Daniel H. Herring
"the Heat Treat Doctor"
2015

ATMOSPHERE HEAT TREATMENT

ATMOSPHERES | QUENCHING | TESTING

VOLUME II

ATMOSPHERE HEAT TREATMENT

ATMOSPHERES | QUENCHING | TESTING

VOLUME II

DANIEL H. HERRING

BNP Media II, LLC | CUSTOM MEDIA GROUP
Troy, Michigan

Copyright 2015 by Daniel H. Herring

All rights reserved. No part of this book may be reproduced or transmitted in any form or by any means without permission in writing from the publisher. For information contact BNP Media II, LLC, 2401 W. Big Beaver Road, Suite 700, Troy, MI 48084.

Printed in the United States of America

Published in 2015 by BNP Media II, LLC, Troy, MI 48084

LIMIT OF LIABILITY/DISCLAIMER OF WARRANTY: The publisher and the author make no representations or warranties with respect to the accuracy or completeness of the contents of this work and specifically disclaim all warranties, including without limitation warranties of fitness for a particular purpose. No warranty may be created or extended by sales or promotional materials. The advice and strategies contained herein may not be suitable for every situation. This work is sold with the understanding that the publisher and author are not engaged in rendering legal, accounting, or other professional services. If professional assistance is required, the services of a competent professional person or professional engineer should be sought. Neither the publisher nor the author shall be liable for damages arising herefrom. The fact that an organization or website is referred to in this work as a citation and/or a potential source of further information does not mean that the author or the publisher endorses the information the organization or website may provide or recommendations it may make. Further, readers should be aware that internet websites listed in this work may have changed or disappeared between when this work was written and when it is read.

Atmosphere Heat Treatment: Atmospheres, Quenching, Testing
Daniel H. Herring
Includes indexes
ISBN 978-0-692-51299-9

Project management by Melanie Kuchma
Art direction and book jacket design by Shannon Shortt
Book designed by Shannon Shortt

Photos reprinted with permission

DEDICATION

To Jeanne…

…my loving wife, soulmate and inspiration for all I do!

To the four Daniel's (George, Hal, Andrew, Quintin)…

…may they stay forever young!

To the readers…

…they say that if you love your job, you never really work a day in your life. For us, the heat-treatment industry is that labor of love. Enjoy the knowledge that this book brings!

> "Do not judge yourself and do not judge others at all. Do not be at a point of judgment, merely be in your discernment that your consciousness may transform and you will walk into your full power. From this place you may create anything."
> – Archangel Metatron, Metatron This Is The Clarion Call

FOREWORD

Worldwide manufacturing is faced with many well-documented challenges across disciplines. Heat treating in particular faces some unique challenges as the labor force changes, leaving gaps of experience on the shop floor around the globe. Practical experience coupled with the operation of equipment and technology used in atmosphere heat treatment will be essential for manufacturers and heat treaters alike.

Atmosphere heat treatment covers a broad spectrum of applications and experience. Finished parts heat treated under atmosphere shape our lives and change how we do things. With precision heat treatment, finished parts meet or exceed engineering designs. *Atmosphere Heat Treatment, Volume II* provides the insight and know-how that allows readers to use techniques and tools for properly changing the metallurgical properties of parts to meet design specifications.

Over the last 20 years, new technologies have enabled heat treaters to become even more efficient at utilizing equipment to maximize throughput without compromising quality. While the use of technology has grown in this industry, so has the need to learn and understand the fundamentals of heat-treatment processes. As a provider of technology used in all areas of heat treatment, Super Systems Inc. (SSi) had the opportunity to collaborate with Dan Herring on many areas of this book. It has been a pleasure to work with Dan on this project knowing there will be numerous heat treaters, engineers, business owners, operations managers and executives finding valuable – and often indispensable – information in this book for years to come.

In *Atmosphere Heat Treatment, Volume II*, Dan balances essential concepts with discussions of complex topics in a clear, relatable manner. Readers will find a level of detail suited to their needs, whether they are new to atmosphere heat treatment or a seasoned veteran. Dan's years of experience in all aspects of heat treatment come to the forefront in this book. The result is a pivotal work covering important examples of past, present and future methods of atmosphere heat treatment.

Super Systems Inc.
Cincinnati, Ohio

PUBLISHER'S PREFACE

Atmosphere Heat Treatment, Volume II is the second in a two-volume set, and both are companions to *Vacuum Heat Treatment*. All three have been authored by Daniel H. Herring and published by *Industrial Heating* with you, the reader, in mind. Volume II puts the fundamentals discussed in Volume I into practice.

"Atmosphere heat treatment" is far more than just the interactions of furnace atmospheres with the parts being treated. Thermal processing, in general, encompasses every heat-treatment process used; the equipment in which these processes – heating, cooling and quenching – are carried out; pre/post-heat-treatment operations (including testing); maintenance; safety; and conservation. This book offers an experience-based perspective on what has been found to be important for all types of atmosphere heat treatments. I can't think of a more cost-effective way to tap the mind of The Heat Treat Doctor.

Like its predecessors, this is a practical book. It is not too technical nor is it devoid of pertinent technical content. It is intended to help individuals who work with atmosphere furnaces operate them more effectively and profitably. It is theory in action from a talented teacher who knows his material and is able to communicate it clearly in a sensible and interesting way.

Also like its predecessors, *Atmosphere Heat Treatment, Volume II* differs from other technical works in that it is a commercial venture. *Industrial Heating* asked many of the companies that have benefited from Dan's experience to help us underwrite a portion of the publishing costs. Without being a distraction, you will see advertisements from these sponsors throughout the book. Please make every effort to support these companies. They are leaders in the atmosphere thermal-processing industry, and they understand that a well-informed industry makes better purchasing decisions.

As the technical editor on this project, it's hard for me to be objective. This book will be a great resource for years to come, and it is an effort undertaken by many. I'd like to offer my sincere appreciation to everyone who contributed to this project.

- ❖ The entire heat-treat community has supported this effort in large and small ways. Individuals, companies, technical societies (ASM International and MTI in particular) and universities all contributed.
- ❖ Bill Mayer, *Industrial Heating*'s managing editor, held it all together. Without his dedicated efforts, we would certainly not have stayed on schedule.

- ❖ Shannon Shortt took care of all of the design and artwork.
- ❖ Melanie Kuchma and Chris Wilson from orangetap, BNP's custom-media group, for managing the project from start to finish.
- ❖ Kathy Pisano, Rick Groves, Steve Roth and Hamilton Pearman – *Industrial Heating*'s outstanding sales team – for arranging many of the sponsorships.
- ❖ The many sponsors of this work (see a complete sponsor list on page xvii). Without you, we wouldn't have this book.
- ❖ Dan Herring, "The Heat Treat Doctor,"® for his ideas, excellent composition and work with innumerable companies and individuals to pull it all together.

We hope you enjoy and benefit from this work. Combine it with *Atmosphere Heat Treatment, Volume I* as your go-to reference for all things atmospheric.

Reed Miller
Associate Publisher/Editor, *Industrial Heating*

PREFACE

The trilogy is complete. The two books that comprise *Atmosphere Heat Treatment*, along with our companion work, *Vacuum Heat Treatment*, present much of what one needs to know about what heat treatment is, how we perform it and what outcome we can expect.

Volume II focuses on furnace atmospheres, quenching practices, testing, safety, conservation, maintenance and specification compliance, whereas Volume I emphasized fundamental principles, materials, metallurgy, applications and equipment. Together, these volumes provide a comprehensive resource on the subject of atmosphere technology as conducted in furnaces and ovens. They also provide unique insights into industry practices and the challenges heat treaters face every day.

While the subject of atmosphere heat treatment is vast, the selection of topics and the depth of coverage are intended to provide the reader with a solid understanding of the technology. In addition, you will come away with valuable insights into the methods, processes and procedures being used in the heat treatment of ferrous and nonferrous materials. Once again, the incredible willingness of those in our industry to share hard-earned knowledge, coupled with my own desire to pass this information on to future generations, is the driving force that has inspired the writing of this book – with a goal of informing, educating, questioning the status quo and challenging the curiosity of the next generations entering our industry.

Whether you are a heat treater, metallurgist, design engineer, supervisor, quality-control engineer, manufacturing engineer, manager or senior management, this book will serve as an invaluable guide and learning tool. What differentiates these books from others is the desire not only to explain what the subject is but also to provide the "why" as to a particular method of heat treatment in simple, everyday language.

To repeat what we have said before, heat treatment is a very "hands-on" science – we roll up our sleeves and get our hands dirty to make it happen. As such, common sense is an important ingredient for success. That does not mean that scientific principles are not at the core of what we do. Coupled with curiosity and hard work, we can not only anticipate the outcome of a heat-treatment operation but take pride in our ability to shape that outcome. Our goal is to manage the heat-treatment process and offer our customers the highest-quality products produced by any technology. We practice our science with eyes wide open and strive to be the absolute best at our craft. This book will help us achieve these goals.

Atmosphere Heat Treatment is all about giving the reader a resource that will endure the test of time. It also provides valuable knowledge into the types and control of furnace atmospheres in use, quenching methods and practices and ways to ensure we meet the highest quality standards. We provide the process and application knowledge needed to do the job – whether you are a CEO, owner or company president; manufacturing, process or quality engineer; department manager; or a heat-treat supervisor or furnace operator curious to learn more about the equipment they operate and the processes they run. The reader will come away with useful information and practical tips about the process and equipment variables that must be controlled, monitored and recorded.

Another unique aspect of this book is the many pictures and illustrations contained herein. The author is incredibly grateful to all of the many companies that provided them. There were literally thousands of images from dozens and dozens of companies to select from, and the choices are my way of providing additional insight to the reader. Study them carefully, and ask yourself questions about what you see. In many cases, these images are as important as the written words.

Within each chapter you will find units of measure expressed in both the metric and English system. It should be noted that the temperature conversions are approximate – by design, as the heat-treatment industry does not often use the same temperature conversion as the scientific community. If exact conversions are necessary, the English unit(s) should be the starting point.

Any errors or omissions in the book are those of the author alone. Your feedback and comments are highly encouraged. Send them directly to the author's principle e-mail address: dherring@heat-treat-doctor.com. I will be sure to acknowledge them.

Finally, you may be asking yourself how valuable a technical reference this can be given the fact we have sponsors and promotional pieces strategically placed throughout the book. This is a unique approach; one the publisher and I feel provides the reader with yet another technical resource, which is why we have chosen this path. Original equipment manufacturers, heat-treat shops and suppliers of components or ancillary products are valuable sources for information and help. Rely on them to supplement what is presented in these pages.

No work is the effort of just one person alone. Literally hundreds of people have contributed in large and small ways, and it is impossible in this short space to list them all. You know who you are and know that you have my eternal gratitude.

There are a few people, however, who must be mentioned. Don Bowe, D. Scott Mackenzie and George Totten stand out for their keen knowledge, technical brilliance and true passion for the industry. A special "thank you" goes out to Christopher Wilson, Melanie Kuchma, Shannon Shortt and the rest of the BNP Media staff; to my editors, Bill Mayer and Reed Miller; and to my son Tim for helping in the book's preparation. Finally, thanks to my lovely wife and partner, Jeanne, for her continued support and keen mind. She has been an invaluable asset asking intriguing questions, offering excellent suggestions and making improvements to my poor attempt at expressing the English language in clear and concise terms. Love you!

So, where do we go from here? The quest for knowledge never ends. As such, there is a need for new books to provide greater depth into specific process applications. "The Doctor" hears the clarion call.

Daniel H. Herring
"The Heat Treat Doctor"®
Chicago, Ill.
2015

CONTENTS

DEDICATION v
FOREWORD vi
PUBLISHER'S PREFACE vii
PREFACE ix
SPONSOR ACKNOWLEDGMENTS xvii
COLOR IMAGE SECTION (MIDDLE) .... A

CHAPTER 9: FURNACE ATMOSPHERES
9.1 | OVERVIEW 1
9.2 | TYPES AND CHARACTERISTICS 7
9.3 | THEORY 15
9.4 | CHARACTERISTICS AND APPLICATIONS (PART ONE) 27
9.5 | CHARACTERISTICS AND APPLICATIONS (PART TWO) 45
9.6 | GENERATED ATMOSPHERES 77
9.7 | CONTROLS AND SENSORS 107

CHAPTER 10: QUENCHING AND QUENCHANTS
10.1 | PRINCIPLES OF QUENCHING 127
10.2 | TYPES OF QUENCHANTS (PART ONE) 135
10.3 | PERFORMANCE VARIABLES IN OIL AND POLYMER QUENCHANTS ... 163
10.4 | TYPES OF QUENCHANTS (PART TWO) 173
10.5 | MANAGING DISTORTRION 191
10.6 | CONSIDERATIONS IN QUENCH-TANK DESIGN 205

CHAPTER 11: HEAT-TREATMENT SPECIFICATIONS, COMPLIANCE
11.1 | THE ROLE OF TRAINING 219
11.2 | AUDITS 229
11.3 | ACCREDITATION 241
11.4 | SYSTEM ACCURACY TESTS AND TEMPERATURE UNIFORMITY SURVEYS ... 257
11.5 | INSTRUCTIONS AND BLUEPRINT REQUIREMENTS 277

CHAPTER 12: TESTING

12.1	RATIONALE	**291**
12.2	HARDNESS AND MICROHARDNESS TESTING	**303**
12.3	MECHANICAL TESTING	**323**
12.4	METALLURGICAL EVALUATION	**341**
12.5	FAILURE ANALYSIS	**355**
12.6	NONDESTRUCTIVE TESTING METHODS	**369**

CHAPTER 13: MAINTENANCE REQUIREMENTS AND PRACTICES

13.1	PLANNED PREVENTIVE-MAINTENANCE PROGRAMS	**381**
13.2	MAINTENANCE OF ATMOSPHERE FURNACE COMPONENTS	**391**
13.3	MAINTENANCE TECHNIQUES FOR ATMOSPHERE FURNACE SYSTEMS	**409**
13.4	MAINTENANCE OF QUENCHANTS AND QUENCH BATHS	**439**
13.5	MAINTENANCE OF SENSORS AND CONTROLS	**459**
13.6	FURNACE TESTING AND TROUBLESHOOTING	**477**
13.7	PROCESSING ISSUES AND THEIR CONTRIBUTION TO MAINTENANCE	**497**

CHAPTER 14: HEALTH, SAFETY, ENERGY AND THE ENVIRONMENT

14.1	HEALTH	**511**
14.2	SAFETY	**519**
14.3	ENERGY AND ENVIRONMENTAL ISSUES IN THE SHOP	**527**
14.4	IDENTIFYING SOURCES OF POTENTIAL QUENCH-OIL FIRE HAZARDS	**545**
14.5	FURNACE WATER-COOLING SYSTEMS	**551**

CHAPTER 15: ANCILLARY TOPICS OF INTEREST

15.1	THE ROLE OF MANAGERS, SUPERVISORS AND OPERATORS	**563**
15.2	LEAN MANUFACTURING AND LEAN HEAT TREATMENT	**569**
15.3	TOTAL COST OF OWNERSHIP	**579**
15.4	THE ROLE OF METALLURGY IN ENGINEERING AND MANUFACTURING	**589**
15.5	THE ROLE OF HEAT TREATMENT IN REVERSE ENGINEERING	**595**
15.6	EMBRITTLEMENT AND CORROSION	**603**
15.7	APPLICATIONS FOR HEAT-RESISTANT MATERIALS	**627**
15.8	HEAT-TREATMENT REQUIREMENTS FOR THE FASTENER INDUSTRY	**635**

CHAPTER 16: REFERENCE MATERIALS

16.1 | HEAT-TREATMENT TERMINOLOGY **653**
16.2 | REFERENCE LIBRARY FOR HEAT TREATERS **683**
16.3 | METALLURGICAL SAMPLE PREPARATION **699**
16.4 | USEFUL PROPERTIES, TABLES AND CHARTS **721**

INDEXES

EQUATION **749**
TABLE **751**
FIGURE **754**
SUBJECT/TERMINOLOGY **767**

SPONSOR ACKNOWLEDGMENTS

SPONSOR	LEVEL	PAGE
Super Systems Inc.	Platinum	6, 125, 475
Lindberg/MPH	Gold	134, 390
Yokogawa	Gold	255, 275
AFC-Holcroft	Silver	162
Aichelin USA	Silver	218
Dry Coolers	Silver	561
Eurotherm by Schneider Electric	Silver	256
Gasbarre Products	Silver	203
GeoCorp Inc.	Silver	437
Heatbath/Park Metallurgical	Silver	188
Ipsen	Silver	228, 302, 577
Linde	Silver	14, 74
Safe Cronite	Silver	633
SECO/WARWICK	Silver	75
Solar Atmospheres	Silver	543
Struers	Silver/Bronze	354
Bodycote	Bronze	
Houghton International	Bronze	171
Phoenix Heat Treating	Bronze	190
Specialty Steel Treating	Bronze	588
Surface Combustion Inc.	Bronze	105
Accurate Steel Treating	Iron	240
Air Products	Iron	26
AmeriKen	Iron	568
Chemtool Incorporated	Iron	172
Derrick Company	Iron	189
Engineered Abrasives	Iron	290
hti	Iron	189
I Squared R Element	Iron	408
Met-Tek Inc.	Iron	240
Nitrex	Iron	76
Orton Ceramic Foundation	Iron	276
Schunk Graphite	Iron	276
South-Tek Systems	Iron	44
United Process Controls	Iron	25

ATMOSPHERE HEAT TREATMENT

ATMOSPHERES | QUENCHING | TESTING

VOLUME II

9 | FURNACE ATMOSPHERES

9.1

OVERVIEW

A critical consideration in heat treatment is the type, consistency and control of the furnace atmosphere. Many component parts subjected to heat treatment need to be protected against exposure to undesirable surface reactions during processing. Selection of a furnace atmosphere often depends on the heat-treatment process and in some instances is influenced by the type of equipment being used.

Purpose

The purpose of a furnace atmosphere varies with the desired end result of the heat-treatment process. In general, furnace atmospheres are designed to be:

- *Passive (or chemically inert/neutral)* to the metal surface; that is, to protect the material being processed from chemical reactions that could occur on their surfaces (e.g., oxidation or carburization in many cases).
- *Reactive (or chemically active)* to the metal surface; that is, to allow the surface of the material being processed to change (e.g., adding or depleting carbon, adding nitrogen or oxygen).

Let's take a look at several examples to better understand a furnace atmosphere's role in the heat-treatment process.

EXAMPLE 1: WHEN THE ROLE OF THE FURNACE ATMOSPHERE IS TO BE CHEMICALLY INERT

High-carbon SAE 52100 steel retaining rings (Fig. 9.1.1) are neutral hardened (i.e., neither carburized or decarburized) at 845°C (1550°F) in an endothermic-gas atmosphere and held at a 1.00% carbon potential (the same as the base material) by enrichment-gas additions so that the surface of the rings will not be subjected to decarburization (the removal of carbon from the surface of the steel) or carburization (the addition of carbon into the surface of the steel). By the use of this type of heat treatment, a retaining ring can be stamped to close tolerances and then hardened to prevent excessive wear in service while maintaining an excellent surface finish.

FIGURE 9.1.1 | Neutral-hardened retaining rings

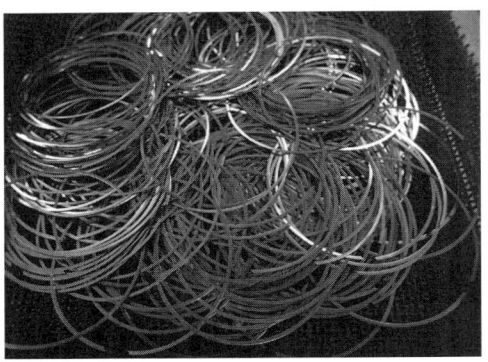

EXAMPLE 2: WHEN THE ROLE OF THE FURNACE ATMOSPHERE IS TO BE CHEMICALLY ACTIVE

Low-carbon SAE 12L14 steel bearing races (Fig. 9.1.2) are case hardened (gas carbonitrided) at 845°C (1550°F) in an endothermic-gas atmosphere and enriched to a carbon potential of 1.00% with natural gas and ammonia to provide a source of both carbon and nitrogen for absorption into the surface of the steel. By the use of this type of heat treatment, a shallow case is imparted onto the surface of the bearing race to prevent excessive wear in service while maintaining an excellent surface finish.

FIGURE 9.1.2 | Case-hardened bearing races (courtesy of Bodycote Inc.)

Note the similarities in the heat-treatment parameters (temperature, basic atmosphere type, the fact that a hydrocarbon enrichment gas was added) and desired end use in each of these examples. The choice of material, 52100 versus 12L14, was based on the design and service demands of the product as well as the cost of the raw material and associated manufacturing steps required, all of which dictate the type of heat treatment to be used. Once all of these variables are known, the purpose of the atmosphere (inert or active) was determined.

Historical View

In the early days of the heat-treatment industry not much thought was given to the atmosphere in which the work was heated. In some instances, such as the processing of aluminum and aluminum alloys, this was of little consequence because of the protective oxide already present on the surface of the aluminum. For steel parts, however, heavy scaling was an unavoidable and generally unacceptable consequence.

Also at that time, the majority of heat-treatment furnaces had intermittently welded shells with loosely fitted doors. Electric furnaces either ran an air atmosphere or had gas curtains to supply an air-gas mixture (Fig. 9.1.3). Most gas-fired furnaces used products of combustion as their atmosphere. By controlling the air/gas ratio at the burners so that a richer ratio is obtained, some degree of protection against oxidation was possible, albeit still inadequate. This, however, introduced other problems such as poor heating and bad temperature uniformity characteristics.

FIGURE 9.1.3 | "Old Number One," the first controlled-atmosphere furnace sold to American industry – "Certain Curtain" furnace May 11, 1927 (courtesy of C.I. Hayes, a Gasbarre Furnace Group Company)

As time passed and better surface quality was demanded, furnace atmospheres were introduced to protect against the effects of surface oxidation. This resulted in furnace designs incorporating gas-tight shells, tight-fitting doors, atmosphere inlets and outlets, retorts and indirect heating of gas-fired furnaces (via radiant tubes and/or muffles). This allowed the "protective" atmosphere to produce "clean" (if not bright) surfaces on the work being treated.

Modern View

As manufacturing methods have evolved, so too has atmosphere heat treatment. Real-time in-situ sensor technology that started in the late 1960s and early 1970s laid the groundwork for better predictability and reproducibility in atmosphere processing. With increased productivity demand came the need for higher quality in heat-treat processes now performed at facilities around the world. Today, engineering specifications call for tighter tolerances than achievable in even the recent past. These tolerances are only made possible by modern sensor technology – the resulting ability to measure atmosphere conditions in real time – and computerized controls that respond to changes quickly when needed.

Acceptance of sensor technology in atmosphere heat treatment was soon followed by commercialization. Accurate sensor technology led to closed-loop process control driving accuracy and predictability in metallurgical results. New methods of verification provided even greater levels of confidence and more assurances of repeatability in the process.

Demonstrating acceptance of new technologies, industry specifications now designate measurement and verification steps and provide guidance on expected results. Paper chart recording, common throughout our heat-treat history, has given way to complete data-acquisition systems that provide proof of process and centralized control. Automated-recipe controllers track operation, maintenance and uptime. They also provide built-in redundancy and, with technological advancements, can allow for complex multi-loop recipe control in addition to single-parameter loop control.

Manufacturing has driven the heat treater to constantly strive for continuous improvement while lowering overall cost. Process control – by means of technology and procedures – is what heat treaters are using to maximize uptime, increase production and deliver the highest quality to customers.

Types of Surface Reactions

When a metallurgical change to the surface of a component part during heat treatment does not alter its microstructure (i.e., when the surface is essentially unchanged), we talk about the atmosphere being neutral or inert. Annealing or normalizing of a 0.40% carbon steel at 800°C (1475°F) in an endothermic-gas atmosphere at a +4.5°C (+40°F) dew point or age hardening of aluminum in air would be examples.

When a metallurgical change to the surface of a component part during heat treatment alters its microstructure, we talk about the atmosphere having an active species or being reactive. The two most common reactions of this type are oxidation/reduction (Section 9.4) and carburization/decarburization (Section 9.4).

Summary

While the subject of furnace atmospheres involves considerable knowledge of chemical reactions, what is most important for the heat treater to understand is that we are able to predict and control the effects of these reactions on the materials being processed. When doing so, we assume the furnace atmosphere to be under equilibrium conditions, although in reality we know the furnace atmosphere is constantly changing. Thus, an understanding of the interactions of the gas species allows us to anticipate when these reactions will occur, and we can control the reactions of greatest importance using modern gas-analysis equipment.

REFERENCES

1. Herring, Daniel H., "Selection and Use of Furnace Atmospheres," *Industrial Heating* webinar, 2011
2. Hotchkiss, A.G. and H.M. Webber, *Protective Atmospheres*, John Wiley & Sons Inc., 1953
3. Donald Bowe, Air Products & Chemicals Inc. (www.airproducts.com), technical and editorial review, private correspondence
4. James Oakes, Super Systems Inc. (www.supersystems.com), technical and editorial review, private correspondence

SuperSystems
incorporated

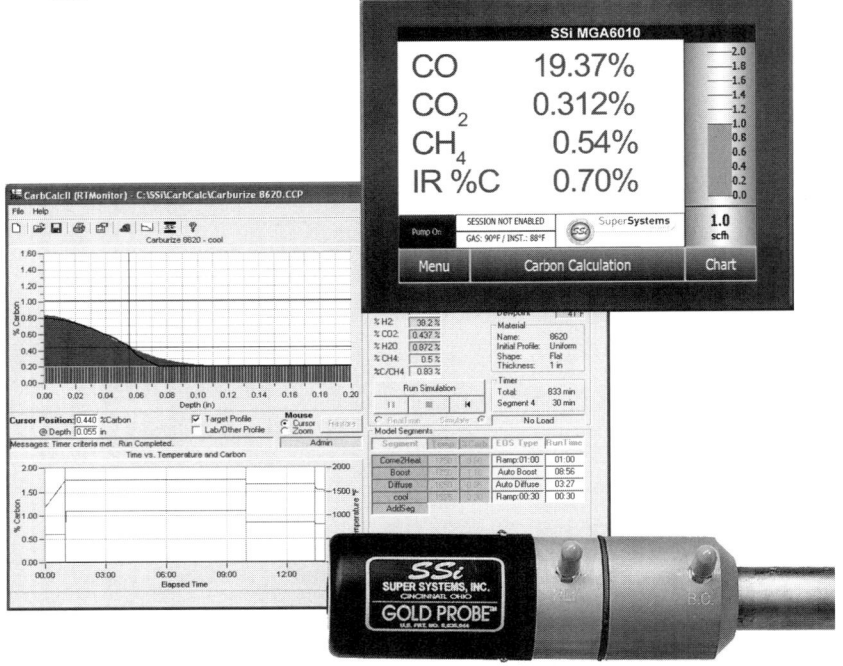

Industry Experts in Control and Measurement of Heat Treating Atmospheres

www.supersystems.com

9.2

TYPES AND CHARACTERISTICS

Many types of furnace atmospheres are available for use in heat treatment (Table 9.2.1). It is natural, therefore, to ask what the most common furnace atmosphere is. The answer is air. Often, nothing more is needed. When an air atmosphere is used, such as in a low-temperature tempering operation, the final condition of the material's surface, or skin, is not considered important. For many other applications, however, protection of the part surface is of primary importance.

Types of Furnace Atmospheres

A great many furnace atmospheres are in use throughout industry. The proper selection of a furnace atmosphere depends on a large number of factors, including but certainly not limited to:

- Type of material being processed
- Quality of the atmosphere
- Availability
- Cost
- Type of equipment in use

Volume Requirements

During operation, the flow rate of a protective atmosphere required for safe use in a particular heat-treating furnace depends to a great extent on the:

- Type and size of furnace
- Presence or absence of doors and/or curtains
- Room environment (especially drafts, truck bay doors, etc.)
- Size, loading, orientation and nature of the work being processed
- Surface cleanliness

TABLE 9.2.1 | Common types of furnace atmospheres (listed alphabetically)

Type	Chemical symbol	Remarks
Air		Typically used when oxidation or decarburization of the part surface is not a factor or when subsequent operations such as machining or shot blasting will be performed.
Argon[a]	Ar	A non-reactive, truly inert gas.
Blended atmosphere		An atmosphere produced from a mixture of hydrogen (100-0%) and nitrogen (0-100%).
Carbon dioxide	CO_2	A common constituent in generated atmospheres.
Carbon monoxide	CO	A common constituent in generated atmospheres.
Custom (special) blends		Combinations of nitrogen and other gases or liquids such as hydrocarbons, alcohols or monoaromatic hydrocarbons (e.g., benzene, toluene).
Generated atmospheres[b]		Atmosphere produced by a gas generator. Examples include endothermic (RX®)[c], exothermic (DX®)[c] and dissociated ammonia.
Helium[a]	He	A non-reactive, truly inert gas.
Hydrocarbon gases		Typically used as additions or enriching gases to furnace atmospheres. Common types include methane (CH_4), propane (C_3H_8) and in some instances butane (C_4H_{10}).
Hydrogen[d]	H_2	A reducing constituent of many furnace atmospheres used to aid in heat transfer and react with any oxygen present.
Nitrogen[e]	N_2	A blanketing or purging gas. Not inert in all applications (e.g., processing of stainless steels).
Oxygen	O_2	Oxidizing to the metal surface at elevated temperatures.
Products of combustion		An atmosphere produced from combustion of a hydrocarbon fuel gas and air. It typically consists of high amounts of carbon dioxide and water vapor.
Steam	H_2O	Water vapor. Most often used to impart a protective oxide layer on a component part (Section 5.12).
Sulfur dioxide	SO_2	An atmosphere used for special heat treatment (e.g., magnesium alloys).
Synthetic endothermic atmosphere[b]		An atmosphere produced by nitrogen and methanol (methyl alcohol).
Vacuum		An "atmosphere" produced by removing air from a sealed vessel.

Notes:

[a] Argon is often used for atmosphere processing when the presence of nitrogen may cause unwanted surface reactions. Argon (and helium) are often associated with vacuum furnaces being used for partial pressure or quenching.

[b] Endothermic gas and nitrogen/methanol systems are the most common atmosphere types for hardening and case-hardening operations.

[c] RX® and DX® are registered trademarks of Surface Combustion Inc.

[d] Hydrogen can hydride certain metals (e.g., Al, Ti).

[e] Nitrogen can nitride certain metals under certain processing conditions (e.g., stainless steels, Ti, Ta).

In all cases, the manufacturer's recommendations should be reviewed and followed since they would have taken these factors into account during the design of the equipment. If the furnace is older, has been relocated or is running a different process application, however, volume requirements may have changed.

To purge air out of a furnace prior to introduction of a combustible furnace atmosphere requires a minimum of five volume changes of the chamber (Table 9.2.2, Figure 9.2.1). This is to ensure that the oxygen content of the chamber is below 1% prior to the introduction of the atmosphere as required by NFPA 86. Confirmation that a furnace has been purged involves sampling of the atmosphere with an oxygen analyzer and continuing to purge with inert gas until two consecutive readings indicate that the oxygen content is below 1% by volume. A vacuum purge is an acceptable alternative if the initial room air within the furnace is pumped out with a mechanical pump to a vacuum level of 0.1 torr (100 microns or 0.1 mm Hg).

TABLE 9.2.2 | Volume changes required for safe purging of furnaces

Number of volume changes	Percentage of air still remaining
0.1	90.48
0.2	81.87
0.3	74.08
0.5	60.65
1.0	36.79
2.0	13.53
3.0	4.98
4.0	1.83
5.0	0.67

FIGURE 9.2.1 | Theoretical purge-down curve[1]

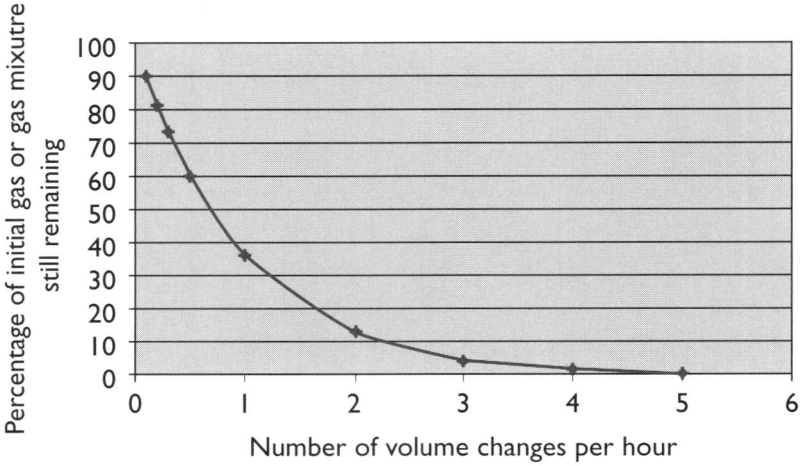

Classification of Blended Atmospheres

Blended (aka synthetic) atmospheres are those supplied from industrial gas sources or those involving on-site storage/supply systems. Examples include: nitrogen/hydrogen, nitrogen/methanol and nitrogen/hydrocarbon. The most common nitrogen/hydrogen blends are 97/3, 95/5 (forming gas), 90/10 and 25/75 (dissociated-ammonia equivalent). In many cases, nitrogen is combined with generated gas to produce the final furnace atmosphere (Table 9.2.3). Blended gases can also be used as substitutes for generated atmospheres (Table 9.2.4).

TABLE 9.2.3 | Examples of blended and diluted atmosphere types and compositions

Atmosphere	Type	%H_2	%N_2	%CO	Dew point, °C (°F)
Hydrogen	Pure	100	0	0	-70 to -85 (-95 to -120)
Nitrogen/dissociated ammonia	Diluted	10	90	0	-40 to -50 (-40 to -58)
Nitrogen/endothermic	Diluted	12	82	6	< 0 (32)
Nitrogen/hydrogen	Blended	3-10	97-90	0	-50 to -60 (-58 to -76)
Nitrogen	Pure	0	100	0	-68 to -85 (-90 to -120)

TABLE 9.2.4 | Common blended atmosphere substitutions for generated atmospheres

Blended atmosphere	Generated atmosphere
Hydrogen in nitrogen (1-3%)	Exothermic (lean)
Nitrogen (100%)	Exothermic (lean)
Hydrogen in nitrogen (4-19%)	Exothermic (rich)
Nitrogen/methanol (5-10%)	Exothermic (rich)
Nitrogen/hydrocarbon	Exothermic (rich)
Nitrogen/methanol	Endothermic
Nitrogen/hydrocarbon	Endothermic
Nitrogen/endothermic dilution	Endothermic
Hydrogen (100%)	Dissociated ammonia
Hydrogen (75%)/nitrogen (25%)	Dissociated ammonia
Nitrogen/dissociated ammonia dilution	Dissociated ammonia

Classification of Generated Atmospheres

According to the relative amounts of the individual gases produced, the American Gas Association (AGA) classifies generated atmospheres (Table 9.2.5). The compositions and characteristics of the most common variants (Table 9.2.6) are of greatest interest to heat treaters.

TABLE 9.2.5 | Classification of generated furnace atmospheres[3]

Class	Base type	Subclasses	Description
100	Exothermic[a]	101 (lean) 102 (rich)	An atmosphere created from the products of partial or complete combustion of an air-gas mixture in a water-cooled combustion chamber. Units operating rich require a catalyst.
200	Nitrogen	201 (lean prepared) 202 (rich prepared) 207 (201+ hydrocarbon) 208 (202 + hydrocarbon) 223 (201 + steam[b]) 224 (202 + steam[c])	A prepared atmosphere using an exothermic base with a large percentage of the carbon dioxide and water vapor removed.
300	Endothermic[a]	301 (lean) 302 (rich) 323 (gas-air-steam[d]) 325 (gas-air-steam[e])	An atmosphere created by partial reaction of an air-gas mixture in an externally heated, catalyst-filled chamber.
400	Charcoal	400	Uncommon today. Formed by passing air over a bed of incandescent charcoal.
500	Exothermic-endothermic	501 (lean) 502 (rich)	Uncommon today. An atmosphere created by complete combustion of a mixture of gas and air, removing a large percentage of the water vapor, and re-forming most of the carbon dioxide to carbon monoxide by reaction with fuel gas in an externally heated catalyst reactor.
600	Ammonia	600 (raw) 601 (dissociated) 621 (lean combusted) 622 (rich combusted)	Any atmosphere created using ammonia as the primary constituent, including nascent (raw) ammonia, dissociated ammonia, or partially or completely combusted ammonia with a large percentage of the water vapor removed.

Notes:
[a] Exothermic reactions are heat-producing. Endothermic reactions require heat to promote the reaction. The composition of the atmospheres produced can be changed in a number of ways. Varying the air/gas ratio or using a different feedstock (e.g., natural gas or propane) will cause the chemistry of the gas to change.
[b] A catalyst is used to convert CO.
[c] A catalyst is used to convert CO.
[d] A catalyst is used to convert CH_4.
[e] A catalyst is used to convert CO.

TABLE 9.2.6 | Composition of selected generated furnace atmospheres[3,4]

Description	Nominal composition, vol. %					Dew point, °C (°F)
	N_2	H_2	CO	CO_2	CH_4	
Exothermic (lean)	86.8	1.2	1.5	10.5		[a]
Exothermic (rich)	71.5	12.5	10.5	5.0	0.5	[a]
Endothermic (neutral or lean)	45.1	34.6	19.6	0.4	0.3	-7 to +10 (+20 to +50)
Endothermic (carburizing or rich)	39.8	38.7	20.7		0.8	-4 to -20 (+25 to -5)
Exothermic/endothermic (lean)	63.0	20.0	17.0			-56 (-70)
Exothermic/endothermic (rich)	60.0	21.0	19.0			-45 (-50)
Dissociated ammonia	25.0	75.0				[b]

Notes:
[a] Dew point is approximately 5.5°C (10°F) higher than the cooling-water temperature measured at the heat exchanger.
[b] Dew point is a function of incoming ammonia supply purity; typically, -51°C (-60°F) or drier.

Summary

By discussing the most common types of furnace atmospheres, we are better prepared to talk about the chemical reactions of the gas species present in each (Section 9.3), how to select a particular furnace atmosphere for a given application (Sections 9.4, 9.5 and 9.6) and how to measure and control that atmosphere (Section 9.7).

REFERENCES

1. Herring, Daniel H., "Purging of Furnaces: Is It Safe?" *Heat Treating Progress*, September 2002
2. Herring, Daniel H., "Furnace Atmosphere Considerations During Heat Treatment," *Furnaces International*, March/April 2009
3. *Metals Handbook, Volume 2: Heat Treating, Cleaning and Finishing*, ASM International, 1964
4. *ASM Handbook, Volume 4B: Steel Heat Treating Technologies*, ASM International, 2014
5. Hotchkiss, A.G. and H.M. Webber, *Protective Atmospheres*, John Wiley & Sons Inc., 1953
6. Herring, Daniel H., "Furnace Atmospheres," white paper, 2001
7. Donald Bowe, Air Products & Chemicals Inc. (www.airproducts.com), technical and editorial review, private correspondence
8. James Oakes, Super Systems Inc. (www.supersystems.com), technical and editorial review, private correspondence

Achieve the best results in heat treatment.
With our leading solutions.

Gases / Technology / Expertise

→ Tailored process and safety solutions
→ Atmosphere and process control
→ Gas supply solutions
→ Cryo treatment solutions

Linde North America, Inc.
575 Mountain Ave, Murray Hill, NJ 07974
Phone 1-800-755-9277, sales.lg.us@linde.com, www.lindeus.com

Linde North America, Inc. is a member of The Linde Group. Linde is a trading name used by companies within The Linde Group. "Linde" and the Linde logo are trademarks of The Linde Group. © The Linde Group 2015. All rights reserved.

9.3

THEORY

Furnace atmospheres are needed to allow the proper interactions to take place between the part's surface and its process environment. The furnace atmosphere must produce both the desired surface condition and a controllable condition throughout the heat-treatment process. Furnace atmospheres are used for a wide variety of reasons. The most common include:

- ❖ Establishing an atmosphere neutral to the material being processed
- ❖ Purging – prior to the introduction of a flammable atmosphere into a furnace or for removal of a furnace atmosphere
- ❖ Safety purging – in the event of an upset condition (e.g., power failure)
- ❖ Blanketing (aka cover) gas – to eliminate air/oxygen from the furnace environment and/or to help maintain a positive pressure inside the furnace in order to prevent air ingress (e.g., to compensate for door openings or other disturbances due to movement of the workload or to negate the effects of furnace leaks)
- ❖ Establishing an atmosphere "active" to the material being processed (e.g., to provide a source of enrichment – carbon or nitrogen – for case hardening)

Furnace Atmosphere Constituents

It is necessary to understand the most common gaseous constituents in the furnace atmosphere. We will review their reactions with one another and the materials being treated in Sections 9.4 and 9.5. We will focus on iron and steel heat treatment for the purpose of this discussion.

PRINCIPAL GASES AND SOLIDS

Principal gases include: hydrogen, carbon dioxide, carbon monoxide, nitrogen, oxygen and water vapor. Principal solids include: carbon (soot), iron, iron oxide and iron carbide.

Chemical Reactions

Now that we know some of what is in a furnace atmosphere, we need to know what reactions are taking place that we must try to measure and control.

To understand the necessity for and interactions involved between the furnace atmosphere and the parts being treated, it is essential to be familiar with some of the basic chemical reactions, both between the various gases and the materials they are trying to protect or react with at elevated temperatures. It is worth mentioning here that the intent is not to provide a list of chemical equations but a reminder that the (equilibrium) reactions shown are responsible for what happens to the steel during heat treatment. Factors that affect these reactions include:[3]

- ❖ **Time.** Most furnace-atmosphere reactions in heat treatment occur over time. In general, therefore, the shorter the time, the less the effect. For example, the moisture content of an atmosphere for bright annealing of carbon steel in a continuous belt furnace could be as high as 4% without bluing because of short cooling time. This same atmosphere in a bell furnace could result in severe discoloration because of slower cooling rates.
- ❖ **Temperature.** Elevated temperatures tend to cause reactions to occur more rapidly. Changes in temperature or temperature uniformity have their most significant influence on the surface of the metal being treated. For all practical purposes, the temperature of the furnace atmosphere can be assumed to be the same as the metal temperature, but not in all cases (such as the impingement of cold incoming gas on a hot metal surface).
- ❖ **Composition of the atmosphere.** The composition and purity of the gases at elevated temperatures may result in unstable or reactive atmosphere conditions, complicating the task of atmosphere control. For example, the introduction of CO or CO_2 into a N_2/H_2 mixture may cause difficulty in controlling the dew point, or moisture content, of the gas.
- ❖ **Gas ratio.** The overall gas ratio determines the oxidation or carburizing potential of the atmosphere.
- ❖ **Flow rate of the atmosphere.** Increases in the volume turnover (i.e., flow rate based on internal volume) of the furnace atmosphere may result in failure to reach gas equilibrium conditions, making control of the furnace atmosphere unpredictable. Certain processes (e.g., carburizing) require constant circulation of the atmosphere, typically by the use of a fan, to ensure uniformity over all surfaces.

❖ **Materials in contact with the atmosphere.** Certain materials can act as catalysts to different gas species, causing dissociation or other unwanted effects. In a nitriding application, for example, iron (baskets, fixtures, retorts) will dissociate ammonia into nitrogen and hydrogen.

❖ **Impurities.** Examples of impurities include water leaks (moisture), air leaks, oil vapors, chemical residue left on parts or fumes from plating operations.

The chemical reactions that follow fall into several general categories, depending on how they alter the steel's surface (Table 9.3.1).

TABLE 9.3.1 | Chemical reaction categories by gas type

Reaction type	Principal gas species
Neutral	Argon, helium, nitrogen[a]
Oxidizing	Water vapor, carbon dioxide, oxygen
Reducing	Hydrogen, carbon monoxide
Carburizing	Carbon monoxide, hydrocarbon gases
Decarburizing	Water vapor, carbon dioxide, oxygen
Water-gas[b]	Water vapor, hydrogen, carbon monoxide, carbon dioxide
Other	Ammonia, products of combustion, hydrogen sulfide, methanol

Notes:
[a] Nitrogen is not neutral to all materials.
[b] Atmosphere control devices monitor this reaction.

AMMONIA REACTIONS

Dissociation of ammonia (NH_3) starts at temperatures as low as 760°C (1400°F) and is aided by the presence of iron as a catalyst. If the dissociation reaction (Equation 9.3.1) occurs at the surface of a steel part, atomic nitrogen (N) has a choice of either entering the surface or combining with another nitrogen atom to form molecular nitrogen.

9.3.1) $2NH_3 \rightarrow 2N + 6H \rightarrow N_2 + 3H_2$

CARBON DIOXIDE REACTIONS

Carbon dioxide (CO_2) is one of the reaction products when a fuel is burned in air and is present in generated gases such as exothermic and endothermic types. *Carbon dioxide is oxidizing* to iron at elevated temperatures (Equations 9.3.2, 9.3.3). To prevent oxidation, it is necessary to have an excess of carbon monoxide.

Carbon dioxide is also extremely decarburizing (Equation 9.3.4). To prevent decarburization, CO_2 levels must be controlled very closely and must depend on the carbon monoxide content, temperature and the carbon content of the steel.

9.3.2) $CO_2 + Fe = CO + FeO$

9.3.3) $CO_2 + 3FeO = Fe_3O_4 + CO$

9.3.4) $CO_2 + Fe_3C = 3Fe + 2CO$

CARBON MONOXIDE REACTIONS

Carbon monoxide (CO) is highly reducing to steel (Equation 9.3.5). The reversible reaction of CO to form carbon in the form of soot and carbon dioxide (Equation 9.3.6) is of particular interest to the heat treater. CO has a high carbon potential and becomes increasingly more stable at elevated temperatures. It is only at lower temperatures around 480-705°C (900-1350°F) that carbon monoxide will form solid carbon in this so-called carbon-reversal reaction. This results in maintenance issues with gas generators (Section 9.6) and heat-treatment furnaces (Section 13.7).

9.3.5) $FeO + CO = Fe + CO_2$

9.3.6) $2CO = CO_2 + C \text{ (soot)}$

COMBUSTION REACTIONS

Combustion reactions involve the partial or total burning of a hydrocarbon fuel (e.g., natural gas or propane) and air (Equations 9.3.7, 9.3.8). Additional gas reactions (Equations 9.3.9, 9.3.10) involve the reaction between the hydrogen or carbon monoxide present and oxygen.

9.3.7) $CH_4 + \text{air} (2O_2 + 8N_2) \rightarrow CO_2 + 2H_2O + 8N_2$

9.3.8) $2C_3H_8 + \text{air} (3O_2 + 11.4N_2) \rightarrow 6CO + 8H_2 + 11.4N_2$

9.3.9) $2H_2 + O_2 = 2H_2O$

9.3.10) $2CO + O_2 = 2CO_2$

HYDROCARBON REACTIONS

Methane (CH_4) and other hydrocarbons (e.g., propane/butane) are contributors to the source of carbon monoxide and hydrogen needed for carburizing. It has been shown[4] that the reaction of carbon monoxide and hydrogen (Equation 9.3.11) is the rate-determining reaction in carburizing atmospheres.

9.3.11) $CO + H_2 \rightarrow C + H_2O$

Furthermore, when steel absorbs carbon, hydrogen (H_2) is the byproduct (Equation 9.3.12). Hydrogen concentrations higher than equilibrium, however, will tend to decarburize.

9.3.12) $CH_4 + 3Fe = Fe_3C + 2H_2$

Nascent (freshly formed) carbon in the form of soot can deposit on the steel surface when hydrocarbons dissociate at elevated furnace temperatures (Equation 9.3.13). This free carbon is not a good source of carbon for carburizing. Therefore, hydrocarbon additions must be controlled for best results. Soot acts as an insulator or barrier through which carbon from the atmosphere must try to diffuse, which slows the carburizing rate.

9.3.13) $CH_4 \rightarrow C\ (soot) + 2H_2$

HYDROGEN SULFIDE REACTIONS

Hydrogen sulfide (H_2S) is usually found in trace amounts in furnace atmospheres manufactured from "dirty" fuel gases. It may interfere with the carburizing process if present in too large a quantity. It is also very harmful to the furnace interior because it reacts with and deteriorates nickel-based alloys, forming a low-melting Ni-S eutectic around 635°C (1175°F). Filtering a gas containing high percentages of H_2S through steel wool separates out the harmful sulfur constituent.

HYDROGEN REACTIONS

Hydrogen is highly reducing (Equation 9.3.14) *and decarburizing* (Equation 9.3.15) to steel when present in concentrations higher than equilibrium.

9.3.14) $FeO + H_2 = Fe + H_2O$

9.3.15) $Fe_3C + H_2 = 3Fe + CH_4$

METHANOL REACTIONS

Methanol/nitrogen mixtures (Equation 9.3.16) or methanol only (Equation 9.3.17) are used to produce either a neutral or carburizing atmosphere.

9.3.16) $\quad 2CH_3OH + 2N_2 \rightarrow 2CO + 3H_2 + 2N_2$ (synthetic endo)

9.3.17) $\quad CH_3OH \rightarrow CO + 2H_2$

NITROGEN REACTIONS

Atomic nitrogen (N) does not normally occur in a furnace atmosphere unless it is purposely introduced by the addition of ammonia (NH_3). Most atomic nitrogen immediately reacts with itself to re-form molecular nitrogen, which is then inert to iron.

OXYGEN REACTIONS

Oxygen is oxidizing (Equations 9.3.18-9.3.20) *and decarburizing* (Equation 9.3.21) to steel even in trace amounts. The type of iron oxide formed depends, among other variables, on the temperature (Section 5.10). Oxygen can also react with carbon (soot) present in the furnace (Equation 9.3.22). Oxygen (O_2) should be avoided if the steel's surface is to be kept clean during heat treatment and free of decarburization.

9.3.18) $\quad O_2 + 2Fe = 2FeO$

9.3.19) $\quad 3O_2 + 4Fe = 2Fe_2O_3$

9.3.20) $\quad 2O_2 + 3Fe = Fe_3O_4$

9.3.21) $\quad O_2 + Fe_3C = 3Fe + CO_2$

9.3.22) $\quad O_2 + C = CO_2$

WATER-VAPOR REACTIONS

Of all the furnace-atmosphere chemical reactions that occur, the water-gas reaction (Equation 9.3.23) is arguably the most important.

9.3.23) $\quad CO_2 + H_2 = CO + H_2O$

Water vapor (in the ppm level) and carbon dioxide are both in this equation, and we can use this fact to control the carbon potential of a furnace atmosphere.

Dew-point analyzers look at the H_2O/H_2 ratio in the water-gas reaction. Infrared analyzers and oxygen probes look at the CO/CO_2 ratio in the water-gas reaction.

This equation represents the key reaction for gas atmospheres because it controls the reactants formed on each side of the equation. The equal sign indicates chemical equilibrium, meaning the reaction can go either way to form carbon monoxide and water vapor or to form carbon dioxide and hydrogen. It depends on the relative percentages of each in the furnace atmosphere.

Finally, *water vapor is a strongly decarburizing gas* (Equation 9.3.24). Therefore, a constituent such as carbon dioxide will have a tendency to form water vapor, which is why carbon dioxide must be closely controlled. In addition, the carbon monoxide and hydrogen must be present in amounts to satisfy the equilibrium condition at processing temperature to prevent decarburization by water vapor.

Water vapor and carbon dioxide oxidize and decarburize steel. Hydrogen is formed when water vapor oxidizes iron. To prevent oxidation and to keep iron bright, therefore, a definite excess of H_2 over H_2O vapor is required.

9.3.24) $Fe_3C + H_2O = 3Fe + CO + H_2$

Ellingham-Richardson Diagram

Changes to a metal's surface that occur by oxidation or reduction depends on the equilibrium ratios of the partial pressures of the gases involved – namely the ratio of hydrogen to water vapor (pH_2/pH_2O), the ratio of carbon monoxide to carbon dioxide (pCO/pCO_2) and the partial pressure of oxygen (pO_2) that will be in equilibrium with a metal oxide.

The Ellingham-Richardson diagram (Fig. 9.3.1) is derived from thermodynamic calculations and expresses the oxidizing or reducing potential of different materials at different temperatures. It is important for the heat treater to understand that this diagram indicates the probability that the reaction will occur for a given set of conditions, and one assumes that the reaction speeds up as temperature increases. The four main uses of this diagram are:[7]

- ❖ To determine the relative ease of reducing a given metallic oxide to a pure metal
- ❖ To determine the partial pressure of oxygen in equilibrium with a metal oxide at a given temperature
- ❖ To determine the ratio of carbon monoxide to carbon dioxide that will reduce the metal oxide to pure metal at a given temperature
- ❖ To determine the ratio of hydrogen to water vapor that will be able to reduce the metal oxide to pure metal at a given temperature

FIGURE 9.3.1 | Ellingham-Richardson diagram[6]

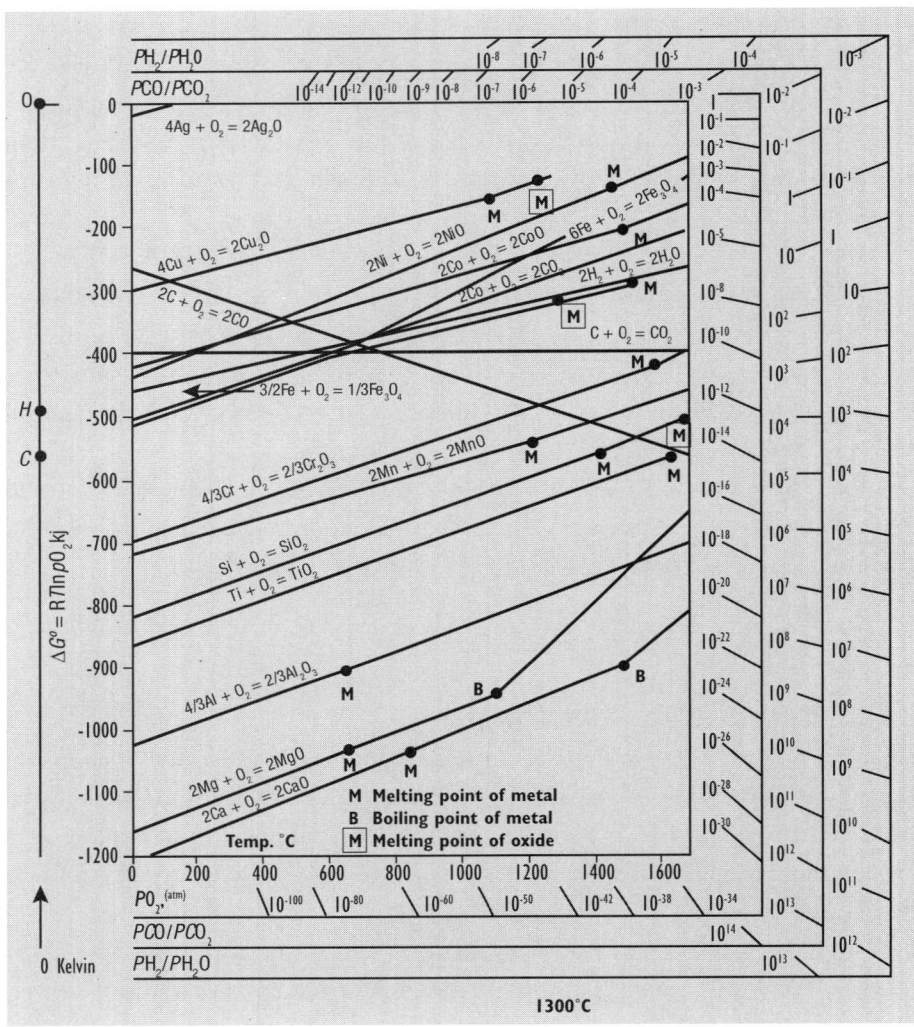

The height of the line above the x-axis is an indication of the stability of the metal oxide as a function of temperature. As one moves toward the bottom of the diagram, the metals become progressively more reactive (and their oxides harder to reduce). For all practical purposes, aluminum oxide cannot be reduced, which is why we can process aluminum alloys in air.

A simplified version (Fig. 9.3.2) allows us to more easily interpret results. For example, we can readily see that copper is easier to reduce at lower temperature and chromium is easier to reduce at high temperature. This can have a significant impact on heat-treatment processes such as annealing of easily oxidized materials, where an atmosphere that was protective at high temperature can easily become oxidizing as the temperature falls.[7]

FIGURE 9.3.2 | Oxidation boundaries for various elements[7] (Copyright Linde AG, courtesy of Linde LLC)

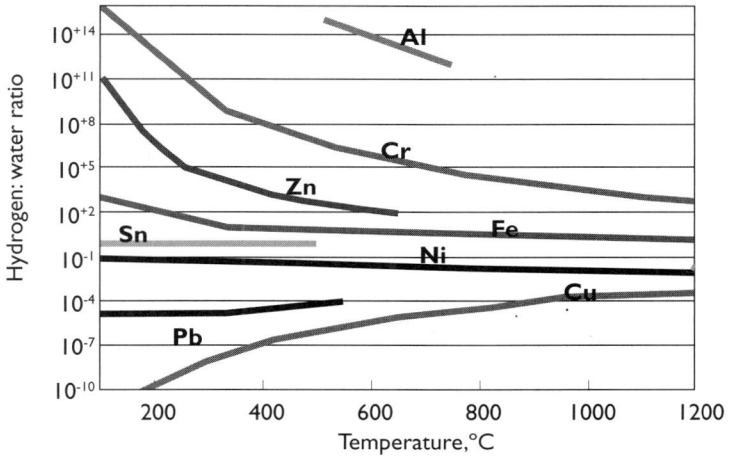

In another example, if we intend to carburize a low-carbon steel at 900°C (1650°F) at a carbon potential of 0.8% C in an endothermic gas atmosphere enriched with propane (producing a 23% CO value), then the CO_2 value would be 0.34%. This yields a CO/CO_2 ratio of 676. If we draw a line on the Ellingham-Richardson diagram between "C" and "676" on the CO/CO_2 scale, we see that it is well below the iron oxide line at 900°C and is, therefore, reducing to iron. It is well above the manganese and silicon lines, however, indicating the internal oxidation (IGO/IGA) cannot be avoided with this atmosphere choice.[7]

The Ellingham-Richardson diagram can also be very helpful when troubleshooting heat-treatment problems. Nickel brazing of stainless steel is an example. This type of brazing is typically run in the temperature range of 925-1200°C (1700-2200°F) depending on the composition of the braze alloy.

One of the elements that we are most concerned about in stainless steel brazing is chromium since the formation of chromium oxides will inhibit braze flow. The diagram shows that a dew point of -37°C (-35°F) is theoretically required at 1000°C (1832°F) with a 100% hydrogen atmosphere to prevent the oxidation of the chromium. We know a dew point of at least -50°C (-58°F) is required to produce good joints. This is because the diagram only considers thermodynamics and not kinetics. The reduction of chromium takes place slowly over time, so it is better to have an atmosphere with a higher driving force (lower dew point) to do the job in a much shorter time.[7]

Copper-Steel-Stainless Steel Test

There is a simple test that lets heat treaters differentiate between an air leak and a water leak in a furnace. The test also provides information about the relative severity of the leak. It can be used with furnaces running endothermic, dry exothermic, nitrogen, nitrogen-hydrogen, dissociated ammonia or hydrogen atmospheres.

The test involves gathering relatively thin, clean (i.e., bright) samples made of steel, copper and (if available) 300-series stainless steel. A short length of copper tubing, sanded if necessary to produce a shiny surface, is often used for the copper sample. If possible, the furnace atmosphere should be switched to nitrogen. The furnace temperature is then lowered to 980-1010°C (1800-1850°F), and the samples are allowed to remain in the furnace for at least 20 minutes.

Caution: Pure copper melts at 1083°C (1981°F), so care must be taken to not approach this temperature. See Table 13.6.2 for results and examples.

Summary

While it is important to understand that chemical reactions occur in the furnace atmosphere, knowing which gas constituents will be oxidizing or reducing – causing decarburization or carburization to occur – is the heat treater's ultimate goal. This allows us to choose atmospheres and design control schemes to ensure the planned reactions take place.

REFERENCES
1. *ASM Handbook, Volume 4B: Steel Heat Treating Technologies*, ASM International, 2014
2. *Metals Handbook, Volume 2: Heat Treating, Cleaning and Finishing*, ASM International, 1964
3. Hotchkiss, A.G. and H.M. Webber, *Protective Atmospheres*, John Wiley & Sons Inc., 1953
4. *Furnace Atmospheres No. 1: Gas Carburizing and Carbonitriding*, Special Edition Booklet, Linde, 2007
5. Plicht, Guido and Rob Edwards, "N_2-Nitrogen On-Site Generation for Heat Treatment of Aluminium," *Metallurgia*, September 2002
6. Stempo, Michael J., "The Ellingham Diagram: How to Use it in Heat-Treat-Process Atmosphere Troubleshooting," *Industrial Heating*, April 2011
7. Stratton, Paul, "Ellingham Diagrams – Their Use and Misuse," *International Heat Treatment and Surface Engineering*, Volume 7, Issue 2, June 2013
8. Donald Bowe, Air Products & Chemicals Inc. (www.airproducts.com), technical and editorial review, private correspondence
9. James Oakes, Super Systems Inc. (www.supersystems.com), technical and editorial review, private correspondence

- **Industrial gas solutions**
- **Applications equipment**
- **Technical expertise**

Nitrogen • Hydrogen • Argon • Helium • Oxygen

Improve part quality and lower overall costs.
Contact us at 800-654-4567 or 610-706-4730.

tell me more
airproducts.com/mp

© Air Products and Chemicals, Inc., 2014 (37353)

9.4

CHARACTERISTICS AND APPLICATIONS (PART ONE)

The types of furnace atmospheres in common use in the heat-treatment industry can be divided into four broad-based categories:

- ❖ Atmospheres for preventing oxidation and/or reducing oxides
- ❖ Atmospheres for preventing decarburization
- ❖ Atmospheres for case hardening (carbonitriding, carburizing, nitriding and nitrocarburizing) and carbon restoration
- ❖ Atmospheres for special applications

We will review each.

Atmospheres for Preventing Oxidation and/or Reducing Oxides

Many people expect the outcome of a heat-treatment operation to produce "clean" and/or "bright" surfaces. However, these terms are highly subjective and difficult to define in a universal way. Instead, it is better to think in terms of a part's surface either being changed (i.e., metallurgically altered) or remaining essentially unchanged. Tempering in air is an example of the latter. A process that will produce a thin oxide layer on the part is often deemed acceptable for the intended service application. If a change does occur, under a given choice of furnace atmospheres, it must be evaluated to determine the impact of the process on the end result.

Oxidation/Reduction Reactions

Changes to a metal's surface can occur by oxidation (Equation 9.4.1) or by reduction (Equation 9.4.2) as a function of the gas species involved.[4,5] Therefore, the oxidizing or reducing potential always depends on the equilibrium ratios of the partial pressures of the gases, namely pH_2/pH_2O and pCO/pCO_2. The Ellingham-Richardson Diagram (Section 9.3) expresses these oxidizing or reducing potentials for different materials at different temperatures.

Air from leaks in piping and flanges, water vapor from water-cooled seals and housings, and leaky radiant tubes are examples of sources of oxygen, water and carbon dioxide.

9.4.1a) $\quad Me + ½O_2 \rightarrow MeO$

9.4.1b) $\quad Me + H_2O \rightarrow MeO + H_2$

9.4.1c) $\quad Me + CO_2 \rightarrow MeO + CO$

Reduction reactions occur due to the presence of favorable conditions or constituents in the atmosphere, such as hydrogen or carbon monoxide.

9.4.2a) $\quad MeO \rightarrow Me + ½O_2$

9.4.2b) $\quad MeO + H_2 \rightarrow Me + H_2O$

9.4.2c) $\quad MeO + CO \rightarrow Me + CO_2$

where "Me" represents a metal in Equations 9.1.4a-9.4.2c

Oxidation is a process in which oxygen combines with the metal to form a metal oxide on the surface. This can range from a thin, tightly adhered layer to one that consists of a thick, loose scale. Rusting of steel is a classic example of the effects of oxygen on iron. The effects of oxidation are accelerated with increasing temperature, and negating this effect is one of the principle reasons for the use of furnace atmospheres.

Reduction – the opposite of oxidation – is a process in which a metal oxide is reduced to its pure metallic state. Copper brazing or powder-metal sintering in a hydrogen or hydrogen/nitrogen atmosphere are examples of processes that rely on oxide-free surfaces to promote joining or bonding of metallic surfaces together.

Science tells us that oxidation/reduction (redox) reactions involve the exchange (i.e., transfer) of electrons between the metal and atmosphere that surrounds it. Oxidation is the loss of electrons (or an increase in oxidation state), while reduction is the gain of electrons (or the decrease in oxidation state).

A simple example is what will happen to iron (steel) as a function of temperature and oxygen content of the atmosphere (Fig. 9.4.1). Looking at the right-hand curve (which gives the equilibrium relations between iron, iron oxide, carbon monoxide and carbon dioxide), we see that the area to the right of the curve is in the oxidizing range, and the area to the left of the curve is in the reducing range. Given equal amounts of CO and CO_2, therefore, it should be possible to heat treat iron to approximately 538°C (1000°F) without discoloration. If the temperature is increased to 593°C (1100°F), oxidation will occur.[3]

The other curve provides the relationship between water vapor and iron. The area to the left of the curve indicates the reducing reaction, and the area to the right indicates the oxidation reaction. It can be seen from the curves that a gas containing 5% water vapor and 95% hydrogen (vertical line a-a') will not oxide steel at any temperature, and iron oxide will tend to be reduced. If we have a gas with 3.7% water vapor and 15% hydrogen (ratio=0.25) at a temperature of 650°C (1200°F), we enter the oxidizing region (vertical line b-b'). This tells us that it is impossible to slow cool steel in this atmosphere without oxidation taking place.[3]

FIGURE 9.4.1 | Theoretical equilibrium relations between iron and iron oxide when in contact with carbon monoxide and carbon dioxide or hydrogen and water vapor at heat-treating temperature[3]

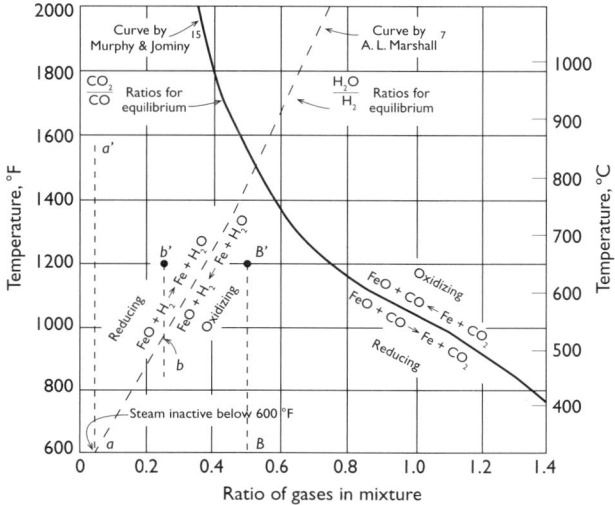

Let's review each of the common gases used for preventing oxidation and/or reducing oxides.

HYDROGEN

Hydrogen is the gas that comes immediately to mind when thinking about preventing oxidation or reducing metal oxides. Hydrogen is the lightest gas (about 1/15 the weight of air), having a gaseous specific gravity of 0.0695 and a boiling point of -252.8°C (-423°F) at atmospheric pressure. It's a colorless, odorless, tasteless, flammable gas found at concentrations of about 0.0001% in air (by volume). Hydrogen is produced by several methods, including steam/methane reforming, dissociation of ammonia, electrolysis and recovery from byproduct streams during chemical manufacturing and petroleum re-forming. Hydrogen is stored and transported as either a gas or a cryogenic liquid.

Hydrogen is flammable and burns with an almost invisible bluish flame. Its auto-ignition temperature is 574°C (1065°F) when mixed with air or oxygen. The flammability range is 4-74% in air. A full tertiary diagram (Fig. 9.4.2) for hydrogen, oxygen and nitrogen includes the flammability envelope for ambient conditions. Mixtures inside the envelope are flammable. Temperature corrections (Fig. 9.4.3) have been determined, assuming oxygen is the contaminant.

FIGURE 9.4.2 | Tertiary diagram for hydrogen, oxygen and nitrogen (with flammability envelope for ambient conditions)[6]

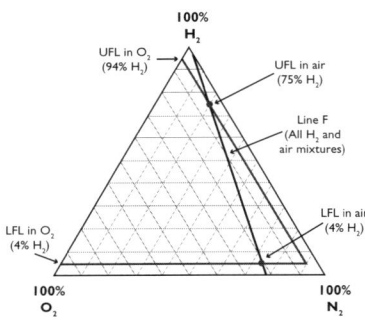

FIGURE 9.4.3 | Portion of right side of tertiary diagram for hydrogen, oxygen and nitrogen at elevated temperatures[6]

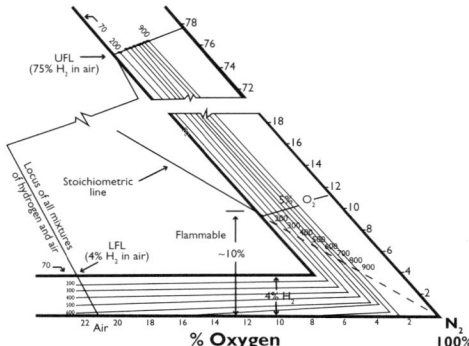

Hydrogen's thermal conductivity (Table 9.4.1) is about 10 times that of endothermic gas. It is an excellent conductor of heat, so furnace designs must take this into consideration with respect to shell temperature. Also, furnaces with high-hydrogen atmosphere contents can run hotter, so this must be taken into account when designing process recipes. Hydrogen sensors should be located at the high point in most shops (e.g., ceiling areas) and tied in with automatic exhaust fans (with explosion-proof motors). These sensors are triggered at a predetermined hydrogen value, typical anywhere below 25% of the lower flammability limit. In addition, they are often tied into a solenoid valve to close the supply line.

Hydrogen is used in heat treating primarily for its properties as a reducing gas.

TABLE 9.4.1 | Selected properties of gases used for furnace atmospheres[3]

Gas type	Specific gravity	Thermal conductivity (relative to air)	Thermal content, MJ/m^3 (BTU/cf)
Air	1.000	1.000	0 (0)
Argon	1.379	0.745	0 (0)
Carbon dioxide	1.527	0.590	0 (0)
Carbon monoxide	0.968	0.959	11.98 (300)
Dissociated ammonia	0.295	5.507	9.08 (244)
Endothermic gas (dry)	0.622	3.228	7.07 (190)
Endothermic gas (wet)	0.798	2.663	5.25 (141)
Exothermic gas (lean)	1.030	0.994	0.19 (5)
Exothermic gas (rich)	0.858	1.878	3.57 (96)
Hydrogen	0.069	7.010	12.10 (325)
Methane	0.554	1.127	37.72 (1,013)

NITROGEN

Nitrogen is the most common purging or blanketing gas used in heat treatment and is considered to be inert to most materials. As a function of temperature, however, nitrogen is reactive to certain materials, especially alloys containing chromium, molybdenum, titanium and niobium (columbium).

The principal reason for nitrogen's popularity is cost. What is important for the heat treater to remember is that nitrogen is nonreactive to the air or oxygen present in or entering the furnace. As such, nitrogen must "push" oxygen molecules out of the furnace rather than react with them. Furnace pressure and atmosphere flow, as well as the tightness of the furnace or oven, become important considerations when attempting to run clean, or "bright," work.

Nitrogen makes up 78.03% of air (by volume), has a gaseous specific gravity of 0.967 and a boiling point of -195°C (-320.5°F) at atmospheric pressure.

It is colorless, odorless and tasteless. Commercially, nitrogen is produced by a variety of air-separation processes, including cryogenic liquefaction and distillation, adsorption separation and membrane separation. Nitrogen can also be supplied from nitrogen generators (Fig. 9.4.4) using either membrane (Fig. 9.4.5) or adsorption (Fig. 9.4.6) technology.

A membrane "mechanically" separates nitrogen from oxygen and other molecules but does not involve a chemical process. The basic components of the system are an air-supply source (e.g., air compressor), dryer, filters, pre-storage tank, nitrogen membrane and nitrogen storage tank.

A pressure swing adsorption system (PSA) is a vessel filled with a carbon molecular sieve (CMS) pressurized with air. Oxygen, carbon dioxide, carbon monoxide and certain other molecules (e.g., ammonia) are captured while the nitrogen is drawn off into a receiving tank. Depressurizing the sieve bed then flushes the trapped gases and regenerates the CMS, which is ready for more air. The basic components of the system are an air-supply source, dryer, filters, pre-storage tank, dual CMS tanks (one active, one discharging byproducts), nitrogen storage tank and gas-analysis equipment (e.g., oxygen, dew point).

Gas purity with these systems can vary from 0.001-5% oxygen and is a function of the flow rate and size of the system.

FIGURE 9.4.4 | Typical nitrogen generator system (courtesy of South-Tek Systems)

a.

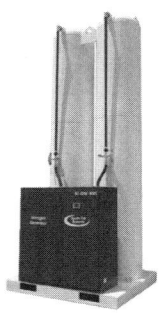

b.

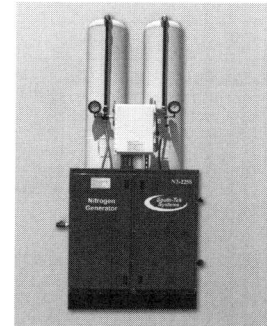

a. Typical mounting arrangements: tank mounted, compact skid, skid mounted (left to right)
b. Nitrogen generator with output ranges from 15 m³/hour (532 scfh) at 99.999% purity to 170 m³/hour (6,042 scfh) at 99% purity

FIGURE 9.4.5 | Schematic representation of membrane technology (courtesy of South-Tek Systems)

a.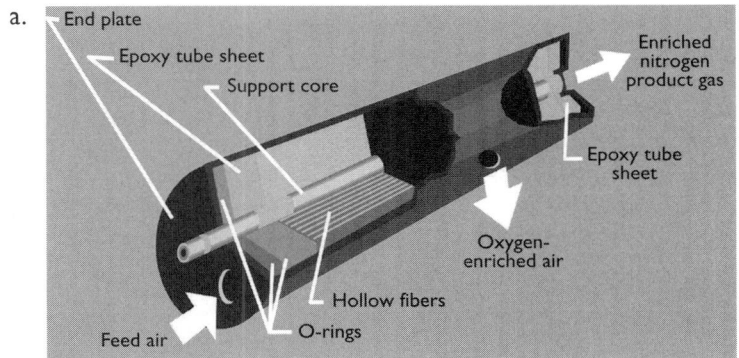

b.

c.

a. Basics of membrane technology
b. Cut-away view of a membrane unit
c. Installed membrane unit

FIGURE 9.4.6 | Schematic representation of PSA adsorption technology

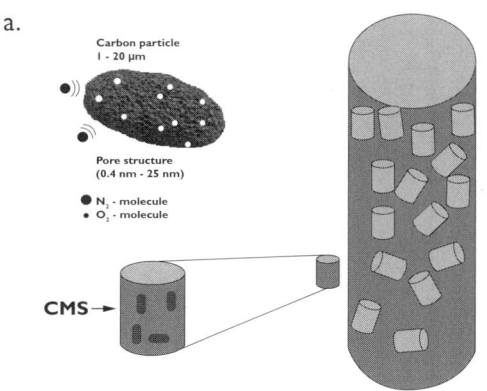 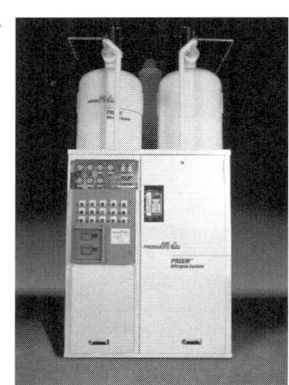

a. Sieve bed (courtesy of South-Tek Systems)
b. Working PSA unit (courtesy of Air Products and Chemicals Inc.)

Nitrogen used for atmosphere furnace applications is, in most cases, adequately supplied by industrial-grade gas, often either generated on site or supplied from a cryogenic source (Fig. 9.4.7). The typical impurity levels in the specification for industrial-grade nitrogen are 10 ppm oxygen (maximum) and a minimum dew point of -68°C (-90°F) or lower (3.4 ppm by volume). The actual levels are usually in the 2 ppm range for both oxygen and water vapor. It is often the case that there is more "pick up" of impurities in the piping to the equipment than in the supply product itself, primarily from leaks.

FIGURE 9.4.7 | Typical cryogenic nitrogen storage system (courtesy of Air Products and Chemicals Inc.)

9 | FURNACE ATMOSPHERES

ARGON

Argon is heavier than air and a truly inert gas. This makes it ideal for the processing of materials to prevent hydrogen embrittlement, which could occur with titanium, tantalum, niobium and zirconium alloys or nitriding of stainless steels. Argon is more expensive than nitrogen and is used primarily in applications where its benefits outweigh its cost and other disadvantages (such as the fact that it will tend to accumulate in pits, basements and other low areas of the shop). As such, oxygen monitors should be placed near the floor to detect safe levels of oxygen when argon is in use.

Argon is a monatomic, chemically inert gas comprised of slightly less than 1% air (by volume). Its gaseous specific gravity is 1.38, and its boiling point is -185.9°C (-302.6°F). Argon is colorless, odorless, tasteless, noncorrosive, nonflammable and nontoxic. Commercial argon is the product of cryogenic air separation, where liquefaction and distillation processes are used to produce a low-purity crude argon product that is then purified to the commercial product.

CARBON DIOXIDE

Carbon dioxide is much heavier than air, having a specific gravity of 1.53 (air =1.0). It is a nonflammable, colorless, odorless gas found in air at concentrations of about 0.03% (by volume). Although not an inert gas, carbon dioxide is nonreactive with many materials (e.g., copper). Carbon dioxide does not oxidize iron or most alloying elements in steel (e.g., Cr, Mn, V, Si, Al, Zn) at elevated temperature (Fig. 9.4.1). Oxygen (typically 0.03% maximum) and nitrogen are the most common impurities. Carbon dioxide is used for fire prevention and fire extinguishing as well as cold treating (e.g., dry ice).

HELIUM

Helium is used in atmosphere heat treatment only for specialized applications due to cost. If used, it is often recycled. Helium is lighter than air and a truly inert gas. Helium is a chemically inert gas, the second-lightest gas behind hydrogen and has a gaseous specific gravity of 0.138. It's a colorless, odorless, tasteless gas with a boiling point of -268.9°C (-452.1°F) at atmospheric pressure. Helium is present in air at a concentration of 0.0005% (by volume). The principal supply source is certain natural gas deposits where the crude helium is extracted from the natural gas stream and then purified. Helium can be stored and shipped either as a compressed gas or cryogenic liquid.

GENERATED ATMOSPHERE

Dissociated ammonia, exothermic gas and endothermic gas can be used to prevent oxidation for most materials and in some cases reduce oxides. These will be discussed in detail in Section 9.6 with several comments appropriate here.

Dissociated ammonia is flammable and about one-third as heavy as air. It can be used as a hydrogen substitute in applications where nitrogen and residual ammonia (50-250 ppm) are acceptable. Typical uses include sintering of powder metal, copper brazing, bright annealing and dilution of nitriding atmospheres for the purpose of compound-layer (white-layer) compositional control.

Exothermic gas is inexpensive (arguably the least expensive gas other than air) with good versatility due in part to the fact that it can be produced in a lean or rich ratio (Table 9.2.6). Rich exothermic gas is combustible, often burning with what has been described as a "wispy" flame. As such, proper atmosphere flow rates and furnace pressures are necessary.

Exothermic gas is mildly reducing due to its hydrogen and carbon monoxide content. Copper brazing, hardening, sintering and bright annealing are typical examples. It will, however, decarburize medium- to high-carbon steels (unless water vapor and carbon dioxide are removed). Lean exothermic gas is noncombustible. It can be used for purging or blanketing as well as annealing of brass and bronze and silver brazing. Dryers and chillers can be used to lower the dew point from +38°C (+100°F) to +15°C (+60°F) or -40°C (-40°F), if required. It can also be used for fire prevention and fire extinguishing as a lower-cost alternative to carbon dioxide.

Due to its high carbon monoxide and hydrogen contents, endothermic gas is flammable and burns with what is often described as a robust flame. It is a popular choice when hardening, brazing, annealing, normalizing or case hardening.

Both exothermic and endothermic gas compositions vary by ratio.

Atmospheres for Preventing Decarburization

The common choices of atmosphere for preventing decarburization include hydrogen, nitrogen, dissociated ammonia, purified exothermic gas (i.e., CO_2-free) and endothermic gas. The hydrogen content of the atmosphere plays a significant role in the depth of the decarburized layer (Fig. 9.4.8).

FIGURE 9.4.8 | Surface decarburization as a function of hydrogen content of the atmosphere[10] (Copyright Linde AG, courtesy of Linde LLC)

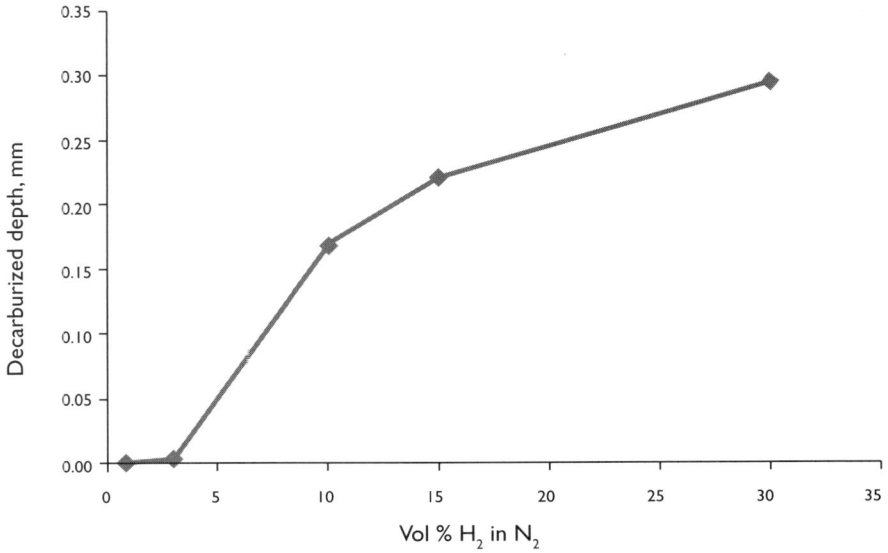

Examples of the effect of these gases on common steels will be shown for exothermic (Fig. 9.4.9) and endothermic (Fig. 9.4.10) gas. Samples in purified exothermic show a slight increase in weight (indicating carbon pickup), while samples run in unpurified gas show a sharp decrease in weight (indicating decarburization). By contrast, samples run in endothermic gas under controlled conditions show considerable increase in carbon but within acceptable limits for neutral hardening. Reference 3 provides additional details on the test parameters.

Finally, Gonser[7] showed the need for purified exothermic gas to prevent surface decarburization (Fig. 9.4.11). In this study, steel samples containing 0.10-1.0% C were run at 950°C (1740°F) for three hours in exothermic gas with 9% hydrogen and 9% carbon monoxide. In curve 1 (5.3% CO_2), all 0.10% C samples decarburized. By contrast, curve 2 (1.4% CO_2) shows that carbon steels of 0.4% C and less picked up carbon while above 0.4% C decarburized. Curve 3 (0.9% CO_2) showed little change in the low- to medium-carbon steels but some improvement for high-carbon steels. Curve 4 (0.2% CO_2) avoided decarburization for steels up to 1.1% C. Note that at 1.1% CH_4 the atmosphere remained neutral up to 0.7% C when compared to curves 1 and 2, which were neutral to only a 0.5% C steel.[3]

FIGURE 9.4.9 | Results of weight-change tests using an atmosphere of purified (CO_2-free, dried) and unpurified exothermic gas[3]

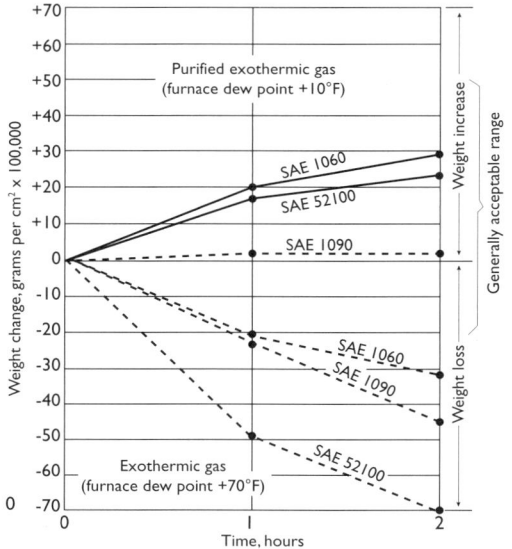

FIGURE 9.4.10 | Results of weight-change tests using an atmosphere of (dry) endothermic gas[3]

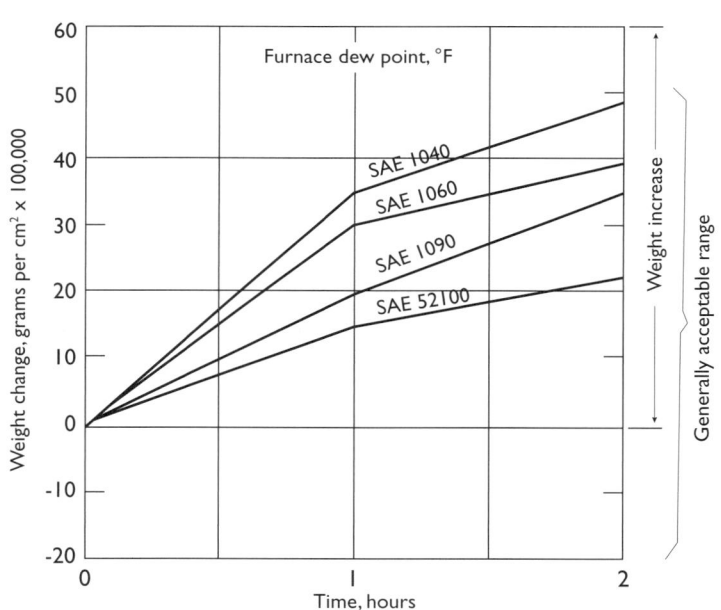

FIGURE 9.4.11 | Gonser's curves showing slight carburizing action of purified exothermic gas (curve 5) compared to increasing decarburizing results (curves 4-1) with increasing CO_2 content[3]

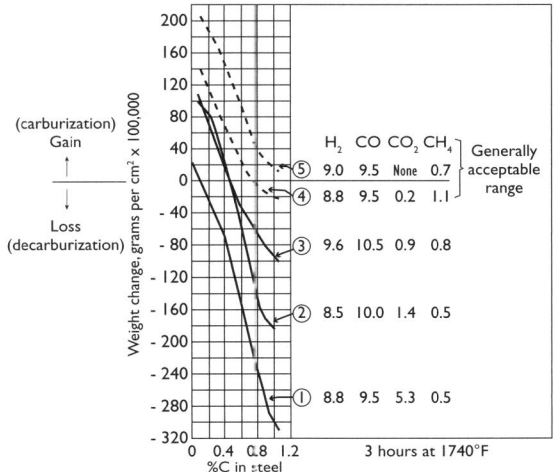

Atmospheres for Case Hardening and Carbon Restoration

Case hardening (Section 5.5) is most commonly performed in an endothermic gas atmosphere or a methanol or nitrogen/methanol atmosphere acting as a "carrier gas" neutral to the surface of the steel enriched at the furnace with a hydrocarbon gas (e.g., methane or propane).

CARBURIZATION/DECARBURIZATION REACTIONS

Changes to a metal's surface can also occur due to the pickup of carbon (carburization) or loss of carbon (decarburization).

Carburization occurs when carbon atoms are deliberately added at the steel surface from the furnace atmosphere. The carburizing potential in the furnace is determined by the atmosphere gas composition. Since surface carbon concentration and carbon flux from the atmosphere to the steel surface changes with time, maintaining a constant atmosphere carbon potential requires continuous adjustment of the carbon setpoint.[3]

Decarburization occurs when carbon atoms at the steel surface interact with the furnace atmosphere and are removed from the steel as a gaseous phase. Carbon from the interior will then diffuse outward toward the surface; i.e., carbon diffuses from a region of high concentration to a region of low concentration, and the decarburization process continues. Because the rate of carbon diffusion increases with temperature when the structure is fully austenitic, the depth of total decarburization will increase as the temperature increases above the Ac_3. For temperatures in the two-phase region – between the Ac_1 and Ac_3 – the process is more complex. The diffusion rates of carbon in ferrite and austenite are different, and they are influenced by temperature and composition. Decarburization (Fig. 9.4.12) is a serious problem because the surface properties will be inferior and result in poor wear resistance and low fatigue life.[4]

FIGURE 9.4.12 | Section of fastener thread showing decarburization, 50X (courtesy of Struers)

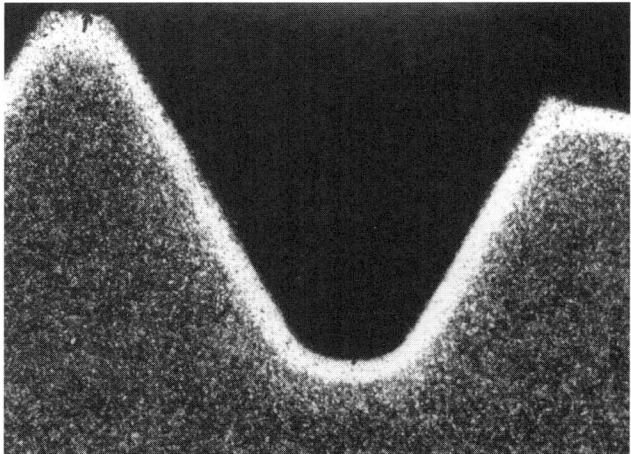

Carburizing/decarburizing reactions consider the effects of furnace-atmosphere composition on the carbon content of the steel in question as a function of temperature. In many cases, we not only want to prevent oxidation from occurring but control the carbon content such that carbon is neither added to nor taken away from the surface of the steel (e.g., neutral hardening) or carbon is deliberately added to the surface (e.g., carburizing).

This will be explained in more detail in Section 9.5, but it suffices for illustrative purposes here showing regions where carburizing conditions can exist for a steel of specific carbon content.

To ensure that reversible reactions with CO_2 and iron carbide remain at equilibrium, a fixed relationship must exist between CO and CO_2 (Fig. 9.4.13). This relationship is a function of temperature and carbon content. For example, a CO/CO_2 ratio of 0.4% C at 900°C (1650°F) is neutral but carburizing to steel at 0.2% C and decarburizing to steel at 0.8% C. If the same ratio was maintained but the temperature was increased, the 0.4% C would fall into the decarburizing range. If the temperature was decreased, it would fall into the carburizing range.[3]

When hydrocarbon additions are made, the ratio of CH_4 to H_2 (Fig. 9.4.14) becomes an important process variable. Any gas mixture that contains more hydrogen than the equilibrium amount will tend to decarburize. As the temperature is raised, small amounts of CH_4 will result in a carburizing atmosphere.[3]

FIGURE 9.4.13 | Carburizing or decarburizing reactions of CO and CO_2 and their relation to carbon in the steel at different steel temperatures[3]

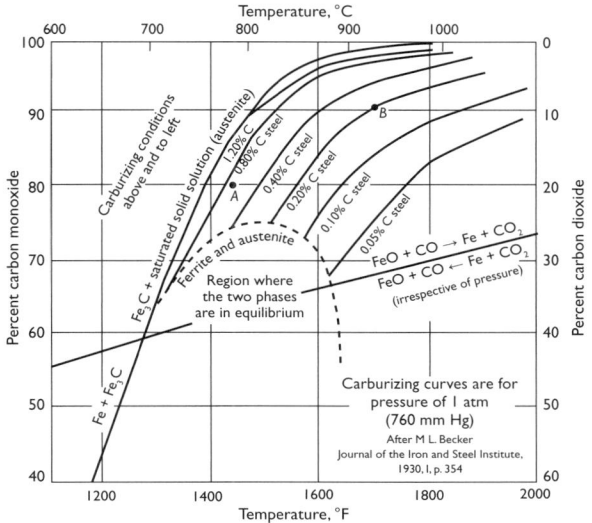

FIGURE 9.4.14 | Carburizing/decarburizing equilibrium of CH_4 for various carbon steels[3]

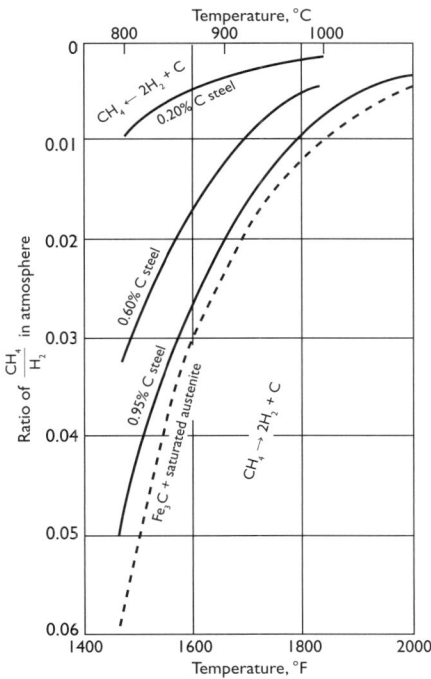

Carbon restoration differs considerably from carburizing in that the carbon level of the decarburized surface layer is less than the core carbon content of the material. The success of the process depends on establishing exact equilibrium between the carbon potential of the atmosphere and the carbon level of the steel prior to decarburization (Fig. 9.4.15).

FIGURE 9.4.15 | Typical carbon restoration cycle with carbon (shaded area) added[3]

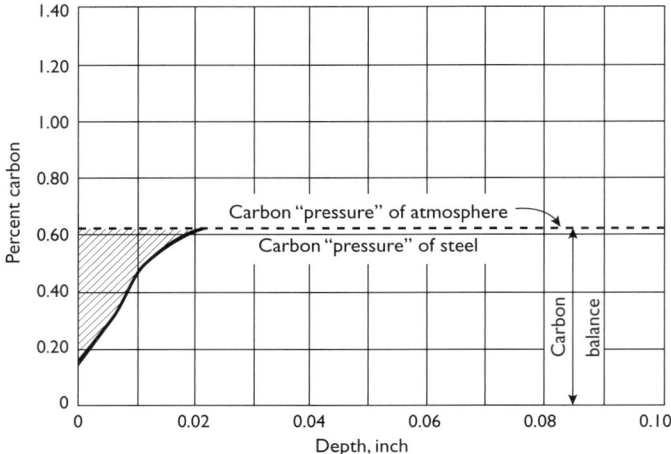

Atmospheres for Special Applications

Designing furnace atmospheres for processing exotic materials or running special processes requires knowledge of the tolerable surface reactions allowed before specific atmospheres and gas characteristics/compositions can be selected. The use of argon for the processing of titanium fasteners is one such example. The type of furnace (e.g., shaker-hearth, rotary-drum) is an important consideration as is the proper flow rate and volume turnover to maintain an oxygen-free atmosphere.

In another example, -98°C (-144°F) ultra-low dew point, impurity-fee hydrogen is required for processing electronic components, where high dielectric strength and electrical resistivity of certain solder alloys is of principal importance.[10]

In many instances, vacuum heat treatment[1] is preferred to avoid unwanted surface reactions. Even here, however, choices of partial pressure and quench gas must be carefully considered.

Summary

The type of furnace atmosphere is a critical decision that depends not only on the characteristics of the atmosphere but on the heat-treatment process selected, the equipment used and (to some extent) on the end-use service application. The wrong choice or misapplication of a furnace atmosphere can result in disastrous consequences.

REFERENCES

1. Herring, Daniel H., *Vacuum Heat Treatment*, BNP Media, 2012
2. Herring, Daniel H., "Selection and Use of Furnace Atmospheres," *Industrial Heating* webinar, 2011
3. Hotchkiss, A.G. and H.M. Webber, *Protective Atmospheres*, John Wiley & Sons Inc., 1953
4. Andersson, Rolf, Torsten Holm, Sören Wiberg and Anders Åstrom, "Furnace Atmospheres No. 4: Brazing of Metals, Special Edition," Booklet, 2005
5. Plicht, Guido and Rob Edwards, "N2-Nitrogen On-Site Generation for Heat Treatment of Aluminium," *Metallurgia*, September 2002
6. Dwyer Jr., John, James G. Hansel and Tom Philips, "Temperature Influence on the Flammability Limits of Heat Treating Atmospheres," Air Products & Chemicals, 2003
7. Gosner, Bruce W., "The Status of Prepared Atmospheres in the Heat Treatment of Steel," *Industrial Heating*, December 1939
8. Karabelchtchikova, Olga, "Fundamentals of Mass Transfer in Gas Carburizing," PhD Dissertation, November 2007
9. Vander Voort, George, Decarburization, VAC AERO International, 2011
10. "Furnace Atmospheres No. 8: Sintering of Steels," Special Edition Booklet, Linde, 2011
11. Cook, B.A., I.E. Anderson, J.L. Harringa and R.L. Terpstra, "Effect of Heat Treatment of the Electrical Resistivity of Near-Eutectic Sn-Ag-Cu Pb-Free Solder Alloys," *Journal of Electronic Materials*, Vol. 31, No. 11, 2002
12. Donald Bowe, Air Products & Chemicals Inc. (www.airproducts.com), technical and editorial review, private correspondence
13. James Oakes, Super Systems Inc. (www.supersystems.com), technical and editorial review, private correspondence

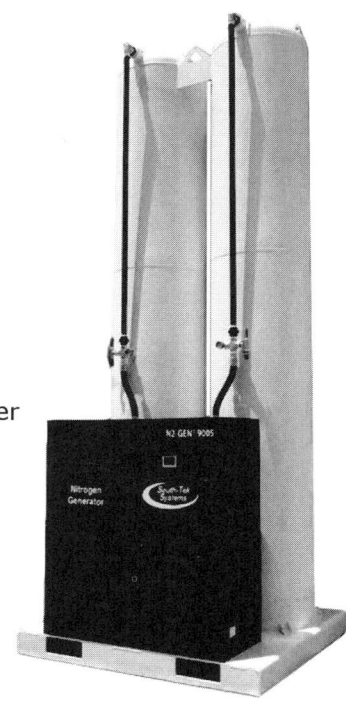

9.5

CHARACTERISTICS AND APPLICATIONS (PART TWO)

The choice of furnace atmosphere (Fig. 9.5.1) for any given process application is a function of the material, processing temperature and time, and the nature of the atmosphere (active or passive) to achieve the desired surface characteristics as well as properties (metallurgical, mechanical, physical). The type of equipment also plays a role in some instances. Our purpose here is to review the major applications and the factors that influence atmosphere selection.

FIGURE 9.5.1 | Variables influencing choice of atmosphere[3] (Copyright Linde AG, courtesy of Linde LLC)

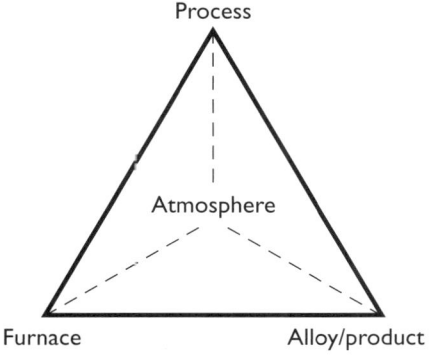

Annealing

Annealing (Section 5.2) involves heating a ferrous or nonferrous alloy above its critical temperature and holding at processing temperature followed by cooling at a controlled (slow) rate. Annealing is performed for such purposes as reducing hardness, improving machinability, facilitating cold working, producing a desired microstructure (for subsequent operations) and reducing internal stresses. It is also carried out for obtaining desired mechanical, physical or other properties. Cooling is often performed in the furnace.

ATMOSPHERES

Various types of furnace atmospheres are used for annealing (Table 9.5.1) and are primarily determined by the desired surface condition of the product being processed. For example, 100% hydrogen is used worldwide by steelmakers for batch bright (subcritical) annealing of semi-finished products such as strip and wire. Improved surface finish and dramatically reduced cycle times have been reported[12] over traditional 10% H_2/N_2 (HNX) atmosphere due to the greater thermal conductivity of pure H_2.

Decarburization-free annealing is always conducted in a very dry, oxygen-free atmosphere (e.g., hydrogen, nitrogen/hydrogen or nitrogen).

TABLE 9.5.1 | Atmosphere choices for annealing

Material	Atmosphere choices
Low-alloy steel	Exothermic, endothermic, nitrogen-methanol, nitrogen/hydrogen, nitrogen, dissociated ammonia, hydrogen, vacuum
Stainless steel (austenitic)	Dissociated ammonia, hydrogen, vacuum
Stainless steel (ferritic)	Dissociated ammonia, hydrogen, vacuum
Stainless steel (duplex)	Dissociated ammonia, hydrogen, vacuum
Stainless steel (precipitation hardening)	Dissociated ammonia, hydrogen, vacuum
Tool steels (M, T series)	Endothermic, nitrogen-methanol
Cu alloys	Nitrogen, hydrogen
Al alloys	Air, nitrogen, products of combustion (limited)
Ni alloys	Dissociated ammonia, hydrogen
Ti alloys	Vacuum, argon, helium

APPLICATIONS

Metal tube and pipe are typical examples of products that are annealed to improve forming and machining. Manufacturers often dictate system requirements in the case of stainless steel and exotic alloys. Typical requirements may include the following variables.

- ❖ Tube outside-diameter size range: 4.75-100 mm (0.1875-4.00 inches)
- ❖ Tube wall size range: 0.4-5.1 mm (0.016-0.200 inch)
- ❖ Alloy: 304/304L, 316/316L, 321, 455, 625, 17-7 PH, 718, 21-6-9, AL6XN
- ❖ Furnace capacity: 680 kg/hour (1,500 pounds/hour)

- ❖ Furnace operating temperature: 815-1175°C (1500-2150°F)
- ❖ Cooling rate: 1040 to 315°C (1900 to 600°F); 65 to 205°C/min (150 to 400°F/ min), variable
- ❖ Pyrometry must meet or exceed AMS 2750, furnace class 2, ±5.5°C (±10°F)
- ❖ Dew point and oxygen (0-100 ppm) at multiple locations
  - Atmosphere supply (H_2, N_2, Ar)
  - Hot-zone muffle
  - Transition zone
  - Fan cool
  - Cooling zone (two locations)
  - Exit purge
- ❖ Water (temperature, pressure, flow): fan cool (entry and exit of heat exchanger)
- ❖ Furnace type mesh-belt conveyor with muffle
- ❖ Atmosphere requirements: nitrogen, hydrogen, argon

Annealing atmospheres for carbon-steel tubes often require the added condition of carbon control (Fig. 9.5.2) so that the atmosphere will consist of nitrogen, hydrogen and carbon monoxide.

FIGURE 9.5.2 | Atmosphere control based on gas analysis – nitrogen and endothermic gas[3] (Copyright Linde AG, courtesy of Linde LLC)

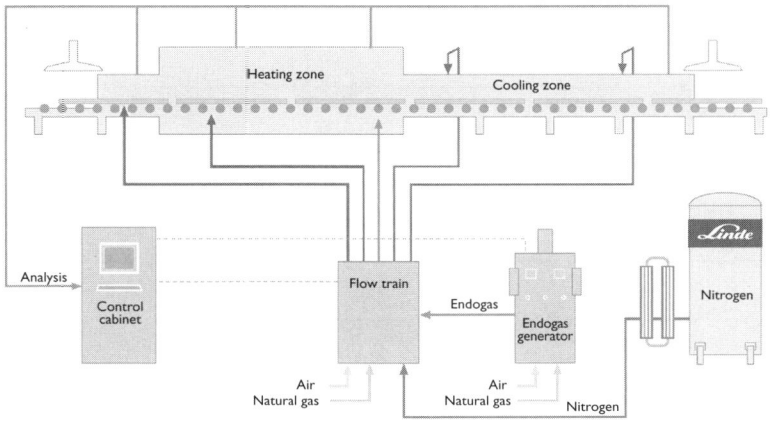

Bell annealing provides another example of the importance of gas analysis throughout the cycle and adequate gas flow during the period where lubricant burn-off (vaporization) takes place (Fig. 9.5.3).

FIGURE 9.5.3 | Furnace atmosphere analysis during heating of coil stock in a bell furnace[3] (Copyright Linde AG, courtesy of Linde LLC)

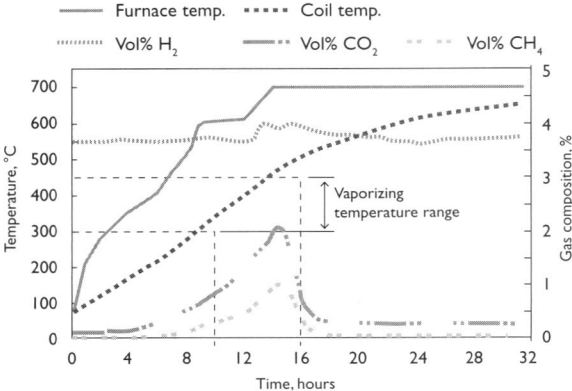

EQUIPMENT

Annealing furnaces are designed to provide a controlled cooling rate from elevated temperature (Table 9.5.2) in order to achieve specific properties (microstructural, metallurgical) in the product being processed.

TABLE 9.5.2 | Selected annealing temperatures[3]

Material	Annealing process temperature, °C (°F)		
	Recrystallization	Stress relief	Solution/aging
Low-alloy steel	650-705 (1200-1300)	550-650 (1020-1200)	
Stainless steel (austenitic)	950-1150 (1740-2100)		
Stainless steel (ferritic)	650-1050 (1200-1920)		
Stainless steel (duplex)	950-1175 (1740-2150)		
Stainless steel (precipitation hardening)			955-1040/480-620 (1750-1900/900-1150)
Tool steels (M, T series)		600-750 (1110-1380)	
Cu alloys	250-825 (480-1520)	150-350 (300-660)	[a]
Al alloys	260-480 (500-900)	200-350 (390-660)	[a]
Ni alloys	750-1200 (1380-2200)	480-1200 (900-2190)	
Ti alloys	650-900 (1200-1650)	480-815 (900-1500)	

Note:

[a] Alloy system dependent

9 | FURNACE ATMOSPHERES

The most common types of continuous-annealing furnaces are the horizontal mesh-belt conveyor (Fig. 9.5.4) and the roller hearth (Fig. 9.5.5). They are typically used for annealing of tubular products, bar, strip and wire coils. Pusher furnaces (Fig. 9.5.6) are also used, particularly for improved machinability via an isothermal annealing process. Finally, tube (strand) annealing furnaces process wire, thin tubular products and strip. Inclined mesh-belt (humpback) conveyor furnaces are also used, albeit in specialized applications (e.g., stainless steel).

The most common types of batch-annealing furnaces are bell furnaces (Fig. 9.5.7), which are designed to process wire coils and flat strip. Box (Fig. 9.5.8), car-bottom and tip-up furnaces (Fig. 9.5.9) are typically used for processing large/heavy loads.

FIGURE 9.5.4 | Typical mesh-belt conveyor furnace (courtesy of Air Products and Chemicals Inc. and St. Marys Pressed Metals)

FIGURE 9.5.5 | Typical roller-hearth furnace (courtesy of SECO/WARWICK Corp.)

FIGURE 9.5.6 | Typical pusher furnace with entrance vestibule to prevent atmosphere disruption (courtesy of Ipsen)

FIGURE 9.5.7 | Typical bell furnace (courtesy of Lindberg/MPH)

FIGURE 9.5.8 | Typical atmosphere box furnace (courtesy of L&L Special Furnace Co. Inc.)

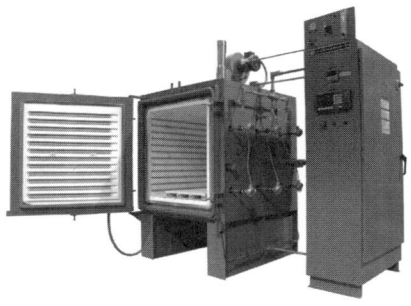

FIGURE 9.5.9 | Typical tip-up furnace (courtesy of J.L. Becker, a Gasbarre Furnace Group Company)

Brazing

Brazing (Section 5.9) is the joining of component parts by flowing a filler metal into the space between them via capillary action. Bonding results from the intimate contact produced by dissolution of a small amount of base metal into the molten filler metal without melting of the base metal. The term brazing is used when the filler metal has a liquidus (melting) temperature of 450°C (840°F) or higher. Brazing is a joining process, not a heat-treatment process per se.

ATMOSPHERES

Most brazing atmospheres are dependent on both the braze alloy selected and base metal involved (Table 9.5.3). The role of the furnace atmosphere is to allow the braze alloy to melt, flow and solidify unimpeded over a narrow range of temperatures. Thus, the atmosphere must be either neutral to the base and filler metal (e.g., argon or nitrogen) or reducing – e.g., pure hydrogen (100%), dissociated ammonia (75% hydrogen) or hydrogen/nitrogen mixtures from 33-90% hydrogen – to assist in eliminating residual surface oxides. Reactions involving oxidation, decarburization, carburization and nitriding must be avoided.

TABLE 9.5.3 | Common furnace atmospheres for brazing[5]

Base metal	Braze alloy	Furnace atmosphere
Aluminum, copper	Al/Si, Cu/Sn/Ni/P	Nitrogen (100%)
Copper, low-carbon steel	Cu, Cu/Ag, Cu/P, Cu/Zn	Exothermic, purified exothermic
Copper, low- and high-carbon steels, low-alloy steels	Cu, Cu/Ag, Cu/P, Cu/Zn	Endothermic
Copper, brass, stainless steel, carbon and low-alloy steels	Cu, Cu/Ag, Cu/P, Cu/Zn, Ni	Dissociated ammonia
Copper, brass, carbon and low-alloy steels, stainless steels	Cu/Ag, Cu/P, Cu/Zn, Cu, Ni	Nitrogen/hydrogen
Copper, carbon and low-alloy steels	Cu/Ag, Cu/P, Cu/Zn, Cu, Ni	Exothermic, endothermic, nitrogen/methanol
Carbon and low-alloy steels	Cu/Ag, Cu/P, Cu/Zn, Cu, Ni	Nitrogen/hydrocarbon
Stainless steels, steel, brass	Cu/Ag, Cu/P, Cu/Zn, Cu, Ni	Hydrogen
Steels containing titanium, zirconium, nitrogen	Cu/Ag, Cu/P, Cu, Zn, Cu, Ni	Argon

APPLICATIONS

Silver brazing of brass and copper HVAC components (Fig. 9.5.10) is a challenge for the furnace atmosphere due to contamination from zinc vapors. Dezincification is a dealloying process brought about by the low vapor pressure of zinc at elevated temperature. A fast belt speed is required to minimize the effects of dezincification and keep the brass reasonably clean. An atmosphere of 28% nitrogen + 72% hydrogen at a dew point of -29°C (-20°F) has been found effective in the temperature range of 705-720°C (1300-1325°F). A heat head is typically run in zone 1. Zinc deposits in the entrance tunnel and cooling jackets require shutdown of the equipment and mechanical cleaning.

FIGURE 9.5.10 | Silver brazing of HVAC components

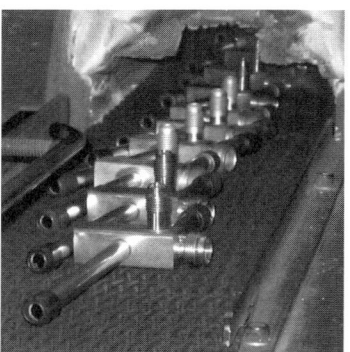

The formula (Equation 9.5.1) for vapor pressure (P) of zinc (in atmospheres) is a function of temperature (T) expressed in degrees Kelvin (K) over the range from melting (693K) to boiling (1177K).[7]

9.5.1) $\ln(P) = -(A/T) + [B \bullet \ln(T)] + C$

where: (for Zn) A = 15,250, B = -1.255 and C = 21.79

At 693K (420°C, 790°F), 1074K (800°C, 1475°F) and 1177K (905°C, 1658°F), the values for P are as follows: P (693K) = 0.00022 atmospheres; P (1074K) = 0.3144 atmospheres; P (1177K) = 0.9617 atmospheres. This illustrates the exponential nature of zinc vaporization as a function of increasing temperature.

Nickel brazing of orthodontic braces (Fig. 9.5.11) can be done in a muffle-style mesh-belt conveyor furnace running dissociated ammonia (75% hydrogen) as long as the dew point is held in the range of -46 to -51°C (-50 to -60°F).

FIGURE 9.5.11 | Nickel brazing of orthodontic braces

9 | FURNACE ATMOSPHERES

EQUIPMENT

Brazing furnaces are designed to produce a specific thermal profile (Fig. 9.5.12) in order to allow the braze alloy to melt, flow and solidify under controlled conditions. This thermal profile consists of preheating, soaking (dwell) below the melting point of the filler metal, ramping at a controlled rate to temperature, holding at brazing temperature for a prescribed period of time, slow cooling (to "set the braze") and cool down to handling temperature. Brazing temperature is selected based on the type of braze alloy (filler metal) chosen (Table 9.5.4).

FIGURE 9.5.12 | Typical brazing cycle[5] (Copyright Linde AG, courtesy of Linde LLC)

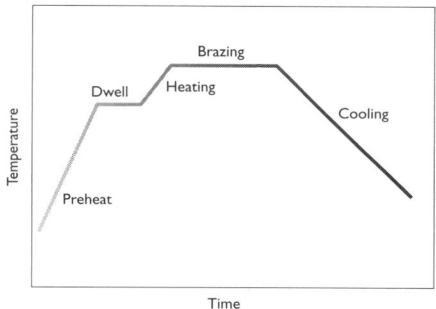

TABLE 9.5.4 | Common brazing temperatures[5]

Braze alloy (filler metal)	AWS designation	Brazing temperature, °C (°F)
Al-Si	BAlSi	555-650 (1030-1200)
Au-Ni	BAu	890-1230 (1645-2245)
Cu-X	BCu	1095-1150 (2000-2100)
Cu-Zn	RBCuZn	910-980 (1670-1800)
Cu-P	BCuP	690-925 (1275-1700)
Cu-Ag	BAg	620-980 (1150-1800)
TM-Si-B	BNi	925-1200 (1700-2200)
(Co, Cr)-Si, B	BCo	1175-1245 (2150-2275)

The most common types of continuous brazing furnaces are horizontal mesh-belt conveyor furnaces (Fig. 9.5.13) in an open chamber or muffle configuration and inclined (humpback) mesh-belt conveyor furnaces (Fig. 9.5.14) utilizing a muffle. They are typically designed to allow the product to move in a continuous fashion through the furnace. Pusher furnaces and roller-hearth furnaces are also used when higher production is warranted.

The most common types of batch furnaces for annealing are box furnaces, often equipped with a retort (Fig. 9.5.15) so that heating and cooling is performed in a controlled-atmosphere environment.

FIGURE 9.5.13 | Typical horizontal mesh-belt conveyor furnace for brazing (courtesy of Abbott Furnace Company)

FIGURE 9.5.14 | Typical inclined (humpback) conveyor furnace (courtesy of C.I. Hayes, a Gasbarre Furnace Group Company)

FIGURE 9.5.15 | Typical box furnace with retort (courtesy of Nabertherm)

9 | FURNACE ATMOSPHERES

Carburizing/Carbonitriding

Carburizing and carbonitriding (Section 5.5) are case-hardening processes in which a workload is brought to austenitizing temperature and allowed to dwell at temperature for a prescribed period of time in which carbon (during carburizing) or carbon and nitrogen (during carbonitriding) transfer from the atmosphere to the surface of the part. Parts are then either rapidly quenched or slow cooled (for subsequent manufacturing operations) prior to reheating in a neutral atmosphere and quenched (Fig. 9.5.16).

FIGURE 9.5.16 | Case hardening of steel by austenitizing and quenching[3] (Copyright Linde AG, courtesy of Linde LLC)

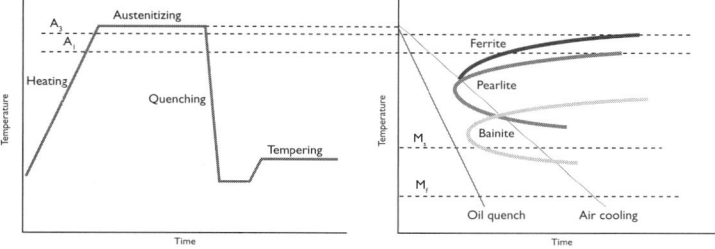

ATMOSPHERES

The furnace atmosphere used for carburizing involves a carrier gas plus a hydrocarbon enrichment gas, typically natural gas or propane. Different carbon-concentration profiles in the part can be achieved by varying the carbon potential of the gas during the carburizing cycle.

Two common carburizing methods are used. The first of these is single-stage carburizing (aka the constant carbon-potential method) in which one fixed carbon potential is held throughout the cycle (e.g., 0.8% C). The other method is two-stage carburizing (aka boost/diffuse carbon-potential method) in which a higher carbon potential (typically near the limit of saturation of carbon in austenite) is used for a portion of the cycle. A lower carbon potential is used in the latter portion of the cycle to allow diffusion of carbon away from the surface (e.g., 1.1% C boost, 0.80% C diffuse).

It is also important to note that, primarily for distortion control during quenching, the temperature at the end of the carburizing cycle is often lowered (aka the drop-temperature method) followed by a soak at the lower temperature. Carbon control of the workload during the time when the temperature is dropping becomes an important process variable to control.

Finally, some shops add ammonia near the end of the carburizing cycle (typically in the last 30 minutes) to diffuse a small amount of nitrogen into the surface of the steel, primarily for a slight surface-hardness improvement. The carbon potential of the atmosphere will change during this period because of the addition of another gas species.

Carrier gases such as endothermic gas and nitrogen/methanol are common throughout the heat-treatment industry. Each will be discussed briefly.

Endothermic Gas

Endothermic-gas generators (Section 9.6) are a common sight in the heat-treat shop. The main components of an endothermic-gas generator (Fig. 9.5.17) are relatively simple and consist of:
- ❖ Heated reaction retort with catalyst
- ❖ Air-gas proportioning control components
- ❖ Pump to pass the air-gas mixture through the retort
- ❖ Cooler to freeze the reaction and prevent soot formation

Endothermic gas (also called Endo or RX® gas) is produced when a mixture of air and fuel is introduced into an externally heated retort at such a low air-to-gas ratio that it will normally not burn. The retort contains an active catalyst that is needed for cracking the mixture. Leaving the retort, the gas is cooled rapidly to avoid carbon re-formation (in the form of soot) before it is sent into the furnace. The endothermic-gas composition (Table 9.5.5) varies by volume depending on the type of hydrocarbon gas feedstock, air/gas ratio and relative humidity in the room.

Endothermic gas is used for neutral hardening and as a carrier gas for gas carburizing and carbonitriding. It is generally produced so that its composition is chemically inert to the carbon level of the surface of the steel, and it can be made chemically active (i.e., higher carbon potential) by the addition of enrichment (hydrocarbon) gas. This is usually done at the furnace.

Endothermic-gas composition (Section 9.6) varies as a function of the fuel source (e.g., natural gas or air) and the relative humidity of the incoming air.

TABLE 9.5.5 | Typical compositional ranges for endothermic gas

Gas constituent	Percentage, % (based on natural gas)	Percentage, % (based on propane)
N_2	40.9	40.9
CO	19.6	23.3
CO_2	0.4	0.1
H_2	38.9	35.5
CH_4	0.2	0.2
Dew point	-7 to +10°C (+20 to +50°F)	-23 to -26°C (-10 to -15°F)
(Air/gas) ratio	2.6:1	7.8:1

FIGURE 9.5.17 | Endothermic-gas generator schematic

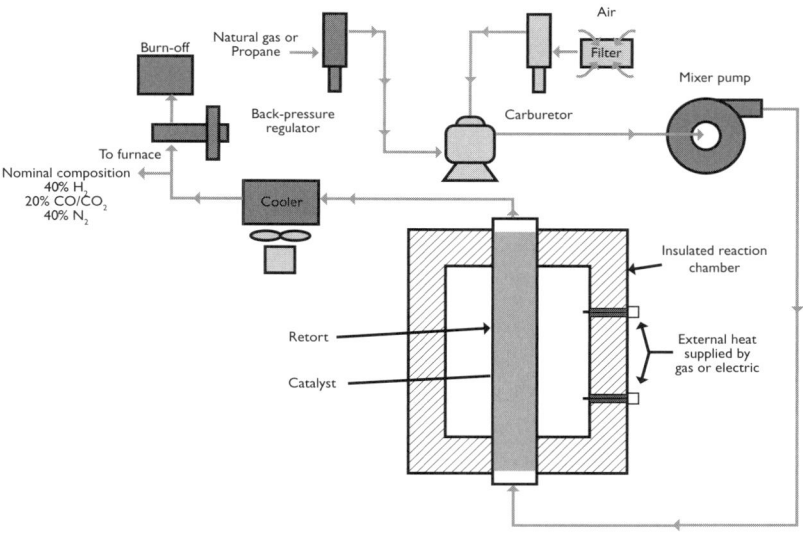

The most common problems with endothermic-gas generator systems are failure to properly maintain the generator, variability in air/gas ratio and running volumes in excess of rated capacity.

Nitrogen/Methanol

Nitrogen/methanol (Section 16.4) is a blended-gas atmosphere used to produce the carrier gas, which is subsequently enriched by a hydrocarbon addition. In some instances, especially outside North America, methanol (only) is used to generate a carburizing atmosphere.

A nitrogen/methanol system (Fig. 9.5.18) produces an "endothermic equivalent" (aka synthetic Endo) gas atmosphere and is obtained by cracking liquid methanol (methyl alcohol) and combining it with nitrogen (Equation 9.5.2), using a typical blend of 40% nitrogen and 60% dissociated methanol (Section 16.4).

9.5.2) $2CH_3OH + N_2 \rightarrow 2CO + 2H_2 + N_2$

This chemical reaction typically takes place inside the furnace as the liquid methanol and gaseous nitrogen are metered in through a special injector (aka sparger), which atomizes the liquid and sprays it into the chamber, usually onto a hot target such as the furnace fan. Complete cracking of methanol into CO and H_2 only occurs if the temperature is above 705-800°C (1300-1475°F). The equivalent of 4 kW of heat is required per 3.8 liters (1 gallon) to crack the methanol.

FIGURE 9.5.18 | Nitrogen/methanol system[3] (Copyright Linde AG, courtesy of Linde LLC)

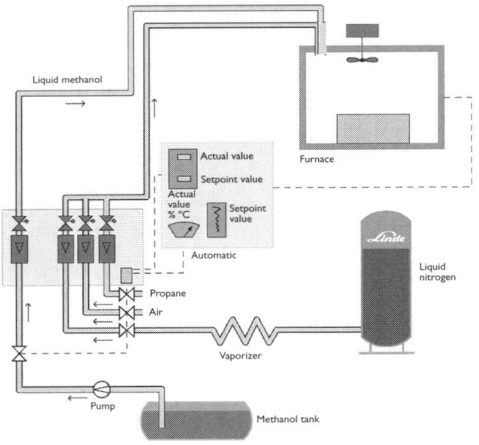

The amount of gaseous methanol produced from liquid is based on the fact that 3.8 liters/hour (1 gallon/hour) of methanol produces approximately 6.5-6.8 m^3/hour (228-241 cfh) of dissociated methanol. A 40% nitrogen/60% dissociated methanol atmosphere is typical for gas carburizing. (Fig. 9.5.19).

The most common problems with nitrogen/methanol systems have to do with the failure to properly vaporize and dissociate (atomize) the methanol. For example, large droplets do not properly decompose, resulting in furnace-control difficulties. One telltale sign of this is sooting of the workload in the area of the injector due to low-temperature dissociation of the methanol. Also, methanol is corrosive to nickel alloys used for internal furnace components (e.g., fans, radiant tubes, belts, etc.)

FIGURE 9.5.19 | Resulting gas composition upon cracking of methanol in an atmosphere containing 40% nitrogen and 60% cracked methanol[3] (Copyright Linde AG, courtesy of Linde LLC)

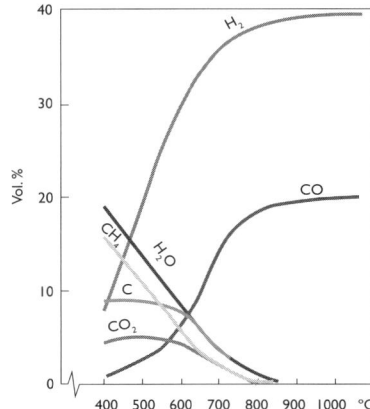

APPLICATIONS

The applications for gas carburizing are almost endless and extremely diverse. Carburizing of gears (Fig. 9.5.20) and shafts (Fig. 9.5.21) in an endothermic-gas atmosphere followed by oil quenching are common examples of the case-hardening process.

FIGURE 9.5.20 | Carburized gears (courtesy of Metals Engineering Inc.)

FIGURE 9.5.21 | Carburized shafts (courtesy of Lawrence Industries Inc.)

For example, ring gears of SAE 8620H were carburized in an electrically heated pit furnace running an endothermic-gas atmosphere and enriched with natural gas. The gross load weight was 390 kg (860 pounds) with a net load of 230 kg (510 pounds). The carburizing process (Fig. 9.5.22) took place at 925°C (1700°F). The carburizing atmosphere was 2.8 m³/hour (100 cfh) enriched with 0.34 m³/hour (12 cfh) of natural gas.

The enriching gas was turned on when the load reached carburizing temperature and was turned off after three hours, or 57% of the total time at temperature. The load was removed after 5.4 hours to a cooling pit, where it was slow cooled. The furnace pressure was maintained at 85-110 Pa (0.34-0.44 inch w.c.), and the dew point at the generator was maintained at -4.4 to -3°C (+24 to +26°F). The 5.08-hour cycle produced a case depth of approximately 1.65 mm (0.065 inch) and an effective case depth (0.40% C) of 1.08 mm (0.042 inch). The surface carbon content was 0.95%.[8]

FIGURE 9.5.22 | Ring-gear processing cycle[8]

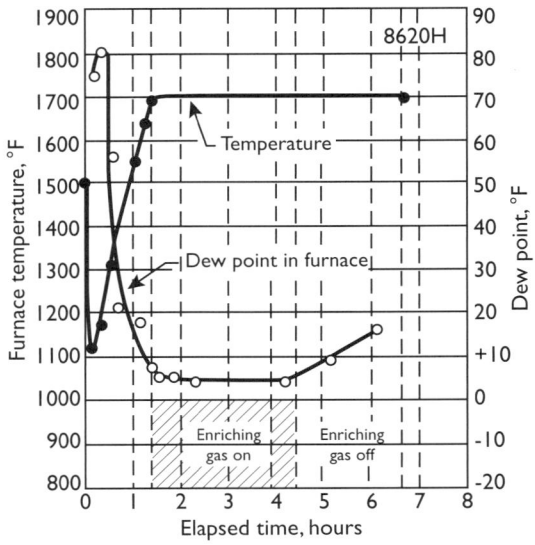

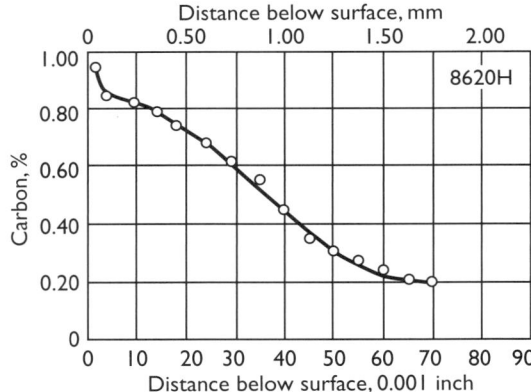

By comparison, SAE 8620H shafts were carburized in an electrically heated pit furnace running an endothermic-gas atmosphere and enriched with natural gas. The gross load weight was 258 kg (569 pounds) with a net load of 145 kg (320 pounds). The carburizing process (Fig. 9.5.23) took place at 925°C (1700°F). The carburizing atmosphere was 2.8 m³/hour (100 cfh) enriched with 0.25 m³/hour (9 cfh) of natural gas. The load was cooled to 845°C (1500°F) prior to oil quench. The enriching gas was turned on when the load reached carburizing temperature and was turned off after 2.83 hours, or 56% of the total time at temperature.

The furnace pressure was maintained at 97-122 Pa (0.39-0.40 inch w.c.), and the dew point at the generator was maintained at -4.4 to -2.8°C (+24 to +27°F). The 5.08-hour cycle produced a case depth of approximately 1.65 mm (0.065 inch) and an effective case depth (0.40% C) of 1.08 mm (0.042 inch). The surface carbon content was slightly below 0.8%.[8]

9 | FURNACE ATMOSPHERES

FIGURE 9.5.23 | SAE 8620H shaft processing cycle[8]

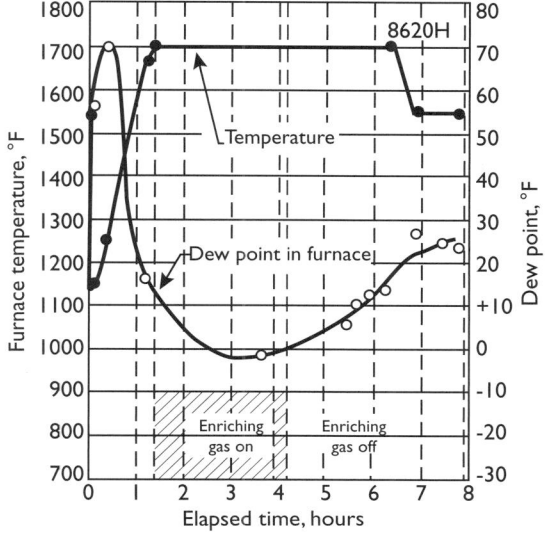

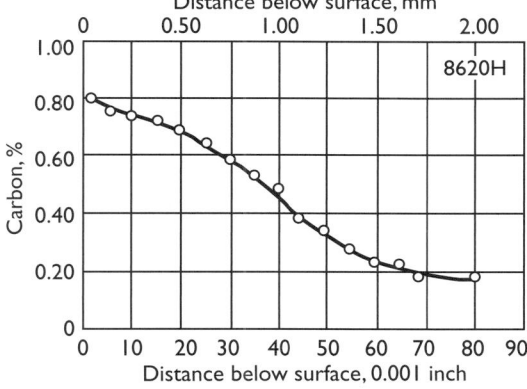

EQUIPMENT

While the equipment used for case hardening is quite diverse – from box furnaces and mesh-belt conveyors to rotary retort, rotary drum and shaker-hearth types – three styles dominate: pit-style (Fig. 9.5.24) typically used for either large (long) parts, heavy loads and/or deep case depths; integral-quench furnaces (Fig. 9.5.25) for diversity in the mid-size workload range; and pusher furnaces (Fig. 9.5.26) for their high-volume throughputs. Many of the furnaces that perform carburizing can also perform neutral hardening.

FIGURE 9.5.24 | Pit furnace system for carburizing wind-energy gears (courtesy of Aichelin Heat Treatment Systems)

FIGURE 9.5.25 | Integral-quench-style furnace (courtesy of Surface Combustion Inc.)

FIGURE 9.5.26 | Pusher furnace (courtesy of Aichelin Heat Treatment Systems)

Hardening

Hardening (Section 5.4) is a process that increases the hardness of a material by a suitable heat treatment, usually involving heating and (rapid) cooling. It does not produce a case on the surface of a part. Hardening in an atmosphere furnace has been referred to by several names, including direct hardening, neutral hardening, quench hardening and surface hardening. The atmosphere carbon potential is neutral to the carbon level in the part. We will refer to it as neutral hardening for the purposes of this discussion.

ATMOSPHERES

The role of a furnace atmosphere for neutral hardening is to prevent surface reactions from taking place, either from the surface of the component part to the atmosphere or from the atmosphere to the part surface. When hardening steel, the water-gas reaction (Equation 9.3.23) tells us that we need to control the atmosphere composition with respect to CO, CO_2, H_2O and H_2 and establish an atmosphere that is neither decarburizing nor carburizing. Hence, the carbon potential of the atmosphere must be carefully controlled.

It is worth noting that heat treaters talk in terms of carbon potential, while the scientific community refers to carbon activity (γ). The carbon potential (C_p), which is equal to the carbon content in weight percent in a binary Fe-C system, can be related to the molar fraction x_c (Equation 9.5.3), which in turn is related to the carbon activity (Equation 9.5.4).[3]

$$9.5.3) \quad x_c = \frac{\dfrac{C_p}{12.01}}{\dfrac{C_p}{12.01} + \dfrac{(100 - C_p)}{55.85}}$$

$$9.5.4) \quad a_c = \frac{e x_c}{(1 - 2 x_c)}$$

where:

$$e = \exp\left\{ \frac{\dfrac{5115.0 + 8339.9 x_c}{(1 - x_c)}}{T - 1.9096} \right\}$$

and T is temperature in Kelvin

APPLICATIONS

Parts that have been carburized and slow cooled, such as gears and bearings, are subsequently neutral hardened (i.e., reheated in an atmosphere neutral to the carburized surface carbon) for quenching. Component parts that are prone to distortion are often processed in this manner.

In another example, bars from 75-300 mm (3-12 inches) diameter manufactured of SAE 4140, 4150 and 4340 are heated in a gas-fired roller-hearth furnace and discharged to a spray water quench (Fig. 9.5.27) for hardening, meeting ASTM A434 (Standard Specification for Steel Bars, Alloy, Hot-Wrought or Cold Finished, Quenched and Tempered) at a production rate of 7,250 kg/hour (16,000 pounds/hour). In this instance, the products of combustion create an acceptable furnace atmosphere.

FIGURE 9.5.27 | 200-mm-diameter (8-inch-diameter) bar of SAE 4150 entering spray water quench for hardening

EQUIPMENT

Equipment for neutral hardening spans the gamut of furnace types from box, pit and integral-quench-style furnaces to rotary-hearth (Fig. 9.5.28), rotary-drum, shaker-hearth and pusher furnaces. Many of the furnaces that perform neutral hardening can also perform case hardening.

FIGURE 9.5.28 | Rotary-hearth furnace for neutral hardening (courtesy of Surface Combustion Inc.)

Nitriding/Nitrocarburizing

Nitriding and nitrocarburizing (Section 5.5) are case-hardening processes in which nitrogen (for nitriding) or nitrogen and carbon (during nitrocarburizing) are added to the surface of the steel to develop certain enhanced properties

(Table 9.5.6). There are two principal transfer mechanisms at work, namely:
- ❖ Gas-to-solid transfer from the active furnace atmosphere to the steel surface
- ❖ Diffusion – first through the compound layer, then into the diffusion layer

TABLE 9.5.6 | Relationship between part properties, steel selection and nitriding/nitrocarburizing parameters[9]

Properties	Static and fatigue strength		Contact load fatigue	Abrasive wear resistance	Adhesive wear resistance	Corrosion resistance
	High	Moderate				
Steel type	High-alloy and nitriding steels	Low-alloy steels, cast irons, sintered steels	High- and low-alloy steels	High- and low-alloy steels, cast irons, sintered steels		
Process	Gas nitriding	Nitrocarburizing	Gas nitriding and nitrocarburizing			Nitrocarburizing + post-oxidation
Compound layer	Minor influence		High hardness, high ε-content			Dense, post-oxide
Diffusion layer	High hardness and depth	Moderate hardness and depth	High hardness and depth	Minor influence		

ATMOSPHERES

The atmospheres for nitriding and nitrocarburizing, while somewhat similar, are in fact quite different and need to be discussed separately.

Nitriding

Nitriding is performed either in a raw-ammonia atmosphere (aka single-stage nitriding) or an atmosphere of raw ammonia followed by dilution with either dissociated ammonia, nitrogen or hydrogen (aka two-stage nitriding). If nitrogen is used as a dilution gas, the compound-layer (white-layer) composition will be different. It is also more difficult to control using nitrogen dilution than dilution with dissociated ammonia. Hydrogen has also been used as a dilution gas so that the nitrogen activity can be varied over a wider range. The most common way to determine the nitriding potential is by gas analysis using ammonia or hydrogen analyzers.

Nitrocarburizing

Several different furnace-atmosphere combinations have been used for nitrocarburizing (Table 9.5.7) to produce an atmosphere of approximately 20-50% ammonia, 2-20% carbon dioxide, 2-38% hydrogen and the balance nitrogen.

TABLE 9.5.7 | Compositions of nitrocarburizing atmospheres

Gas composition choices
60% ammonia – 40% endothermic gas
50% ammonia – 50% endothermic gas
35% ammonia – 5% CO – 60% nitrogen
33% ammonia – 5% CO – 62% nitrogen
28% ammonia – 4% CO – 68% nitrogen
50% ammonia – 40% endothermic gas – 10% air

APPLICATIONS

Nitralloy 135M is an example of material that is commonly gas nitrided. The procedure necessary to develop optimum core properties prior to nitriding is to normalize then austenitize at 940°C (1725°F) and then quench and temper for final core hardness (Table 9.5.8). The goal is to temper 10°C (50°F) below the final nitriding temperature. Single-stage nitriding (Fig. 9.5.29) or two-stage nitriding (Fig. 9.5.30) can then be performed to achieve the desired case depth. A comparison to several other nitridable materials (Fig. 9.5.31) is also provided.

TABLE 9.5.8 | Nitralloy 135M core hardness as a function of tempering temperature[10]

Tempering temperature, °C (°F)	Yield strength, MPa (ksi)	Tensile strength, MPa (ksi)	% Elongation, 50 mm (2 inches)	% RA	Hardness, HRC	Hardness, Brinell
538 (1000)	1,251 (181.5)	1,420 (206)	13	46	45	415
593 (1100)	1,138 (165)	1,248 (181)	15	54	41	368
650 (1200)	972 (141)	1,096 (159)	17	56	36	320
705 (1300)	862 (125)	1,000 (145)	20	64	32	285

FIGURE 9.5.29 | Typical depth-hardness curves for Nitralloy 135M nitrided at 525°C (975°F) with 30% dissociation rate by the single-stage process[10]

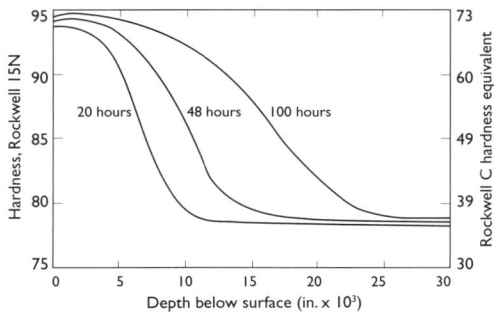

FIGURE 9.5.30 | Typical depth-hardness curves for Nitralloy 135M nitrided by the two-stage process[10]

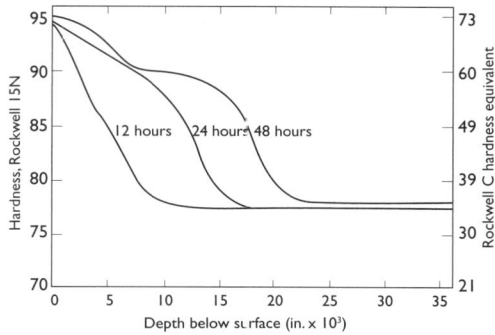

FIGURE 9.5.31 | Comparison of nitrided 135M with nitrided SAE 4140 and SAE 4340[10]

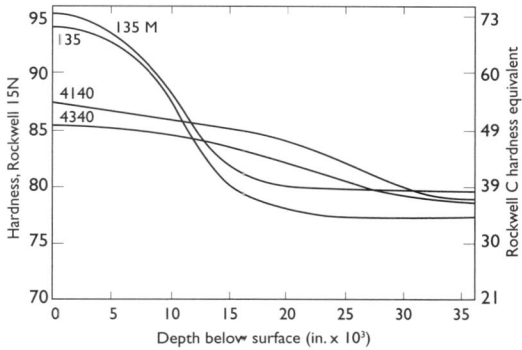

In another example, a load of piston plates (Fig. 9.5.32) is ferritic nitrocarburized at 580°C (1075°F) for 90 minutes in an atmosphere of 35% endothermic gas and 65% ammonia. Parts are subsequently oil quenched. Targeted hardness is 400 HV50 at a case depth of 0.01 mm (0.0004 inch).

FIGURE 9.5.32 | Load of piston plates after ferritic nitrocarburizing

EQUIPMENT

While nitriding can be performed in a number of different furnace types, pit furnaces (with or without retorts) dominate the landscape (Fig. 9.5.33). The internal materials of construction, including liners and/or retorts, fans, circulation shrouds and gas inlet pipes, are an important consideration because the alloy selection will influence the dissociation rate. Inconel® 600 has been found to work best and have excellent life. This is followed (with increasing dissociation rates) by Inconel 601®, 330, 304 and carbon steel. Some integral-quench furnaces are also being used, primarily for nitrocarburizing.

FIGURE 9.5.33 | Pit nitriding furnaces (courtesy of Nitrex Metal Inc.)

Sintering

Sintering (Section 5.11) is the diffusion bonding of adjacent powder particle surfaces in a green compact, resulting in a change to the part microstructure that influences its mechanical properties. Other common names and related subjects are cold/hot isostatic pressing, liquid-phase sintering and metal injection molding. Sintering is not a heat-treating process per se.

ATMOSPHERES

Atmosphere requirements in a sintering furnace vary considerably depending on the type of furnace (e.g., mesh-belt conveyor, pusher, walking beam), style of furnace (batch or continuous) and if delubrication (aka delub, dewax) is required to be performed. The basic atmosphere requirements for a continuous mesh-belt furnace (Fig. 9.5.34) are as follows.

- ❖ In the delubrication zone, a high dew-point atmosphere in the range +10°C to +20°C (+50°F to +70°F) is generated by air or wet-gas additions to aid in lubricant removal.

9 | FURNACE ATMOSPHERES

❖ In the sintering zone, a low dew-point atmosphere in the range of -29°C to -40°C (-20°F to -40°F) aids in oxide reduction to promote bonding of the powder-metal (PM) particles together.
❖ In the post-cool zone or in some cases the sintering zone (optional), carbon control of certain materials prevents surface decarburization.
❖ In the cooling zone, sufficient gas flow is needed to prevent oxidation. O_2 levels in commercial practice often run in the 10-50 ppm range (maximum). The goal is to attain the lowest practical level of oxygen.

FIGURE 9.5.34 | Role of a sintering furnace atmosphere by location inside the furnace

		Preheating Zone	Hot zone	Slow cooling zone	Water cooling zone
Atmosphere functions →	• Convey heat quickly and uniformly • Burn and sweep out lubricants to front exit	• Reduce surface oxides • Carbon diffusion	• Copper melting, coating or infiltrating • Bonding • Carbon control	• Carbon control • Cooling rate control	• Cooling • Prevent oxidation or controlled light oxidation
Atmosphere composition →	• Lightly oxicizing	• Highly reducing • Neutral to carbon	• Reducing • Neutral to carbon preferred	• Reducing • Neutral to carbon preferred	• Convey heat quickly and uniformly • Slightly reducing or neutral or slightly oxidizing
Temperature range (steel) →	• 425-650°C (800-1200°F)	• 659-1040°C (1200-1900°F)	• 1040-1120°C (1900-2050°F)	• 1120-815°C (2050-1500°F)	• 815°C (1500°F) to ambient

The most common sintering atmospheres are mixtures of nitrogen/hydrogen or dissociated ammonia diluted by nitrogen additions. Ratios vary from as low as 5-7% to as high as 20-30% hydrogen. Stainless steels and some tool steels are often processed in 100% hydrogen as are metal-injection-molded (MIM) parts. They are commonly run in either pusher furnaces or vacuum furnaces operating with a partial pressure of hydrogen.

APPLICATIONS

Automotive applications (e.g., anti-lock brakes, mirror mounts) dominate the demand for PM components, making up over 70% of the PM marketplace. Most of this production is either iron, steel or stainless steel. Nonautomotive examples of PM components are gears and pinions for hand and power tools, carbide inserts and cutters, firearm components, medical and orthodontic devices, magnets, appliance parts and lock components.

PM valve-seat inserts (Fig. 9.5.35) are an example of an automotive engine component that is often sintered in a mesh-belt conveyor furnace running an atmosphere of 7% hydrogen/93% nitrogen. Valve-seat inserts provide a wear-resistant sealing surface for the valve beyond the parent metal of the cylinder head. Valve-seat inserts are typically produced from highly alloyed steel, including tool steels and/or composite-phase materials, to improve temperature and wear resistance. Important properties for valve-seat inserts include not only temperature and wear resistance but thermal conductivity, hot hardness, corrosion resistance and impact resistance, depending on the specific engine-system design.

FIGURE 9.5.35 | Sintering of valve-seat inserts (courtesy of Federal Mogul Corp.)

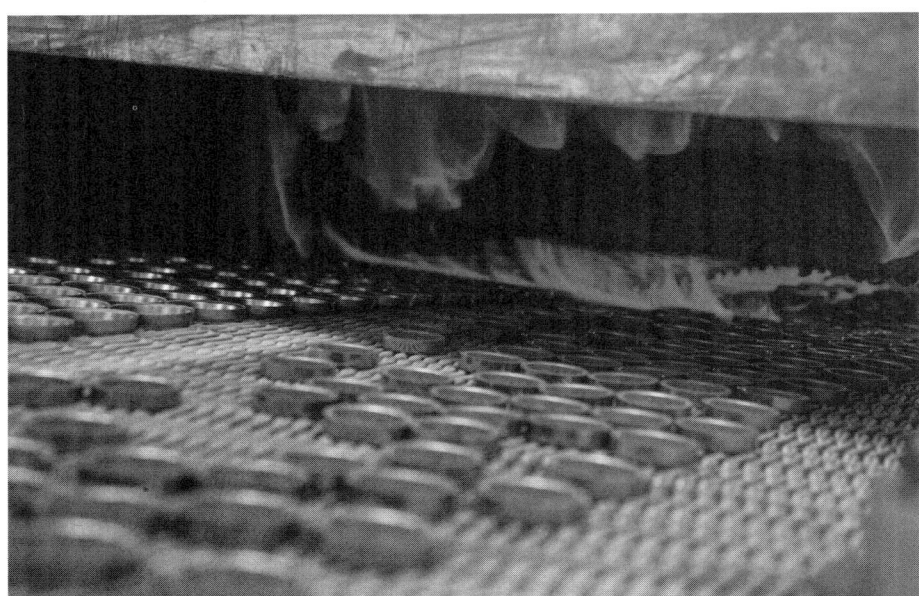

EQUIPMENT

Sintering performed in a mesh-belt conveyor furnace requires the control of a number of process and equipment variables, including: material, density, lubricant, belt speed, belt loading, furnace-door height, temperature uniformity across the belt, temperature profile along the length of the furnace, entrance tunnel, front and rear curtain integrity, atmosphere inlet type (dump tube, spray bar), location and direction (angle of gas entry in relation to the moving belt). All of these parameters help determine the atmosphere volume and composition.

Sintering furnaces process PM components made from a wide variety of materials (Table 9.5.9). The operations can be done in batch or continuous furnaces, the latter being more common in the industry due to production demands. Sintering furnaces are often matched to the output of one or two presses so that green compacts are produced and immediately sintered.

TABLE 9.5.9 | Sintering temperature and equipment type by material

Material	Sintering temperature range, °C (°F)	Most common type of equipment
Aluminum	550-650 (1020-1200)	Alloy mesh-belt, inclined humpback conveyor
Bronze	750-900 (1380-1650)	Alloy mesh-belt conveyor
Iron	1200-1350 (2200-2460)	Mesh-belt conveyor, pusher, vacuum, walking beam
Stainless steel	1200-1350 (2200-2460)	Ceramic mesh-belt conveyor, pusher, walking beam
Stainless steel	1095-1150 (2000-2100)	Alloy mesh-belt or ceramic mesh-belt conveyor, pusher
Steel	1120-1200 (2050-2200)	Alloy mesh-belt or ceramic mesh-belt conveyor, inclined humpback conveyor, pusher, walking beam
Tool steel	1150-1350 (2100-2460)	Pusher, vacuum
Tungsten carbide	1550-1650 (2820-3000)	Vacuum
Uranium oxide	1600-1700 (3000-3100)	Pusher, walking beam

Sintering typically involves conventional or double "press and sinter," copper infiltration and/or liquid-phase sintering. In most cases, the following steps are involved: delubrication/preheating, sintering and cooling (slow or fast depending on the material).

The goals of sintering are to produce: a good bond between PM particles to a targeted density; rounded pores; minimal oxidation; good dimensional control; and adequate mechanical properties such as strength and corrosion.

The most common type of continuous sintering furnace is a horizontal mesh-belt conveyor furnace (Fig. 9.5.36) using either alloy belts for conventional sintering temperatures or ceramic belts for temperatures over 1175°C (2150°F). These types of furnaces are used for a variety of sintering processes primarily involving automotive components and industrial/commercial products. Pusher furnaces and walking-beam furnaces (Fig. 9.5.37) are also used principally for large or heavy components, parts made of stainless steel or tool steel, or when processes exceed 1175°C.

The most common types of batch sintering furnaces are bell and single-chamber vacuum (Fig. 9.5.38). They are often used to process stainless steels and metal-injection-molded parts.

FIGURE 9.5.36 | Typical horizontal mesh-belt conveyor sintering furnace (courtesy of Sinterite, a Gasbarre Furnace Group Company)

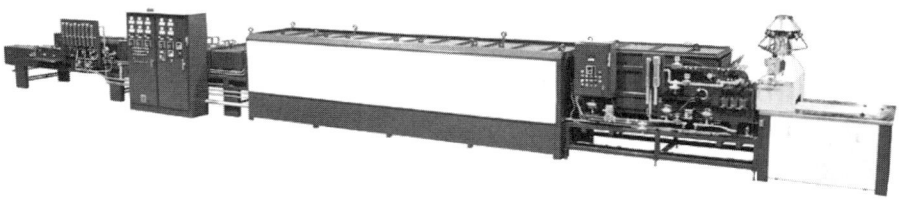

FIGURE 9.5.37 | Typical walking-beam furnace (courtesy of Lindberg/MPH)

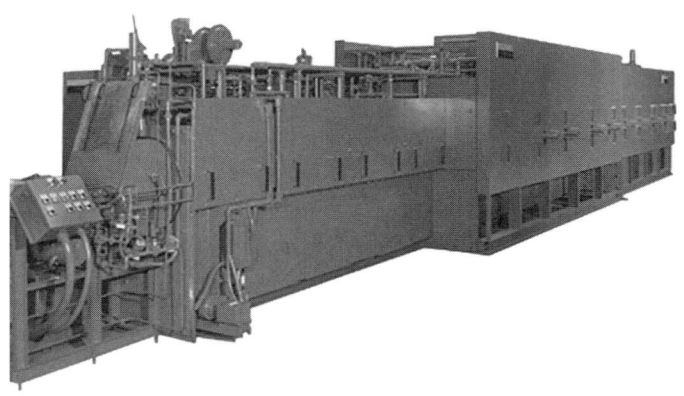

FIGURE 9.5.38 | Typical single-chamber batch vacuum furnace (courtesy of Elnik Systems)

9 | FURNACE ATMOSPHERES

Summary

We have looked at the furnace atmosphere requirements for a number of common heat-treatment processes along with application examples and a discussion of the typical equipment in which the process can be run. The proper selection of a furnace atmosphere must take into consideration many variables, which include: the material being run, the desired properties one wishes to achieve, production demands, loading, handling, thermal profile (heating rate, temperature, cooling rate, cooling medium), equipment and production cost.

REFERENCES

1. Herring, Daniel H., *Atmosphere Heat Treatment, Volume I*, BNP Media Group 2014
2. Andersson, Rolf, Torsten Holm, Thomas Mahlo and Sören Wilberg, "Furnace Atmospheres for Tube Annealing," Special Edition Brochure, Linde, 2004
3. "Furnace Atmospheres No. 2: Neutral Hardening and Annealing," Special Edition Brochure, Linde, 2005
4. "Furnace Atmospheres No. 1: Gas Carburizing and Carbonitriding," Special Edition Brochure, Linde, 2007
5. Andersson, Rolf, Torsten Holm, Sören Wilberg and Anders Åstrom, "Furnace Atmospheres No.4: Brazing of Metals," Special Edition Brochure, Linde, 2005
6. Wikipedia (www.wikipedia.org)
7. Gaskell, David R., *Introduction to Metallurgical Thermodynamics*, McGraw-Hill Book Company, 1981
8. *Carburizing and Carbonitriding*, ASM International, 1977, pp. 111-115
9. "Furnace Atmospheres No. 3: Gas Nitriding and Nitrocarburizing," Special Edition Brochure, Linde, 2009
10. Bever, Michael B. and Carl F. Floe, *Case Hardening of Steel by Nitriding*, Source Book on Nitriding, ASM International, 1977, pp. 125-143
11. "Furnace Atmospheres No. 8: Sintering of Steels," Special Edition Brochure, Linde, 2011
12. "Heat Treatment of Metals," Wolfson Heat Treatment Centre, 1990.1, pp. 1-4
13. Donald Bowe, Air Products & Chemicals Inc. (www.airproducts.com), technical and editorial review, private correspondence
14. James Oakes, Super Systems Inc. (www.supersystems.com), technical and editorial review, private correspondence

Achieve the best results in heat treatment.
With our leading solutions.

Gases / Technology / Expertise

→ Tailored process and safety solutions
→ Atmosphere and process control
→ Gas supply solutions
→ Cryo treatment solutions

Linde North America, Inc.
575 Mountain Ave, Murray Hill, NJ 07974
Phone 1-800-755-9277, sales.lg.us@linde.com, www.lindeus.com

Linde North America, Inc. is a member of The Linde Group. Linde is a trading name used by companies within The Linde Group. "Linde" and the Linde logo are trademarks of The Linde Group. © The Linde Group 2015. All rights reserved.

9 | FURNACE ATMOSPHERES

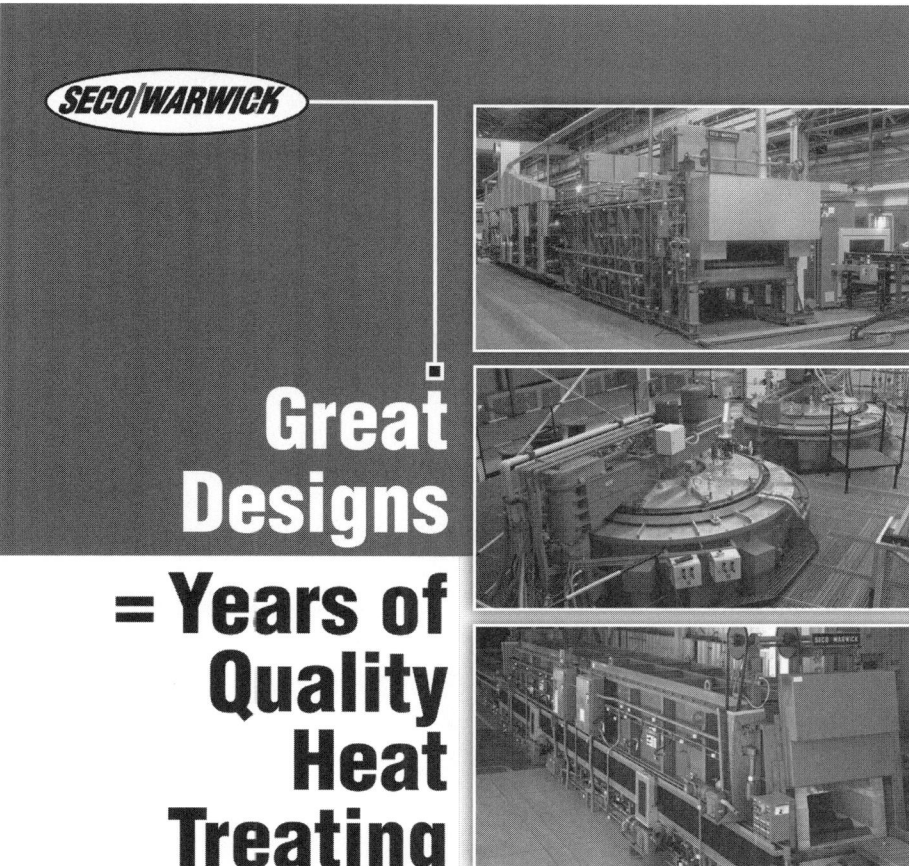

Great Designs = Years of Quality Heat Treating

When you look closer at SECO/WARWICK's atmosphere furnace designs, you discover custom engineered equipment that meets your production requirements at a reasonable cost.

Most heat treating processes can be accomplished in a variety of batch and continuous equipment designs and we have the technical experience to select the most innovative design to produce repeatable, quality heat treatment results for years of reliable service.

If you are looking for the equipment value and energy savings only atmosphere furnaces can offer, take a look at SECO/WARWICK on our website, www.secowarwick.com, or call our Thermal Process Group at 814-332-8400.

SECO/WARWICK
Meadville, PA 16335
814-332-8400 · Fax 814-724-1407
info@secowarwick.com · secowarwick.com

USA · Poland · India · Brazil · China

9.6

GENERATED ATMOSPHERES

Gas generators are a common sight in most heat-treatment shops. The purpose of this section is to provide a basic understanding of the gases they produce, how they operate and the applications for which they are best suited.

Dissociated Ammonia

Ammonia dissociators have been supplied commercially since the 1930s. Originally, dissociated ammonia was considered for use as a furnace atmosphere when a low-cost alternative to pure (100%) hydrogen could be tolerated. More recently, the need for dissociated ammonia (DA) has been expanded to include a low-cost source of hydrogen in nitrogen atmospheres and as a replacement for other generated atmospheres. In addition to high purity, DA is stable and easy to control.

GAS CHEMISTRY

On heating to temperatures above 480°C (900°F), ammonia (NH_3) will begin to dissociate into its component gases, namely nitrogen and hydrogen. Ammonia will dissociate completely around 650°C (1200°F) in the presence of an active catalyst such as iron or nickel. The use of higher dissociation temperatures – in the range of 870-1010°C (1600-1850°F) – in combination with a catalyst will speed up this dissociation process.

The DA gas reaction (Equation 9.6.1) produces an atmosphere of 75% hydrogen and 25% nitrogen (Fig. 9.6.1) plus a small percentage of undissociated ammonia, typically in the range of 0.005% (50 ppm).

9.6.1) $2NH_3 \rightarrow N_2 + 3H_2$

FIGURE 9.6.1 | Dissociated-ammonia generator gas chemistry

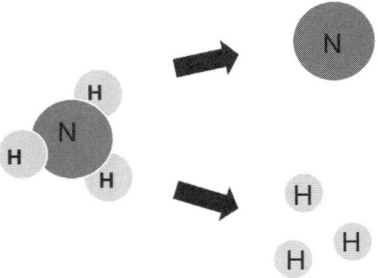

In order to generate 1.3 m³/hour (45 cfh) of DA at standard pressure and temperature, 0.65 m³ (22.5 cfh) of raw ammonia is required. This doubling of gas volume is the result of the chemical reaction responsible for the formation of DA. Equation 9.6.1 shows that two moles of ammonia will dissociate into four moles of reaction products, producing an atmosphere of three parts (75%) hydrogen and one part (25%) nitrogen. The volume of DA produced is double that of the ammonia volume supplied.

Since this reaction takes place at relatively low temperatures, a very stable atmosphere composition is produced with very low residual ammonia – no more than 0.025% (250 ppm). In general, the dew point of the incoming ammonia supply is drier than -54°C (-65°F), which produces a DA atmosphere of the same relative dew point.

EQUIPMENT

Ammonia dissociators (Fig. 9.6.2) are available in standard sizes ranging from 0.7 m³/hour (25 cfh) to 170 m³/hour (6,000 cfh). Larger units are normally custom designs.

FIGURE 9.6.2 | Typical dissociated-ammonia generator (courtesy of C.I. Hayes, a Gasbarre Furnace Group Company)

9 | FURNACE ATMOSPHERES

FIGURE 9.6.3 | Schematic of a dissociated-ammonia generator (courtesy of C.I. Hayes, a Gasbarre Furnace Group Company)

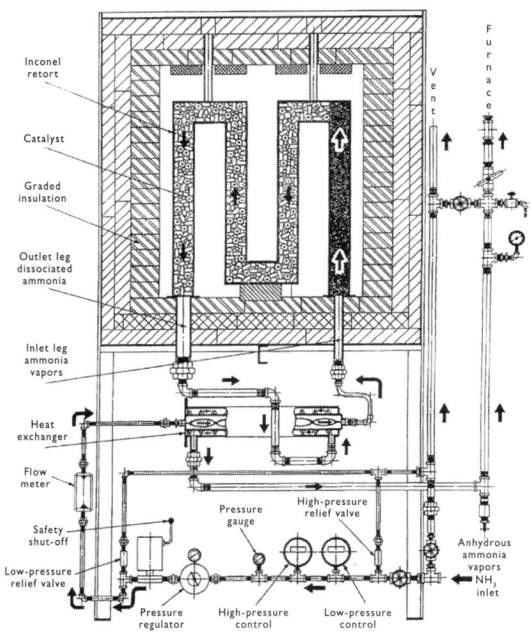

FEATURES

The components that make up this equipment (Fig. 9.6.3) can be described as follows.

Retort

The basic ammonia dissociator consists of a heated chamber housing an alloy retort or catalyst chamber, typically either cylindrical or "U" shaped and usually 75-150 mm (3-6 inches) in diameter. Testing has shown[4] that a 100-mm-diameter (4-inch-diameter) design optimizes uniform heat distribution throughout the catalyst bed.

Since the gas flowing through the retort is under pressures as high as 200 kPa (30 psig), both the material and method of retort construction must be carefully considered. Inconel 600® is the preferred retort material, providing superior life versus cost.

Catalyst

A retort design (Fig. 9.6.4) that optimizes dissociation typically uses three different catalyst types. Iron is the best catalyst, but it will nitride in the presence of undissociated ammonia and begin to expand (grow), which leads to severe surface cracks. To avoid this concern, nickel-impregnated ceramic spheres are used in the first one-third of the catalyst bed. Catalyst A, a large-diameter sphere, begins the dissociation process. Catalyst B, nickel-impregnated ceramic spheres 50% smaller than catalyst A, provides the surface area necessary for proper dissociation. Catalyst C, sintered (pure) iron rings, provides both large surface area and enough dwell time to ensure dissociation rates to 0.0050% (50 ppm) residual ammonia.

When larger-diameter retorts are supplied, it is not uncommon to use an alloy tube within the center of the catalyst bed to occupy this volume or to use a gas preheating arrangement, which finds tubing wrapped around the exterior of the retort prior to gas entry in some cases. Neither of these alternatives is as efficient as the use of a small, multiple-pass retort design. This is critical since improper temperature or time result in an end product that will contain excessive amounts of undissociated ammonia.

FIGURE 9.6.4 | Retort with catalyst (courtesy of C.I. Hayes, a Gasbarre Furnace Group Company)

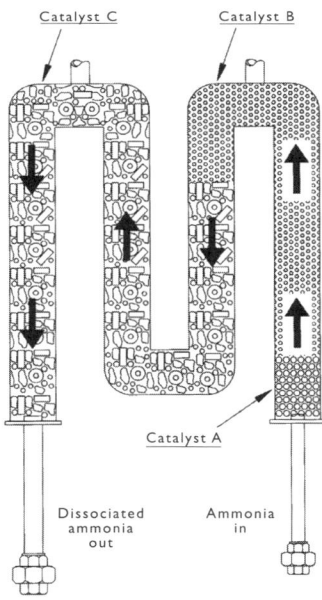

Heating Chamber

The actual heating chamber of the dissociator is refractory-lined and can be gas-fired or electrically heated. The heat storage in the refractory walls aids in maintaining a consistent temperature gradient throughout the retort. In many systems, silicon-carbide Globars® or Starbars® are provided to ensure maximum heat-transfer efficiency by close coupling the heat source to the retort. Silicon-carbide heating elements are superior to metallic elements in this application since they resist oxidation. The use of silicon-controlled rectifiers or saturable core reactors ensures maximum element life.

Piping
Heavy-wall (i.e., schedule-80) black-iron piping and fittings must be used on all ammonia and DA lines. In addition, all valves and components must be either black iron or stainless steel. No copper or copper-bearing materials (e.g., brass) are permitted. Incoming anhydrous ammonia is supplied to the dissociator in a gaseous state from a liquid supply source, typically a storage tank equipped with a vaporizer (Fig. 9.6.5). Although cylinders can be used, high-volume systems require storage tanks located outside of the factory. The incoming ammonia is supplied at inlet pressures of 140-200 kPa (20-30 psig) and then passes through a pressure regulator, shutoff valve and flowmeter prior to entry into a gas heat exchanger. A DA flowmeter on the outlet side of the retort is often provided downstream.

FIGURE 9.6.5 | Ammonia tank installation (courtesy of Airgas)

Low and high gas-pressure controls are set to shut off the system (in the case of low pressure) or vent the incoming gas (in the case of high pressure). Vent lines should be piped out of the building or bubbled through water to prevent ammonia discharge into the air. A nitrogen purge of the retort should be initiated in the case of loss of gas pressure or power failure.

Heat Exchanger

One of the unique features of the ammonia dissociator is the use of a dual-purpose gas heat exchanger. The incoming (cold) ammonia gas is passed inside a tube bundle to be preheated by the (hot) DA having just exited the retort. In the process of heat exchange, the hot gas is cooled so both incoming and exiting gas supplies benefit from this design.

MAINTENANCE

The operational temperature of most ammonia dissociators is around 955°C (1750°F). This is a compromise between the decomposition rate of the ammonia, which increases with increasing temperature, and equipment life.

One of the biggest advantages of ammonia dissociators is that they require minimal maintenance. The temperature-control instrumentation should be checked periodically and thermocouples replaced on a scheduled basis. The heating chamber should be inspected annually, if possible, and heating elements or burners should be replaced or adjusted as needed. Safety equipment should be inspected and tested on a routine basis. Nitrogen purging is required in case of power failure to dilute and vent the mixture in the retort.

Perhaps the single biggest concern is pressure drop through the catalyst bed. This should be carefully monitored since it is an indicator of the catalyst-bed condition. Under normal operating conditions, outlet pressures of 75-100 kPa (10-15 psig) are common. If inlet pressure must be increased to maintain outlet pressure or flow, this is a sign that the catalyst is deteriorating. The retort catalyst tends to disintegrate over time, mainly due to problems directly associated with ammonia purity. This causes the pressure differential across the dissociating chamber to increase. An advantage of using a sintered-iron catalyst is that it will not crumble and pack together, and ablation (i.e., erosion due to hot gases) will not cause a loss of nickel content (since most impregnation techniques are surface treatments) and, therefore, dissociation ability.

OTHER TECHNICAL CONSIDERATIONS

Several other factors are important when considering the use of an ammonia dissociator. These include supply requirements, the presence of impurities and nitrogen pickup.

Supply Requirements

One of the most important considerations in the use of DA generators is the incoming (raw) gas purity. Anhydrous ammonia for heat-treatment applications is available in three basic grades (Table 9.6.1), with metallurgical grade preferred.

TABLE 9.6.1 | Anhydrous ammonia product specifications

Constituent	Commercial grade, %	Refrigeration grade, %	Metallurgical (premium) grade, %
Ammonia (minimum)	99.5	99.990	99.995
Moisture	[a]	0.005	0.0033
Oil	[a]	0.005	0.0020

Note:
[a] Total impurities (moisture/oil) = 0.50%

Impurities and Their Effects

Two problems with incoming ammonia supplies are common. They relate to either a failure to properly vaporize (i.e., liquid ammonia is ingested directly into the dissociator) or the presence of contaminants in the gas stream (oils or water).

Metallurgical, or premium-grade, ammonia is preferred for best results, maximum equipment life and highest gas purity. Most users do not recognize that oil as an impurity can be drawn into the retort and contaminate (poison) the catalyst by reacting with the nickel coating, thereby destroying its ability to promote dissociation. Ammonia supply tanks should be refilled when they still have 20-30% remaining. Under no circumstances should the ammonia supply tank be allowed to run dry. In addition, the vaporizer should be drained on a schedule of once approximately every 2,250-4,500 kg (5,000-10,000 pounds) of ammonia put through the unit to remove any traces of oil that tend to collect.

In situations where lesser grades of ammonia must be supplied, such as overseas installations, activated charcoal and porostone filters on the upstream side of the dissociator can be supplied to remove oil. Under these conditions, a properly functioning ammonia dissociator will produce a 99.95% dissociated product having a dew point of -46 to -51°C (-50 to -60°F).

Water vapor is another concern in many process applications. Molecular sieve dryers on the downstream side of the gas stream are used to lower the gas dew point prior to injection into the furnace. These sieves will not only absorb moisture but also undissociated ammonia up to approximately 4% of their weight. The relationship between dew point and moisture content of DA (Table 9.6.2) determines the level to which absorption is needed. The use of a molecular sieve dryer can provide residual ammonia levels as low as 1-10 ppm.

TABLE 9.6.2 | Relationship between dew point, moisture content and percent dissociation

Dew point, °C (°F)	H_2O, ppm by volume	H_2O, %	Dissociated ammonia, %
-73 (-100)	7	0.0007	99.9993
-68 (-90)	15	0.0015	99.9985
-65 (-85)	25	0.0025	99.9975
-62 (-80)	31	0.0031	99.9969
-59 (-75)	45	0.0045	99.9955
-57 (-50)	62	0.0065	99.9938
-54 (-65)	90	0.0090	99.9910
-51 (-60)	119	0.0119	99.9881
-48 (-55)	150	0.0150	99.9850
-46 (-50)	200	0.0200	99.9800
-43 (-45)	275	0.0275	99.9725
-40 (-40)	398	0.0398	99.9602
-37 (-35)	550	0.0550	99.9450
-34 (-30)	702	0.0702	99.9298

Nitrogen Pickup

In certain situations, both ammonia and (atomic) nitrogen can cause nitrogen pickup on the surface of a material. Equation 9.6.1 can be rewritten as Equation 9.6.2.

9.6.2) $\quad 2NH_3 \rightarrow 2N + 6H \rightarrow N_2 + 3H_2$

The intermediate step shown involves the formation of atomic nitrogen and atomic hydrogen. If this reaction occurs at the surface of the material exposed to the DA atmosphere, nitrogen pickup (nitriding) will occur. This is one of the main reasons why the amount of residual ammonia must be kept to an absolute minimum and why proper ammonia-dissociator design is so important in critical process applications. DA should be used as a substitute for hydrogen or other gases only if the parts being processed are not susceptible to nitriding. Similarly, molecular nitrogen will break into its atomic constituents and has been known to nitride materials such as stainless steels above 1040°C (1900°F). Proper understanding of the furnace atmosphere is critical to its proper application.

APPLICATIONS

The applications for DA are quite diverse and include the following.

Annealing

One of the primary uses for DA (Fig. 9.6.6) is bright annealing, which will provide a highly reducing and oxygen-free atmosphere for processing stainless steel; brass and bronze alloys; copper and copper alloys; beryllium copper and beryllium nickel; nickel and nickel alloys; and silicon iron. Strip, tubing, wire, stamped and formed parts run in DA are common. In addition, annealing of metal powder is commonly performed in DA.

FIGURE 9.6.6 | Oxidation/reduction diagrams for 18% chromium stainless steel at 1095°C (2000°F)[8]

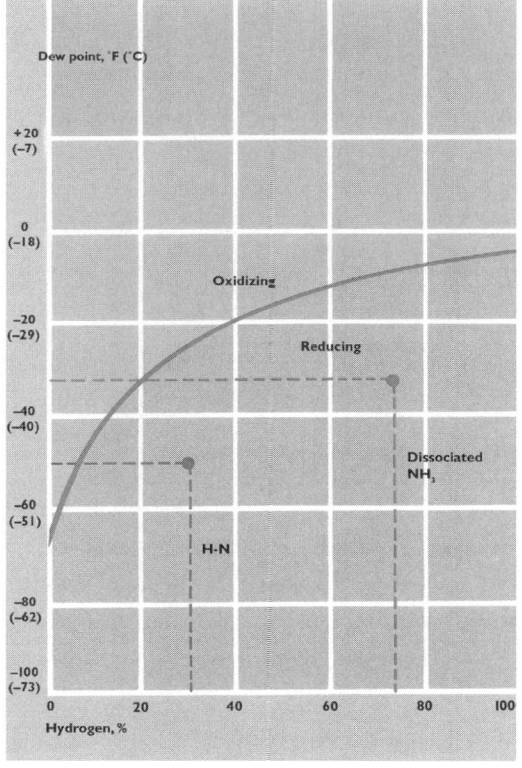

Brazing

One of the most important factors in the wetting stage of the brazing process is the cleanliness of the surface to be wetted. Oxide or carbon (soot) layers inhibit wetting and spreading of the molten filler metal. The main purpose of any protective atmosphere in brazing is to remove any oxide and prevent carbon deposition in the joint area and expose clean base metal.

Good wetting and spreading of the liquid filler metal on the base metal are necessary because the mechanics of the process demand that the filler metal be brought smoothly, rapidly and continuously to the joint opening. If the conditions within the capillary space of the joint do not promote good wetting, capillary attraction will be inhibited and metal will not be drawn into the space. In fact, capillary attraction is the result of the same forces that make for good wetting – the adhesion force of closely spaced parallel solid surfaces for a liquid are greater than the cohesive forces of the liquid. The imbalance of these forces causes the liquid (including molten metal) to flow between the closely spaced surfaces, even against the flow of gravity. Clearly, the joint surfaces must provide close parallel surfaces free of contamination.

By virtue of its high hydrogen content and low dew point, DA is a highly reducing atmosphere that can control the formation of oxides, reduce oxides present, and present metallurgically clean surfaces at the onset and propagation of the wetting action. Thus, it promotes wetting and braze-alloy flow. An additional benefit of the high hydrogen content in DA is an improvement in thermal conductivity, which significantly aids both part heating and temperature uniformity. This also contributes to quality brazing.

Hardening
Clean (aka bright) hardening of martensitic stainless steels and tool steels is performed in a DA atmosphere, often in a retort or bell furnace. Some tempering operations for tool steel also employ a DA atmosphere to avoid any surface oxidation.

Sintering
DA is typically used as a source of hydrogen (up to 10%) in sintering furnace atmospheres composed primarily of nitrogen. Iron and carbon steels, iron-copper and copper steels, iron-nickel and nickel steels, low-alloy steels, brass and bronze powders, stainless and tool steels can all be processed in blended nitrogen/DA atmospheres.

Other Applications
Many other thermal processes involve the use of DA either as a stand-alone gas or in combination with other gases in a furnace atmosphere. Applications include metallizing of ceramics, glass-to-metal sealing, degasifying metals, hydro-generation, ore refining, purification of gases and reduction processes. In addition, DA can be used as a nitrogen source, as an ultra-pure hydrogen source, as a forming gas or as a dilutant gas to control the percentage dissociation in nitriding. Work has also been done with DA as a carrier gas for carburizing and carbonitriding.

SAFETY

Anhydrous ammonia has a very pungent smell, detectable by the average human nose around 53 ppm. It is a toxic, flammable and colorless gas with a specific gravity of 0.5971 as compared to air, which is 1.0. A stream of the pure ammonia gas will not support its own flame in air at room temperature. Ammonia/air mixtures, however, are flammable in the range of 16-25% ammonia in air. They will typically burn but not explode. There are rules, regulations and codes involved with transportation and storage of anhydrous ammonia.

The fact that ammonia has a pungent odor means that detection is rapid, and the source can be isolated and repaired immediately. For small leaks, litmus paper moistened with water is used since ammonium hydroxide – a base – is formed when ammonia comes in contact with water. The litmus paper will turn color (blue litmus paper turns red in the presence of a base). Gas sniffers (detectors) are also available.

If ammonia odor is detected on the downstream side of the dissociator, either the flow of ammonia through the retort has been increased to an extent that the retort is overloaded or the retort temperature has dropped, perhaps in conjunction with too much flow. This most often occurs if the unit is improperly set – trying to produce 110 m³/hour (4,000 feet³/hour) out of a unit designed for 55 m³/hour (2,000 cfh).

DA is readily flammable, with a specific gravity of 0.296 (compared to air at 1.0). Equipment provided for the use of DA must have NFPA 86 safety and purging equipment.

Exothermic (DX®) Gas

Exothermic gas (aka exo) is a popular atmosphere choice from both a cost and flexibility standpoint. Exo is used for a variety of applications, such as a source of inert gas for purging/blanketing gas (lean) or to create a furnace atmosphere with up to 12% hydrogen (rich).

GAS CHEMISTRY

The exothermic-gas reaction, which happens in two steps (Equation 9.6.3), produces an atmosphere of nitrogen, carbon dioxide and water vapor with varying percentages of carbon monoxide, hydrogen and methane (Fig. 9.6.7).

9.6.3)   1st step: $CH_4 + Air\ (2O_2 + 7.6N_2) \rightarrow CO_2 + 2H_2O + 7.6\ N_2$

            2nd step: $2CH_4 + Air\ (O_2 + 3.8N_2) \rightarrow 2CO + 2H_2 + 3.8\ N_2$

FIGURE 9.6.7 | Exothermic-gas generator chemistry

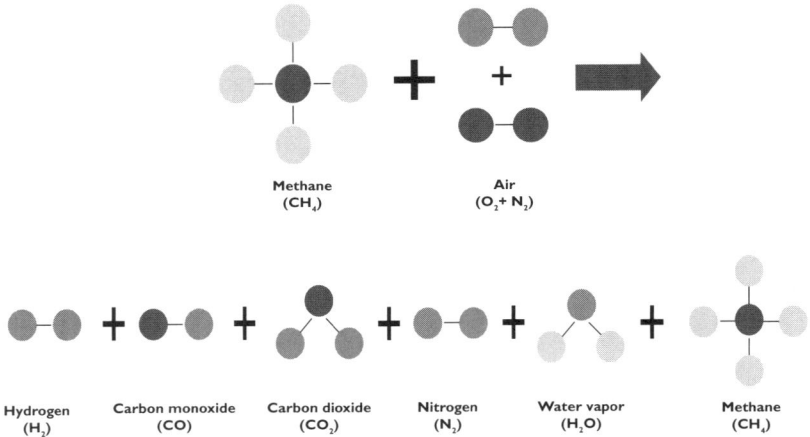

EQUIPMENT

An exothermic-gas generator (Fig. 9.6.8) consists of several basic components: a gas mixer, burner, horizontal combustion chamber and heat exchanger (Fig. 9.6.9). The products of combustion of a fuel (e.g., natural gas) and air are combined at different air/gas ratios (Fig. 9.6.10) to create the atmosphere. The reaction produces heat. As such, these generators typically have water-cooled combustion chambers.

FIGURE 9.6.8 | Typical exothermic-gas generator (courtesy of C.I. Hayes, a Gasbarre Furnace Group Company)

9 | FURNACE ATMOSPHERES

FIGURE 9.6.9 | Schematic of an exothermic-gas generator[11]

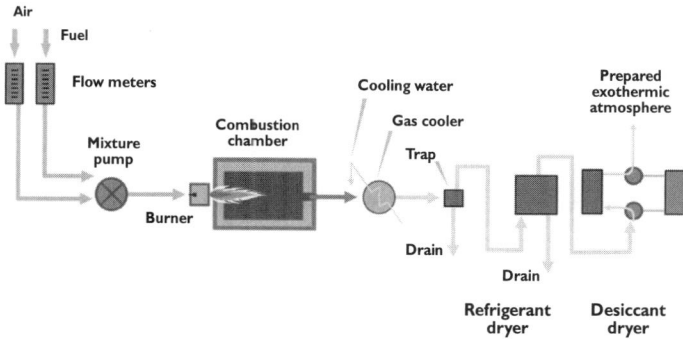

FIGURE 9.6.10 | Exothermic-gas composition as a function of air/gas ratio (courtesy of SECO/WARWICK Corp.)

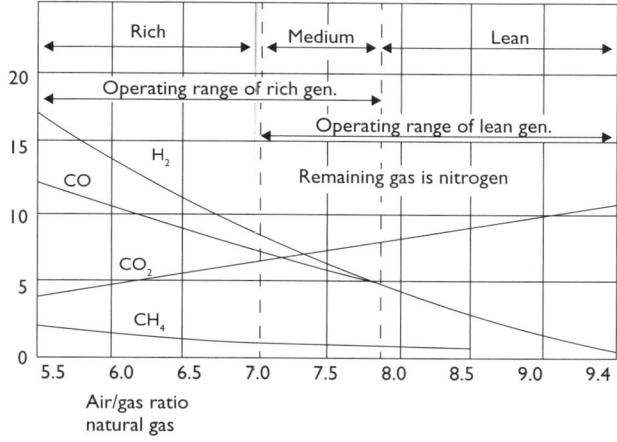

SIZES AND TYPES

Exothermic-gas generators produce either a lean or rich atmosphere composition (Table 9.6.3) depending on the chosen air/gas ratio. Generator capacities vary from 14-1,415 m³/hour (500-50,000 cfh) to larger. Depending on the application, the gas produced is cooled either by direct water spray into the gas stream or indirectly through a shell-and-tube heat exchanger. Outlet pressures of approximately 0.01-0.014 MPa (1.5-2.0 psig) are common, but output gas pressures of up to 1 MPa (150 psig) and higher can be designed.

Lean exothermic gas is produced from air/gas ratios of approximately 7.35:1 to 11:1. A typical lean generator gives off 85,460 kJ (81,000 BTU) per 28.3 m³/hour (1,000 cfh) produced atmosphere.

Rich exothermic gas is produced from air/gas ratios of approximately 5.5:1 to 8.2:1. A typical rich generator only gives off 11,605 kJ (11,000 BTU) per 28.3 m³/hour (1,000 cfh) of products.

TABLE 9.6.3 | Typical exothermic-gas composition and specifications[9]

	Lean gas atmosphere	Rich gas atmosphere	Cogenerated atmosphere
Carbon dioxide	5.2 to 10.5%	8.5 to 5.2%	11.8%
Carbon monoxide	0.01 to 3.0%	5.0 to 10.0%	0.01 to 3.0%
Hydrogen	0.01 to 3.0%	6.0 to 12.0%	0.01 to 3.0%
Methane	0.0 to 0.2%	0.0 to 0.1%	0.0 to 0.2%
Nitrogen	Balance	Balance	Balance
Oxygen	0.2 to 0.01%	< 50 PPM	0.2 to 0.01%
Typical delivery pressure	2,490-3,490 Pa (10" to 14" w.c.)	3,490 Pa (14" w.c.)	3,490 Pa (14" w.c.)
Dew point	38-43°C (100-110°F)	38-43°C (100-110°F)	38-43°C (100-110°F)
Air/gas ratio	7.0:1 to 10:1	5.5:1 to 7.8:1	

Note:

[a] Gas analysis depends on fuel used (natural gas assumed above).

Although hydrocarbon gases such as natural gas, propane and butane are the most common, virtually any combustible fuel can be used, including coke-oven gas, MAP gas, and liquid fuels such as kerosene, alcohols and fuel oil. Conversions between gases require only the changing of a burner and flowmeters.

FEATURES

The components that make up this equipment (Fig. 9.6.11) can be described as follows.

FIGURE 9.6.11 | Cross-sectional view of an exothermic-gas generator (courtesy of C.I. Hayes, a Gasbarre Furnace Group Company)

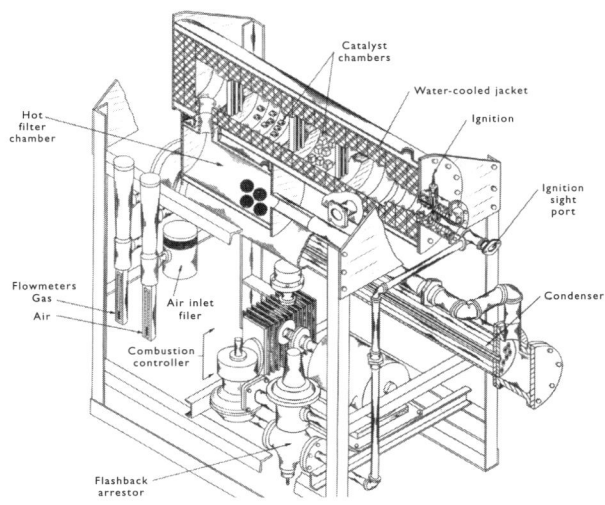

Combustion Chamber

Conventional lining of insulating firebrick and hard firebrick are traditionally used due to their heat-retention properties at lower flame temperatures (air/gas ratios in the rich range). Ceramic fiber capable of withstanding mixture temperatures to 1537°C (2800°F) reportedly offers a number of benefits, including rapid heating. Some designs do not use any insulation in the combustion chamber.

Catalyst

The use of a catalyst in the combustion chamber of a rich exothermic-gas generator is common to retain temperature and ensure complete cracking of the reaction products. Catalyst use in lean operation is often optional. This catalyst is similar to that used in nickel-impregnated endothermic gas generators consisting of 25-mm (1-inch) cubes. Untreated insulating firebrick can also be used.

The gas, upon completion of the combustion process and now depleted of oxygen, is cooled to about 50°C (120°F) and is then ready for introduction into the process. Normally, the final dew point of the process gas is approximately +5.5°C (+10°F) above the cooling-water temperature measured at the heat exchanger. With the addition of gas dryers and refrigerated coolers, however, dew points as low as -40°C (-40°F) are possible. Purifiers are available to reduce oxygen or carbon monoxide to less than 10 ppm so that lean exothermic gas can be used (in some cases) as a source of nitrogen.

In addition to the standard components found on an exothermic gas generator, several other optional features are available and will be discussed here.

Combustible Analyzers

Today's gas analyzers measure a number of variables. These include oxygen, carbon monoxide and carbon dioxide, total combustibles, temperature and even net combustion efficiency. Many of these systems are equipped with alarms and 4-20 mA signals, and they can be easily integrated into a control scheme. Outputs connected to motorized valves maintain a desired gas analysis based on analyzer setpoint. They provide a quick and easy way to fine-tune generators and furnaces or other industrial combustion processes.

Gas Coolers and Refrigerant Dryers

Most gas coolers have the capacity to cool and dry the rated gas flow from +40°C (+105°F) down to +4.5°C (+40°F) dew point. Hot-gas bypass systems in some designs allow the dryer to run from 100% of full rating down to approximately 60%, or the dryer can be bypassed completely. They are built in a variety of capacities from 14 to 283 m^3/hour (500 to 10,000 cfh). Refrigerant dryers can then reduce the +4.5°C (+40°F) gas to -40°C (-40°F) or lower. These refrigerant dryers are charged with environmentally friendly refrigerants.

MAINTENANCE

Most of the maintenance problems can be traced to poor water quality (i.e., water high in dissolved minerals and hardness values greater than 7 grains/gallon) and corrosion failures of the water jacket. Water-treatment options have made great strides in eliminating problems with water jackets and heat exchangers. Symptoms often include an uncontrollably high dew point. Other causes for high dew point can be traced to faulty flowmeters and air infiltration into the combustion chamber.

Sooting of the combustion chamber and catalyst bed, especially at low air/gas ratios, and soot collecting on the tubes in the heat-exchanger tube bundle are other maintenance issues. Shell-and-tube heat exchangers should provide effective cooling to within 7°C (15°F) of the cooling-water temperature.

OTHER TECHNICAL CONSIDERATIONS

Newer exothermic-gas generators are being equipped with additional, noteworthy features.

Mixer System Technology

Exothermic-gas systems (Fig. 9.6.12) using advanced software packages combined with regenerative blowers are now capable of delivering "flow on demand" throughout the working range of the generator. Ratio deviation alarms can be provided to confirm that the desired gas mixture is being produced. These types of systems typically replace existing carburetors/pumps with a gas-injection system. These systems can be adapted for low-pressure gas supplies or very high turndowns.

FIGURE 9.6.12 | Exoinjector® fuel-injection gas mixing system (courtesy of Atmosphere Engineering Company)

Cogeneration Technology

Some generators combine the ability to produce exothermic gas with that of generating steam or hot water (Fig. 9.6.13). This cogeneration technology – operated independently or integrated into an existing system – adds flexibility, saves energy and reduces cost. To do this, the standard water-cooled combustion chamber is replaced with a commercially available fire tube boiler. The efficient use of heat recovery is reported to lower operating costs by as much as 50%.

FIGURE 9.6.13 | Exothermic-gas generator with cogeneration capability (courtesy of Modern Equipment Co. Inc.)

APPLICATIONS

Exothermic gas can be used in a variety of industries, including heat treating, melting, petrochemical and food. The uses of exothermic gas in heat treatment include:

- ❖ Purging and nonflammable blanketing during various processes (lean)
- ❖ Bright annealing of nonferrous alloys such as aluminum and copper (lean or slightly rich)
- ❖ Processes requiring low oxygen percentage in the gas (lean)
- ❖ Bright annealing of low-carbon steel (rich)
- ❖ Normalizing (lean or rich)
- ❖ Silver brazing of nonferrous alloys (lean)
- ❖ Bluing (lean)
- ❖ Clean annealing of silicon steels (rich)
- ❖ Annealing (rich)
- ❖ Copper brazing (rich)
- ❖ Brazing of dissimilar metals onto low-carbon steels (rich)
- ❖ Reduction of surface oxides on metals (rich)

SAFETY

Perhaps the most important safety feature for use with exothermic gas systems is that of a carbon monoxide monitor (Section 14.2). There are a number of excellent panel-mounted and portable carbon monoxide sensors available, and monitoring the generator area is strongly recommended. There are also a variety of other sensors that look at a wide range of combustible and toxic gases as well as oxygen deficiency.

Endothermic (RX®) Gas

Endothermic gas (aka endo) is used for neutral hardening and as a carrier gas for gas carburizing and carbonitriding. Endo is generally produced so that its composition is chemically inert to the surface of the steel and can be made chemically active by the addition of enrichment (hydrocarbon) gas, which is usually done at the furnace.

GAS CHEMISTRY

Endothermic gas is produced when a mixture of air and fuel is introduced into an externally heated retort at such a low air/gas ratio that it will normally not burn. The retort contains an active catalyst that is needed for cracking the mixture. Leaving the retort, the gas is cooled rapidly to avoid carbon re-formation (in the form of soot) before it is sent into the furnace.

The endothermic-gas reaction (Equations 9.6.4, 9.6.5) produces an atmosphere of nitrogen, hydrogen and carbon monoxide with varying percentages of carbon dioxide, water vapor and residual methane (Fig. 9.6.14).

9.6.4) $\quad CH_4 + air\ (2O_2 + 8N_2) \rightarrow CO_2 + 2H_2O + 8N_2$

9.6.5) $\quad 2C_3H_8 + air\ (3O_2 + 11.4N_2) \rightarrow 6CO + 8H_2 + 11.4N_2$

The endothermic-gas composition (Table 9.6.4), by volume, varies depending on the type of hydrocarbon-gas feedstock. The use of a nickel-based catalyst makes the reaction easier. The nickel attracts the hydrogen atoms of the methane, which attach to the catalyst. The oxygen molecules approach and are attracted to the carbon atoms. The carbon atoms combine with the oxygen atoms to form carbon monoxide (CO). The hydrogen atoms combine to form H_2 and are released from the nickel attraction. The now-available nickel attracts new methane to continue the reaction (cracking) process. After the passage of the air-gas mixture over the catalyst, the reaction is frozen by chilling the gas rapidly to 315°C (600°F) in either an air-cooled or water-cooled heat exchanger.

FIGURE 9.6.14 | Endothermic-gas generator chemistry (courtesy of Surface Combustion Inc.)

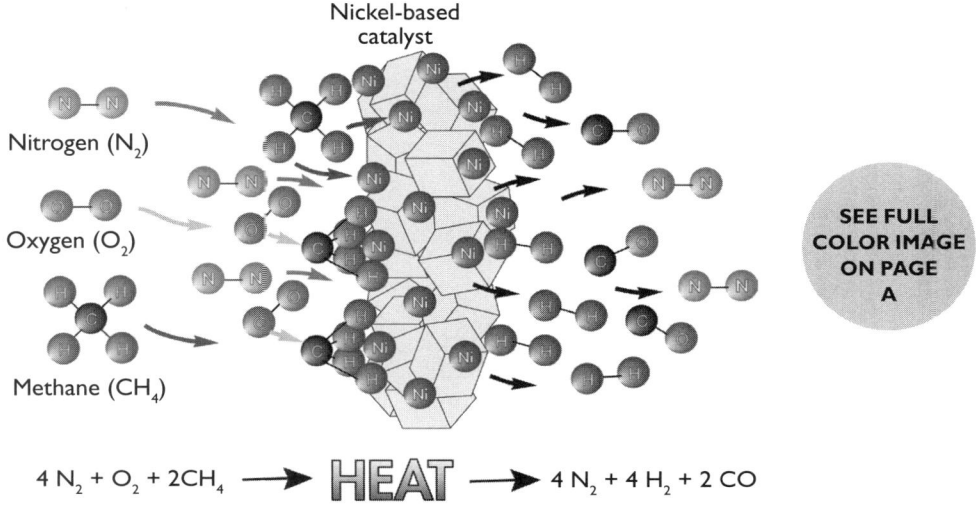

$$4 N_2 + O_2 + 2CH_4 \longrightarrow \text{HEAT} \longrightarrow 4 N_2 + 4 H_2 + 2 CO$$

TABLE 9.6.4 | Typical compositional ranges for endothermic gas

Gas constituent	Percentage (based on natural gas)	Percentage (based on propane)
N_2	40.9%	40.9%
CO	19.6%	23.3%
CO_2	0.4%	0.1%
H_2	38.9%	35.5%
CH_4	0.2%	0.2%
Dew point	-7 to +10°C (+20 to +50°F)	-23 to -26°C (-10 to -15°F)
Air/gas ratio	2.6:1	7.8:1

EQUIPMENT

Endothermic-gas generators are comprised of several basic components: a gas mixer, burner, combustion chamber and heat exchanger. They are available in single-retort (Fig. 9.6.15) and multiple-retort (Fig. 9.6.16) designs. The products of combustion of a fuel (e.g., natural gas) and air are combined at slightly different air/gas ratios, typically between 2.5:1 and 3.5:1 to create the atmosphere. The reaction requires heat to proceed, so these generators usually have heated combustion chambers.

FIGURE 9.6.15 | Single-retort endothermic-gas generator

FIGURE 9.6.16 | Modular multi-retort endothermic-gas generator (courtesy of Surface Combustion Inc.)

FEATURES

Endothermic-gas generators are common in the heat-treat shop. The main components of an endothermic-gas generator (Fig. 9.6.17) are relatively simple and consist of:

- ❖ Heated reaction retort with catalyst
- ❖ Air-gas proportioning control components
- ❖ Pump to pass the air-gas mixture through the retort
- ❖ Cooler to freeze the reaction and prevent soot formation
- ❖ Fire-check valve (to prevent backfire)
- ❖ Burn-off vent
- ❖ Thermocouples (control, over-temperature, recording)

FIGURE 9.6.17 | Endothermic-gas generator schematic piping arrangement[11]

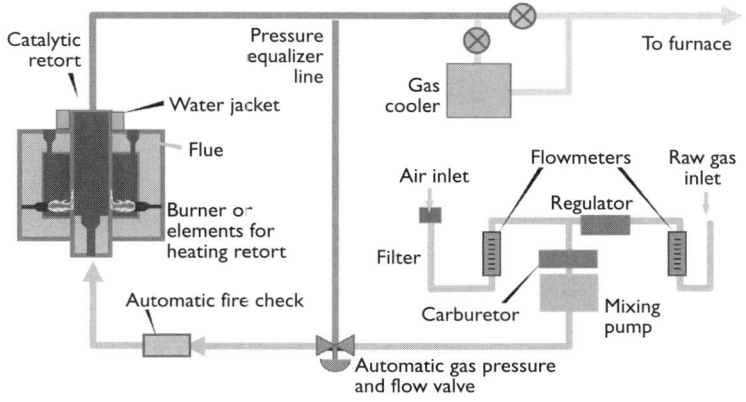

Retort
The retort for an endothermic-gas generator is typically a cast alloy, with HU (38% Ni, 18% Cr) and HK (20% Ni, 25% Cr) being common. In some instances, retorts are fabricated of Inconel 600® (preferred) or made of silicon carbide.

Retorts in most industrial generators are either thin and long or thick and short. They vary in diameter from about 150 to 300 mm (6 to 12 inches). To avoid issues with a cold center in the catalyst bed in larger-diameter designs, either the inlet pipe runs down through the center of the retort or the space is occupied by a closed-end pipe, typically 50-75 mm (2-3 inches) in diameter.

Catalyst
For economic reasons only, manufacturers have gone away from supplying pure-nickel shot as a catalyst and currently utilize refractory cubes typically 25 mm (1 inch) in size coated with 3-7% nickel sulfate ($NiSO_4$). Smaller-sized 17.5 mm (11/16 inch) cubes and spheres of 19 mm (0.75 inch) diameter have also been used, but pressure through the catalyst bed must be monitored due to increased packing density. The use of a refractory catalyst necessitates the use of a smaller-diameter retort to ensure both proper heat distribution throughout the catalyst bed and adequate dwell time at temperature for complete dissociation.

Mixer
A mixing pump and carburetor (optional) control the air/gas ratio of an endothermic-gas generator in the range of 2.5:1 to 3.5:1. In some generator designs, air/gas ratios have been known to run as low as 2.0:1. Most generators run a dew point in the range of +4.5 ± 0.2°C (+40 ± 2°F) to minimize maintenance concerns (e.g., sooting). Some heat treaters prefer to run generators in the

range of -1.1 ± 0.2°C (+30 ± 1°F) for carburizing applications. Final adjustments to the atmosphere are typically made at the furnace.

Similar to their exothermic-gas generator counterparts, endothermic generators can be equipped with advanced software packages combined with regenerative blowers capable of delivering flow-on-demand throughout the working range of the generator (Fig. 9.6.18).

FIGURE 9.6.18 | Endoinjector® fuel-injection gas mixing system mounted on an endothermic-gas generator (courtesy of Atmosphere Engineering Company)

Insulation
The heating chamber of an endothermic-gas generator can either be refractory-lined or lined with ceramic-fiber insulation. Modular designs and clamshell-style chambers (to facilitate retort removal) housing single retorts are more popular than the past practice of using large heating chambers housing multiple retorts.

Heating Source
Endothermic-gas generators are either gas-fired or electrically heated. If gas-fired, ring burners or combustion burners are commonplace. Electrically heated units either use nickel-chromium or silicon-carbide heating elements. If metallic, they often include the addition of rare-earth elements (e.g., Hf, Y) to extend operating life in air. Power is regulated by on/off or proportional control (zero-fired or phase-angle-fired SCRs). The typical operation temperature of an endothermic-gas generator is 1010-1095°C (1850-2000°F), depending on design. Most units run at 1065°C (1950°F).

Thermocouples
Most generators run either type-K, type-N or type-S thermocouples.

Firecheck
The firecheck (Fig. 9.6.19) is a safety device designed to prevent backfire into the

incoming gas supply line. The functionality of the firecheck should be checked every six months or more frequently if recommended by the manufacturer. The sad reality is that most heat treaters and plant maintenance personnel don't understand its function and never make sure that it is operating properly.

FIGURE 9.6.19 | Firecheck on an endothermic-gas generator

APPLICATIONS
The uses of endothermic gas (Table 9.6.5) include the following applications:
- ❖ Carburizing
- ❖ Bright hardening
- ❖ Sintering
- ❖ Brazing
- ❖ Carbonitriding
- ❖ Carbon restoration
- ❖ Neutral hardening

TABLE 9.6.5 | Applications for endothermic gas

Application				
Steel type	Low carbon (to 0.20%)	Medium carbon (0.20 to 0.60%)	High carbon (above 0.60%)	Special steels and irons
Process	Carburizing Carbonitriding Brazing Sintering	Clean hardening Carburizing Carbonitriding Bright annealing Carbon restoration Brazing	Bright annealing Clean hardening	Bright annealing Clean hardening

MAINTENANCE
Maintenance on endothermic-gas generators includes the following general areas:
- ❖ Test indicators of catalyst sooting
  - Small ratio adjustments do not result in a change of dew point
  - Very high dew-point readings

- Methane (CH_4) higher than 0.5%
- When operating between +30 to +40°F, the CO_2 should be approximately equivalent to the dew point divided by 100 (0.30-0.40% CO_2).

❖ Daily checklist
- Check the temperature-control instrumentation for proper operating temperatures.
- Check for proper flow and pressure of the generated atmosphere.
- Check for proper inlet air/gas ratios.
- Check either the gas analysis or the dew point of the unit. Make sure that manual and automatic readings coincide. Recalibrate automatic gas analyzers.
- Check that the floats in the gas flow tubes are free and operating.
- Check that the compressor is operating and functional.
- Check that the gas cooler is operating. If installed, check the temperature of the exiting gas to confirm that the carbon-reversal reaction is not occurring (and that soot is not being formed) on gas discharge from the generator to the furnace. If the system is water-cooled, check sight drains or temperature gauges (or both) to confirm proper water flow, pressure and temperature.
- Check that there are no leaks from any of the joints on the process retort, particularly at the point of entry of the process gas from the compressor.
- Check the heating chamber and visually confirm it is incandescent. If gas-fired, check the combustion equipment (including pilots, spark igniters and flame rods) for proper operation. Check burners for proper ignition and combustion characteristics. If electrically heated, check the current draw on the heating elements.
- Make sure atmospheric burners or pilots (or both) are protected from drafts.
- Check the burn-off stack to confirm ignition of flammable atmosphere gases.
- Monitor the carbon monoxide (CO) level in the immediate area of the generator. Confirm it is < 0.01%.
- Check for proper operation of the exhaust hoods and stacks.
- Check for excessive temperature in all areas of the generator.
- Check hand valves, manual dampers, secondary air openings or adjustable bypasses, valve motors and control valves for smooth action, proper position and adjustment.
- Check all pressure switches for proper pressure settings.

- Check blowers, compressors and pumps for unusual noise or vibration.
- Check belt tension.
- Check for evidence of any damage from any cause.

❖ Weekly checklist
- Burnout/regenerate the catalyst as per the recommended manufacturer's instructions and at the frequency recommended by the manufacturer.
- Remove the air filter from the compressor; clean and/or replace.
- Once the burnout/regeneration is complete, start the gas-making procedure. Check either the gas analysis or the gas dew point.
- Make sure the flame-sensing equipment is in good condition, properly located and free of foreign debris. Clean the burner flame rod. Check ignition-spark electrodes for proper operation and gap.
- Test thermocouples and lead wire for shorts and loose connections. Check protection tubes for sagging, cracks and proper insertion depth.
- Test visible and audible alarm systems for proper functionality.
- Remove the air filter from the compressor; clean and/or replace.
- Check ignition-spark electrodes for proper operation and gap.
- Remove the floats from the flowmeter glass tube, clean the internal and external surfaces of the flowmeter, and reassemble.
- Check the thermocouples for calibration.
- Check the gas pressure of the gas at the compressor.
- Check the instrumentation for calibration. This means temperature as well as gas analysis or dew point.
- Test interlock sequences of all safety equipment. Manually make each interlock fail, noting that related equipment closes or stops as specified by the manufacturer.

❖ Monthly checklist
- Test pressure-switch settings by checking switch movements against pressure settings and comparing with actual impulse pressure.
- Inspect all electrical devices for proper current and voltage and be sure that all electrical contacts and switches are functioning properly.
- Clean or replace the air blower filter.
- Clean any filters or strainers.
- Inspect burners and pilots.
- Check ignition cables and transformers.

- Test automatic and manual turndown equipment.
- Test pressure-relief valves; clean as necessary.
- Check back-pressure regulators; inspect and clean/replace diaphragms.
❖ Quarterly checklist
    - Inspect the catalyst. If necessary, fill to the recommended mark or replace.
    - Inspect and clean the burners. Check the gas train for functionality.
    - Remove the gas delivery line from the generator to the furnace and clean. There may be soot present if there have been any problems with the gas cooler.
    - Check all safety solenoids and safety controls.
❖ Semiannual checklist
    - Inspect the retort, refractory, heat exchangers, refrigerators, dryers and other accessories; repair or replace as necessary.
    - Lubricate the instrumentation, valve motors, valves, blowers, compressors, pumps and other components.
    - Test instrumentation; clean slide wires and electrical components.
    - Test flame-safeguard units.
    - Burn out carbon in the retort(s).
    - Check for plugging of hot pipes, tube bundles and jacketed pipes.

PROBLEMS

The most common problems experienced with endothermic-gas generators involve the following:

❖ Efficiency of the endothermic reaction is thermally dependent. The entire atmosphere must reach a minimum temperature in order for the gas to be completely reacted. If the temperature is not reached, no reaction will occur.

❖ If the gas coolers are not operating efficiently, sooting will occur due to the carbon reversal reaction. This typically takes place outside the retort, which causes a restriction in the outlet piping and the piping to the furnace. This then leads to back-pressure and a loss of flow. Soot can accumulate in a gas cooler in a matter of minutes, which is why ratio control is so important.

❖ The catalyst is often a nickel-impregnated refractory chosen for its capability to support itself at the required high operating temperature, withstand catalyst regeneration cycles and maintain physical stability in the presence of the reaction products, mostly carbon monoxide. If the temperature is not high enough and the gas is not completely reacted, then sooting in the catalyst bed will result. The catalyst becomes ineffective once it starts to soot, and the gas composition will drift and produce higher percentages of methane, carbon dioxide and water vapor.

SAFETY

Like any other piece of atmosphere equipment that uses a combustible gas, great care must be taken when both starting up and operating endothermic-gas generators. When starting up a new generator or after a shutdown, the unit should be raised to its operating temperature slowly to reduce the risk of thermal shock and the potential for cracking of the furnace refractories and the process retort. Under no circumstances should you consider putting gas or any other combustible gas mixture into the process retort under 760°C (1400°F). Otherwise, a serious explosion will most likely occur and can result in serious injury or death and significant damage to the equipment.

Summary

Each generated atmosphere has its own particular uses and alternatives, particularly in the form of blended nitrogen/hydrogen atmospheres. The decision to use a gas generator often comes down to cost and convenience. For this reason, a good understanding of the economics of any gas choice is needed. For their part, gas generators have a long track record of success in a diverse number of applications with problems being well understood and solvable on the shop floor.

REFERENCES
1. Herring, Daniel H., *Atmosphere Heat Treatment, Volume I,* BNP Media, 2014
2. O'Neill, Raymond G. and Daniel H. Herring, "The Use of Dissociated Ammonia in Heat Treating Applications," *Industrial Heating*, August 1998
3. "The Use of Dissociated Ammonia in Industry," Booklet No. TBK-1, LaRoche Industries Inc.
4. Diman, William, "Dissociated Ammonia," C.I. Hayes, white paper
5. Herring, Daniel H., "Technical Considerations in the Use of Dissociated Ammonia or Nitrogen/Propane as Brazing Furnace Atmospheres," C.I. Hayes, white paper
6. Herring, Daniel H., "Heat Treating and Atmosphere Generation," ICGCI II, Institute of Gas Technology, 1996
7. Snell, C.V., "Effective Use of Dissociated Ammonia," *Metals and Alloys*, 1945

8. Ellison, Thomas, Robert H. Shay and Kerry R. Berger, "Selecting Atmosphere Compositions For Bright Annealing Stainless," *Metal Progress*, June 1983
9. Herring, Daniel H., "Exothermic Gas Generators: Forgotten Technology?" *Industrial Heating*, June 2005
10. *Steel Castings Handbook, Supplement 9, High Alloy Data Sheets, Heat Series*, Steel Founder's Society of America, 2004
11. *Metals Handbook, Volume 2: Heat Treating, Cleaning and Finishing, 8th Edition*, ASM International, 1964
12. "Metal Minutes," SECO/WARWICK Corp.
13. Donald Bowe, Air Products & Chemicals Inc. (www.airproducts.com), technical and editorial review, private correspondence
14. James Oakes, Super Systems Inc. (www.supersystems.com), technical and editorial review, private correspondence
15. Donald Bystricky, Modern Gas Equipment Co. (www.moderneq.com), private correspondence
16. Jason Jossart, Atmosphere Engineering Co. www.atmoseng.com), private correspondence
17. Michael Schmidt, SECO/WARWICK Corp. (www.secowarwick.com), private correspondence
18. DX® and RX® are registered trademarks of Surface Combustion Inc.

9 | FURNACE ATMOSPHERES

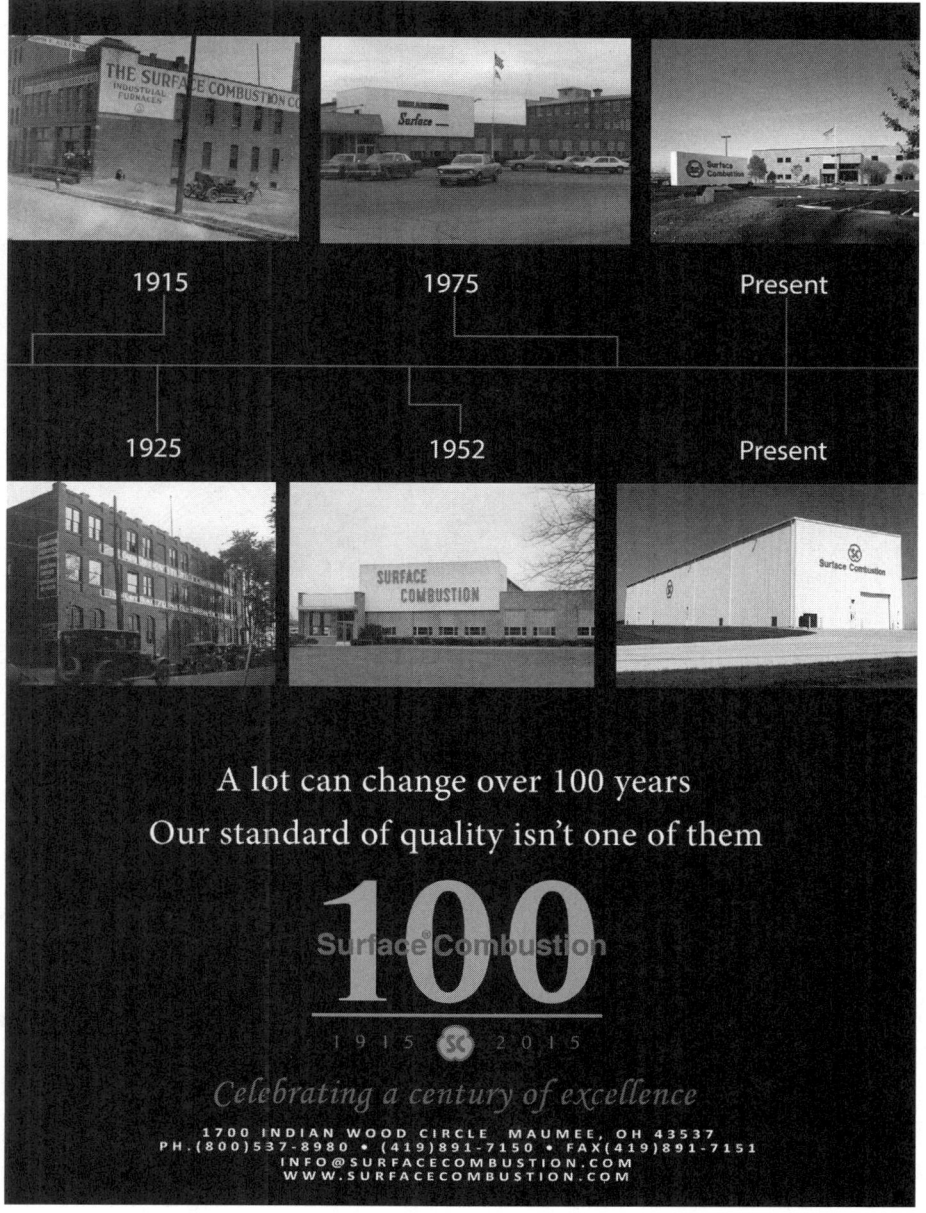

ATMOSPHERE HEAT TREATMENT

105

9.7

CONTROLS AND SENSORS

The composition of a furnace atmosphere is constantly changing. Therefore, we must use measurement and control devices to ensure the proper atmosphere chemistry necessary to meet the metallurgical quality and the desired mechanical/physical properties throughout the workload.

The trend today is to use multiple measurement tools to obtain the most accurate snapshot of the atmosphere in real time. In addition, the industry is moving toward automated control systems.

Manual or automatic control?

In one form or another, atmosphere control has been around for decades, and the control method may be categorized as either manual or automatic. There are benefits and limitations to both (Table 9.7.1).

In a manual control scheme, success depends on the vigilance of the people operating the equipment. They must rely on their experience coupled with trial and error and historical records/information.

A typical example of this type of control would be as follows. A furnace operator sets up a load, charges it into the furnace and manually adjusts temperature and atmosphere parameters. As process values approach setpoint, the operator begins periodically checking the atmosphere with sensor devices (e.g., dew-point instruments) and logging values on a heat-treat form of some design.

If process deviation occurs, the operator intervenes and attempts to make adjustments "on the fly" based on experience. At the end of the process, the heat treater checks the load to confirm it does not need to be reworked or scrapped. Essentially, operators learn what worked and what did not on a case-by-case basis over time, adding to their knowledge base. For key parameters or data records, a form, recorder or printer would provide a paper copy to be filed.

Automated control differs from manual control mainly in that sensors, computers and specially designed software are used during the heat-treatment process to monitor, control and record data about the furnace atmosphere. Load-entry software can even be used to control the timing of a charge and electronically record all load parameters set up by the operator. Output is set and changed by a process controller designed specifically for heat-treatment processes or a PLC (Section 4.7).

If the controller detects a process deviation or a condition likely to result in one, it can make adjustments or generate a process alarm if operator intervention is needed. Information is delivered through mobile devices to operators, supervisors or even maintenance based on an escalation and profile of the alarm and recipient. Data logging is electronic. Logged data points can be accessed at any point in the future.

TABLE 9.7.1 | Benefits and limitations of manual and automated control[4]

	Manual control scheme	Automated control scheme
Benefits	• Lower up-front cost • Initial training is easier • Benefits of "manufacturing wisdom" of experienced line operators	• Significantly reduced need for operator intervention • Greater control precision • Ability to customize recipes and alarms • Ability to automate maintenance alerts • Significant cost savings possible over time due to reduced scrap and rework • Reduced possibility of operator error • Opportunity to replace obsolete older equipment for which replacement parts may not be available or which may not be in compliance with industry requirements
Limitations	• Reduced ability to respond to possible deviation conditions quickly • Higher amount of scrap and rework • Much higher possibility of operator error • Limits on alarm complexity	• Higher up-front cost • Reduced ability to apply "manufacturing wisdom" that comes with operator experience on the line • Additional knowledge/training requirements for operation

Measurement Devices and Sensors

Monitoring and/or controlling the heat-treatment process are accomplished by use of sensors and measurement devices. The most common include:

- ❖ Oxygen (carbon) probes
- ❖ Non-dispersive infrared analyzers (NDIR) – single or multiple gas
- ❖ Dew-point analyzers
- ❖ Oxygen analyzers (Section 13.5)
- ❖ Combustion analyzers (Section 13.5)

Whether neutral or case hardening, annealing or normalizing, a number of variables determine how well a furnace does its job. It is critical to the process that we control the percentage of carbon dioxide, oxygen and water vapor throughout the entire cycle, as well as the ratio of enriching gas (or air) to carrier gas.

9 | FURNACE ATMOSPHERES

For example, surface carbon can be controlled within ±0.05% C during carburizing by measuring one or more of the gases mentioned above and adjusting the addition gases (hydrocarbon and/or air) accordingly.

OXYGEN PROBES

The oxygen (aka carbon) probe (Fig. 9.7.1) is an in-situ device that looks similar to a thermocouple for measuring temperature and typically sits inside the furnace and/or inside the generator (above the catalyst bed or in a separate heated "well" into which the furnace atmosphere is pumped). In whatever location, the oxygen probe measures minute changes in oxygen concentration of the furnace atmosphere.

A difference in partial pressure of oxygen in the furnace atmosphere and the partial pressure of oxygen in the room air induces a voltage across the electrodes in the probe. At any given temperature, there is a known relationship between the probe millivolt output and the oxygen potential of the atmosphere. The oxygen potential can be directly related to the carbon potential (Section 16.4). Hence, monitoring the furnace temperature and the probe output can control the carbon potential of the furnace atmosphere.

FIGURE 9.7.1 | Oxygen (carbon) probes and recommended mounting arrangement (courtesy of Super Systems Inc.)

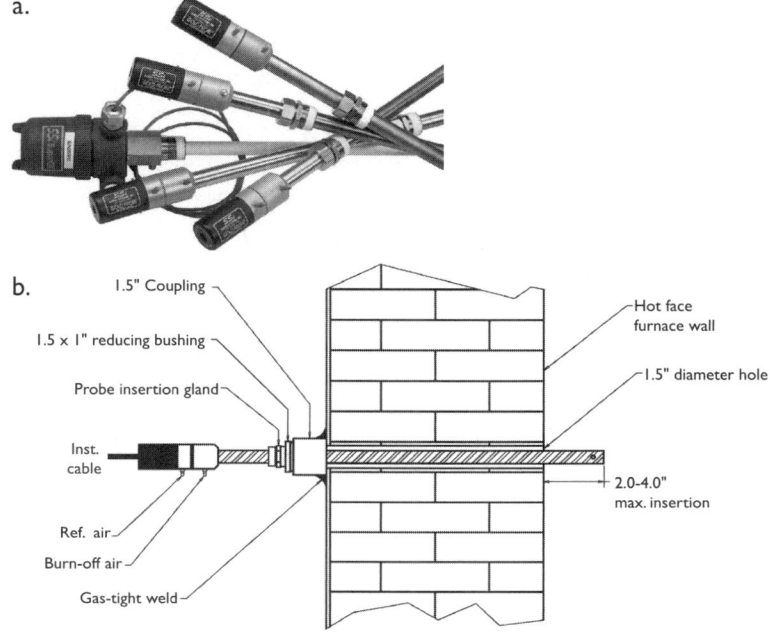

a. Family of oxygen (carbon) probes

b. Oxygen-probe mounting arrangement

The oxygen probe uses a conductive ceramic sensor, which is most often manufactured from zirconium oxide (Zr_2O_3). The operating range of the probe is normally 650-980°C (1200-1800°F). Oxygen probes can be used for a variety of atmosphere compositions, but they need to be calibrated for the specific one in use. They are fast-response devices and subject to contamination by carbon, zinc and certain stop-off paint vapors (Section 7.3). The presence of ammonia will shorten the life of the probe when used in carbonitriding applications.

An oxygen probe in a carburizing atmosphere must incorporate periodic air burnouts (Table 9.7.2). The carburizing process in use will determine the burnout frequency.

TABLE 9.7.2 | Recommended burnout frequency[5]

Atmosphere type	Frequency of sensor burnout, hours	Duration of sensor burnout, seconds[a]	Flow of burnout air, m³/hour (cfh)
Neutral	24	90	0.28 (10)[b]
High carbon	8 to 12	90	0.28 (10)[b]

Notes:
[a] Recommended burnout time. Maximum time should not exceed 120 seconds before allowing the system to re-stabilize.
[b] Minimum recommended flow rate.

A burnout consists of at least 0.28 m³/hour (10 cfh) of air piped to the burn-off fitting on the head of the probe. Room air or filtered combustion air are most commonly used. It is important not to use compressed air due to water and oil contamination that can damage the oxygen probe. The carbon controller should either control the frequency and duration of the burnout or shut off the gas additions in order to prevent excessive gas from compensating for the flow of air to the probe. Burnout flow and duration recommendations vary by manufacturer based on sheath diameter and tip design.

It is good practice never to exceed 90 seconds of air addition at any one time to avoid overheating the tip of the probe. A consistent way to verify a correct burnout is to monitor the millivolts of the carbon controller during the burnout phase. If a proper burnout is taking place, the output will drop below 200 millivolts. This can also vary based on the circulation in the furnace and the probe placement.

A possible side effect of extended burnout duration is oxidation of the tip of the sensor. This problem can manifest itself in higher-than-normal millivolt values over the remaining life of the sensor, which will require a lower CO factor setting for the same calculation of carbon potential. Consideration should be made for the duration of the burnout based on the carbon level in the furnace.

The current trend is to use oxygen probes in combination with three-gas analysis (CO, CO_2, CH_4) equipment (Fig. 9.7.2) to calibrate the probe to a known CO value and monitor the amount of free hydrocarbon gas in the furnace atmosphere.

FIGURE 9.7.2 | Combination furnace oxygen probe and infrared control (courtesy of Super Systems Inc.)

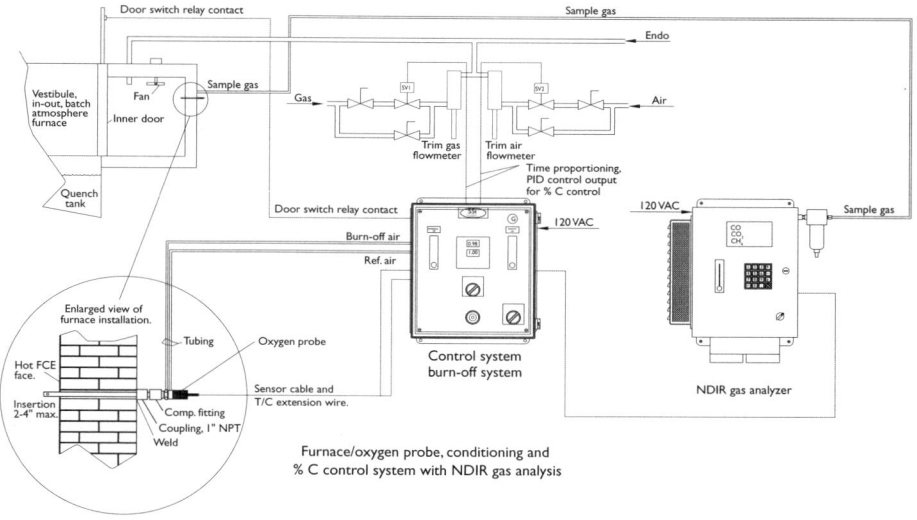

Measuring an Endothermic Atmosphere in a Furnace[5]

Carbon (oxygen) probes, shim stock or carbon analyzers (Section 13.6) and dew point are typically used to measure the carbon potential of the endothermic atmosphere in a heat-treatment furnace. An infrared analyzer can also be used. The CO_2 value – similar to water vapor and oxygen – varies based on furnace temperature and carbon potential but will typically lie between 0.25 and 0.50%. Multiplying the CO_2 value by 100 can roughly approximate the dew point (in Fahrenheit). Charts and tables can also be used (Section 16.4).

If a radiant tube in a gas-fired furnace is leaking, the leak (oxygen and moisture) will dilute the atmosphere carbon potential in the heat chamber when the furnace goes to high fire. As a result, the carbon controller will compensate by calling for more enriching gas to increase the carbon potential coincidental with high fire. A "saw-tooth" trend appears on the recorder. Setting the control to low fire will eliminate the saw tooth if the tube is cracked and leaking.

Oxygen probes and infrared analyzers are often used in conjunction when monitoring and controlling a nitrogen-methanol atmosphere. Using NDIR, values between 14-28% CO have been reported (target 20%). CO_2 values typically range from 0.10-0.80%, and free methane can be as high as 8% depending on the carbon setpoint. All testing should occur without hydrocarbon (e.g., natural or propane) gas additions. For nitrogen/methanol systems, the CO percentage is largely determined by furnace temperature and the relative flows of nitrogen and methanol.

Measuring an Endothermic Atmosphere in a Generator[5]

Another popular trend, especially in the control of endothermic-gas generators, is to use an oxygen probe (Fig. 9.7.3) in combination with a dew-point analyzer.

Using an oxygen sensor to measure and control the dew point in an endothermic generator is done either in-situ, in a thermal well or by use of an oxygen probe with an integral thermal well/sheath arrangement. The millivolt signal can be converted to a dew-point value (Section 16.4) or by calculation in some of the more advanced dew-point control systems. This calculation uses the millivolts generated by the probe, the hydrogen factor of the controlling instrument and the temperature of the oxygen sensor. The temperature is required for the calculation, but the dew point of the gas is not temperature-dependent.

Oxygen sensors are often run at 1040°C (1900°F), although they will provide the same reading when operating at 815°C (1500°F). The generator gas exiting the retort is sent through a heat exchanger to freeze the composition. As long as the sensor is accurately measuring the millivolts of the gas, the temperature of the sensor can be as low as 593°C (1100°F).

Changes in sensor location have occurred over the years. Initially, a sensor would be mounted on the top of a retort in an air-cooled, fabricated fixture to measure the oxygen. Then a ceramic reheat well mounted through the sidewall of the generator was used. Now a modified sheath with an integral reheat well makes the installation much easier. The sheath and the integral well are aluminized prior to assembly so that the nickel in the sheath material does not react with the endothermic gas. This is especially important between 705 and 480°C (1300 and 900°F), where the endothermic reaction will reverse over time if nickel is present.

Lambda-style probes like those used in an automobile engine are also used in generator applications. While less expensive, they lack the long-term stability of zirconia technology and must be re-oxidized to avoid drift.

FIGURE 9.7.3 | Endothermic generator dew-point control scheme via oxygen probe (courtesy of Super Systems Inc.)

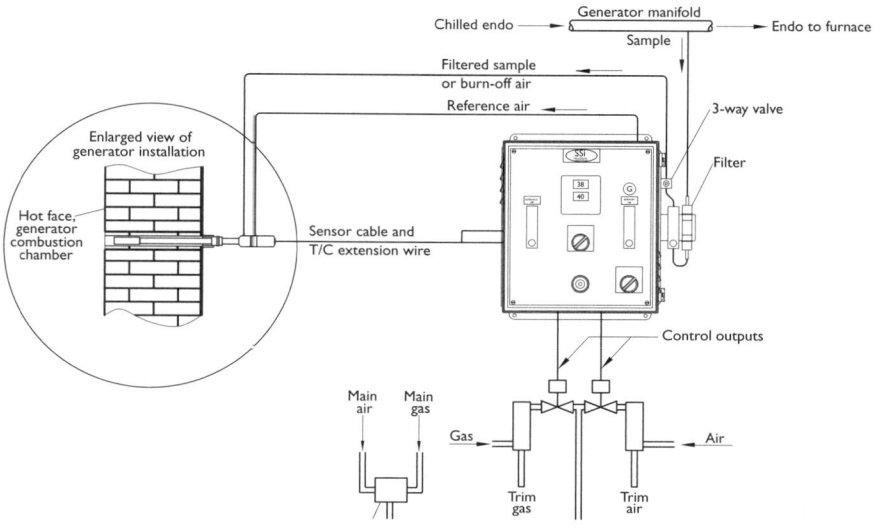

INFRARED (NDIR) ANALYZERS

Infrared analysis (Fig. 9.7.4) uses light in the infrared spectrum to analyze a gas sample and determine the percentage of that constituent in the furnace atmosphere. Single-gas (carbon monoxide) or multiple-gas analyzers detect the percentage of these gases in the furnace atmosphere.

FIGURE 9.7.4 | Infrared measurement of a gas (courtesy of Super Systems Inc.)

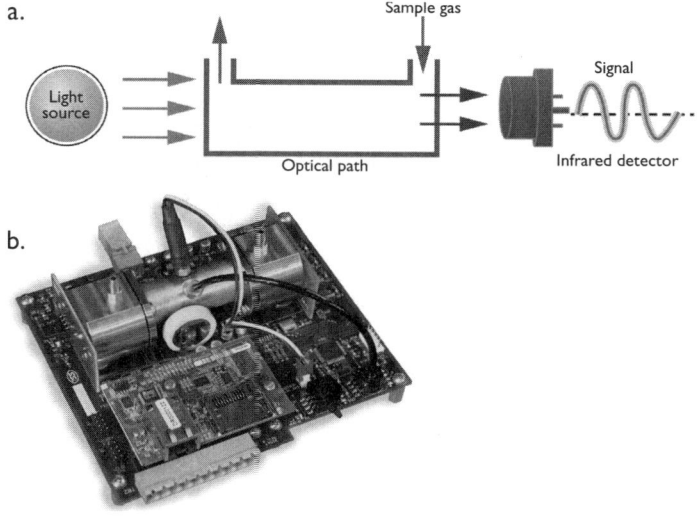

a. Principle of operation

b. Board-mounted IR sensor

The amount of carbon dioxide in the furnace is another indirect way of measuring the carbon potential of the atmosphere. Charts/tables can be used to correlate CO_2 readings with dew point or millivolt signals from oxygen probes.

Today, three-gas infrared analyzers (Fig. 9.7.5) are popular, and they are used to monitor the atmosphere produced by generators (Table 9.7.3) and furnaces (Fig. 9.7.6). The analyzers work on the principle that individual gases absorb infrared radiation of very specific wavelengths. The amount of absorption increases with gas concentration. The unit operates under the principle that a gas sample passes through a cell, where a light emits infrared energy of known wavelength. The sensor converts measured infrared energy into an electrical signal. These values are usually compared to the values obtained with a reference gas. Infrared analyzers are known for their fast response and are easily calibrated.

FIGURE 9.7.5 | Infrared gas analyzers (courtesy of Super Systems Inc.)

a.

b.
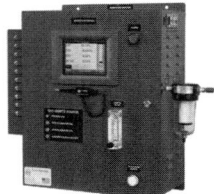

a. Portable unit

b. Panel-mounted unit

TABLE 9.7.3 | Typical field data for an operating endothermic-gas generator[1]

Constituent	First quarter	Second quarter	Third quarter	Fourth quarter
% CO	19.02	19.66	19.32	19.21
% CO_2	0.260	0.252	0.254	0.257
% CH_4	0.07	0.08	0.09	0.09
Generator dew point, °C (°F)	+4 (+39)	+4 (+39)	+4 (+40)	+4 (+39)
Dew point at furnace inlet, °C (°F)	+3 (+37)	+3 (+37)	+3 (+37)	+3 (+37)
Zonal dew point (Z2-Z4), °C (°F) [3]	+4.5-5.5 (+40-42)	+4.5-5.5 (+40-42)	+4.5-5.5 (+40-42)	+4.5-5.5 (+40-42)

Notes:

[1] 85 m³/hour (3,000 cfh) output

[2] Natural gas feedstock

[3] Neutral hardening

9 | FURNACE ATMOSPHERES

FIGURE 9.7.6 | Furnace infrared (three-gas) control scheme (courtesy of Super Systems Inc.)

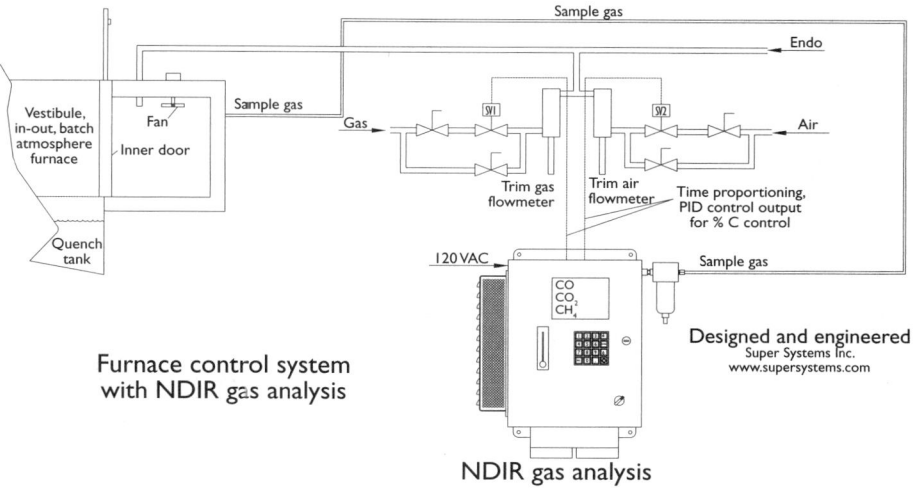

Dew Point

Dew point is defined as the temperature that water vapor starts to condense. In simplest terms, a dew-point analyzer (Fig. 9.7.7) measures the amount of water vapor present in the furnace atmosphere (Table 9.7.4). This information can then be used to determine the carbon potential of the atmosphere (Table 9.7.5). It is considered an indirect-measurement technique if it involves pulling a gas sample from the furnace into the instrument.

FIGURE 9.7.7 | Dew-point analyzers (courtesy of Super Systems Inc.)

a.

b.

a. Portable, Model DP-2000

b. Panel-mounted, Model DPC 2530

TABLE 9.7.4 | Typical dew-point levels

Dew point, °C (°F)	Water vapor (ppm)
+8 (+46)	10,590
-4 (+25)	4,320
-18 (0)	1,240
-40 (-40)	127
-68 (-90)	3.4

Note: 1% moisture = 10,000 ppm water vapor = +4.4°C (+40°F) dew point

TABLE 9.7.5 | Dew point vs. surface carbon at selected temperatures

Dew point, °C (°F)	815°C (1500°F)	870°C (1600°F)	925°C (1700°F)
-1.1 (+30)	1.10	0.80	0.55
+4.4 (+40)	0.85	0.60	0.40
+10 (+50)	0.60	0.40	0.27

If performed properly, dew point is a simple and accurate atmosphere measurement technique and indicates the condition inside the atmosphere generator or heat-treat furnace. It is important to keep the sample port (Fig. 9.7.8) free of soot (the capability to rod out the port should be included in the design), and a method of cooling the gas sample should be provided (by use of a small heat exchanger or coiling the copper/stainless tubing into loops to allow greater dwell time of the gas).

FIGURE 9.7.8 | Dew-point sample port on a continuous mesh-belt furnace

Dew point will help tell you if the reaction is stable or unstable (constant dew point or changing dew point over time). It can tell you when the catalyst bed in your endothermic-gas generator is starting to soot and if there is a water leak, an air leak or non-uniformity ("breathing") of the atmosphere inside your furnace.

The types of dew-point analyzers available include capacitance sensors, chilled mirrors and older-model fog chambers (Alnor®, Dew Cup). Condensation is a problem for all dew-point devices if the sample temperature is less than the dew point of the gas.

One solution is to heat trace the sample lines. If the ambient temperature exceeds 40°C (105°F), as it often does in heat-treat shops, the analyzer will not consistently give accurate readings unless special precautions are taken, such as keeping the unit in the office and bringing it out on the shop floor for only a short period of time in hot conditions.

INTERPRETATION OF DATA

In order to interpret furnace-atmosphere data correctly, it is important to collect data in real-time. However, it is of critical importance to understand the exact furnace operating conditions at the time the data was collected (e.g., zone temperatures and gas flows, furnace pressure, exhauster settings, fan rotation and speed, etc.).

In more than one instance, data was analyzed and adjustments made based on improper sample port locations, instrumentation that was not properly calibrated, filters that were dirty and/or sample ports not extending fully into the furnace chamber. Mistakes such as these can be devastating to both equipment and part quality.

Process Control

In its basic form, process control applied to heat treatment is controlling temperature to maintain a desired setpoint. Process control can then be extended to other variables and, in its most complex form, to a lights-out operation where all mechanical, electrical and control functions are being handled by instrumentation. We will consider some simple examples.

GENERATOR CONTROL

For gas generators, maintaining the quality of the gas (i.e., temperature, gas composition, dew point) using microprocessor controls and sensors to regulate the addition of enriching gas or dilution air is vital. Control systems can utilize fuel injection, which performs the mixing and control in one package with very little adjustment or maintenance required (Section 9.6).

CARBURIZING CONTROL

Carburizing requires precise management of its variables to achieve the desired metallurgical results in the parts being processed. This includes such items as heat-up and soak times, temperature, carbon potential (single or boost/diffuse), cool down and quench (Fig. 9.7.9).

The carburizing atmosphere is typically controlled by use of an oxygen probe. The carbon algorithm (using sensor millivolts and furnace temperature) used when determining carbon potential available in the furnace atmosphere assumes a consistent gas composition of the enriching gas coming into the building. The equation also assumes a known CO value. If the composition of the enriching gas or air changes, it not only influences the quality of the endothermic gas but also changes the amount of additive or enriching gas. The calculation of surface carbon from the millivolts and temperature monitored by the probe could be off. The automation component comes in when the calculation in the controller is automatically adjusted based on the three-gas calculation of carbon potential, thus providing a more accurate reading of carbon. This is commonly referred to as IR compensation.

Periodic verification of the process is typically performed using dew point, shim stock or portable gas analyzers. If possible, shim stock (the only true test of carbon available in the furnace atmosphere) should always be used for verification.

FIGURE 9.7.9 | Carburizing recipe-control scheme (courtesy of Super Systems Inc.)

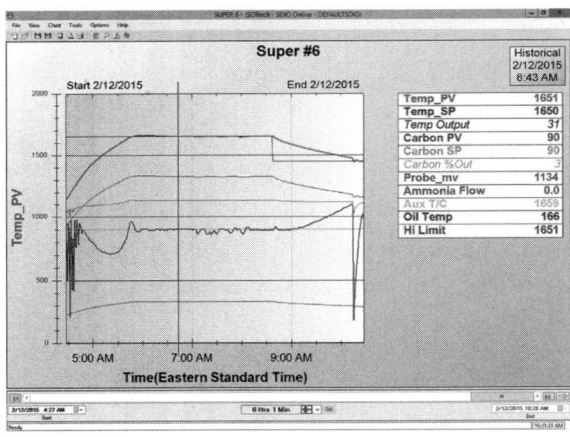

NITRIDING CONTROL

Nitriding (Fig. 9.7.10) is another process that requires precise management of its variables. To achieve this, multi-variable control systems (Fig. 9.7.11) monitor temperature, atmosphere, nitriding potential, flow rate and pressure to provide both safe and precise control of compound (i.e., white) layer and diffusion-zone variations in material. Load and furnace types are also controlled automatically.

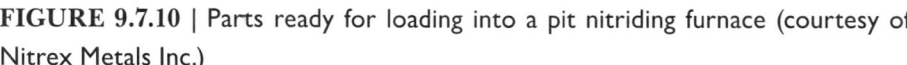

9 | FURNACE ATMOSPHERES

FIGURE 9.7.10 | Parts ready for loading into a pit nitriding furnace (courtesy of Nitrex Metals Inc.)

FIGURE 9.7.11 | Nitriding control scheme (courtesy of Super Systems Inc.)

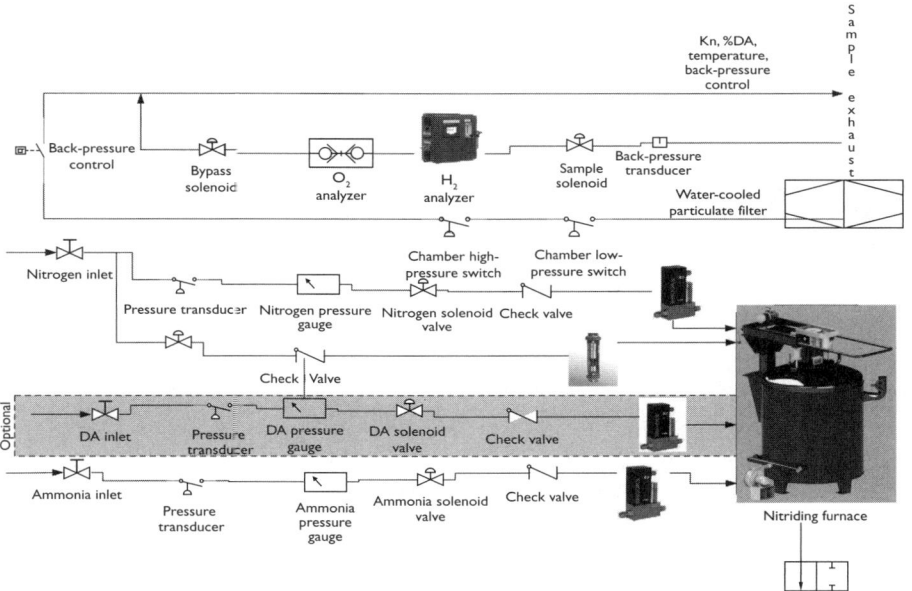

Automated control for both nitriding and nitrocarburizing is based on temperature, time, flow and measurement of hydrogen, oxygen, carbon monoxide (CO) and CO_2. In nitriding, hydrogen is the process variable that must be measured as a result of the reaction of NH_3 during dissociation (Fig. 9.7.12).

FIGURE 9.7.12 | Thermal-conductivity cell for measuring hydrogen (courtesy of Super Systems Inc.)

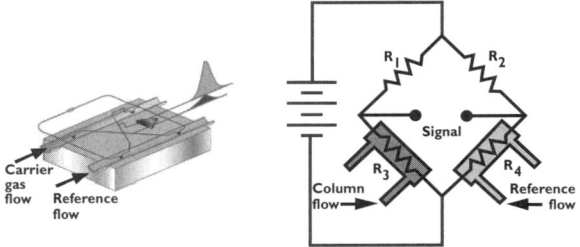

It is possible to define the material, group and class. Determination of the desired K_N (Equation 9.7.1, Table 9.7.6) and K_C based on the required epsilon layer and allowable porosity can be found in AMS 2759/12A (Table 9.7.7).

9.7.1) $$K_N = \frac{p(NH_3)}{p(H_2)}$$

where: p = gas partial pressure

TABLE 9.7.6 | Relationship between dissociation and nitriding potential (K_N)

Percentage dissociated ammonia	Nitriding potential, K_N
1.33	986.67
2.67	344.13
6.67	83.48
13.33	27.41
20.00	13.77
26.67	8.20
33.33	5.33
40.00	3.65
46.67	2.58
50.00	2.18
53.33	1.84
60.00	1.33
66.67	0.94
73.33	0.65
80.00	0.43
86.67	0.25
93.00	0.11
100.00	0.00

TABLE 9.7.7 | K_N and K_C values

Material	Process temperature[a] °F	Process temperature[a] °C	Process time, hours	Class 1 Not exceeding 15% of white-layer thickness K_N min	Class 1 K_N max	Class 1 K_C min	Class 1 K_C max	Class 2 Above 10% but not exceeding 50% of white-layer thickness K_N min	Class 2 K_N max	Class 2 K_C min	Class 2 K_C max
Group 1[b]	1040	560	3-6	2.13	2.41	0.57	0.69	2.48	2.68	0.49	0.54
	1075	579	2-5	1.50	1.60	1.10	1.22	1.68	1.78	0.86	0.94
Group 2[c]	980	527	6-30	4.51	5.55	0.16	0.24	6.03	7.10	0.09	0.13
Group 3[d]	1060	571	3-10	1.82	2.10	0.76	0.99	2.22	2.64	0.48	0.68

Notes:

[a] Temperatures shown are not firm requirements. Once a temperature is selected, however, both potential values shall be within specified limits for the given temperatures.

[b] Group 1: HLSA, carbon steels

[c] Group 2: 4140, 4340, Nitralloy 135M

[d] Group 3: Cast iron

FERRITIC NITROCARBURIZING (FNC) CONTROL

Ferritic nitrocarburizing (FNC) control (Fig. 9.7.13) is similar to nitriding but with the addition of a carbon-bearing gas. Hydrogen is one of a number of variables that must be considered when controlling FNC. With the use of CO_2 (common in pit-type furnaces), there is a reaction with hydrogen that produces CO to create carbon activity, K_C (Equation 9.7.2).

$$9.7.2) \quad K_C = \frac{p(CO_2)\, p(H_2)}{p(CO)\, p(H_2O)}$$

where: p = gas partial pressure

FIGURE 9.7.13 | Nitrocarburizing control scheme (courtesy of Super Systems Inc.)

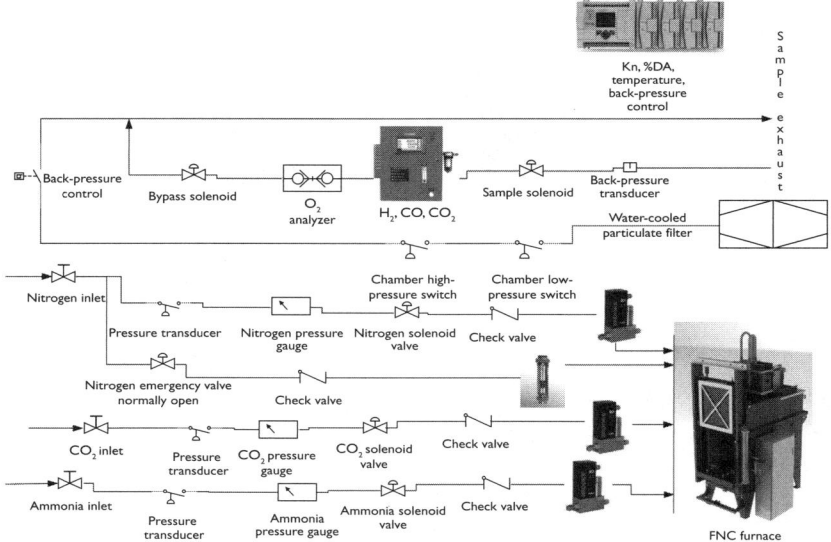

CO and CO_2 are both reliably measured with infrared technology. It should be noted that infrared cell technology must be compatible with NH_3 and water for the FNC process. Oxygen is measured using zirconia or lambda-probe technology. As with nitriding, it is important to meter the gas flows when using nitrogen and any other diluents. When using endothermic gas with 40% hydrogen present, it is important to meter and compensate for the gas flow and hydrogen content. With these variables being considered, K_N and K_C are calculated to allow for more precise control of the nitrogen and carbon compound layers.

Sample lines and filters to the H_2 analyzer and oxygen cell must be maintained on a regular basis. Calibration of the instrumentation and analyzers must be completed at manufacturer-required intervals. In addition, the mechanical assembly of the system must take into account the need for filtration and ease of maintenance. For example, sample systems must be supplied with drip legs in order to remove any moisture that might be generated during the pre-oxidation process.

Filtration and heat tracing must be included in the exhaust piping when nitriding stainless steel and using a pre-activation process (typically involving polyvinylchloride). Lines and filters must be cleaned for each run. Filtration systems should be designed with pressure-indication and automatic-bypass modes. FNC processes, in particular, are more challenging due to the formation of ammonium carbonate in the exhaust.

Maintenance

Preventive-maintenance programs (Section 13.1) are common to all heat-treatment operations, but nowhere is it more important than for gas-sampling systems. By using existing inputs and outputs on controllers and PLCs, operators can set specific timers and counters. These are extremely useful when determining when certain maintenance tasks should be planned or considered. With an established baseline, quick evaluation of real-time versus historical information can identify problem situations. Taken a step further, notification software can be used to automate emails and text messages to key personnel.

Here are some quick tips and items that need to be routinely maintained.

- ❖ Sampling lines
    - Install, inspect and replace the line. Analyze filters as necessary.
    - Verify sample flow (many analyzers are flow-dependent).
    - Flowmeters at the source or analyzer.
    - Provide a way to clean out the sample port (i.e., rod out if carbon buildup occurs).
    - Don't over-pressurize/back-pressure the analyzer as many cells are dependent on flow and pressure. Install metering valves if necessary.
    - Purge the sample lines and analyzer with nitrogen when not in use.
- ❖ Analyzers
    - Zero the infrared and thermal-conductivity analyzers with nitrogen.
    - Span with a gas of known concentration. Purchase a cylinder filled with a known gas composition.
    - Span with outside air or a moisture generator.
    - Span with lithium-chloride or potassium-chloride salt solutions.
    - Some units (Alnor® dew pointers) require factory calibration only.
- ❖ Oxygen probes
    - Check surface-carbon values against shim stock.
    - Measure temperature near the probe.
    - Measure and correct for CO content of the furnace atmosphere.
    - Confirm reference air and burnout air are being supplied unimpeded.
    - Confirm proper probe location and insert depth.
    - If hot insertion or removal is required, follow manufacturer's recommendations with respect to prescribed speed of movement of the probe. Slower is always better.

Summary

While manual control methods are highly effective when properly employed and continue to be used throughout the heat-treatment industry, the future lies in the precision and customizability of automated atmosphere-control methods. By utilizing specialized process controllers and software, these systems help enhance productivity through real-time monitoring and control, which leads to reduced scrap and frees up operators for other tasks.

Automated systems also form the basis for electronic data acquisition, which improves process traceability and makes it much easier to analyze data in making decisions for improving processes. Preventive-maintenance programs often grow from automated methods, reducing unplanned downtime. Instrumentation upgrades associated with automation bring several additional benefits, including:

- ❖ Increased focus on operation-specific features such as carburizing, vacuum heat treating and nitriding
- ❖ Recipe control and management, leading to fewer entry points and operator requirements
- ❖ Potential for greater operating ranges due to the ability to manage multiple parameters at the same time

Clearly, as technology improves, processes improve. And while the initial cost of control upgrades may seem significant, these upgrades will pay for themselves many times over, and customer satisfaction will grow.

REFERENCES

1. Herring, Daniel H., "Selection and Use of Furnace Atmospheres," *Industrial Heating* webinar, 2011
2. Herring, Daniel H., "Understanding Furnace Atmospheres, Atmosphere Operation and Atmosphere Safety," Heat Treating Hints, Vol. 1, No. 7
3. Thompson, Stephen, "A Practical Approach to Controlling Gas Nitriding and Ferric Nitrocarburizing (FNC) Processes," *Industrial Heating*, December 2013
4. Oakes, James and Jeremy R. Merritt, "Automated Control of Heat Treating Processes: Technology, Data Acquisition, Maintenance and Productivity Gains," *Industrial Heating*, August 2014
5. Fincken, Robert, "Protective Atmospheres, Measurement Technologies and Troubleshooting Tools," Conference Proceedings, Furnaces North America 2014
6. "Process Automation," Super Systems Inc., white paper
7. Donald Bowe, Air Products & Chemicals Inc. (www.airproducts.com), technical and editorial review, private correspondence
8. James Oakes, Super Systems Inc. (www.supersystems.com), technical and editorial review, private correspondence
9. Thomas Philips, Air Products & Chemicals (www.airproducts.com), private correspondence

9 | FURNACE ATMOSPHERES

Super**Systems**
incorporated

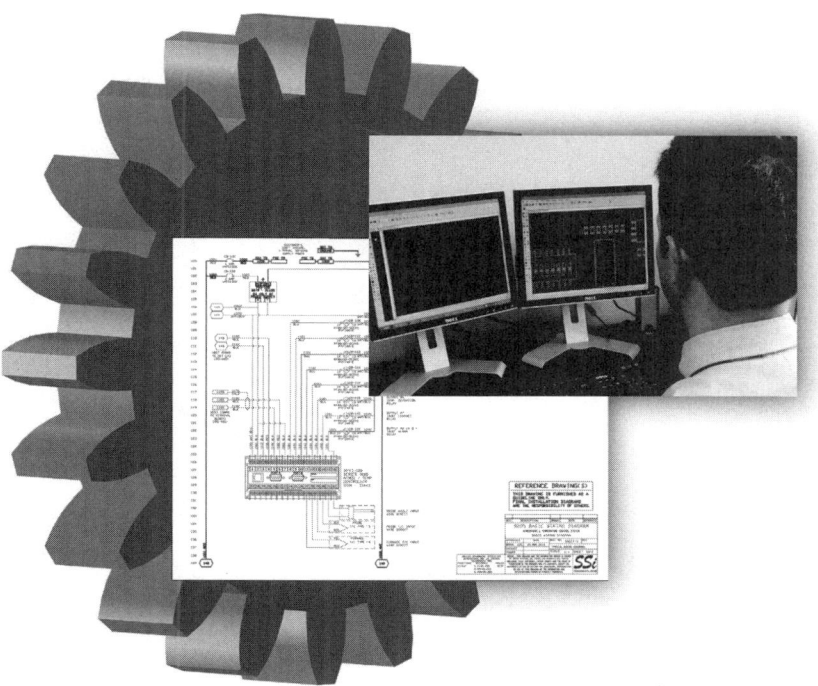

Automation and Controls Engineering for Heat Treatment

www.supersystems.com

10 | QUENCHING AND QUENCHANTS

10.1

PRINCIPLES OF QUENCHING

Heat-treatment operations fall into two broad-based categories: processes that soften a material (e.g., annealing, normalizing, stress relief) and processes that harden a material (e.g., through or direct hardening, case hardening). While there are important differences in each of these processes, what is perhaps most noticeable to the heat treater are the differences in quenching (cooling) rates from elevated temperature.

In general, a slow quench (i.e., slow cooling rate) produces a soft material with a primarily ferrite and pearlite microstructure. This category of processes is intended to remove stresses, refine the grain structure and put the material in a workable condition for subsequent manufacturing operations. By contrast, a rapid quench (i.e., fast cooling rate) produces a primarily martensite or bainite microstructure. This category of processes is intended to increase strength, improve surface hardness and wear resistance, and increase toughness and resistance to impact.

Since quenching is multifaceted, the selection of the quenchant and the design of the quenching system become important factors in the success of this portion of the heat-treatment operation. In addition, managing distortion often becomes an overriding consideration. As such, there is a delicate balance between achieving desired properties and maintaining dimensional control of the component part.

Quenching Choices

The method of quenching (cooling) depends on the hardenability of the steel, the shape of the part and the properties to be imparted to the steel. Quenchants (Fig. 10.1.1) are either liquids or gases. This includes brine/caustic, water, oils, polymers and molten salt as well as air (still or moving), nitrogen, argon, hydrogen, helium and gas mixtures. Each has characteristics that make them ideal for certain applications and poor choices for others.

FIGURE 10.1.1 | Common quenchant choices and their effect

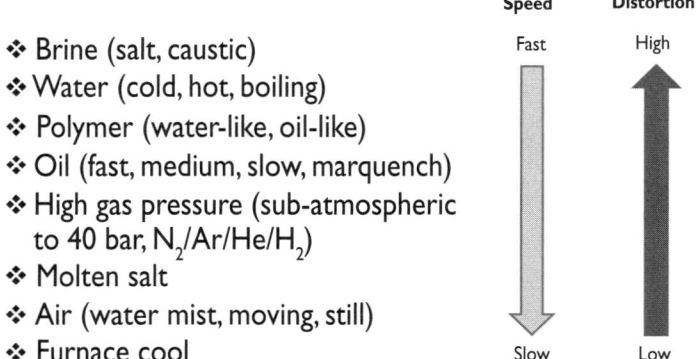

- Brine (salt, caustic)
- Water (cold, hot, boiling)
- Polymer (water-like, oil-like)
- Oil (fast, medium, slow, marquench)
- High gas pressure (sub-atmospheric to 40 bar, N_2/Ar/He/H_2)
- Molten salt
- Air (water mist, moving, still)
- Furnace cool

A key difference between quenching in a liquid and quenching in a gas lies in the different mechanisms involved in their heat-transfer characteristics. Most liquids (Fig. 10.1.2a) – such as water, polymer or oil – have distinct boiling points and, thus, different heat-transfer mechanisms (and rates) at various temperature stages. In general, liquids have three distinct heat-transfer phases: a vapor blanket or "film" boiling stage, a nucleate or "bubble" boiling stage and a convection cooling stage. By contrast, heat transfer takes place by convection only for gaseous media (Fig. 10.1.2b).

Quenching in a gaseous medium can be performed while the parts cool in the furnace (e.g., annealing), with the parts being slow cooled in still air (e.g., normalizing), with parts buried in sand or ash, and by fog quenching using a water mist or using pressure-quenching techniques from sub-atmospheric pressure to 20+ bar (atmospheres).

Quenching by a liquid is typically done either by immersion or spray quenching. In many cases, agitation is employed; in other cases, it is avoided. Quench baths are heated in some instances, while they are not in others. The time to quench a part can vary from several seconds to many minutes (e.g., austempering). Quenching can be continuous or interrupted, and it can be performed manually or automatically.

Quench systems may be part-specific, designed specifically for a given application or designed to provide quenching for a wide variety of parts. All of these factors, and more, mean that it is challenging to predict or manage the quenching process. For all our hope that quenching is a science (and in many ways it is), one must remember that quenching is still an art in many ways and unique to the task at hand. As such, common sense and empirical knowledge become important factors in the process. The key is to document these aspects of the process so that the knowledge is not lost over time.

FIGURE 10.1.2 | Heat-transfer characteristics in different quench media (courtesy of ALD Vacuum Technologies GmbH)

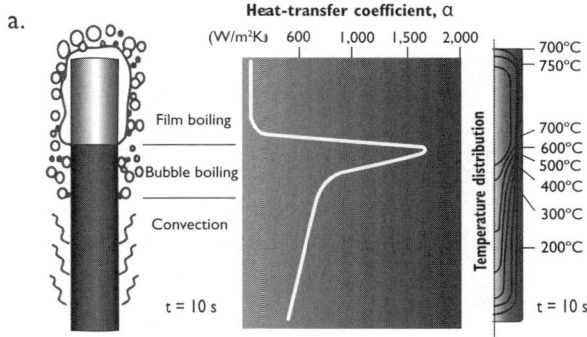

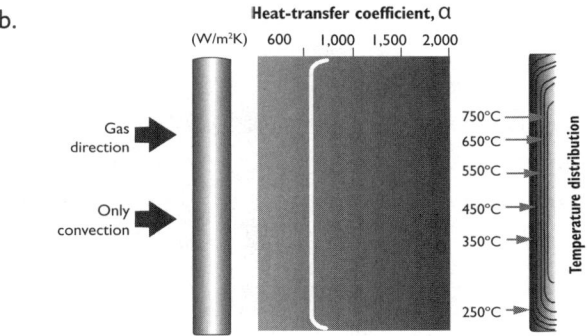

a. Liquid quenching

b. Gaseous quenching

The choice of a quenchant often comes down to the rate of heat extraction by the quenching medium. The design of the quench tank should circulate the quench medium in such a way as to achieve the balance between properties and distortion (dimensional change). The selection of a quenchant medium depends on the:

- ❖ Hardenability of the particular alloy
- ❖ Section thickness (ruling section) and shape
- ❖ Cooling rates required to achieve the desired microstructure (and mechanical properties)

To many heat treaters, when someone refers to "quenching a part," the inference is rapid cooling from austenitizing temperature to produce a desired microstructure or from solution heat-treatment temperature in the case of non-ferrous materials to set up the age-hardening process. When one refers to "cooling a part," the implication is that a slow cooling rate is desired.

Selecting the proper quench medium is critical to avoiding problems such as distortion or cracking. From the hardening or case-hardening temperature, for example, the various types of quenching methods employed may include direct quenching, time quenching, interrupted quenching, selective quenching and spray or fog quenching.

Tempering (Section 5.8) or aging (Section 6.2) is generally required after quenching. In the case of steel, tempering transforms martensite to tempered martensite and helps to balance strength and ductility properties in the material. The tempering process also improves the toughness of the material. Tempering should be carried out within a short time after the quenching operation and as soon as the parts have cooled to 50-80°C (120-180°F).

For medium- and high-hardenability steels (e.g., 4140 or 4340) "temper immediately" often refers to having these materials placed into a heated tempering furnace within 15-30 minutes of quenching. Aging causes precipitation to occur for precipitation-hardening alloys (e.g., certain steels, aluminum). The degree of stable equilibrium for a given grade is a function of time and temperature. Aging creates a change in the microstructure by precipitation of a secondary phase, allowing it to recover from the metastable condition produced by quenching.

Fundamental Quenching Relationship

A fundamental relationship exists between the cooling rate, metallurgical transformation and the internal (residual) stress state of a given material (Fig. 10.1.3). The goal of quenching is not only to achieve the desired microstructure, properties and dimensional state of the material being processed but also to be able to repeat the heat treatment nearly identically time after time. To accomplish this, factors such as loading arrangement, integrity and stability of the quenchant (i.e., quenching intensity) and transfer times to the quench bath must be optimized and consistent.

FIGURE 10.1.3 | Relationship between cooling rate, metallurgical transformation and internal stress (adapted from references 3 and 4)[2]

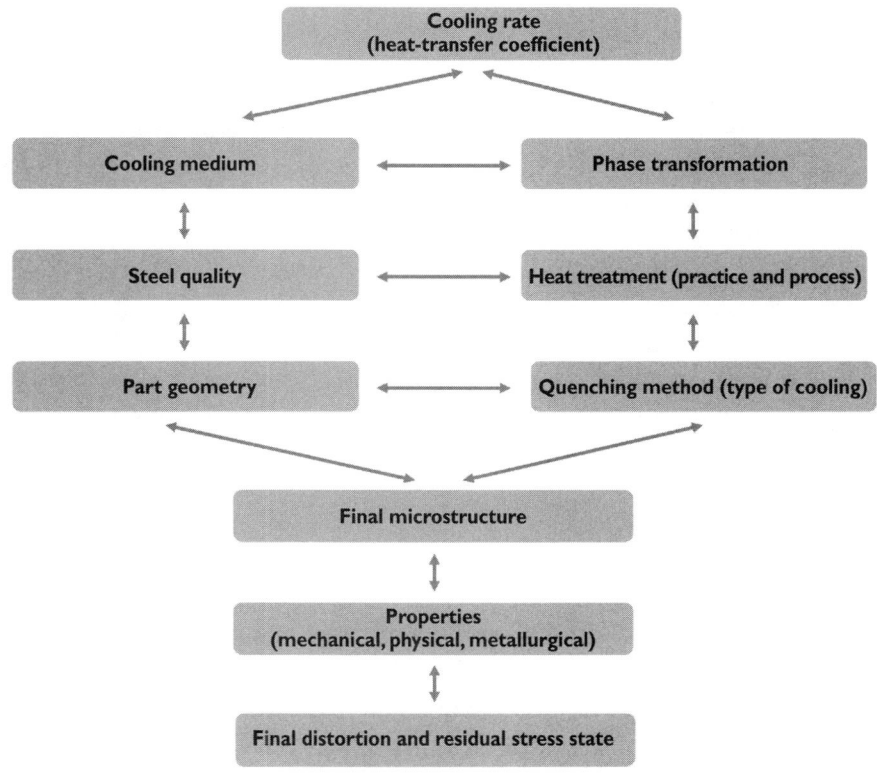

Quenching Variables

Processing of component parts with minimal part distortion has been identified as one of the top industry needs for the 21st century (Section 1.3). As such, understanding the quenching process has been the focus of considerable attention by the scientific community (Fig. 10.1.4) and the subject of considerable effort of modelers looking at component parts' dimensional changes, residual-stress state and microstructural transformations.

The principal variables involved include understanding the process to be run, the furnace the process will be run in, the component parts (including what they are made of and how they will be loaded), the controls in place, the maintenance of the quenchants and even how the results will be measured. Of all these, the human element (e.g., the operator, the engineer) must not be overlooked. Quenching, despite the scientific principles employed and engineering analysis conducted, is and will remain a process controlled by those performing the work.

FIGURE 10.1.4 | Ishikawa diagram for quenching[6]

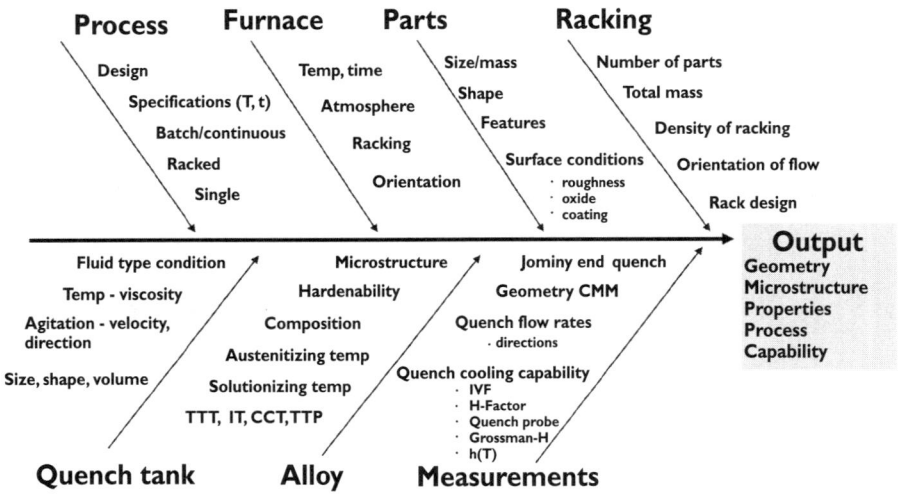

Summary

For quenching to be successful, it is important that the heat treater recognize that many factors influence the outcome. These include the desired outcome of the quenching process (i.e., specification requirements), material and material state, part geometry (size, weight, shape), loading, quenchant type, equipment in which the quenching process will take place, and the variables introduced into the quenching process from external sources (e.g., operator intervention, maintenance, assumptions made). Quenching is such a critical step in the heat-treatment process that a considerable investment in time is needed to ensure it is well understood and managed.

Finally, there is a huge amount of valuable literature on the subject of quenching, some of which is referenced in Section 16.4.

REFERENCES

1. Herring, Daniel H., *Atmosphere Heat Treatment, Volume I*, BNP Media, 2014
2. Herring, Daniel H., "A Review of Gas Quenching from the Perspective of the Heat Transfer Coefficient," *Industrial Heating*, February 2006
3. *Steel Heat Treatment: Metallurgy and Technologies, 2nd Edition*, George E. Totten (Ed.), CRC Press, 2007
4. Herring, Daniel H., "Quenching & Managing Distortion: An Overview," *Industrial Heating* webinar, 2010
5. Dr. D. Scott Mackenzie, Houghton International (houghtonintl.com),

technical and editorial contributions, private correspondence
6. Dr. George E. Totten, G.E. Totten & Associates (getottenassociates.com), technical and editorial contributions, private correspondence
7. Professor Richard D. Sisson Jr., private correspondence

10.2

TYPES OF QUENCHANTS (PART ONE)

Quenching is a vital part of all heat-treatment processes. Quenchants are employed to transform the microstructure of a component part for the purpose of producing desired properties. The quenchant choices are quite diverse and include:

- Brine/caustic
- Water
- Polymer
- Oil
- High gas pressure
- Molten salt
- Air
- Furnace cool

We will briefly review oil and polymer quenchants here and defer the other quenchants to the second part of our discussion (Section 10.4).

Once a quenchant is chosen, recognize that it may be many years before it is changed, which makes the initial choice critically important. The key to the selection process is to find a balance between speed of transformation and distortion (i.e., dimensional change). In other words, the role of the quenchant is to help manage the internal-stress state produced in a component part during the quenching process.

With regard to product selection, the heat treater must carefully consider factors such as material, geometry, required end-use properties and loading arrangement. One must also be aware that applications may (and often do) change over time, so flexibility (i.e., the ability to change quenching parameters) must be part of the selection process.

Other considerations include the type of equipment being employed, furnace atmosphere, load transport to quench (method and time), operating temperature of the bath, how the quenching operation will be performed (manual or automatic), skill of the operators, heating and cooling method of the quenchant, etc. For these reasons, a "universal" quenchant – one that can be used on every part for every application – seldom exists. A quenchant is almost always a compromise, with the goal being consistency and repeatability. In addition, quenchants must be constantly monitored and maintained over their lifetime.

Quenching in Oil

Oil quenching is one of the most common methods used for heat treatment of component parts. Many parts rely on oil quenching to achieve consistent and repeatable mechanical and metallurgical properties and predictable distortion patterns. The reason oil quenching is so popular is due to its excellent performance results and stability over a broad range of operating conditions.

Oil quenching facilitates hardening of steel by controlling heat transfer during quenching, and it enhances wetting of steel during quenching to minimize the formation of undesirable thermal and transformational gradients, which may lead to increased distortion and cracking. For many, the choice of oil is the result of an evaluation of a number of factors, including:[1]

- ❖ Economics/cost (initial investment, maintenance, upkeep, life)
- ❖ Performance (cooling rate/quench severity)
- ❖ Minimization of distortion (quench system)
- ❖ Flexibility (controllable cooling rates)
- ❖ Environmental issues (recycling, waste disposal, etc.)
- ❖ Simplicity of use (ease of control and monitoring)

Oil quenching in atmosphere furnaces, given its ability to control many of the quenching variables, offers an attractive alternative to other technologies. All oil-quench furnaces must meet NFPA 86 standards.

One approach to minimizing distortion is to alter the flow characteristics, whether by providing variable-speed agitation, delaying the onset of agitation or adding directional pipes with supplementary flow.

Commercially available oils have their compositions specially blended to take into consideration the widest possible range of quenching applications (Tables 10.2.1, 10.2.2).

TABLE 10.2.1 | Quench-oil selection guide (part 1)[4]

Product information[a]		Application process			Hardenability[c]			Typical physical properties	
Name	Oxidation resistance[b]	Type	Max temp., °C (°F)	Quench speed	Low	Medium	High	Viscosity, cSt @ 40°C	Flash, COC °C
Houghto-Quench 100	C	Cold	82 (180)	Slow	NR	S	R	22	177
Voluta C201	C	Cold	82 (180)	Slow	NR	S	R	17	210
Houghto-Quench 3420	C	Cold	82 (180)	Slow	NR	S	R	25	182
Soluble Oil #2	B	Cold	82 (180)	Slow	NR	S	R	22	210
Houghto-Quench 105	B	Cold	82 (180)	Slow	NR	S	R	22	179
Houghto-Quench 3470	B	Cold	82 (180)	Medium	S	R	S	16	174
Vac Quench #7490	B	Cold	82 (180)	Medium	S	R	S	28	199
Houghto-Quench 3430	A	Cold	82 (180)	Medium	S	R	S	17	177
Houghto-Quench G	A	Cold	82 (180)	Medium	S	R	S	20	177
Klen Quench R	B	Cold	82 (180)	Fast	R	S	NR	18	174
Voluta C400	A	Cold	82 (180)	Fast	R	S	NR	28	206
BioQuench 700	C	Cold	82 (180)	Fast	R	S	NR	38	324
Houghto-Quench 3440	A	Cold	82 (180)	Fast	R	S	NR	17	168
Houghto-Quench K	A	Cold	82 (180)	Fast	R	S	NR	15	174
Voluta C302	A	Cold	82 (180)	Fast	R	S	NR	18	177
Dasco LPA-15	B	Cold	82 (180)	Fast	R	S	NR	17	182
Dasco LBA-15	B	Cold	82 (180)	Fast	R	S	NR	17	182
Mar-Temp 705-T	A	Hot	163 (325)	Slow	NR	S	R	161	243
Voluta VH401	B	Hot	163 (325)	Slow	NR	S	R	278	296
Mar-Temp 705	A	Hot	163 (325)	Slow	NR	S	R	157	232
Mar-Temp 2500	B	Hot	205 (400)	Slow	NR	S	R	582	288
Mar-Temp 250	C	Hot	121 (250)	Medium	S	R	S	53	218
Mar-Temp 355	A	Hot	135 (275)	Medium	S	R	S	74	229
Mar-Temp 3530	A	Hot	177 (350)	Medium	S	R	S	123	232
Mar-Temp 755	A	Hot	163 (325)	Medium	S	R	S	161	243
Mar-Temp 2525	B	Hot	205 (400)	Medium	S	R	S	568	288
Mar-Temp 2565	A	Hot	205 (400)	Medium	S	R	S	526	288
Q. 3082-64-01	B	Hot	177 (350)	Fast	R	S	NR	128	262
Dasco MBA-60	B	Hot	135 (275)	Fast	R	S	NR	70	227
Dasco MPA-60	A	Hot	135 (275)	Fast	R	S	NR	84	232

Notes:

[a] Other manufacturers produce products with similar properties.

[b] A = best; C = worst

[c] NR = not recommended; S = satisfactory; R = recommended

TABLE 10.2.2 | Quench-oil selection guide (part 2)[4]

Product information[a]			Typical cooling-curve data (°C or °C/s)							
Name	CR max, °C/s	TCR max, °C	Time to temperature, seconds				Cooling rate at temperature, °C/s			
			600°C	500°C	400°C	300°C	600°C	500°C	400°C	300°C
Houghto-Quench 100	48.3	505.6	10.8	14.2	17.8	30.0	16.6	48.0	12.4	6.2
Voluta C201	50.0	505.6	10.7	14.2	17.7	30.1	16.5	50.0	13.8	6.5
Houghto-Quench 3420	52.2	518.3	10.6	13.8	17.6	30.6	15.7	49.4	12.1	5.8
Soluble oil #2	53.3	551.1	11.2	13.2	17.4	30.4	30.7	45.8	10.8	5.8
Houghto-Quench 105	60.6	542.8	10.6	12.6	16.6	29.2	25.0	50.1	10.9	6.0
Houghto-Quench 3470	67.2	545.6	10.0	11.6	13.6	22.6	52.5	61.2	31.7	6.2
Vac Quench #7490	72.8	586.1	9.6	11.0	14.0	26.0	73.0	52.9	18.0	5.5
Houghto-Quench 3430	93.9	631.1	7.2	8.6	11.0	21.8	90.4	59.0	22.8	5.8
Houghto-Quench G	96.1	616.7	7.2	8.4	10.6	21.0	94.2	66.4	27.7	5.8
Klen Quench R	101.7	616.1	6.8	8.0	10.2	20.0	98.9	67.7	26.7	5.6
Voluta C400	98.9	626.7	6.8	8.0	10.5	22.0	98.7	71.5	23.8	6.0
BioQuench 700	106.0	745.6	3.0	5.5	12.4	25.9	67.3	27.7	10.0	14.4
Houghto-Quench 3440	103.3	614.4	7.2	8.4	10.4	19.0	100.2	68.3	30.4	5.8
Houghto-Quench K	105.6	627.8	7.2	8.4	10.2	17.8	99.3	66.8	33.6	5.8
Voluta C302	106.1	601.7	8.1	9.1	10.9	18.2	105.2	86.0	40.5	6.2
Dasco LPA-15	107.8	637.2	6.6	7.8	9.8	18.8	101.5	66.7	31.9	5.6
Dasco LBA-15	115.6	607.2	6.2	7.4	9.6	19.2	103.8	59.8	29.6	5.8
Mar-Temp 705-T	67.8	610.0	7.0	8.8	15.0	33.0	67.5	37.6	6.6	3.7
Voluta VH401	65.6	638.9	6.8	9.5	19.9	37.9	67.3	60.5	7.2	4.0
Mar-Temp 705	68.3	618.9	7.4	9.4	16.4	33.2	66.3	31.9	6.7	3.9
Mar-Temp 2500	72.8	676.1	5.0	7.6	17.0	36.0	54.7	24.6	6.6	3.6
Mar-Temp 250	75.6	556.7	9.0	10.4	13.4	28.4	71.8	54.1	15.8	3.9
Mar-Temp 355	75.6	562.2	8.2	9.6	12.2	26.0	69.7	57.3	21.2	4.1
Mar-Temp 3530	76.7	615.6	6.6	8.2	12.8	31.0	72.7	40.7	10.1	3.6
Mar-Temp 755	77.2	605.6	7.8	9.4	13.6	30.8	75.7	43.6	10.4	3.8
Mar-Temp 2525	77.2	676.7	5.2	7.6	17.4	36.0	57.7	24.9	6.6	3.5
Mar-Temp 2565	83.3	642.8	6.2	8.0	11.6	29.0	74.4	40.7	14.8	3.3
Q. 3082-64-01	88.9	643.9	5.4	7.0	11.2	28.8	77.6	42.9	12.2	3.4
Dasco MBA-60	91.7	717.2	4.2	6.4	15.2	32.8	61.4	26.2	6.6	3.5
Dasco MPA-60	97.2	683.9	4.4	6.0	10.6	28.6	79.4	43.0	8.1	3.8

Note:

[a] Other manufacturers produce products with similar properties.

COOLING-RATE CHARACTERIZATION

Measuring the efficiency, or speed, of an oil quench can be done in one of two ways: by measuring the oil's hardening power (i.e., its ability to harden a steel) or by measuring the cooling ability of the liquid. This second method is popular since cooling ability is independent of steel selection (e.g., composition and grain size) and because it provides information about the oil itself independent of its end-use application.

The preferred test method today is cooling-curve analysis (ISO 9950 or ASTM D 6200), which involves a laboratory test using a 12-mm-diameter (0.475-inch-diameter) nickel-alloy probe for the determination of the cooling characteristics of industrial quenching oils (Figure 10.2.1). The test is conducted in non-agitated oils. Thus, it is able to rank the cooling characteristics of the different oils under standard conditions and provide information on the cooling pathway, which must be known if the ability of quench oil to harden steel is to be determined.

FIGURE 10.2.1 | Typical cooling curves and cooling-rate curves for new oils[7]

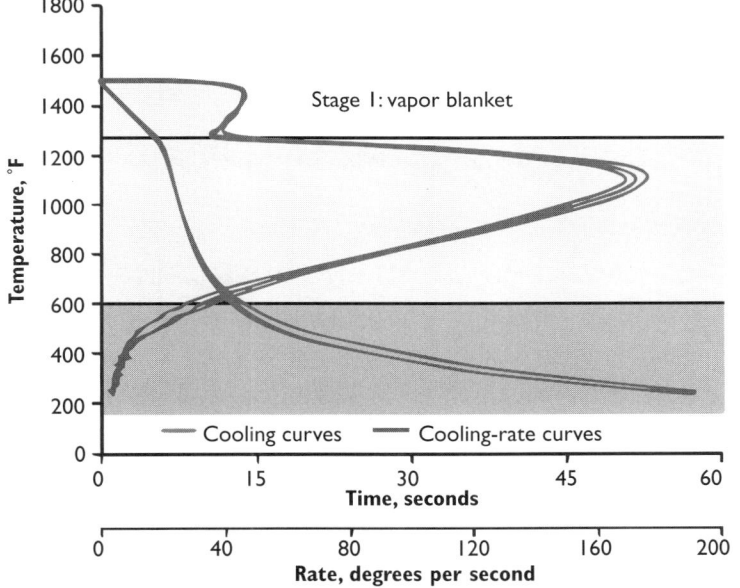

Older methods, such as the GM Quench-O-Meter (ASTM D3520 – withdrawn) or the hot-wire test, are still commonly referred to by the heat-treatment industry. The GM Quench-O-Meter method, for example, measured the overall time to cool a 22-mm (7/8-inch) nickel ball from 885 to 355°C (1625 to 670°F), while the hot-wire test is influenced by the heat-extraction rate of the oil at temperatures close to the melting point of Nichrome, about 1510°C (2750°F).

Oils are generally classified by their ability to transfer heat. They are given a speed rating as fast, medium, slow or marquench oils (Table 10.2.3). Fast oils (7-9 seconds) are used for low-hardenability alloys, carburized and carbonitrided parts, and large cross sections that require high cooling rates to produce maximum properties. Medium oils (10-14 seconds) are typically used as a quench medium for high-hardenability steels and for many general-purpose applications. Slow oils (15-20 seconds) are used where hardenability of steel is high enough to compensate for the slow cooling aspects of this medium.[3] It is also important to recognize that each oil has an operating-temperature range that optimizes its performance. Finding this out from the manufacturer will enhance its performance.

TABLE 10.2.3 | Classification of quench oils[8]

Type	GM Quench-O-Meter rating, seconds[a]	Equivalent cooling rate, °C/s (°F/sec)
Fast oil	7-9	> 74 (165)
Medium oil	10-14	52-74 (125-165)
Slow oil	15-20	32-52 (90-125)
Marquench oil	18-25	32-52 (90-125)

Note:

[a] The GM Quench-O-Meter test does not provide information concerning the cooling pathway to the Curie point.

MECHANISMS FOR HEAT REMOVAL DURING QUENCHING

The mechanism of cooling a component part of a given material in oil or any liquid is largely dependent on geometry and loading, which often dictates the requirements of the quench system.

Traditionally, we talk about three distinct stages of cooling (Fig. 10.2.2). The first stage of quenching is called the vapor-blanket (or film-boiling) stage. It is characterized by the Leidenfrost phenomenon, which is the formation of an unbroken vapor blanket that surrounds and insulates the workpiece. It forms when the supply of heat from the surface of the part exceeds the amount of heat that can be carried away by the cooling medium. The stability of the vapor layer, and thus the ability of the oil to harden steel, is dependent on the metal's surface irregularities; oxides present; surface-wetting additives, which accelerate the wetting process and destabilize the vapor blanket; and the quench oil's molecular composition, including the presence of more volatile oil degradation byproducts.[8] In this stage, the cooling rate is relatively slow in that the vapor envelope acts as an insulator, and cooling is a function of conduction through the vapor envelope.

The second stage of cooling is known as the vapor-transport (or nucleate-boiling or bubble-boiling) stage. It is during this portion of the cooling cycle that the highest heat-transfer rates are produced and the greatest amount of distortion occurs. The point at which this transition occurs and the rate of heat transfer in this region depend on the oil's overall composition (base oil, speed improvers and antioxidant package).[8]

It begins when the surface temperature of the part has cooled enough so that the vapor envelope formed in stage one collapses. This results in violent boiling of the quenching liquid, and heat is removed from the metal at a very rapid rate, largely due to heat of vaporization. The boiling point of the quenchant determines the conclusion of this stage. Size and shape of the vapor bubbles are important in controlling the duration of this stage.

The third stage of cooling is called the convective (or liquid) cooling stage. The cooling rate during this stage is slower than that developed in the second stage and is exponentially dependent on the oil's viscosity, which will vary with the degree of oil decomposition. Increasing oil decomposition will result initially in a reduction of oil viscosity followed by increasing viscosity as the degradation process increases. Heat-transfer rates increase with lower viscosities and decrease with increasing viscosity.[9]

This final stage begins when the temperature of the metal surface is reduced to the boiling point (or boiling range) of the quenching liquid. Boiling stops below this temperature, and slow cooling takes place by conduction and convection. The difference in temperature between the boiling point of the liquid and the bath temperature is a major factor influencing the rate of heat transfer in liquid quenching. Viscosity of the quenchant also plays a major role in the cooling rate in this stage.

FIGURE 10.2.2 | Three stages of liquid quenching[11] (courtesy of Houghton International)

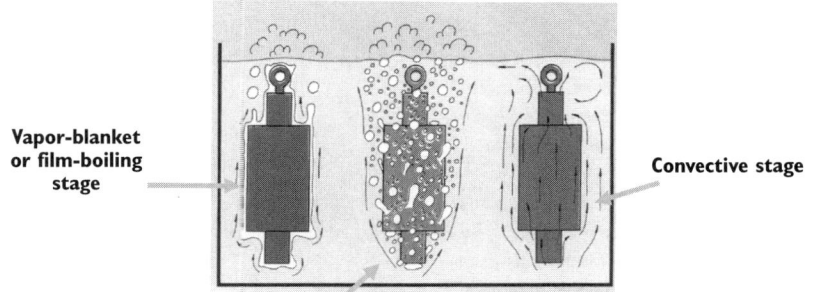

According to an investigation by Kobasko,[12] there are four modes of heat transfer around a part being quenched: shock-film boiling, full-film boiling, nucleate boiling and convection (Fig. 10.2.3a). For any vaporizable liquid there are two critical heat-flux densities at quenching, qcr_1 and qcr_2 (Fig. 10.2.3b). The first critical heat-flux density, qcr_1, is the heat-transfer rate necessary to form the compact vapor film around the part.

Due to the large temperature difference between the very hot surface of the part and the quench-bath temperature, the liquid layer in contact with the hot part's surface heats up to boiling temperature in about one-tenth (1/10) of a second. First, small vapor bubbles occur followed by larger bubbles that grow in size and number until they detach from the part surface, forming the vapor blanket. This is the shock-film boiling stage. The second critical heat-flux density, qcr_2, is the minimal heat flux at which the transition from full-film boiling to nucleate boiling occurs. According to Tolubinski,[12] the critical heat-flux density qcr_1 is given as Equation 10.2.1.

10.2.1) $qcr_1 = 7r \, (af\rho'\rho'')^{0.5}$

where:
   r = heat of vaporization
   a = $\lambda/\rho c$ = thermal diffusivity
   f = frequency of vapor bubble detachment
   ρ' = density of the liquid
   ρ'' = density of the liquid's vapor

FIGURE 10.2.3 | a. Four modes of cooling during quenching; b. Critical heat-flux densities[12]

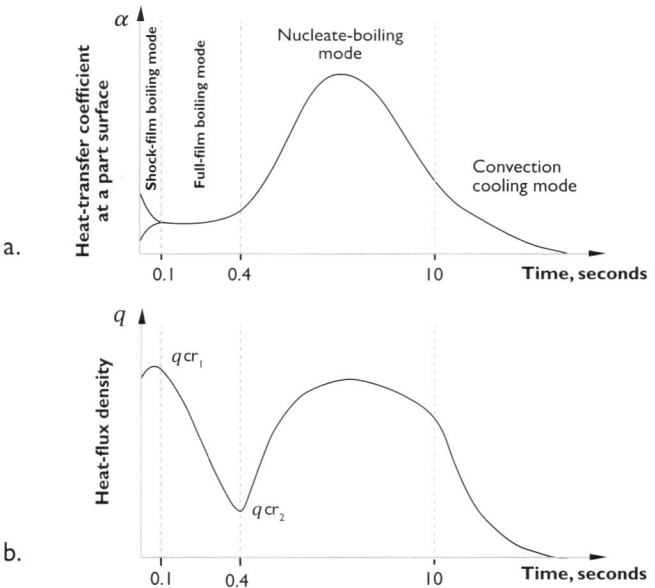

The only variable in Equation 10.2.1 on which qcr_1 depends is "f," the frequency of vapor bubbles detaching from the surface. In other words, a compact vapor film around the part will be formed only if the critical heat-flux density,

q_{cr_1} is attained (which depends on the number of vapor bubbles and the frequency with which they detach from the part's surface).

In addition, these stages of cooling may not occur at all points on a part at the same time (Fig. 10.2.4). As the internal heat moves to the surface, differences in heat rejection may vary based on the surface configuration. Consequently, the need for a uniform and controllable agitation of liquid over the part surface is imperative. Controlled movement of the quenching liquid is vital because it causes an earlier mechanical disruption of the vapor blanket in the first stage and produces smaller, more frequently detached vapor bubbles during the vapor-transport cooling stage. Agitation constantly imparts a cooler liquid to the part surface, which provides a greater temperature difference that allows for improved heat rejection.

FIGURE 10.2.4 | The influence of part geometry on cooling[13]

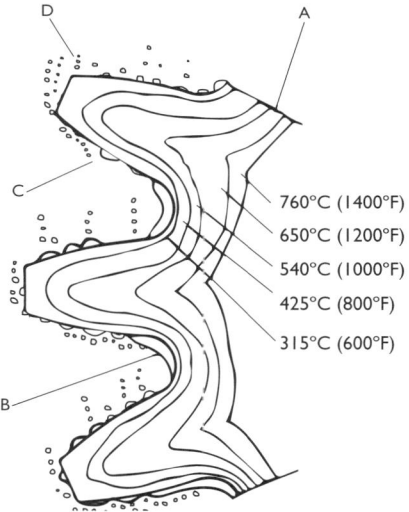

A represents the flow of heat from the hot core of the gear. Temperature and flow rate vary with time.

B represents the vapor-blanket stage, which still exists due to a large heat source and poor agitation.

C represents trapped vapor bubbles, restricted from freely escaping and condensing.

D represents vapor bubbles allowed to freely escape and condense.

PROPERTIES OF AN IDEAL QUENCHING MEDIUM

The ideal quenching medium is one that would exhibit high initial quenching speed in the critical hardening range (through stage one and two) and a slow final quenching speed through the lower temperature range (stage three).

Therefore, the ideal quenchant is one that exhibits little or no vapor stage, a rapid nucleate-boiling stage and a slow rate during convective cooling.

The high initial cooling rates allow for the development of full hardness by getting the steel past the "nose" of the isothermal transformation diagram (quenching faster than the so-called critical transformation rate) and then cooling at a slower rate beginning at the time the steel is forming martensite. This allows stress equalization, so distortion and cracking are reduced. As such, the first criterion any quenchant must meet is its ability to approach the ideal-quenching condition.

When conventional quenching oils are used, the duration of stage one is longer, the cooling rate in stage two is considerably slower and the duration of stage three is shorter. As such, the "quenching power" of oil is far less drastic than that of water. Water and water solutions exhibit high initial cooling rates. Because of water's low boiling point, this fast cooling persists until the steel is cooled to below 150°C (300°F) – well below the martensite start temperature for most commercial steels. Since most steels have formed or are forming martensite by this point, stresses are given little time to equalize. Therefore, water is typically limited to simple shapes or low-hardenability materials.

Oil has a major advantage over water due to its higher boiling range. A typical oil has a boiling range of 230-480°C (450-900°F). This causes the slower convective cooling stage to start sooner, enabling the release of transformation stresses. As a result, oil is able to successfully quench intricate shapes and high-hardenability alloys.

Oil has a proportional drop in viscosity as it is heated. This allows the quenchant to move more freely, which increases the tendency to break the vapor-blanket layer. The nucleate-boiling stage is not drastically altered by changes in bath temperature. The cooling rate in the convective stage of an oil quench will slow as the bath temperature increases. This is advantageous for obtaining a slower rate of cooling through the austenite-to-martensite transformation range. In general, as the temperature of a specific quenching oil increases, the overall quenching rate increases.

Practical heat-transfer coefficient (α) values in the 1,000-2,500 W/m²·K range can be achieved depending on oil characteristics and degree of agitation. Peak values of α in the cooling range of oil are 4,000-6,000 W/m²·K, or a cooling rate greater than 100°C/s (180°F/sec).

AMOUNT OF OIL REQUIRED

Three general rules (aka rules of thumb) are considered to be highly important in oil quenching, and all are considered necessary in order to successfully oil quench.

- ❖ The first of these is the "1 gallon per pound of steel" rule. The gross weight (i.e., total load weight) should not exceed the number of gal-

10 | QUENCHING AND QUENCHANTS

lons of oil in the quench tank; that is, 8.4 L/kg (1 gallon/pound) of quenchant is required.

❖ The second is "the total temperature rise" rule. In a properly designed integral-quench tank, the quench-temperature rise should not exceed 22°C (40°F). For open tanks, this value is dependent on the application and both the gross load weight and the mass of the component part(s) being quenched (Fig. 10.2.5).

❖ The third is the maximum-use temperature of a quench oil below the flash point (Table 10.2.4). The temperature safety factor, or cushion, should be at least 38-65°C (100-150°F) below the flash temperature. In addition, there must be an adequate amount of oil above the height of the tallest part or highest point of the load, typically 150-300 mm (6-12 inches) or more.

FIGURE 10.2.5 | Relationship of the volume of oil and the weight of the work quenched in relation to the temperature rise[16]

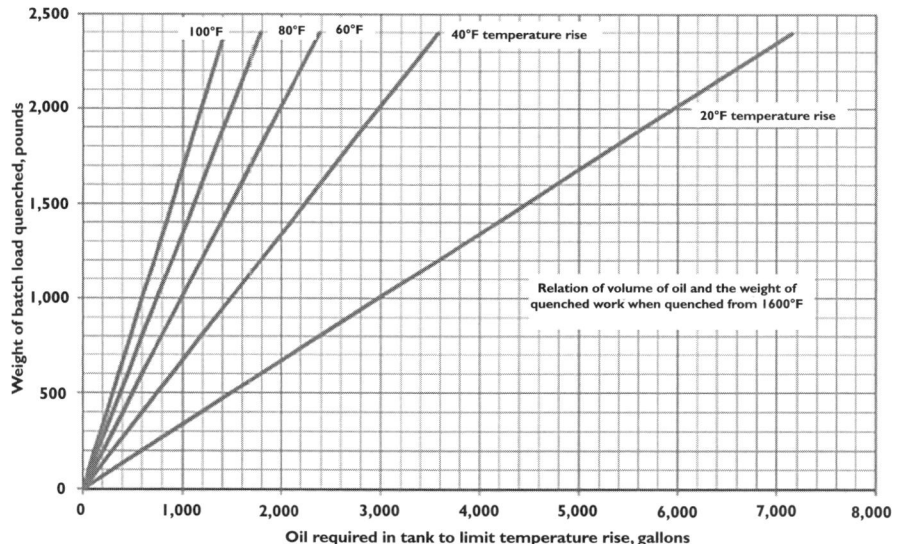

TABLE 10.2.4 | Example of a calculation for the maximum oil operating temperature[16]

Descriptor	Temperature, °C	Temperature, °F
Flash temperature	176	350
Temperature cushion	55[a]	100
Temperature rise on quenching	39[b]	70
Maximum recommended oil operating temperature	82	180

Notes:
[a] An oil temperature of 350-100°F = 250°F (121°C) corresponds to a 55°C decrease in temperature from the 176°C value.
[b] An oil temperature of 250-70°F = 180°F (82°C) corresponds to a 55°C + 39°C = 94°C decrease in temperature from the 176°C value.

One must also allow adequate time for the quench system to recover back to the initial quench conditions before the onset of the next quench operation.

Appropriate cautions apply when using these rules, namely that quenching is highly dependent on a number of material, process and equipment variables. While these rules of thumb have, in general, been found to keep oil-quench systems safe and properly transform component parts, there are instances where quench-tank design (Section 10.6) plays a significant role.

In one instance, a manufacturer quenching 50 kg/hour (110 pounds/hour) of carbonitrided parts from a mesh-belt conveyor furnace into a 11,350-liter (3,000-gallon) quench tank was having severe distortion and staining problems. It turned out that quenching was taking place only in the (improperly designed) quench chute. So, effectively, only 210 liters (55 gallons) of oil was available to carry out the quenching process.

TYPES OF OIL QUENCHING

Oil quenching can be classified as free or unrestricted quenching (Fig. 10.2.6); restricted quenching (Fig. 10.2.7) by press, roll or plug; quenching of parts in baskets (Fig. 10.2.8); and quenching of parts in fixtures (Fig. 10.2.9).

FIGURE 10.2.6 | Example of free quenching (courtesy of Aichelin Heat Treatment Systems)

10 | QUENCHING AND QUENCHANTS

FIGURE 10.2.7 | Example of restricted (press) quenching

FIGURE 10.2.8 | Example of quenching parts in baskets

FIGURE 10.2.9 | Example of quenching parts on fixtures (courtesy of Safe Cronite)

APPLICATION EXAMPLES

Bearing races (Fig. 10.2.10) of SAE 8627H weighing 3.7 kg (8.1 pounds) each are oil quenched to develop a surface hardness of 58-62 HRC. The net load weight is approximately 420 kg (929 pounds). The heat-treatment sequence for these parts is to carburize to an effective case depth of 1.65 mm (0.065 inch), oil quench, subzero treat and temper to final hardness.

FIGURE 10.2.10 | Load of 140-mm-diameter (5.5-inch-diameter) bearing races (courtesy of Bodycote Inc.)

Structural bolts (Fig. 10.2.11) of SAE 4140 weighing 2.7 kg (6 pounds) each are oil quenched to achieve a surface hardness of 27-35 HRC. The load weight is approximately 545 kg (1,200 pounds). The load was hardened at 845°C (1550°F) for two hours in an endothermic-gas atmosphere held at a carbon potential of 0.40% (to avoid carbon pickup), then quenched and tempered at 425-705°C (800-1300°F) to achieve final hardness.

FIGURE 10.2.11 | Load of structural bolts (courtesy of Desert Fire Heat Treat)

Power-transmission gears (Fig. 10.2.12) of SAE 4620 weighing 14.5 kg (32 pounds) each are oil quenched to achieve a surface hardness of 61-63 HRC. The load weight is approximately 270 kg (600 pounds). The load was carburized at 955°C (1750°F) for 16 hours then diffused at 845°C (1550°F) to achieve an effective case depth of 1.8-2.0 mm (0.070-0.080 inch). Parts were tempered at 175°C (350°F).

FIGURE 10.2.12 | Power-transmission gears (courtesy of Metals Engineering)

Automotive and light-truck hood pivot hinges (Fig. 10.2.13) of SAE 1075 weighing 0.9 kg (2 pounds) each are oil quenched and tempered to achieve a surface hardness of 42-47 HRC. The load weight is approximately 205 kg (450 pounds). The load was austenitized in an endothermic-gas atmosphere with a carbon potential of 0.75% at 860°C (1575°F) then quenched into 65°C (150°F) oil. Parts were tempered at 415°C (775°F).

FIGURE 10.2.13 | Load of pivot hinges for hardening and oil quenching (courtesy of MET-TEK Inc.)

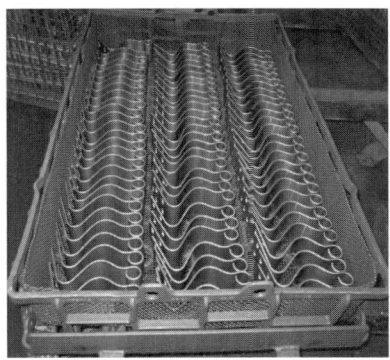

A drive-shaft assembly (Fig. 10.2.14) consists of an SAE 1117 shaft and SAE 1030 laminated-steel gears. The part weight is 0.8 kg (1.8 pounds), and the assemblies are copper brazed prior to carbonitriding. Parts are oil quenched and tempered to achieve a surface hardness of 80 HRC on the shaft. The load weight is approximately 160 kg (350 pounds). The load was austenitized in an endothermic-gas atmosphere with a carbon potential of 0.95% at 870°C (1600°F). The carbon potential was then reduced to 0.75% prior to quenching into 82°C (180°F) oil. Parts are not tempered.

FIGURE 10.2.14 | Drive-shaft assemblies for combination brazing and carbonitriding followed by oil quenching (courtesy of Lawrence Industries Inc.)

Landing-gear components (Fig. 10.2.15) of SAE 300M/4340M material weighing 5.4 kg (12 pounds) each are oil quenched to develop a final hardness of 53-55 HRC. The net load weight is approximately 180 kg (400 pounds). The heat-treatment sequence for these parts is to austenitize at 870°C (1600°F) for 1.5 hours +1 hour/-0.25 hours in an endothermic gas atmosphere at 0.50% C (to avoid decarburization), quench in 38°C (100°F) oil and temper at 300°C (575°F) for four hours.

FIGURE 10.2.15 | Aerospace landing-gear components (courtesy of Pacific Metallurgical Inc.)

ADDITIONAL NEEDS

Water-detection devices and sensors need to be designed to measure the heat transfer inside a quench bath and used to monitor quench intensity in the quench tank in real time as a quality-control tool. This ensures that the quench conditions are known and corrective action or repairs can be implemented before more parts are quenched (Section 13.4).

Quenching in Polymer

Polymer quenching (Fig. 10.2.16) is attractive due in large part to the fact that the quenchant can be made oil-like or water-like. This flexibility in quenching is possible through selection of polymer type, concentration, temperature, agitation and quench-tank design. Polymer has a number of other advantages, including:

- Wide range of achievable metallurgical results
- Flexibility of quenching speed
- Reduction of residual stress, cracking and distortion (in many cases)
- Tolerance to water contamination
- Lower temperature rise (of the bath) during quenching (e.g., almost twice the specific heat capacity of quench oils)
- Minimal drag-out (dependent on polymer type and relative loss)
- Nonflammable (no fire hazard)
- Environmentally friendly

It should be noted, however, that oil-like refers to a similar cooling profile and not limitations or use conditions. For example, you cannot overload the tank with high surface-area product (irrespective of the heat-transfer coefficient) in polymer solutions like you can with oil.

Polymer quenching also has other limitations, including:

- Quench tanks specifically designed for its use
- Tank volume must be 12.7-16.8 liters/kg (1.5-2.0 gallons/pound)
- Maximum allowable temperature rise of the polymer
- Corrosion (if concentration varies) to parts, fixtures and the quench tank
- Cracking/distortion of parts if cooling rate (i.e., speed of quenching) is too fast or heat extraction is non-uniform

FIGURE 10.2.16 | Large mill pinion being lowered into a polymer quench

Polymer quenching is one of the most sought-after solutions for quenching because it has a number of unique properties (Tables 10.2.5, 10.2.6). There are applications in which their use is invaluable and applications in which, for a wide variety of reasons, polymer fails to deliver expected benefits. Factors for successful polymer quenching include:

- ❖ Proper hardenability of the material
- ❖ Achievable properties (mechanical, metallurgical, physical)
- ❖ Correct quench-tank design
- ❖ Proper part temperature and transfer time

TABLE 10.2.5 | Polymer selection guide (part 1)[4]

Product name[a]	Type	Open quench tank		Continuous-belt furnace		Aerospace aluminum (AMS 2770)
		Low hardenability	High hardenability	Low hardenability	High hardenability	
Recommended polymer type		PVP	PVP	PVP/PAG	PVP/PAG	PAG
Aqua-Quench 140	PAG	10-15%	15-30%	10-15%	15-25%	N/A
Aqua-Quench 145	PAG	10-15%	15-30%	10-15%	15-25%	N/A
Aqua-Quench 245	PAG	6-12%	13-20%	5-7%	7-12%	N/A
Aqua-Quench 260	PAG	6-12%	13-20%	5-7%	7-12%	15-40%
Aqua-Quench 260-B	PAG	6-12%	13-20%	5-7%	7-12%	15-40%
Aqua-Quench 261	PAG	6-12%	13-20%	5-7%	7-12%	N/A
Aqua-Quench 251	PAG	5-10%	12-18%	4-6%	6-10%	12-32%
Aqua-Quench 251-B	PAG	5-10%	12-18%	4-6%	6-10%	12-32%
Aqua-Quench 351	PAG	5-10%	12-18%	4-6%	6-10%	N/A
Aqua-Quench 352	PAG	5-10%	12-18%	4-6%	6-10%	N/A
Aqua-Quench 364	PAG	4-8%	10-15%	2-4%	4-8%	N/A
Aqua-Quench 365	PAG	4-8%	10-15%	2-4%	4-8%	N/A
Aqua-Quench 365-H	PAG	4-8%	10-15%	2-4%	4-8%	N/A
Aqua-Quench 3699	PVP/Hybrid	5-7%	10-15%	2-5%	7-10%	N/A
Aqua-Quench 3699J	PVP/Hybrid	5-7%	10-15%	2-5%	7-10%	N/A
Aqua-Quench C	PVP	5-7%	10-15%	2-5%	7-10%	N/A
Aqua-Quench 3600	PEOX	7-12%	15-25%	5-10%	12-15%	N/A
Aqua-Quench 110	ACR	4-6%	8-12%	2-5%	5-8%	N/A

Note:

[a] Other manufacturers produce products with similar properties.

TABLE 10.2.6 | Polymer selection guide (part 2)[4]

Product name	Typical physical properties				
	Viscosity, cSt	Specific gravity @ 15.6°C	pH	Factor	Appearance
Aqua-Quench 140	80	1.06	8.7	2.0	Clear to slightly hazy amber
Aqua-Quench 145	120	1.06	10.3	2.3	Clear to slightly hazy amber
Aqua-Quench 245	195	1.07	10.3	2.0	Clear to slightly hazy amber
Aqua-Quench 260	550	1.09	8.8	2.0	Clear to slightly hazy amber
Aqua-Quench 260-B	550	1.09	8.8	2.0	Clear to slightly hazy amber
Aqua-Quench 261	540	1.09	9.0	2.0	Clear to slightly hazy amber
Aqua-Quench 251	280	1.07	9.5	2.5	Clear to slightly hazy amber
Aqua-Quench 251-B	280	1.07	9.5	2.5	Clear to slightly hazy amber
Aqua-Quench 351	280	1.08	9.0	2.5	Clear to slightly hazy amber
Aqua-Quench 352	350	1.08	8.8	2.5	Clear to slightly hazy amber
Aqua-Quench 364	330	1.08	9.5	2.5	Clear to slightly hazy amber
Aqua-Quench 365	410	1.07	8.6	2.5	Clear to slightly hazy amber
Aqua-Quench 365-H	450	1.09	10.5	2.5	Clear to slightly hazy amber
Aqua-Quench 3699	690	1.02	9.3	5.0-5.5	Slightly hazy, light amber
Aqua-Quench 3699J	1475	1.05	9.9	5.0-5.5	Slightly hazy, light amber
Aqua-Quench C	500	1.04	9.0	5.0-5.5	Clear to slightly hazy amber
Aqua-Quench 3600	72	1.03	9.0	5.5	Clear, reddish-brown
Aqua-Quench 110	560	1.07	9.0	N/A	Almost clear, light amber fluid

Note:
[a] Other manufacturers produce products with similar properties.

Typical design factors to consider when choosing a polymer include:
- ❖ Material chemistry and hardenability
- ❖ Part geometry, mass and physical dimensions
- ❖ Prior (manufacturing) operations/stress state
- ❖ Polymer type (quench characteristics, cooling-curve data)
- ❖ Polymer concentration and temperature

- ❖ Quench-tank volume
- ❖ Grids, baskets, fixtures (material and design)
- ❖ Part load (number of parts, weight, spacing)
- ❖ Quench-tank design factors
  - Height of the polymer over the load
  - Agitation (type, characteristics)
  - Flow velocity through the workload
- ❖ Post-heat-treat operations (if any)

Quench-tank design factors must include consideration of the following:
- ❖ Number of agitators or pumps
- ❖ Location of agitators or pumps
- ❖ Size of agitators or pumps
- ❖ Internal tank baffling (draft tubes, directional flow vanes, etc.)
- ❖ Flow direction
- ❖ Propeller size (diameter, clearance in draft tube) and/or motor horsepower
- ❖ Quench elevator design (flow restrictions)
- ❖ Volume of polymer
- ❖ Temperature of polymer
- ❖ Type of agitators (fixed, two-speed, variable-speed)
- ❖ Maximum (design) temperature rise
- ❖ Heat-exchanger type, size, heat-removal rate (kcal/hour or BTU/hour and instantaneous kcal/minute or BTU/minute)

MECHANISM OF HEAT REMOVAL DURING QUENCHING

The principle of heat extraction with the use of single liquid organic polymers is as follows:
- ❖ When hot metal is immersed in these single polymer quenchants, a film of liquid organic polymer is deposited on the hot metal as it contacts the diluted quenching fluid.
- ❖ The speed of the quenching rate depends on the thickness of the deposited coating.
- ❖ The thickness of this coating is regulated by the concentration of polymer and agitation rates in the quench bath.
- ❖ As the hot metal is cooled to below the separation point of the polymer, the liquid organic polymer drops off the hot metal and again becomes soluble in water.

Quenching characteristics vary with polymer concentration (Fig. 10.2.17).

FIGURE 10.2.17 | Difference in bubble formation due to concentration after 6.5 seconds: a. 10%, b. 22%.[8,9]

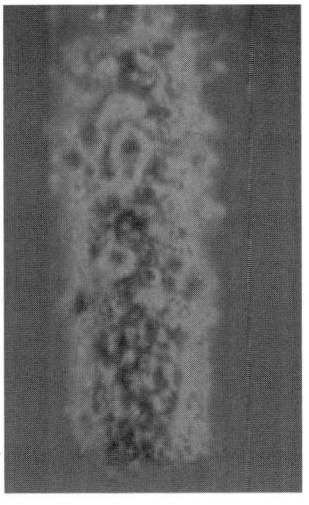

a. b.

AMOUNT OF POLYMER REQUIRED

Three general rules are considered to be highly important in polymer quenching. They are considered necessary in order for a successful polymer quench to take place.

- ❖ The first is the "1.5-2 gallon per pound of steel" rule. The gross weight (i.e., total load weight) multiplied by 1.5-2.0 equals the number of gallons of polymer and water solution needed in the quench tank.
- ❖ The second is the "total temperature rise" rule. For open tanks, this value is dependent on the application and both the gross load weight and the mass of the component part(s) being quenched. When quenching aluminum, the quench temperature should not typically exceed 5.5°C (10°F). For steel, this value can be higher.
- ❖ The third rule is that in no event must the bulk temperature in the tank reach the inverse solubility temperature of the polymer (Fig. 10.2.18). For a typical PAG quenchant, the cloud temperature (aka cloud point; that is, the temperature at which the polymer redissolves in the aqueous solution) is 57-82°C (135-180°F), depending on molecular weight. Thus, the maximum peak operating temperature should be no less than 1.5°C (35°F) below the cloud temperature for effective quenching. Contact the manufacturer for the specific separation temperature of the quenchant.

FIGURE 10.2.18 | Relationship of the polymer volume and the weight of the work quenched in relation to the temperature rise (courtesy of Houghton International)

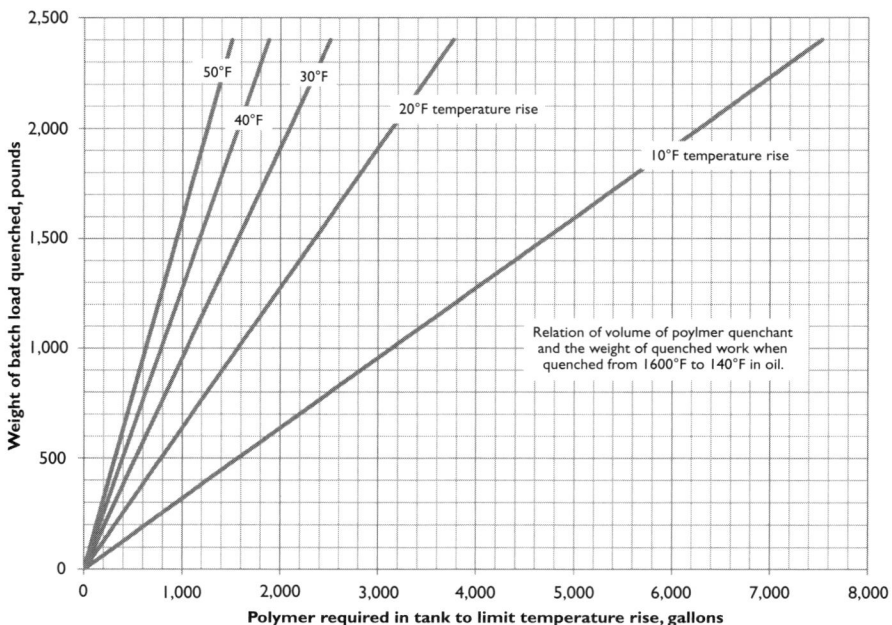

One must also allow adequate time for the quench system to recover back to the initial quench conditions before the onset of the next quench operation.

Again, appropriate cautions apply when using these rules, namely that quenching is highly dependent on a number of material, process and equipment variables. In many instances, such as when quenching aluminum, specifications (e.g., AMS 2770) dictate the maximum bath temperature at the start of quenching and the maximum rate of rise.

TYPES OF POLYMER QUENCHANTS

Various types of polymer quenchants have widely differing properties (Fig. 10.2.19) and application uses. These include:

- ❖ PAG (polyalkylene glycol) and PEG (polyethylene glycol)
- ❖ PVP (polyvinyl pyrrolidone)
- ❖ ACR (sodium polyacrylate)
- ❖ PEO (polyethyl oxazoline)
- ❖ Custom blends

10 | QUENCHING AND QUENCHANTS

FIGURE 10.2.19 | Application uses for polymer quenchants[a] (courtesy of Houghton International)

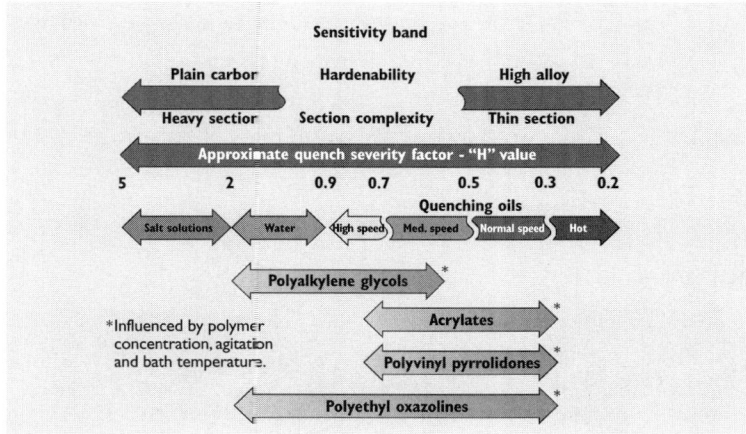

Note:

[a] Ranges shown are for illustrative purposes only.

PAG (Polyalkylene Glycol) and PEG (Polyethylene Glycol)

PAGs and PEGs are used for immersion quenching of steel components and intended to replace water, soluble oil and mineral oil. They are applicable to low-, medium- and high-alloy steels, including carburizing grades and martensitic stainless steels (15-25% concentration). They can be used for the heat treatment of aluminum (16-22% concentration) and for induction hardening followed by spray quenching (5-15% concentration). They leave a tacky residue that must be washed (preferably with ambient-temperature water) or burned off (adequate ventilation must be provided because fumes are acidic) provided the temperature of the subsequent operation is at least 315°C (600°F). If burned off, a potentially black, gummy residue can be left on the parts. This has been known to cause difficulty in subsequent machining and/or grinding operations.

PAGs and PEGs are also inversely soluble. In other words, if the temperature of the bath exceeds about 70°C (160°F), the polymer/water mixture will separate – depending on the polymer (actually, the electrolyte content) – and fall to the bottom or rise to the top of the tank. In point of fact, all water-soluble polymers have a cloud point (i.e., inverse separation point), but some exhibit this above the boiling point of water.

PAGs are widely used for quenching heavy-section castings and forgings; thin-section aluminum sheet; and wrought or cast aluminum alloys such as airframes, engine blocks, cylinder heads and wheels. Other applications include rolled rings, bolts, bearings, crankshafts, fasteners, various agricultural parts, springs, steel bars, steel coils, forgings and other components.

Large aerospace rolled rings up to 6.1 meters (20 feet) in diameter can be quenched in specially designed polymer quench tanks (Fig. 10.2.20). Other factors that determine proper polymer quenching of rolled rings and other components include cost, compatibility with existing equipment, cracking/distortion, ease of cleaning, disposal and safety.

FIGURE 10.2.20 | Large polymer quench tank for rolled rings (internal baffling not pictured)

a. Typical rolled ring quenched in a PAG (courtesy of Houghton International)

b. Polymer quench tank (courtesy of Tenaxol)

a.

b.

PVP (Polyvinyl Pyrrolidone)

PVPs are used when oil-like quenching characteristics are desired. They are applicable to low-, medium- and high-alloy steels, including carburizing grades and martensitic stainless steels. PVPs are widely used in the steel industry for quenching of bars (Fig. 10.2.21), rolled sections and forgings. They can also be used for the heat treatment of aluminum. They can replace water, soluble oil or mineral oil and are typically used in concentrations between 10-20%. PVPs are widely used for quenching many of the same products as PAGs.

FIGURE 10.2.21 | Typical load of bars quenched into a PVP (courtesy of Houghton International)

ACR (Sodium Polyacrylate)

ACRs are a high-speed quenchant with oil-like characteristics similar to medium-speed quenching oils. Concentrations generally do not exceed 15-25%. These products do not exhibit inverse solubility. Instead, they produce high viscosity and polymer-rich layers around the part that reduce the cooling rate during the convection phase, which helps to minimize distortion.

ACRs are critically dependent on polyvalent ion content (i.e., hard metals), which will precipitate the polymer. Also, these products may degrade (shear) over time because they are typically high in molecular weight.

Applications for ACRs include seamless tubing for the oil industry, gas cylinders (Fig. 10.2.22), thin-section alloy-steel crankshafts, high-carbon-chromium grinding balls and certain forgings and castings. They are used for railroad rails and other applications involving pearlitic steels because of their characteristic prolonged first stage of cooling.

FIGURE 10.2.22 | Gas cylinders quenched into an ACR (courtesy of Houghton International)

PEO (Polyethyl Oxazoline)

PEOs are custom-blended chemistries that consist of a mixture of PAGs and ACRs, producing oil-like quenching characteristics. They are used in a wide range of applications, including induction and flame hardening of steel and cast iron (5-10% concentration), high-hardenability steel castings and forgings (5-25% concentration) and wire products (Fig. 10.2.23). A wax-like residue remains on the parts, which can be removed by cold water. Drill pipe, camshafts, crankshafts and automotive gears are other types of products that benefit from the use of PEOs.

FIGURE 10.2.23 | Wire product quenched into a PEO (courtesy of Houghton International)

Summary

Oil and polymer are two of the most widely used fluids for quenching. In order for any quenching operation to be successful, one must carefully match the desired outcome of the process with the type of quenchant best suited for the job. The reason to select one quenchant over another, or to use one method of quenching over another, is often highly subjective and based on items such as historical performance, convenience, cost or flexibility. What is important for the heat treater to understand is that no quenchant can be successful in all applications, and quenching and quenchant variables must be carefully controlled.

REFERENCES
1. Herring, Daniel H., *Vacuum Heat Treatment*, BNP Media, 2012
2. Herring, Daniel H., *Atmosphere Heat Treatment, Volume I*, BNP Media, 2014
3. Herring, Daniel H., "A Review of Factors Affecting Distortion in Quenching," *Heat Treating Progress*, December 2002
4. Wachter, D.A., G.E. Totten and G.M. Webster, "Quenchant Fundamentals: Quench Oil Bath Maintenance," *Advanced Materials & Processes*, 1997
5. Herring, Daniel H., "Oil Quenching Part One: How to Interpret Cooling Curves," *Industrial Heating*, August 2007
6. "Cooling Rate Curves for Various Liquid Quenchants," *Advanced Materials & Processes*, February 1998

7. Totten, G.E., G.M. Webster and D.A. Wachter, "Quenchant Fundamentals: Condition Monitoring of Quench Oils," *Practicing Oil Analysis*, January/February 2003, pp. 26-31
8. Totten, George E., C.E. Bates and N.A. Clinton, *Handbook of Quenching and Quenching Technology*, ASM International, 1993
9. Tensi, H.M. and K. Lainer, "Wiederbenetzung und Wärmeubergand beim Tauchkühlen in Hochleistungsölen," *HTM*, 1997.52, pp. 298-303
10. Herring, Daniel H., *Industrial Heating* Quenching webinar, September 2010
11. Liscic, Bozidar, Sasa Singer and Harmut Beitz, "Dependence of the Heat Transfer Coefficient at Quenching on Diameter of Cylindrical Workpieces," IFHTSE Conference, Rio de Janeiro, 2010
12. *ASM Handbook, Volume 4: Heat Treating*, ASM International, 1991, p. 70
13. *ASM Handbook, Volume 4A: Steel Heat Treating Fundamentals and Processes*, Jon L. Dossett and G.E. Totten (Eds.), ASM International, 2014, pp. 91-157
14. Tensi, H.M., A. Stich and G.E. Totten, "Fundamentals of Quenching," *Metal Heat Treating*, March/April 1995
15. Mackenzie, D. Scott, "Sizing Quench Tanks for Batch Immersion Quenching," Technical Data Sheet, Houghton International
16. Dr. D. Scott Mackenzie, Houghton International (houghtonintl.com), technical and editorial contributions, private correspondence
17. Dr. George E. Totten, G.E. Totten & Associates (getottenassociates.com), technical and editorial contributions, private correspondence

UBQ: Universal Batch Quench Furnace. ▲
E-Z Series Endothermic Generators. ▶
Ultimate in flexibility and versatility.

AFC-Holcroft:
Strength and Innovation since 1916.
Powerful Solutions for the Future.

As a privately owned company with thousands of installations worldwide, AFC-Holcroft is a worldwide leader in the heat treat equipment industry.

- One of the most diverse product lines in the heat treat equipment industry: **Pusher Furnaces, Continuous Belt Furnaces, Rotary Hearth Furnaces, Universal Batch Quench (UBQ) Furnaces and Endothermic Generators.**

- Robust construction and long service life, designed for ease of maintenance.

- Various global facilities in North America, Europe and Asia for fastest local delivery, service and support.

Get in touch with us today to learn more about how we can improve your production processes and how we can give you the edge over the competition.

For further information please visit
www.afc-holcroft.com

AFC-Holcroft USA
Wixom, Michigan
Phone: +1-248-624-8191

AFC-Holcroft Europe
Boncourt, Switzerland
Phone: +41 32 475 56 16

AFC-Holcroft Asia
Shanghai, China
Phone: +86-21-58999100

10.3

PERFORMANCE VARIABLES IN OIL AND POLYMER QUENCHANTS

Oil and polymer quench systems are of keen interest to the heat treater. Some of the key variables to monitor and control with these quenchants will be discussed here, while the actual maintenance checks are covered in Section 13.4.

As part of a sound quality program, it is important to catch problems and perform corrective actions well in advance of quality or production issues. As such, the heat treater and quenchant supplier should work together to maximize the life and performance of each quenching product in use.

Oil Quenchants

For quench oils, routine testing should be done quarterly or even monthly if the oil is being pushed at the upper ends of its performance limits (e.g., high operating temperature, maximum gross furnace loads and high frequency of quenches). Testing should include checks of: viscosity, total acid number (i.e., precipitation number), quench speed, flash point, sludge content and presence of water. Each is briefly explained.

RELATIONSHIP OF PHYSICAL PROPERTIES TO PERFORMANCE[1,2]

Oil is often analyzed to determine its performance characteristics. The testing laboratory issues a report that contains information about the physical-property characteristics of the oil. Here is a listing of various tests and insights into the meaning of the results obtained.[1]

❖ **Viscosity**: Quenching performance is dependent on the viscosity of the oil (Fig. 10.3.1). Oil viscosity changes with time due to degradation (the formation of sludge and varnish). Samples should be taken and analyzed for contaminants, and a historical record of viscosity variation should be kept and plotted against a process-control parameter such as part hardness.

FIGURE 10.3.1 | Viscosity change in martempering oil[1]

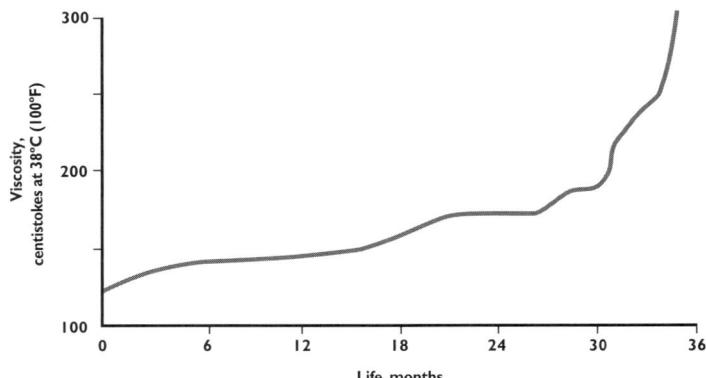

❖ **Water content**: Water from oil contamination or degradation may cause soft spots, uneven hardness, staining and (perhaps worst of all) fires. This is because a volume expansion of approximately 1,700 times occurs when water turns into steam. For example, 3.8 liters (1 gallon) of water turns to 6 m^3 (215 feet^3) of steam. A percentage of water as low as 0.05% is often considered problematic, and anything above 0.1% can cause foaming, fires and explosions.

When water-contaminated oil is heated, a crackling sound may be heard. This is the basis of a qualitative field test for the presence of water in quench oil. The most common laboratory tests for water contamination are either Karl Fisher analysis (ASTM D 1744) or distillation. When water is present, the effect on the cooling curve is to prolong the vapor phase, and the transition to the convection phase is almost nonexistent.

When providing a sample for analysis, remember that water will reside at the bottom of the tank, so it is important to draw a sample from this area with the bath in a quiescent state. Placing a sample in a clear jar and leaving it in a vibration-free area can provide a quick determination if water is present (the oil will float atop the water).

- **Flash point**: The flash point is the temperature where the oil, in equilibrium with its vapor, produces a gas that is ignitable but does not continue to burn when exposed to a spark or flame source. As a general guideline, the maximum-use temperature of quench oil should be 38-65°C (100-150°F) below the flash point of the oil. Finally, changes in the flash point indicate contamination of the oil has taken place.

    There are two types of flash-point values that may be determined: closed-cup or open-cup. The liquid and vapor are heated in a closed system in the closed-cup measurement. Traces of low-boiling contaminants may concentrate in the vapor phase, resulting in a relatively low value. When conducting the open-cup flash point, the relatively low-boiling byproducts are lost during heating and have less impact on the final value. The most common open-cup flash-point procedure is the Cleveland Open Cup procedure described in ASTM D 92. The minimum flash point of oil should be 90°C (160°F) above the oil temperature being used.

- **Oxidation:** This results from buildup of organic acids in the oil and is measured by several methods, including infrared (Fig. 10.3.2). Total acid number (TAN), precipitation number, sludge content and viscosity are indications of oxidation. The cooling curve is affected by an increase in speed for cold oils and a decrease in speed for martempering oils.

    Staining of parts can occur at a TAN greater than 1.5-2.0. Parts meeting hardness and distortion requirements may still be able to be produced, but they will be difficult to clean.

    Oxidation should be carefully monitored in tanks running martempering oil or oils being run above their recommended operating range. Oxidation is detected by infrared spectroscopy. However, the TAN can provide a good indication of the oil's oxidation. Nitrogen blanketing of the oil is one way to reduce both oil oxidation and sludge formation.

- **TAN and neutralization number:** As an oil degrades, it forms acidic byproducts. The TAN indicates the level of oxidation. As it increases, the vapor phase is less stable. The maximum cooling rate also increases, as does the tendency toward staining, and distortion/cracking may become problematic.

    The amount of these byproducts may be determined by chemical analysis. The most common method is the neutralization number, which is determined by establishing the net acidity against a known standard base such as potassium hydroxide (KOH). This is known as the acid number and is reported as milligrams of KOH per gram of sample (mg/g).

FIGURE 10.3.2 | IR spectra – new vs. used oil (courtesy of Noria Corp. and *Machinery Lubrication* magazine)[2]

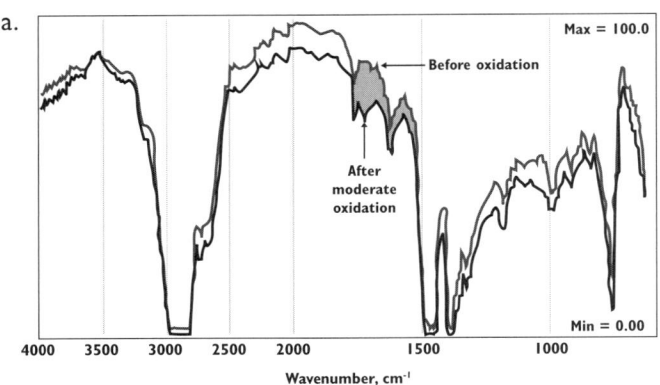

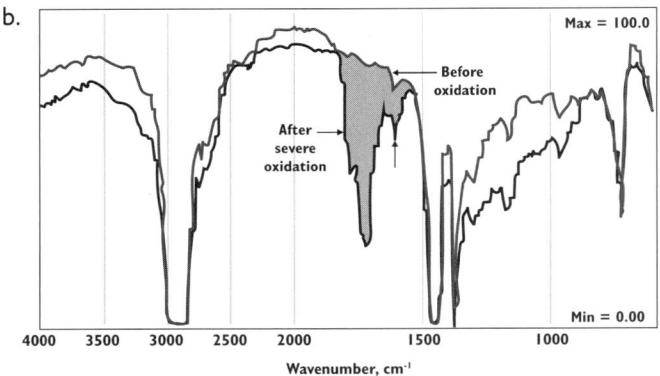

a. Moderately degraded oil

b. Severely degraded oil

❖ **Precipitation number:** This value is an indication of the tendency of the oil to form sludge. The higher the number, the greater the chances parts will stain. The precipitation number is determined using ASTM D 91. The relative propensity of sludge formation of new and used oil may be compared providing an estimate of remaining life. This value can be controlled by filtration.

❖ **Sludge formation:** Sludge is caused by the breakdown of the oil (i.e., polymerization) by localized overheating (frying). Sludge is one of the biggest problems encountered with quench oils. Although other analyses may indicate that a quench oil is performing within specification, the presence of sludge may still be sufficient to cause non-uniform heat transfer and increased thermal gradients, cracking and distortion.

Sludge may also plug filters and foul heat-exchanger surfaces. The loss of heat-exchanger efficiency may cause overheating, excessive foaming and possible fires. Generally, the viscosity increases as the oil degrades. Sludge can be controlled by filtration.

- ❖ **Accelerator performance:** Cooling-curve analysis is the most direct way to evaluate if speed enhancers are required. Testing involves either induction-coupled plasma (ICP) spectroscopy or FTIR (Fourier transform infrared spectroscopy) analysis, coupled with X-ray techniques if organic products or antioxidants are present. When additives (such as metal salts) are used as quench-rate accelerators, their effectiveness can be lost over time by both drag-out and degradation. Their effectiveness can be quantified by performing ICP spectroscopy (a direct analysis for metal ions), and compensating measures (such as the addition of a specific percentage of new accelerator) can be taken.

GM QUENCH-O-METER TEST AND COOLING-CURVE ANALYSIS

The GM Quench-O-Meter test (Fig. 10.3.3) has been used to classify quench oils for nearly 50 years, but it is increasingly being replaced by ISO 9950 or ASTM D6200 cooling-curve-analysis procedures (Fig. 10.3.4) for quench oils. It is important to note that Nadcap, CQI-9 and other audit agencies do not recognize the GM Quench-O-Meter test. The value of cooling-curve analysis is that it can be used to identify cooling variations of quench-oil oxidation over time.

FIGURE 10.3.3 | GM Quench-O-Meter test shortcoming – cooling pathway A, B or C is not revealed (courtesy of Noria Corp. and *Machinery Lubrication* magazine)[2,4]

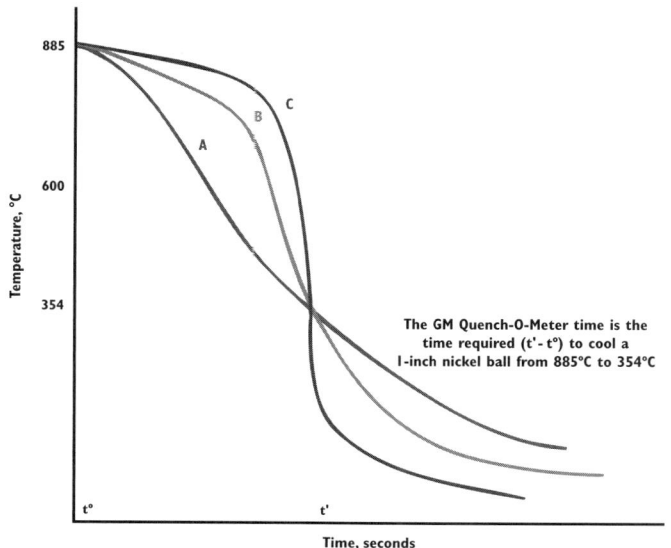

FIGURE 10.3.4 | Value of cooling-curve analysis (courtesy of Noria Corp. and *Machinery Lubrication* magazine)[2,4]

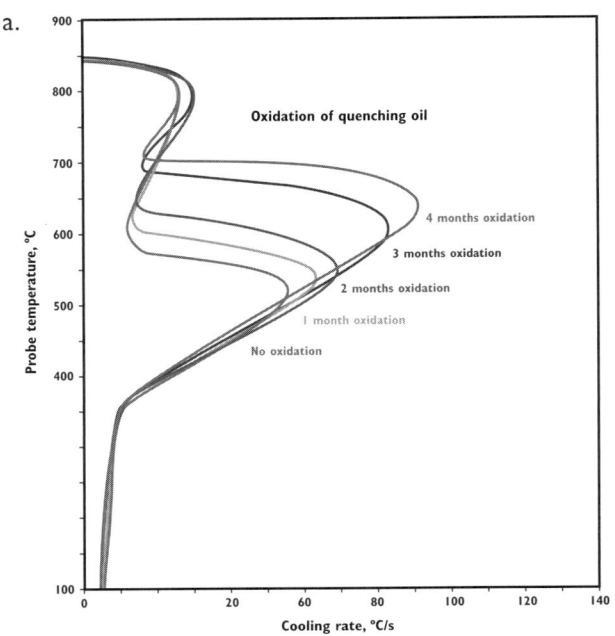

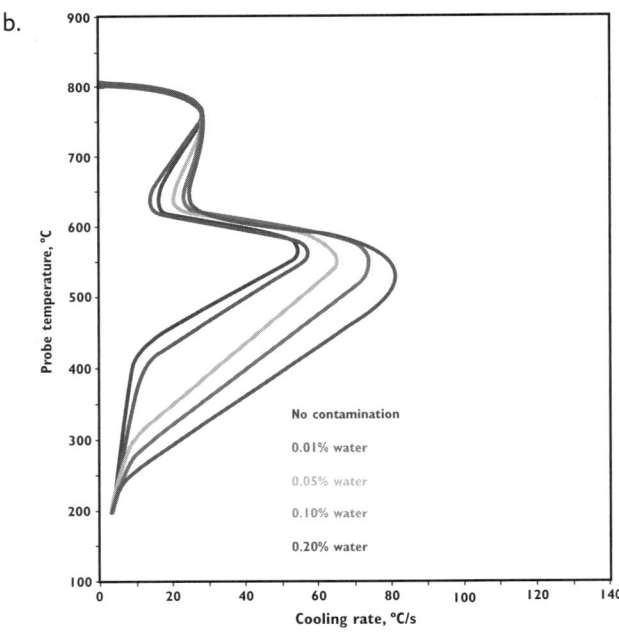

a. Effect of quench-oil oxidation on cooling rate

b. Effect of water contamination on cooling rate

Polymer Quenchants

A quarterly maintenance program is recommended for polymer quench tanks to check for bacteria and mold, concentration (by kinematic viscosity), solids, organic contamination and the presence of proper amounts of rust inhibitors for corrosion control.

Contamination of polymer quench tanks is a common occurrence in the heat-treat shop, from airborne particles (e.g., dust, dirt), the introduction of other fluids (e.g., cutting fluids, rust-preventive oils, hydraulic fluids) or solids (e.g., scale, soot) into the quench system. Since cooling-curve behavior is dependent on contamination, its control is paramount.

As a result, polymer-quenchant concentration should be checked daily, and concentration can be readily controlled on the shop floor by taking refractometer (% Brix) measurements. This is a rapid, easy-to-perform test involving an inexpensive instrument, but it is prone to misinterpretation and contamination. Kinematic viscosity is performed in the laboratory and is the best method to determine concentration since it is relatively immune to contamination. It should be noted that chemical analysis is often used to supplement cooling-curve analysis since it cannot detect all problems, such as foaming in polymer quenchant.

Percentage Brix is the percentage of sucrose (sugar) in water at 20°C (72°F). This provides a direct conversion to refractive index. Several scales are available: 0-8%, 0-20% and 0-32%. Readings are accurate to 0.5% Brix (±0.25% Brix). Instrument errors are only 0.1-0.2% Brix, resulting in good concentration control (e.g., ±1.0% PAG)

Filtration is very important in polymer systems, and sand filtration is often recommended because it is simple, inexpensive and readily maintained. Cartridge and bag filters can be used but are not recommended since they are prone to bacteria formation.

Polymer quenchants are susceptible to microbiological contamination. If the polymer tank begins to stink, this indicates that bacteria and fungus are likely present and can cause the depletion of pH and corrosion inhibitors. Stagnant solutions contribute to localized oxygen depletion, so quenchant should be kept moving and oxygenation can be used. Bacteria can be readily controlled without the use of biocides by keeping the tank clean (i.e., filtered) and having a proper oxygen content.

Summary

A sound maintenance program relies on our ability to check that the variables that impact the ability of the quench system to perform as intended are well controlled. With quenchants, we cannot assume that all is well simply because no problems have been reported. Quench systems are in a dynamic state of continual change, one which we must recognize and verify on an ongoing basis.

REFERENCES

1. Wachter, D.A., G.E. Totten and G.M. Webster, "Quenchant Fundamentals: Quench Oil Bath Maintenance," *Advanced Materials & Processes*, 1997
2. Totten, G.E., G.M. Webster and D.A. Wachter, "Quenchant Fundamentals: Condition Monitoring of Quench Oils," *Practicing Oil Analysis*, January/February 2003, pp. 26-31
3. Houghton International Quenching Presentation, SECO/WARWICK Corp. Furnace & Atmospheres for Today's Technology seminar, June 2014
4. Totten, G.E., C.E. Bates and N.A. Clinton, *Handbook of Quenchants and Quenchant Technology*, ASM International, 1993, pp. 204-206
5. Dr. D. Scott Mackenzie, Houghton International (houghtonintl.com), technical and editorial contributions, private correspondence
6. Dr. George E. Totten, G.E. Totten & Associates (getottenassociates.com), technical and editorial contributions, private correspondence

10 QUENCHING AND QUENCHANTS

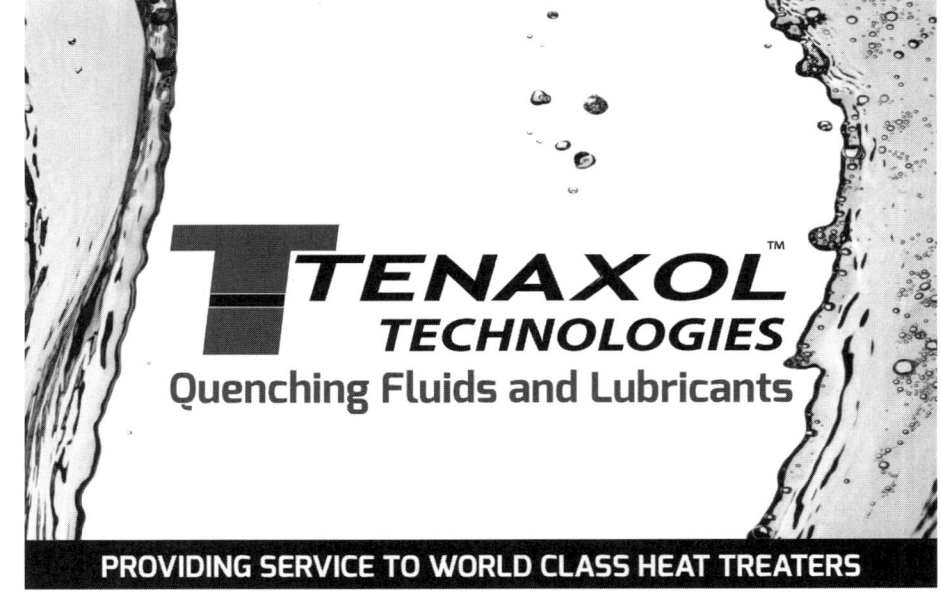

10.4

TYPES OF QUENCHANTS (PART TWO)

We continue our discussion of the various types of quenchants by focusing on air, molten salt, water and brine/caustic solutions.

Quenching in Air

Quenching or cooling in air is considered by many to be the simplest and most straightforward of the quenching processes. When performing these processes, however, one must take into account the temperature of the air, its relative humidity and its flow pattern (i.e., heat-extraction profile) as it passes over the part geometry or through the load in question. The heat-transfer coefficient due to convective heat transfer (Equation 10.4.1) plays an important role in this process.

10.4.1) $Q = h_c \bullet A \bullet (T_{fluid} - T_{object})$

where:
   Q is the heat transferred to the quench medium in watts (Joule/second)
   h_c is the heat-transfer coefficient in W/m²-K
   A is the surface area in m²
   T is the temperature in K

Cooling in air is primarily used to reduce part temperature for handling (Fig. 10.4.1), while air quenching is intended to achieve specific metallurgical and/or mechanical properties (Figs. 10.4.2, 10.4.3). Air quenching can be employed provided the material has enough hardenability to make it practical and if the process allows it (in other words, if surfaces can tolerate oxidation and/or scale).

Types of air quenching include still-air cooling in a controlled environment; room- or outside-air cooling in an uncontrolled environment (in other words, an area subject to drafts, changes in weather conditions and unregulated air movement); or by forced convection (fan) cooling with or without air mist. The amount of distortion (i.e., dimensional change) that can occur after air cooling (Fig. 10.4.4) or air quenching must take into account the prior stress state of the material due to fabrication methods as well as the heat-treat parameters selected.

FIGURE 10.4.1 | Furnace equipped with a fan cooling section (courtesy of SECO/WARWICK Corp.)

FIGURE 10.4.2 | Normalizing large paper rolls using fan-assisted cooling due to part mass (courtesy of Metals Engineering Inc.)

FIGURE 10.4.3 | Air cooling after stress relief in an outdoor environment (courtesy of Derrick Company Inc.)

FIGURE 10.4.4 | Stress relief of 304 and 316 stainless steel plates – distortion improperly blamed on air quenching

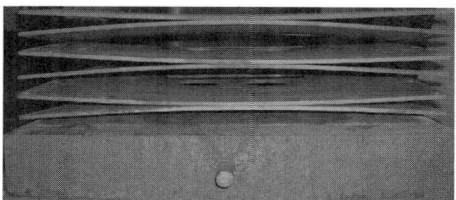

Quenching in Molten Salt

Molten salt is an effective quenchant both for interrupted quenching (e.g., austempering) and for distortion (i.e., dimensional) control. Convective heat transfer (Equation 10.4.2) is allowed to take place in an extremely uniform bath, with typical variation in salt temperature being in the order of 0.56°C (1°F).

$$10.4.2) \quad Q = k \bullet \frac{A \bullet (T_{hotter} - T_{colder})}{L}$$

where:

Q is the heat transferred in watts (Joule/second)
k is thermal conductivity of the material in W/m-K
A is the surface area in m²
$T_{hotter} - T_{cooler}$ is the temperature difference across the material in K
L is the thickness of the material in meters

In interrupted quenching, parts are cooled rapidly from the austenitizing temperature to a point just above the martensite start temperature (M_s), where they are held for a specified time (usually in the order of 12-20 minutes) and then air cooled to room temperature (Fig 10.4.5). The temperature at which the quench is interrupted is usually in the 370-175°C (700-350°F) range for most steel and alloys. Temperature, agitation, water content and residence time are the main variables when quenching in molten salt (Figs. 10.4.6, 10.4.7).

FIGURE 10.4.5 | Austempered stampings exiting the salt quench tank (courtesy of hti)

FIGURE 10.4.6 | Typical cooling and cooling-rate curves for salt quenching: no agitation; no water addition; average cooling rate from 650-260°C (1200-500°F) is 33.6°C/s (60.5°F/sec)[3]

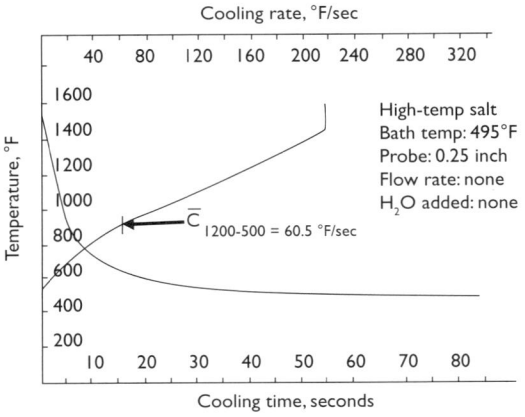

FIGURE 10.4.7 | Effect of agitation and water content on quench severity in molten salt[3]

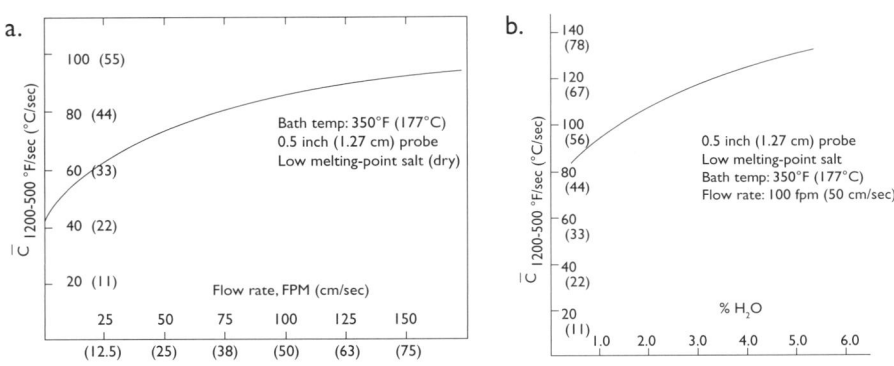

a. Agitation

b. Water content

Quenching salt is a eutectic mixture of nitrates and nitrites of sodium and potassium. There are nearly a dozen compositions available, ranging in melting points from 135-260°C (275-500°F). Selection of a quenching salt primarily depends on the lowest temperatures at which it is going to be operated and its melting point. The difference between the two should preferably be 38°C (100°F) or greater. Once molten, these salts all exhibit nearly the same physical properties (Table 10.4.1).

10 | QUENCHING AND QUENCHANTS

TABLE 10.4.1 | Properties of molten salt

Property	Range of values
Specific gravity	1.84-1.92
Specific heat	0.35-0.40 cal/gram-°C
Thermal conductivity	0.571 W/m-K (0.33 BTU/hour-feet-F)
Heat-transfer coefficient	4.5-16.5 kW/m^2-K (800-2,900 BTU/hour-ft^2-F)
Drag-out rate	50-100 g/m^2 (1-2 pounds/1,000 ft^2)

SALT PROCESSES[2]

The following processes are commonly performed in molten salt.

Austempering and Variations

- Austempering (Section 5.6) is used when a bainitic or bainitic/martensitic structure is desired. Steels that can be austempered include carbon steels such as SAE 1050-1080; steels with 1% manganese; low-alloy steels such as SAE 4150, 4365 and the 5100 series; ductile irons; and gray irons.

    Manipulating agitation can enhance austempering of thin-walled parts such as cylinder liners. To minimize thermal shock, for example, agitation is not used during the first 15-20 seconds of the quench. It is then turned on automatically to complete the standard austempering operation. This procedure helps minimize distortion without any sacrifice in required mechanical properties.

- Austempered ductile iron (Section 5.6) is a process performed in the 230-400°C (450-750°F) range for 0.5-4 hours depending on steel composition and mechanical-property requirements. The microstructure produced is ausferritic rather than bainitic.

- Carboaustempering is a process used on low-carbon-content steels after carburizing to produce a high-carbon bainitic case and either a martensitic or bainitic core depending on steel composition and quench severity. What makes this process unique is that the core becomes hard first, followed by the surface, producing less overall distortion. In addition, while the surface is being austempered, any martensite in the core is being tempered. Parts that are carboaustempered have excellent fatigue strength and wear resistance and have been reported to be dimensionally and functionally superior to carburized and conventionally quenched parts.

- Modified austempering is a modification of austempering in which quench severity is intentionally decreased to force the cooling curve to intersect the pearlite nose of the TTT diagram. Since a mixed microstructure of bainite and pearlite results, hardness is relatively low –

usually in the 30-42 HRC range – but ductility is extremely high and strength is moderately high.

Wire patenting (i.e., passing wire through tubes in a furnace) operating at 970°C (1780°F) is an example of modified austempering. The method should also be considered for carbon steels and heavier sections where ductility greater than that possible by standard austempering is required. Some trial and error will be necessary to develop the optimum cycle for the particular steel, section thickness and property requirements.

Martempering and Variations

- Martempering is usually performed at 175-260°C (350-500°F) depending on steel composition, and it is held long enough for the temperature to equalize throughout the cross section. The part is then removed and cooled in air to room temperature, allowing austenite to transform to martensite. The duration of the treatment is short, typically just a few minutes, depending on section thickness and quench severity. As with conventionally quenched parts, martempered parts must be tempered. Materials suitable for martempering include carbon steels, low-alloy steels and gray cast irons.

  Martempering with two baths is used to increase the depth of hardening in thick sections. The part is first quenched in water or brine for a very short time and then transferred into a martempering bath. Hardening of SAE 52100 steel bearing balls is one such example.

- Modified martempering differs in that the temperature of the quench bath is somewhat below M_s, usually in the 150-175°C (300-350°F) range. This increases quench severity. The process can be used for low-hardenability steels when increasing agitation and water additions are not enough to avoid the pearlite nose of the TTT diagram. It is, therefore, applicable to more steels than conventional martempering.

- Carbomartempered parts tend to be tougher than conventionally carburized and hardened parts. Ferrous metals with low carbon contents can be martempered only after the parts have been carburized or carbonitrided. The greatest advantage, however, is minimum distortion.

  If parts are carburized, they can be quenched directly into salt. If they are carburized in a cyanide salt bath, however, they must first pass through a neutral salt bath to avoid any violent reactions that could result in an explosion.

  The carbomartempering temperature is a compromise between the M_s of the case and that of the core, and it is chosen so as not to

10 | QUENCHING AND QUENCHANTS

jeopardize case hardness. It is generally in the 175-260°C (350-500°F) range, and the duration is usually just a few minutes.

❖ Other variations include the following:
- A process that combines martempering and austempering enables users to realize the maximum benefits of both martensite's hardness and bainite's toughness. Parts are quenched in an austempering bath and held for about half the normal austempering time. They are then removed and cooled to room temperature. The time-temperature cycle can be tailored to provide a range of hardness/toughness combinations that would be unobtainable using either process alone.
- Another variation combines martempering with straightening. Some parts require straightening or re-forming after heat treatment, but the operation is sometimes not feasible after conventional quenching and tempering. Straightening, however, can be performed immediately after the part is removed from the martempering salt while its microstructure is still essentially austenitic.

Quenching in High-Pressure Gas

There is a high degree of interest today to replace liquid-quenching methods with gas-quenching technology. Companion work *Vacuum Heat Treatment*[2] covers this subject in more detail. Atmosphere furnaces are being designed so that they can interface with a pressure-quench system similar in many respects to their vacuum-furnace cousins. In the case of atmosphere furnaces, quenching takes place in a separate section from the heating section.

WHAT IS PRESSURE QUENCHING?

The description most often used for high-pressure gas quenching is "accelerating the rate (speed) of quenching by densification and cooling of gas."[4] One of the many reasons for the intense interest in this quenching technique is related to improved part distortion. A critical concern in using this technology is to avoid sacrifice of metallurgical, mechanical or physical properties; that is, retain the ability to transform a material to a microstructure that is similar, identical or superior to that of a known quenching medium (e.g., oil, polymer or salt).

Pressure Levels

Selecting the optimum gas pressure for quenching is highly dependent on a number of factors, including material, component geometry, loading, net-to-gross load ratio, gas parameters and equipment design (Fig. 10.4.8). Pressure ranges (Table 10.4.2) are typically classified as sub-atmospheric, low, medium, high and ultrahigh pressure irrespective of the type of gas used.

FIGURE 10.4.8 | Variables in high-pressure gas-quenching performance

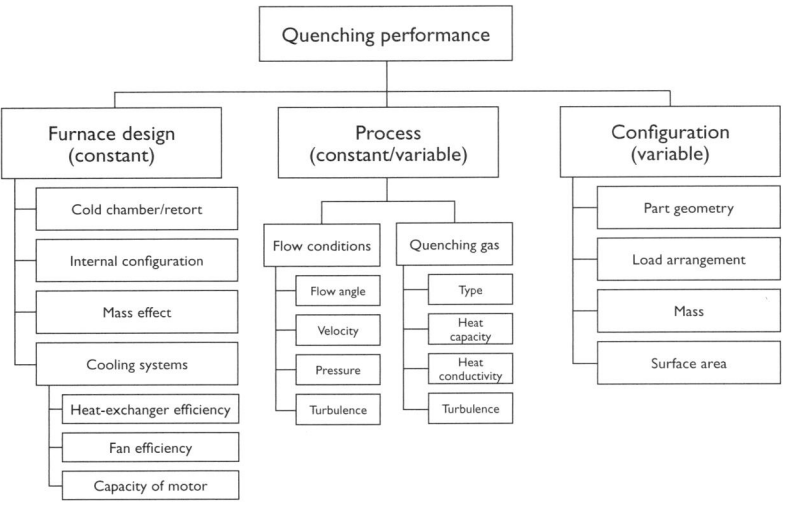

TABLE 10.4.2 | Classification of gas-quenching pressure ranges for atmosphere furnaces[2]

Classification	Pressure range, bar
Sub-atmospheric pressure	-0.67 to -0.17
Low pressure	2-4
Medium (mid-range) pressure	5-9
High pressure	10-20
Ultrahigh pressure	>20

Quenching in Water

Water is one of the most readily available and inexpensive choices for quenching. It has many advantages, including:

- ❖ Environmentally friendly (no fumes or smoke)
- ❖ Nonflammable (no fire hazard)
- ❖ Nontoxic
- ❖ No health or safety hazards, per se

In addition, water has an extremely high thermal-transfer coefficient (both instantaneous and average α-values) and the ability to transform low-hardenability steels. It is also easy to handle parts after quenching (no cleaning required) and promotes scale removal from part surfaces during quenching (i.e., the scale present "pops off" during quenching).

The cooling rate of water is independent of material properties (e.g., thermal conductivity, specific heat) and depends to a great extent on temperature (which controls the cooling rate) and agitation (Table 10.4.3). The cooling rate decreases with increasing water temperature, and the length and stability of the vapor barrier increases. The maximum cooling rate also decreases with increasing water temperature.

TABLE 10.4.3 | Effect of water temperature on cooling rate

Water temperature	Maximum cooling rate	Maximum cooling-rate temperature	Cooling rate at temperature, °C (°F)		
°C (°F)	°C/s (°F/sec)	°C (°F)	@ 705 (1300)	@ 345 (650)	@ 230 (450)
40 (105)	153 (310)	535 (995)	60 (140)	97 (205)	51 (125)
50 (120)	137 (280)	542 (1010)	32 (90)	94 (200)	51 (125)
60 (140)	115 (240)	482 (900)	20 (70)	87 (190)	46 (115)
70 (160)	99 (210)	448 (840)	17 (65)	84 (185)	46 (115)
80 (175)	79 (175)	369 (696)	15 (60)	77 (170)	46 (115)
90 (195)	48 (120)	270 (520)	12 (55)	26 (80)	42 (110)

Water quenching is used in the tubular-products industry (bar, tube, pipe for the water, gas and oil industry) to transform a variety of steels and other materials to meet API (American Petroleum Institute) standards (Fig. 10.4.9).

FIGURE 10.4.9 | SAE 4140 pipe moving into water-spray quench (courtesy of Interpower Induction USA)

Cast aluminum alloys (Fig. 10.4.10) are an example of parts quenched in water or boiling water to reduce the temperature differential and minimize risk of cracking when thin and thick sections exist on the same casting.

Given these advantages, why isn't water used for every quenching application? The answer lies in the limitations that come with a water quench, including:

❖ Corrosive attack (parts, fixtures, quench tank)
❖ Non-uniform quenching characteristics

❖ Distortion/cracking
❖ Microbiological growth

FIGURE 10.4.10 | C356.0 aluminum castings ready for solution heat treatment and water quenching

In addition, although water quenching can properly transform low-hardenability material, it must be done uniformly (Fig. 10.4.11) to prevent unwanted transformation products, cracking and other problems.

FIGURE 10.4.11 | Improper spray-quench coverage and spray parameters resulting in failure to uniformly remove heat, which leads to cracking of the part

Non-uniform quenching characteristics translate to severe limitations on the use of water as a universal quenchant. There are a number of factors that contribute.
❖ Temperature dependence: As the water temperature increases, the vapor phase becomes prolonged and the maximum cooling rate decreases sharply. This can lead to the formation of soft spots on the parts (particularly in induction hardening) and inconsistent hardness.
❖ Vapor-phase stability: The stability of the vapor phase is dependent on

the surface finish of the part. The vapor film lingers on smooth surfaces but breaks up readily with the onset of the boiling stage at sharp corners, rough surfaces, defects or stress risers. This variation in stability produces markedly different cooling rates across the part, resulting in distortion and cracking.
- ❖ High cooling rates: Water exhibits very high cooling rates in the convection phase when compared to other quench media. High cooling rates in the martensite transformation range (M_s to M_f) can result in high residual stresses, excessive distortion and cracking.

These detrimental effects can be minimized by maintaining the water at a low temperature, vigorous agitation (to disperse the vapor blanket), adding an inorganic salt and adding corrosion inhibitors and biocide packages (although this introduces additional complexity).

INTENSIVE WATER QUENCHING

An alternative to conventional water quenching is intensive water quenching (IQ) technology, which relies on highly agitated water applied very uniformly over the part surface contoured to a particular part geometry in single-part-focused high-velocity quenching or in batch loads. Automotive ball studs, cold-work punches, castings, forged rings, coil springs, gear racks, shafts, pinions and dies are typical examples of parts that benefit from this process.

IQ can be thought of as promoting the conditions that lead to maximum surface compressive stresses (Fig. 10.4.12). To properly design a system, it is important to consider: alloy, part shape, cross-section size, minimum quenching/cooling rate and final machining/grinding.

IQ works on most martensitic alloys and carburized parts because of the uniform development of very high surface compressive stresses and the timed interruption of the quench with air cooling when the surface compressive stresses are at their maximum value. Solution treating of stainless steels, titanium and aluminum alloys is another application of the technology.

FIGURE 10.4.12 | Comparison of intensive quenching to other quenching technologies (courtesy of IQ Technologies Inc.)

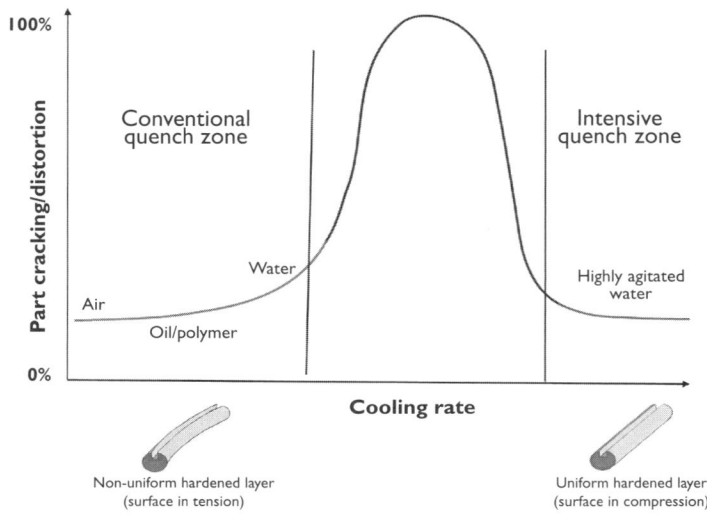

Two IQ techniques are currently used in heat-treatment practice: the batch IQ process (Fig. 10.4.13) is conducted in intensively agitated water tanks, and the single-part processing method is performed in a quench tube or chamber. In batch quenching, the film-boiling stage does not occur. In the higher-velocity, single-part quenching system, both the film- and nucleate-boiling stages are eliminated. Therefore, the highly uniform convection mode of heat transfer takes place almost from the beginning of the quench, reducing distortion.

FIGURE 10.4.13 | Atmosphere integral-quench furnace with a 41,600-liter (11,000-gallon) water tank adapted for IntensiQuench® (courtesy of Euclid Heat Treating Co.)

Quenching in Brine/Caustic

The addition of salts (aka brine) or caustic to water improves the quenching power of water by helping break up the vapor phase, and it prevents water from dissolving gases. These minute crystals are deposited and wet the part surface during quenching. The localized high temperatures cause these crystals to fragment violently. This creates turbulence that destroys the vapor film (Fig. 10.4.14) and produces very high cooling rates (Table 10.4.4). The greater the percentage of brine used, the less air it is able to dissolve and the slower the quenching rate. Brine/caustic quenching (Fig. 10.4.15) is used in many applications when transformation by rapid heat transfer is needed and concerns over cracking are mitigated by material selection and part geometry.

FIGURE 10.4.14 | Effect of crystals on vapor layer[8]

a. 1st stage

b. 2nd stage

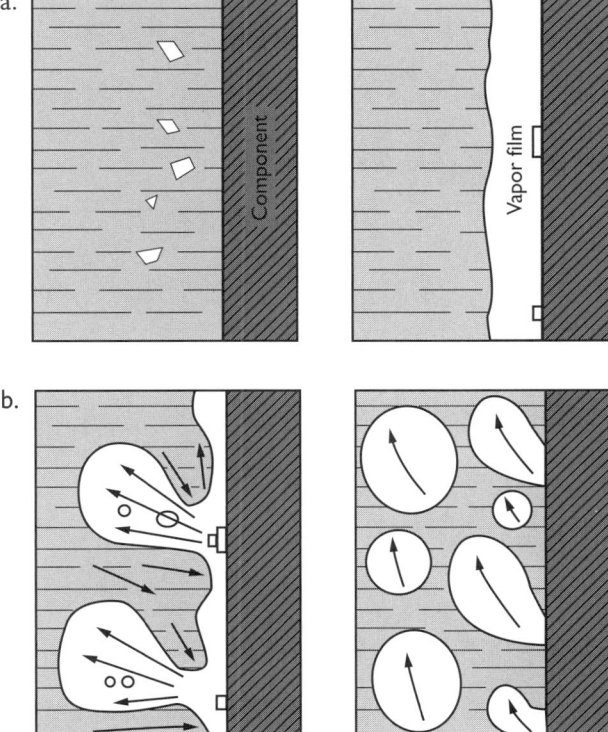

TABLE 10.4.4 | Relative cooling rate as a function of quench medium[5]

Quench medium	Cooling rate (compared to water)	Flash point, °C (°F)	Fire point, °C (°F)
Sodium hydroxide (10%)	2.06		
Brine (10%) at 17°C (65°F)	1.96		
Caustic soda (10%)	1.38		
Water at 17°C (65°F)	1.0		
Oil[a]	0.19-0.44	185-295 (365-565)	205-360 (405-685)
Air, circulated	0.032		
Air, still	0.0152		

Note:

[a] Here the term "oil" covers a number of products and types.

FIGURE 10.4.15 | Load of SAE 4130 cylinder liners being brine quenched (courtesy of Phoenix Heat Treating Inc.)

Rock salt or sodium chloride (NaCl) is used at concentrations of 5-10% for brine quenching, and sodium hydroxide (NaOH) is used at concentrations around 3-5% for caustic quenching. Some in the industry still use the "potato test," which relies on the fact that an uncooked potato will float on the surface when the proper salt concentration is reached, as a very rough approximation of proper brine concentration. The corrosion problems associated with the use of sodium chloride and the toxicological problems associated with the use of sodium hydroxide can be avoided by using certain proprietary blends. Brine solutions are most commonly operated in the 17-38°C (65-100°F) temperature range.

Due to the corrosive nature of brine and caustic, nonferrous materials are not processed in these solutions.

Summary

The various forms of quenching discussed are used throughout the heat-treatment industry for their flexibility and control of quenching parameters. In order for any quenching operation to be successful, one must carefully match the desired outcome of the process with the type of quenchant best suited for the job. The reason to select one quenchant over another or to use one method of quenching over another is often highly subjective and based on items such as historical performance, convenience, cost and/or flexibility. What is important for the heat treater to understand is that no quenchant can be successful in all applications, and quenching and quenchant variables must be carefully controlled.

REFERENCES

1. Herring, Daniel H., *Atmosphere Heat Treatment, Volume I*, BNP Media, 2014
2. Herring Daniel H., *Vacuum Heat Treatment*, BNP Media, 2012
3. Foreman, R.W., "New Developments in Salt Quenching," *Industrial Heating*, March 1993
4. Herring, Daniel H., "A Review of Factors Affecting Distortion in Quenching," *Heat Treating Progress*, December 2002
5. Singh, U.K. and Manish Dwivedi, *Manufacturing Processes*, 2nd Edition, New Age International Limited, 2009, p. 54
6. Kobasko, Nikolai, Michael Aronov, Joseph Powell and George Totten, "Intensive Quenching Systems: Engineering and Design," ASTM International, 2010
7. Aronov, Michael, Nikolai Kobasko and Joseph Powell, "Intensive Quenching of Steel Parts," *ASM Handbook, Volume 4A, Steel Heat Treating Fundamentals and Processes*, 2013
8. *Houghton On Quenching*, Houghton International, p. 13
9. Dubal, Gajen P., "Salt Bath Quenching," *Advanced Materials & Processes*, December 1999
10. Dr. D. Scott Mackenzie, Houghton International (houghtonintl.com), technical and editorial contributions, private correspondence
11. Dr. George E. Totten, G.E. Totten & Associates (getottenassociates.com), technical and editorial contributions, private correspondence

10 | QUENCHING AND QUENCHANTS

10.5

MANAGING DISTORTION

Processing with minimal part distortion has been recognized as a major need for heat treatment in the 21st century (Section 1.6). In one study,[1] it was found that the contribution of the heat-treatment process to the distortion state of a part was as little as 5-10% (Fig. 10.5.1).

FIGURE 10.5.1 | Sources of gear distortion (from Hart.-Tech. Mitt., 43.1, 1988, Carl Hanser Verlag)[1]

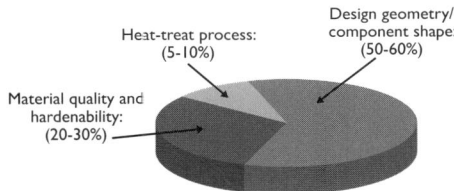

Factors Influencing Distortion

Most, if not all, of the sources of distortion (Fig. 10.5.2) have been identified. They have been and will continue to be the subject of intense study throughout the heat-treatment industry.

FIGURE 10.5.2 | Ishikawa distortion diagram[2]

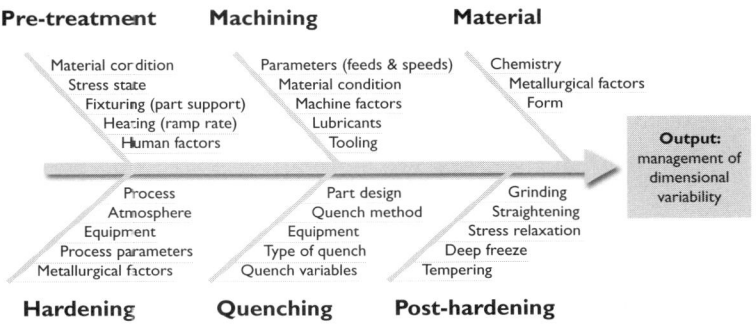

The role each of these factors has on managing distortion is noteworthy. A textbook example is a gear designed for weight reduction achieved using several large holes (Fig. 10.5.3a). After heat treatment, significant distortion resulted. After redesign (Fig. 10.5.3b) using smaller equally spaced holes, the distortion was significantly less. In the real world, poorly manufactured gears (Fig. 10.5.3c) will warp on heat treatment.

FIGURE 10.5.3 | Example of bad and good gear design and a real-world example

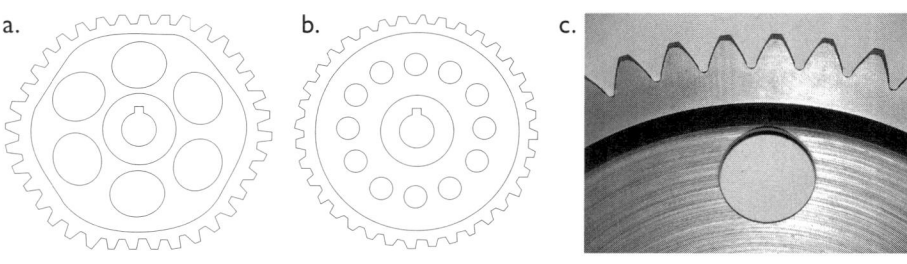

a. Poor design resulting in distortion
b. Good design avoiding distortion
c. Real-world example of improper hole placement leading to distortion of adjacent gear teeth

The higher the hardenability of the material, the more distortion-prone it is. The secret is to use the lowest-hardenability material that will achieve the required core hardness. Normalizing (Section 5.3) is one technique that can counteract some of the factors influencing distortion by helping to negate stresses from mill operations. In addition, other heat-treatment factors also contribute, namely:

- Effect of heating rate
- Temperature variation
- Austenitizing temperature
- Time at temperature
- Choice of equipment
- Basket and fixture design
- Loading arrangement (e.g., part orientation)
- Quenchant technology choice

Another frequently used method to manage distortion due to prior operations (e.g., machining) is to perform a stress relief (Section 5.7) prior to heat treatment. The magnitude of the distortion can be determined by measuring the part before and after stress relief. This can be done after each operation to help understand the contributions of these prior operations to the overall distortion.

Specialized Techniques for Managing Distortion

Several types of specialized quenching methods have been developed in the heat-treatment industry to help manage distortion. Two of them will be discussed.

HOT-OIL QUENCHING

Martempering and marquenching are terms often associated with hot-oil quenching. Formally, martempering (full martempering, true martempering) is a term applied when an austenitized workpiece is quenched into a medium whose temperature is essentially maintained in a bath just above the martensite start (M_s) temperature of the steel and held in that medium until its temperature is uniform throughout – but not long enough to permit bainite (or pearlite) formation – and then allowed to cool in air (Fig 10.5.4a). When the martempering process is applied to carburized material, the controlling M_s temperature is that of the case. As such, this process variation is called marquenching.[3]

A modified form of marquenching or martempering is known as hot-oil quenching, which is designed to take place just below the M_s temperature (Fig. 10.5.4b). The required critical-cooling rate in many steels is such that quenching faster than that possible by marquenching or martempering is necessary to obtain full hardness. It has also been found effective in reducing quench stresses and improving part dimensional stability.

FIGURE 10.5.4 | Comparison of oil-quenching methods[4]

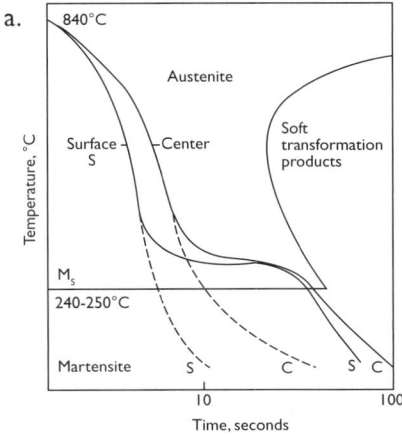
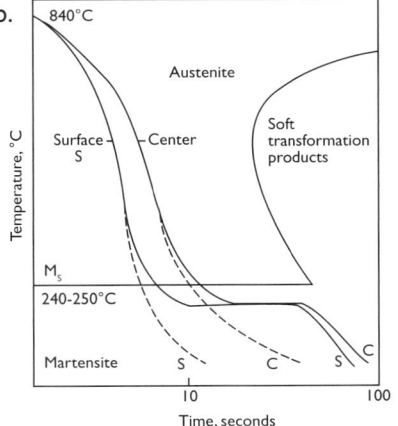

a. Martempering

b. Hot-oil quenching

In carburizing, the core of the component (because it has a lower carbon content) will transform first, while the higher-carbon case will transform last. This contributes to establishing a compressive residual stress state on the surface of the part.

What is hot oil?
Quenching into oil above 100°C (212°F) has traditionally been referred to as hot-oil quenching. Oil temperatures in the 90-205°C (195-400°F) range have been used, with both ends of the spectrum normally reserved for special applications. Hot-oil temperatures in the 105-190°C (220-375°F) range are somewhat common in the heat-treatment industry and take place in either batch or continuous atmosphere sealed-quench furnaces. Since these oils are designed for elevated-temperature service, oxidation can be an issue. Quality hot oils will have a substantial quantity of antioxidant packages to ensure long life at elevated temperatures and repeatable quenching operations.

Part Distortion
Distortion arising from heat treatment can be classified as two types: size distortion due to volume changes resulting from phase transformations (which are generally predictable) and shape change or warpage (which is generally unpredictable). The latter takes many forms, including out-of-round, out-of-flat, bending, bowing, bucking, taper, dishing, canning or closing-in of bores. It can arise either during heating or austenitizing through the relief of internal stresses from prior operations; sagging or creep due to inadequate part support during heating; mechanical damage; non-uniform heating; or during quenching, the result of an imbalance of the internal residual stresses generated. A number of key factors influence these changes, including:

- Steel composition and hardenability
- Component part geometry
- Mechanical handling
- Types of quenchant
- Temperature of quenchant
- Condition of quenchant
- Circulation of the quenchant

In the real world, we find that cooling rates vary widely based on the type, temperature, condition and circulation of the quenchant, which changes daily. Due to the fact that cooling rarely (if ever) occurs uniformly on the component part, we find that different quenchants produce different time durations throughout the three stages of liquid quenching (vapor phase, boiling phase, convective phase) and have a direct bearing on distortion.

10 | QUENCHING AND QUENCHANTS

Operating temperature can have a dramatic influence on distortion. Hence, oils have been developed for martempering and marquenching applications to minimize residual stresses by promoting uniform transformation. Hot oils are generally applied to high-precision engineering components such as thin-section bearing races and transmission gears and shafts requiring critical dimensional control.

The condition of the quenchant, which involves oxidation, contamination and degradation for hot oils, dramatically influences results since viscosity increases. It is not uncommon to filter or centrifuge hot-oil baths daily and replace the oil every 18-24 months. With proper care, filtration and design of the agitation and heating elements, however, longer oil life may be possible.

With hot oil, the cooling rate in the convective phase is increased significantly with agitation. So, excessive agitation is to be avoided to counter distortion. The speed and direction of oil flow over the workpiece can also determine the nature of the distortion. In general, a flow rate of 0.5-1.0 m/sec through the workload area is recommended. The uniformity of the flow around the part is more important than an absolute speed.

Hot-Oil Selection
The choice of hot oil depends on the need and is dependent on factors such as:
- Hardenability of the alloy
- Critical cooling rate and M_s and M_f temperatures as determined from the transformation (TTT) diagram of the steel
- Part austenitizing temperature in relation to oil temperature
- Part and load geometry
- Part cleanliness requirements
- Distortion expectation (allowable dimensional changes)
- Safety

For example, leaner-alloyed parts usually require lower-viscosity martempering oil, often with speed enhancers added. Higher-alloyed parts usually require higher-viscosity oils and certain additives to allow higher oil temperatures where distortion management is maximized. Thin distortion-prone components are best martempered or quenched into the slowest-speed martempering oil, which will produce the required metallurgical properties.

Hints for Extending Oil Life
In general, the hotter the oil's temperature, the lower the resultant distortion and the faster the oil's degradation. Also, the higher the austenitizing temperature from which a part is quenched, the more damage to the oil and the faster

the oil will deteriorate. Other common concerns are oxidation and slug/contamination buildup. These can be minimized to a degree by the addition of antioxidants and with the use of a protective-atmosphere cover (such as a nitrogen blanket) over the oil during heat-up and operation.

Oil without antioxidant additives will give the brightest and most consistent part surface appearance, but it will oxidize rapidly followed by a discoloration of the work. Antioxidant additives will normally produce a consistent surface finish while extending the oil's useful life.

Fresh or makeup oil can be added to further reduce oil degradation. A hidden danger is heating marquench oil up too rapidly, which can also degrade the oil (low-velocity burners or low watt-release resistance heaters should be used). It is recommended that heaters be interlocked with the agitators to prevent the energizing of the heaters without flow around them. Heaters that have an energy density of no greater than 0.0093-0.0155 watt/mm^2 (6-10 watts/inch^2) are recommended. Higher energy density can be used, but the agitation rate must be increased substantially to prevent excessive local heating of the oil and excessive heater surface temperature. The overall time to heat the bath to operating temperature is typically in the 8-12 hour range.

It is important to note that oil capacity is not always an assurance of success. For example, parts run in continuous furnaces discharging parts into quench chutes may see problems with low hardness or staining due to breakdown (fractionation) of the oil in a small localized area, lack of proper heat extraction or poor oil circulation in the quench chute.

Cooling systems should be sized to handle the heat extraction and should be free of copper and other materials known to be catalysts for oxidation of oil products.

Effect of Oil Changes Over Time
As the oil ages, its heat-extraction capabilities will change, influencing the end product gradually over time. During the life of a hot oil, the cooling characteristics change progressively (Fig. 10.5.5). As seen, the vapor-blanket phase disappears and the maximum quench rate increases (and occurs at higher temperature), resulting in a slowing down of the quench rate at lower temperature.

Finally, when changing to marquenching oils, it is important to recognize that the design of the quench tank may have to be changed or optimized for their use. Previous quenchants, sludge and water should be completely removed from the tank prior to introduction of the hot oil. Remember that the oil will expand and its viscosity will change during heating. Protective gas atmosphere over the oil should always be used during initial heat-up, and exposure to air must be minimized at all times.

FIGURE 10.5.5 | Influence of hot-oil age on cooling-rate/temperature profile[4]

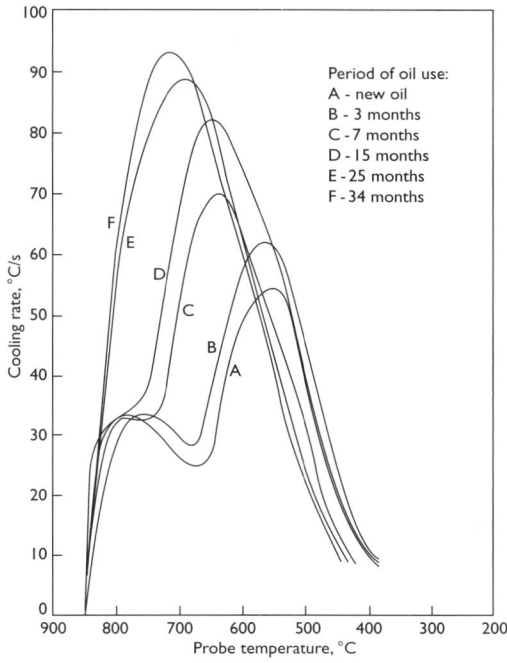

PRESS QUENCHING

Heat treaters know that gears, bearing races and shafts are especially prone to dimensional changes during hardening and quenching, which can cause a number of problems during post-heat-treatment manufacturing operations. Typically, additional stock allowances are needed to compensate for distortion so that parts can be machined to the proper finished dimensions. The objective of press, plug and roll quenching is to hold parts round and flat while they are being cooled, thereby reducing (though typically not eliminating) distortion.

Hardened or case-hardened gears or bearing races are either free or press quenched. The latter involves moving them individually out of the furnace into a quench press (Fig. 10.5.6) by manual or robotic means. The transfer from the furnace to the quench press (with oil or polymer flowing) must be relatively fast, typically under 10 seconds. A high volume of quenchant, uniformly directed to all internal and external surfaces of the parts, is needed. Control of out-of-roundness to within 0.127 mm (0.005 inch) and out-of-flatness to within 0.025 mm (0.001 inch) is common from most industrial presses.

FIGURE 10.5.6 | Typical press-quench operation

Critical Considerations

Consideration and regulation of the following press parameters have been found to help with the final distortion state:

- Transfer time from furnace to press (consistent)
- Manipulator contact (area, duration)
- Part positioning on the die
- Die design
- Die pressure (clamping force, expander force) applied to hold the component
- Quenchant temperature
- Direction of quenchant flow
- Quenchant pressure
- Quantity of quenchant
- Location of points of contact on the component
- Duration of quenching (at various flow rates)
- Flow paths to and through the lower die for the quenchant to reach top and bottom simultaneously
- Die-set maintenance and repair
- "Pulsing" feature (optional)

It is especially noteworthy to mention that old, worn or damaged tooling can be responsible for as much, if not more, distortion than part design or geometry. Quench dies should be routinely inspected for damage or wear and repaired or replaced as necessary. It is also important to check final dimensions after repair since one of the hidden dangers is that of altered or restricted quenchant flow.

The "pulse" feature, which allows a part to expand and contract normally while still controlling shape, is a popular option on presses. Without it, stresses are induced because the part is not allowed to contract and expand. Pulsing reduces the friction caused by constant pressure and clamping on the part as it contracts during cooling. This friction promotes stresses that result in eccentricity and out-of-flatness. Properly applied, the pulsing technique finds the die in contact with the part throughout, but the pressure is released and re-applied every several seconds during the entire quench cycle. The expander pressure is normally not pulsed.

How the Press Works

A typical press-quench system (Fig. 10.5.7) will operate as follows. The component to be quenched is removed from the furnace (usually a continuous-pusher or a rotary-hearth furnace) and placed onto the lower die in the out position. Initiation of the automatic cycle moves the loaded lower-die assembly into the center section of the machine. When this is fully advanced, the upper ram assembly and dies descend, with an expander centering the part just prior to the inner and outer dies locating on their respective pressure points. When the expander and dies are properly located, the ram holding them is latched in the down position and pressure is applied to all three. The inner die, outer die and expander usually have completely independent pressure controls, ranging from 0-107 kN (0-24,000 lbf) on the dies and 0-49 kN (0-11,000 lbf) on the expander. These are often regulated by hydraulic valves and monitored via pressure gauges. When used in a bore, the expander cone pushes out against the segmental lower die in order to hold the bore round and to size. The inner and outer dies help keep the component flat.

Normally, a guard completely encloses the upper dies, forms a quenching chamber and attaches to the upper ram moving up or down with it. When the upper ram latches, the quenchant pump starts. This supplies the chamber with fluid, the volume of which is normally controlled by a series of switches. In a typical press, each switch controls a solenoid valve supplying different flow volumes, often in the range of 190 liters/min (50 gal/min), 380 liters/min (100 gal/min) and 760 liters/min (200 gal/min). Any combination may be used up to a maximum of 1,135 liters/min (300 gal/min). With all switches off, a minimal flow of around 140 liters/min (37 gal/min) is still maintained.

FIGURE 10.5.7 | Quenchant circulation in a typical press quench[8]

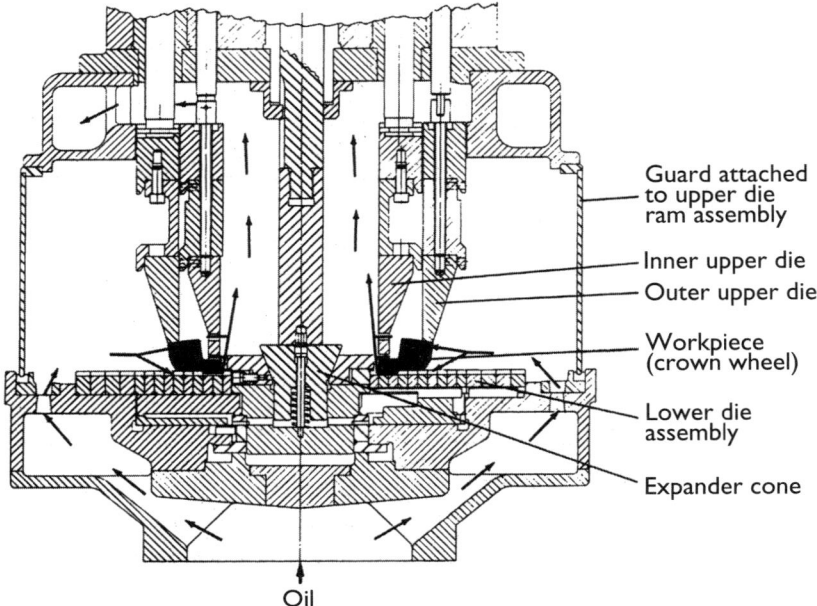

A circulation path within the press is created as the quenchant is pumped into the quench chamber through apertures around the outside diameter or through holes in the lower die, fills the chamber around the component and flows out at the top. The elongated exit apertures may be fully open or closed to restrict flow, depending on requirements. Timers, used in conjunction with flow selector switches, provide control of the duration of flow as well as volume, allowing a variety of flow/duration combinations.

Quality-Assurance Considerations

In addition to the actual press-quenching operation itself, there are several ways to assist in maintaining the parts as distortion-free as possible. Component design and manufacturing methods are the most critical in minimizing distortion, but other factors are also important. In engineering, the design should consider distortion from the initial concept through all phases of production processing, including material considerations (type and source). For example, the forging process should be designed so that material flow minimizes stress patterns. Forgings should be normalized above the austenitizing temperature, and a stress relief at subcritical temperatures is needed after machining.

Part Design

It is important in part design to avoid excessive or abrupt section changes that promote unequal heating and cooling rates. Fixturing and part support at temperature should also be considered during the design phase because creep from poor support is another major cause of distortion. The machinery employed in pressure quenching can normally hold twice the machining tolerances required of one area in its relationship to another. For example, if the relationship between two surfaces on a single plane is ± 0.051 mm (0.002 inch), then the best one can expect to attain after quenching is ± 0.102 mm (0.004 inch).

Part Sampling

It is always a good idea to run a sample lot of parts prior to general production in press quenching. In this way, the degree and nature of the dimensional changes can be observed. Possible changes in manufacturing operations can then be determined. Sampling also enables one to tell if specialized tooling is needed in conjunction with the press-quench machine for size control. Finally, metallurgical checks will allow determination of the optimized part microstructure as a function of setup parameters.

Record Keeping

Press-quench die sets and associated components (plugs, expanders, rolls, etc.) should be marked in a suitable manner to identify them for use with certain part numbers. The press-quench setup data should be stored in a computer database or kept as a written record. This includes: die numbers, plug numbers, quench time, flow rates, ram and expander pressure, pulse (on/off) and other pertinent data.

Summary

The management of distortion is a major driver in the heat-treatment industry. In addition to understanding the process variables, techniques such as martempering, marquenching, hot-oil quenching and press/plug/roll quenching can improve part distortion and minimize rejects when applied properly.

REFERENCES

1. Bergstrom, C M., L-E. Larsson and T. Lewin, "Measurements for Reducing Distortions Induced by Case Hardening," *Hart.-Tech. Mitt.* 43.1 (1988), pp. 36-40.
2. Professor Richard D. Sisson Jr., Worcester Polytechnic Institute, private correspondence
3. Krause, G., *Steels: Heat Treatment and Processing Principles*, ASM International, 1990

4. Hampshire, J.M., "User Experience of Hot Oil Quenching," *Heat Treatment of Metals*, 1984.1, pp. 15-20
5. Conference Proceedings, ASM International 1st International Conference on Quenching and Control of Distortion, Chicago, September 1993
6. Brennan, R.J., "How to Use Martempering Oils for Control of Part Distortion," *Industrial Heating*, January 1993
7. Herring, Daniel H., "Secrets of Effective Hot Oil Quenching," *Industrial Heating*, April 2006
8. Hewitt, G., "The Mechanics of Press Quenching," *Heat Treatment of Metals*, 1979, pp. 88-91
9. Tristano, Michael, "Considerations in Press Quenching of Gears," *Industrial Heating*, July 1981
10. Jones, L.E., "The Fundamentals of Gear Press Quenching," *Gear Technology*, March/April 1994
11. Herring, Daniel H., "Fundamentals of Press Quenching," *Industrial Heating*, December 2006
12. Dr. D. Scott Mackenzie, Houghton International (houghtonintl.com), technical and editorial contributions, private correspondence
13. Dr. George E. Totten, G.E. Totten & Associates (getottenassociates.com), technical and editorial contributions, private correspondence

10.6

CONSIDERATIONS IN QUENCH-TANK DESIGN

The choice of quench tank (type, design, size) and the methods for proper control of quenching variables (e.g., temperatures and flow rates) are a function of many factors, including material, part geometry (and mass), desired properties and the required production rate of the product.

When selecting an oil-quench process, factors include:
- Material – chemistry, hardenability, grain size
- Grids, baskets and fixtures – material and design
- Load configuration – part spacing, orientation, arrangement (load density), etc.
- Load weight (gross, net)
- Maximum quench fixture size, weight, shape
- Part geometry and mass (maximum and minimum part section thickness – whether the component part is uniform in thickness or has thin and thick sections)
- Prior (manufacturing) operations/stress state of the part
- Oil type (quenching characteristics, cooling-curve data)
- Oil speed and viscosity
- Oil temperature
- (Effective) quench-tank volume
- Height of oil over load
- Quench-tank design factors
  - Agitation method (agitators or pumps)
  - Number of agitators or pumps
  - Location of agitators or pumps
  - Size of agitators or pumps

- Type of agitators (fixed, two-speed or variable-speed)
- Propeller size (diameter, clearance in draft tube)
- Internal tank baffling (draft tubes, directional flow vanes, etc.)
- Flow direction
- Flow restrictions (quench elevator and baffling design)
- Volume of oil
- Maximum (design) temperature rise
- Heat-exchanger type, size, heat removal rate (instantaneous and total demand)

❖ Flow velocity through load
❖ Number of furnaces to be served by the quench system
❖ Duty cycle (i.e., the frequency of quenching or time between quenches)
❖ Post-heat-treat operations (if any)

For free quenching, the process may be manual (quenching of individual parts), batch or continuous, interrupted, or directed (i.e., impingement) quenching. In interrupted quenching, the load is pulled up out of the quenchant momentarily to check part temperature so as to determine if additional quenching time is necessary. If it is, the part is lowered back into the bath. In directed quenching, impingement on the part surface by the forceful movement of the quenchant is used. Fixture quenching (Section 10.5) involves restricting the part by some type of mechanical means and forcing quenchant over the part surfaces in an attempt to improve dimensional control.

Quenching of Steel into Oil

Of the variables mentioned, some that affect oil quenching of iron and steel components deserve a more detailed explanation.

QUENCH AGITATION[2,3]

Quench-tank agitation can be provided by various methods, including recirculation pumps, submerged spray, impeller mixers, ultrasonic, air or nitrogen sparge, actual movement of the part itself, or a combination of these methods. Ultrasonic tends to be expensive; pumps often suffer from problems of non-uniform agitation throughout the tank; and air or nitrogen sparge is not generally recommended because it promotes oxidation and may present a safety risk (e.g., quench-tank fires).

Mixers provide good fluid movement but are strongly dependent on the volume of fluid they can turn over and, as such, are related to impeller type and flow pattern. Airfoil-type impellers and marine propellers are commonly used for top-entering, angle-entering and side-entering mixer designs (Fig. 10.6.1).

A streamlined volume pattern can be observed above an open, high-speed impeller in a baffled quench-tank arrangement compared to relatively quiescent or stagnant mixing zones near the bottom of the quench tank. By contrast, the use of a draft tube in a down-flowing system produces a strong flow pattern that will sweep fluid away from the draft tube and the bottom of the tank. Compared to the same system with an open mixer, it results in a significant improvement in the spatial uniformity of flow, especially close to the surface (Fig. 10.6.2).

Draft-tube design (Fig. 10.6.3) is important to the overall performance of the system and should have the following basic characteristics:[2]

- ❖ A down-pumping flow path (to take advantage of the tank bottom)
- ❖ An angle of 30° on the entrance flare (to minimize head loss and establish a uniform velocity profile at the inlet)
- ❖ Liquid coverage over the top of the draft tube of at least one-half of the tube diameter (to avoid flow restriction and disruption of the inlet velocity profile)
- ❖ Anti-cavitation or internal-flow straightening vanes (used to prevent fluid swirl)
- ❖ Proper impeller positioning – both insertion depth into the draft tube (a distance equal to at least one-half of the tube diameter as dictated by the required inlet velocity profile) and diameter (fitting tight enough to prevent fluid flow along the sides of the draft tube)
- ❖ Anti-deflection vanes to compensate for irregular oil movement

FIGURE 10.6.1 | Impeller arrangements and resultant flow patterns[2]

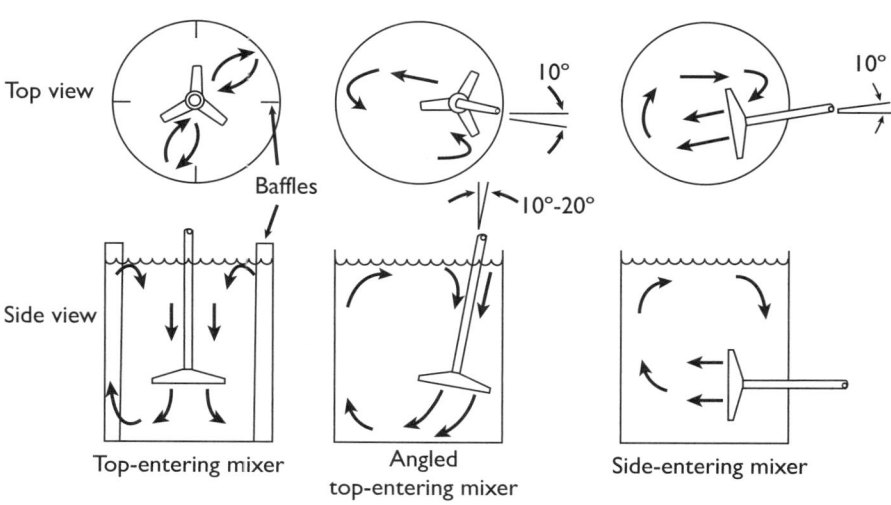

FIGURE 10.6.2 | Fluid-flow patterns comparing open and draft-tube systems[2]

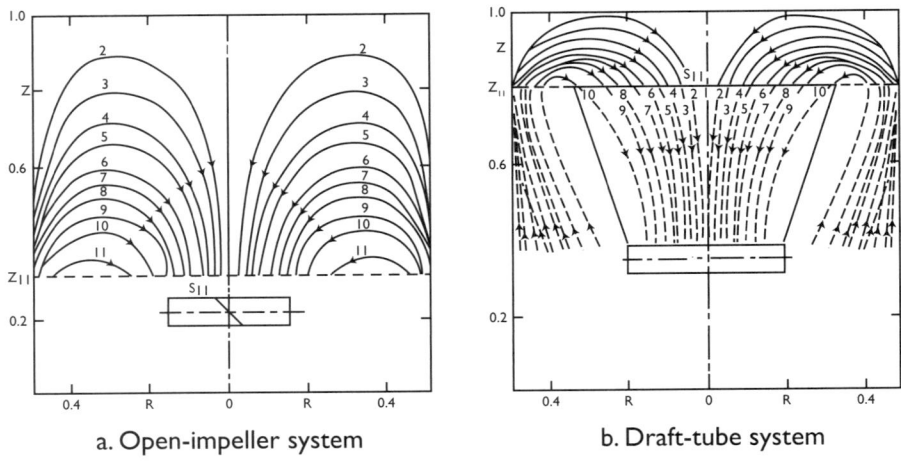

a. Open-impeller system   b. Draft-tube system

FIGURE 10.6.3 | Draft-tube impeller system design[3]

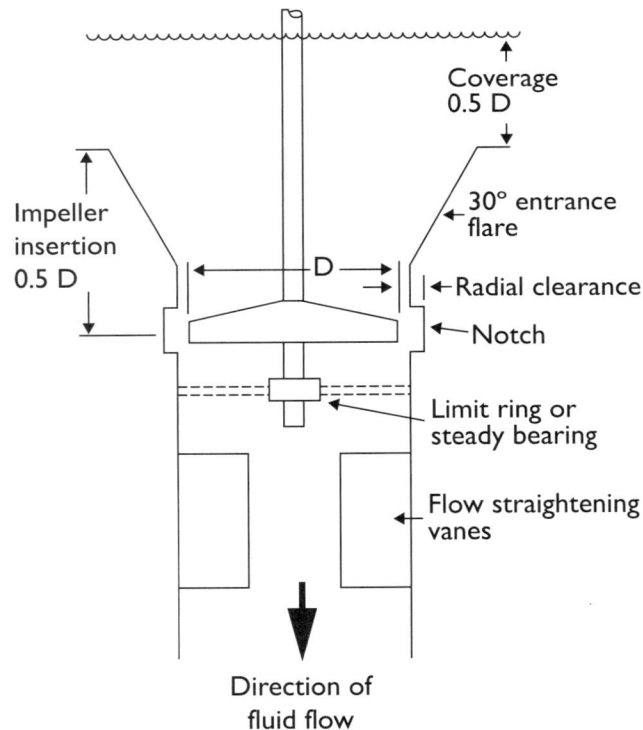

EFFECT OF INCREASING BATH TEMPERATURE

The temperature of the quenchant can have a dramatic influence on the rate of part cooling. Increasing the bath temperature from 21 to 120°C (70 to 250°F) produces the following effects (Section 10.2): slightly faster cooling in stage 1 because the viscosity of the oil decreases; cooling that is only slightly increased in stage 2; and cooling that decreases near the end of the quench in stage 3 because the temperature differential between the bath and the steel is decreased. However, many oils with robust speed improvers are less likely to show dramatic differences in quench speed.

Quenching oils have different characteristics that allow for varying cooling rates as a function of not only their boiling point but also their temperature. This has a direct bearing on properties such as viscosity, conductivity and heat rejection (based on the log mean temperature differential, or LMTD) of the heat-exchange system. Bubble size as well as the conductivity of the vapor barrier in all three stages of cooling are (to a degree) selectable by oil choice.

For most quench oils (other than marquench), the optimum rates of cooling are normally obtained when the bath temperature is between 50-65°C (120-150°F). Properly refined mineral oils are indefinitely stable in this temperature range, and the effect of viscosity is drastically reduced. Various manufacturers usually have an optimum temperature range for their product. The instantaneous rate of rise of the entire quench bath is also important. This is normally design-dependent but usually averages between -7 and 4.5°C (20-40°F).

EFFECT OF INCREASING DEGREE OF AGITATION

Even with a properly selected oil and correct quench-environment design, the stages of cooling previously described will not occur at all points on a part configuration at the same time. Part cooling is a function of the quality of the oil as well as the part's geometry, fixturing and loading techniques. Based on the surface configuration, differences in heat rejection will vary as the part's internal heat moves to the surface. Consequently, the uniform and controllable agitation of clean, properly controlled liquid flow over the part surface is imperative.

Controlled movement of the quenching liquid is vital because it causes an earlier mechanical disruption of the vapor blanket in the first stage and produces smaller, more frequently detached vapor bubbles during the vapor-transport cooling stage. Agitation constantly provides a cooler liquid to the part surface. This affords a greater temperature difference, which allows for improved heat rejection.

Cleanliness of the oil is also important. The oil must be free of particulate materials such as carbon, sludge and water. Carbon is formed after evaporation and fractionation under conditions of insufficient oxygen, or it is introduced by processes such as carburization. Oil breakdown on the part surface may occur if sufficient quenchant agitation is not provided. High levels of fine particulates like soot can decrease the presence of the vapor phase, which can vanish completely in extreme cases.

EFFECT OF QUENCH-TANK DESIGN

Quench-tank design itself (Figs. 10.6.4-10.6.7) is critical to quenching success. Draft tubes are just one component that highlights an often-overlooked aspect of quenching. The limitations imposed by the design of the quench tank can have a significant (negative) effect on the ability of the quench oil, or any quench medium, to perform properly. The volume of oil, localized instantaneous temperature rise of the bath, ability to circulate the quench medium through the load (measured in feet/sec or m/sec), capacity of the heat-exchanger system and overall maintenance of the tank all influence quenching.

A rule of thumb for good quench-tank design (Section 10.2) is that the velocity of the oil up through a workload should be in the range of 0.5-1.0 m/sec (1.65-3.3 feet/sec). While this is often difficult to measure with a workload in place, it can be modeled. In some instances, especially when quenching massive parts where reheating of the surface by heat from the core must be taken into consideration, this value has been increased to as high as 5 m/sec (16.4 feet/sec).

FIGURE 10.6.4 | Pit quench-tank design: top view; 91,000-liter (24,000-gallon) quench tank (courtesy of Aichelin Heat Treatment Systems)

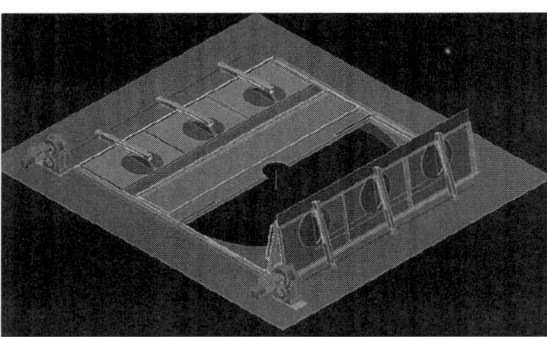
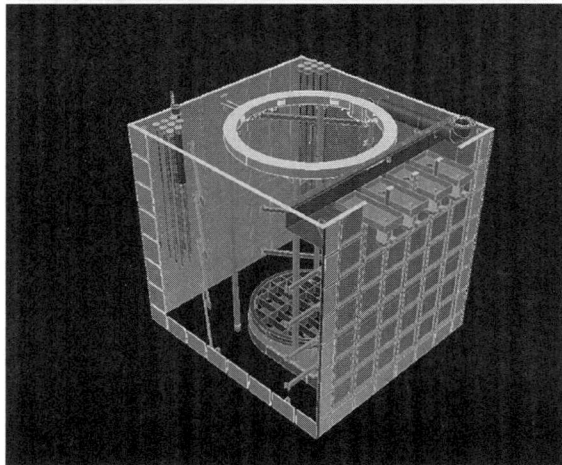

10 | QUENCHING AND QUENCHANTS

FIGURE 10.6.5 | Pit quench-tank design: side view; 91,000-liter (24,000-gallon) quench tank (courtesy of Aichelin Heat Treatment Systems)

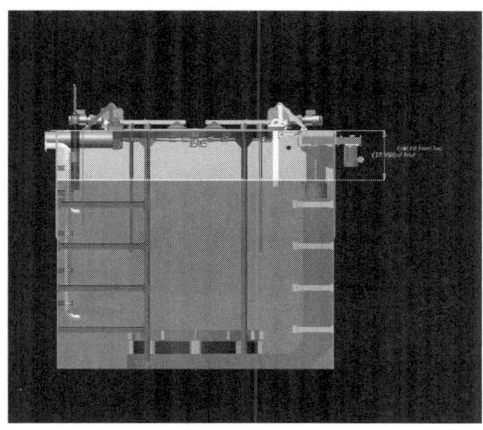

FIGURE 10.6.6 | Pit quench-tank design: draft-tube assembly; 91,000-liter (24,000-gallon) quench tank (courtesy of Aichelin Heat Treatment Systems)

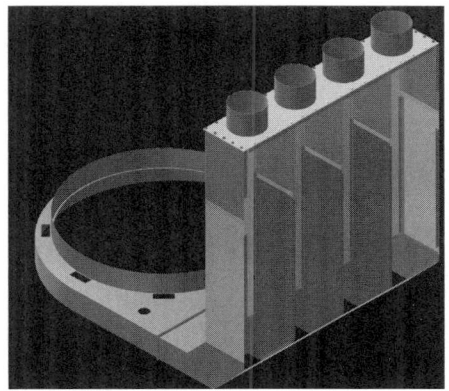

FIGURE 10.6.7 | Pit quench-tank design: deflector vanes under workload; 91,000-liter (24,000-gallon) quench tank (courtesy of Aichelin Heat Treatment Systems)

DISTORTION AND CRACKING

Another important advantage of oil quenching is that it minimizes the tendency to cause distortion and cracking. While better than some mediums (e.g., water or brine), other mediums (such as salt or high-pressure gas quenching) tend to produce less overall distortion. A key consideration, however, is the uniformity/repeatability of the distortion profile, and oil quenching has the ability to produce a very consistent profile.

Fast oils that are highly agitated tend to produce the highest rates of distortion, while slow (marquench) oils tend to minimize distortion. Quenching in still (non-agitated) oil at the beginning of the quench cycle (usually for the first 1-2 minutes) is an often-overlooked technique of distortion control on high-hardenability or sensitive-geometry parts. One method is to use high agitation at the beginning of the cycle followed by low agitation rates toward the end of the cycle. This helps achieve properties at the beginning of the quench while slowing down the rate of the quench as the steel approaches the martensite transformation temperature, thus minimizing thermal gradients.

WATER IN QUENCH OIL

One of the concerns regarding oil quenching is the presence of water in the quench oil. It is dangerous since it will form steam on quenching, resulting in a volume expansion. As the steam bubble rises out of the quench tank, its surface is coated with oil. As it exits from the furnace (usually under extremely high pressure), it is readily ignited. Water contents as low as 600 ppm can cause problems with the formation of soft spots and cracking (often at the same time). The presence of water can lead to the formation of non-martensitic transformation products (Section 2.12) due to an extended vapor phase. Water contents in oil above 1,000 ppm (0.1%) can result in serious risk of fire or explosions (Section 14.4), especially in closed systems such as integral-quench furnaces.

Quenching of Aluminum into Polymer[4,5]

Quenching of aluminum is accomplished by immersing the part(s) into the quench bath. The two most common types of quenchants used are water and polyalkylene glycol, or PAG (Section 10.2). While there are many factors that impact the successful design of a polymer quench system, some of the more important are:

- ❖ Material – chemistry, form
- ❖ Grids, baskets and fixtures – material and design
- ❖ Load configuration – part spacing, orientation, arrangement (load density), etc.
- ❖ Load weight (gross, net)

- Maximum quench fixture size, weight, shape
- Part geometry and mass (maximum and minimum part section thickness – whether the component part is uniform in thickness or has thin and thick sections)
- Prior (manufacturing) operations/stress state of the part
- Polymer type (quench characteristics, heat-transfer rate, cooling-curve data)
- Polymer concentration and separation methods
- Polymer temperature
- Volume of polymer
- (Effective) quench-tank volume
- Height of polymer over load
- Agitation method
- Part racking and baskets
- Transfer speed into the quench
- Quench delay time
- Cooling/heating method of the tank
- Polymer maintenance
- Quench-tank design factors
  - Agitation method (agitators or pumps)
  - Agitation rate
  - Number of agitators or pumps
  - Location of agitators or pumps
  - Size of agitators or pumps
  - Type of agitators (fixed, two-speed or variable-speed)
  - Propeller size (diameter, clearance in draft tube)
  - Internal tank baffling (draft tubes, directional flow vanes, etc.)
  - Flow direction
  - Flow restrictions (quench elevator and baffling design)
  - Volume of polymer
  - Maximum (design) temperature rise
  - Cooling/heating method
  - Heat-exchanger type, size, heat removal rate (instantaneous and total demand)
- Flow velocity through load
- Number of furnaces to be served by the quench system
- Duty cycle (i.e., the frequency of quenching or time between quenches)
- Post-heat-treat operations (if any)

QUENCH FLOW

Quench flow is a process variable that has a strong impact on quenching. It is either measured empirically or modeled, and the maximum flow that should be specified for aluminum batch quenching with water or PAG is on the order of 24-36 cm/sec (0.8-1.2 feet/sec) across the parts. This flow rate may not be achievable, however, especially in large tanks where measured flows in the area of 7-12 cm/sec (0.25-0.40 feet/sec) have been used successfully.[4] As pointed out previously, the overall flow rates are not as important as the uniformity of the flow. Aluminum alloys are sensitive to variations in flow rate due to hydrostatic pressure differences, as well as variations in the heat transfer. This contributes to large thermal gradients, which can cause distortion.

The type of agitation (pump or propeller) should be considered as well. Pumping is versatile and minimizes tank space (sparger pipes, eductors and nozzles can be tucked close to the sidewall or bottom of the tank). However, pumping has relatively low efficiency per liter (gallon) of quench moved compared to other types of agitation devices. The use of an eductor can significantly increase the amount of quench moved inside the tank. The volume goes up by a factor of four, and the velocity goes down with the same factor. The overall effect is that the flow generated is normally sufficient for a proper quench.

Propeller agitation is divided between open-type and inside draft tubes. Open-type propellers are normally located on one or both sides of the workload. They are typically marine-type propellers. The swirling action of the quench when it leaves the propeller tip generates a good nonlinear flow, but it can be uneven and affect as-quenched properties. In many cases, airfoil-style propellers (Table 10.6.1) can avoid this problem by spinning at a fast rate.

TABLE 10.6.1 | Comparison of airfoil-type draft tubes to pump horsepower for different volumes of water[4]

Propeller agitation				Pump agitation		
Airfoil type	HP	RPM	L/m (gpm)	Type	kPa (psi)	HP
13.5	5.5	810	21,200 (5,600)	End suction	138 (20)	75.5
13.5	2.0	520	12,100 (3,200)	End suction	138 (20)	42.6
13.5	1.0	426	11,165 (2,950)	End suction	138 (20)	39.0

Draft tubes are widely used in large open-tank systems. The draft tube consists of an impeller or propeller placed inside a tube. The clearance inside the tube allows for a more controlled and predictable flow. The distance from the fluid to the entrance of the draft tube must be such that cavitation does not occur (in some designs, anti-cavitation baffles are used). Otherwise, foaming and frothing of the polymer will occur, creating bubbles in the quench.

Quenching of Steel into Polymer[6]

Polymer quenchants can also be used in batch integral-quench (Fig. 10.6.8) and continuous (Fig. 10.6.9) furnace systems for ferrous parts. For batch integral-quench systems, a tight inner-door seal is mandatory to prevent water vapor from entering the heating chamber and disrupting the furnace atmosphere dew point. It is also necessary to purge the quench vestibule with nitrogen to prevent entrainment of air into the heating chamber or to prevent an atmosphere collapse that can contribute to an internal explosion in the unit. There have been cases reported of doors blowing off the furnace due to this phenomenon. For these and other reasons, polymer quenching in integral-quench furnaces is not a popular option.

For continuous systems, adequate agitation, new polymer availability in the chute area and a properly designed eductor are critical design features to prevent polymer separation (due to elevated temperature within the quench chute) and negate the effects of water vapor rising up the chute.

FIGURE 10.6.8 | Oil circulation pattern in a batch-style integral-quench furnace[6]

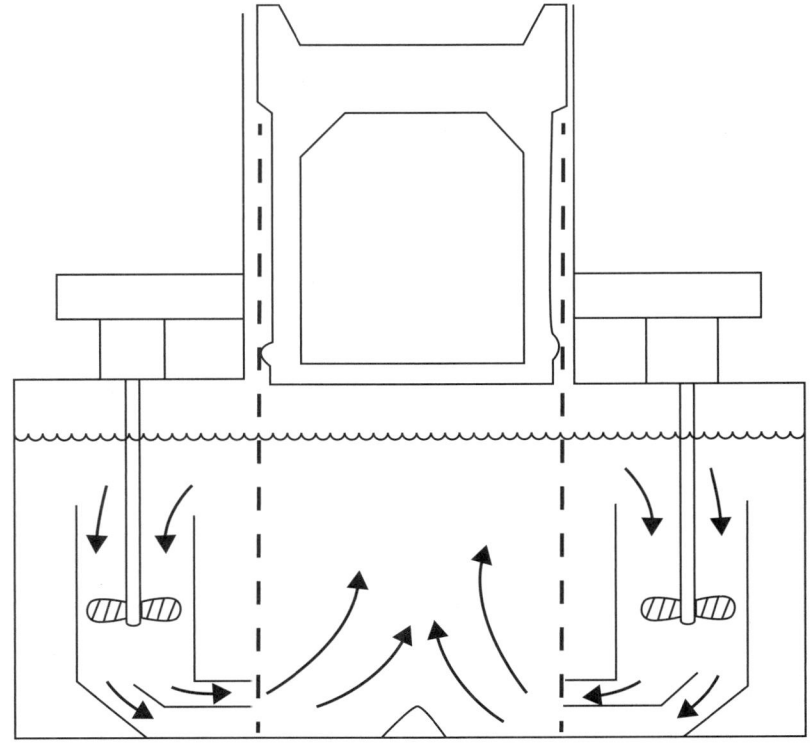

FIGURE 10.6.9 | Quench-chute design for use with continuous furnaces[6]

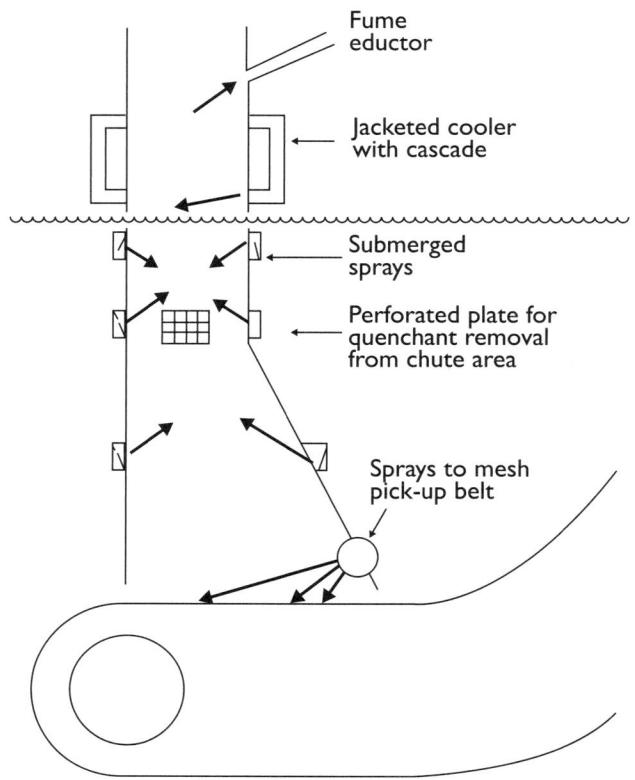

Summary

The heat treater should now have a basic understanding of the factors that influence quench-tank design and how it has a profound effect on the resultant properties and our ability to manage distortion.

REFERENCES
1. Herring, Daniel H., *Atmosphere Heat Treatment, Volume I,* BNP Media, 2014
2. Totten, G.E. and K.S. Lally, "Proper Agitation Dictates Quench Success, Part 1," *Heat Treating,* October 1992
3. Totten, G.E. and K.S. Lally, "Proper Agitation Dictates Quench Success, Part 2," *Heat Treating,* September 1992
4. Mackenzie, D. Scott, "Design of Quench Systems for Aluminum Heat Treating Part I – Quenchant Selection," *Industrial Heating,* June 2006
5. Mackenzie, D. Scott, "Design of Quench Systems for Aluminum Heat Treating Part II – Agitation," *Industrial Heating,* September 2006

6. Totten, G.E. and L.M Jarvis, "How to Effectively Use Polymer Quenchants," *Industrial Heating*, October 1991
7. *Handbook of Quenchants and Quenching Technology*, G.E. Totten, C.E. Bates, and N.A. Clinton (Eds.), ASM International, 1993
8. *Steel Heat Treatment Handbook, 2nd Edition*, George E. Totten (Ed.), CRC Press, 2007
9. Herring, Daniel H., "A Review of Factors Affecting Distortion in Quenching," *Heat Treating Progress*, December 2002
10. Herring, Daniel H., *Industrial Heating* Quenching webinar, September 2010
11. Dr. D. Scott Mackenzie, Houghton International (houghtonintl.com), technical and editorial contributions, private correspondence
12. Dr. George E. Totten, G.E. Totten & Associates (getottenassociates.com), technical and editorial contributions, private correspondence

Need a Retort Pit Furnace? Then You Need

AICHELIN
Heat Treatment Systems

For many heat treatment professionals, the ability to handle added flexibility with varying work pieces and part geometry is critical to the daily work schedule.

We Can Help You!

Aichelin's Vertical Retort Pit Furnace represents the beneficial out growth of years of practical experience and design within this highly demanding field of application.

AICHELIN HTS, Inc.
44160 Plymouth Oaks Blvd., Plymouth, MI 48170
(734) 459-9850

sales@aichelinusa.com | www.aichelinusa.com | www.ema-indutec.de

11 | HEAT-TREATMENT SPECIFICATIONS, COMPLIANCE

11.1

THE ROLE OF TRAINING

The importance of training in the heat-treatment industry cannot be understated. Training is imperative for anyone involved with heat treatment – from company presidents to shop-floor personnel. Training increases understanding at all levels, especially for those in the heat-treat shop. Understanding heat treatment is not only helpful for processing parts, it is useful for other activities as well, such as quoting jobs.

A number of companies and organizations offer various types of training programs to ensure that heat-treat operators, supervisors, maintenance and quality personnel, engineers, managers and just about everyone who is involved in heat treatment knows and understands the subject. Companies want their employees to know as much as possible about what their company does, what products they produce and how these products are manufactured. The heat-treatment process is no exception.

Most training programs are (or should be) designed to meet the requirements of a formal structured program, such as SAE-ARP-1962 (Training and Approval of Heat-Treating Personnel). This document calls out both classroom instruction and OJT (on-the-job training) for heat treaters. One of the most misunderstood aspects of this training program is that the company can customize it to fit their specific training needs. If you haven't designed a training program specifically for your organization, you should consider doing so.

A key element to the success of any training program (Fig. 11.1.1) is to understand how to balance classroom instruction with hands-on shop-floor training by someone skilled in the types of processes and equipment being run.

How We Learn

It is important to recognize that learning continues over one's lifetime and is not just a one-time event. Different people learn in different ways – through books, in formal (classroom) settings, by word of mouth, by reading trade journals (such as *Industrial Heating*), by watching others, from the Internet, by participating in training programs, by trial and error, and by being shown how (and, hopefully, why) something works. This last method is an important element to any training program for heat treatment and what we call hands-on training. This type of training has been found to be highly effective in shortening the learning curve and in breaking bad habits. In addition, recipients gain practical experience on the furnace systems relevant to them.

FIGURE 11.1.1 | Training options[1]

Repetition, understanding and knowing how and why something works is critical. Then practicing it over and over again helps many of us learn. This is especially true in the heat-treat shop. The more an operator understands the process of heat treatment, the easier the training is for that employee. If we can explain the process and then show our operators how to accomplish it, they have a better understanding of what is right and what is wrong.

The Value of Training

In-house training saves time and money for the company that supports it. The travel time and expense of sending multiple employees off-site can be very costly. Also, you can break the training into shorter periods of time to help keep the audience engaged. Most in-house training can be geared more to what the company does instead of the industry as a whole. The trainer can then concentrate on specific subjects/concerns when they are addressing a group from the same company or organization.

The payback for training is tremendous and includes benefits such as:
- Immediate reinforcement of newly acquired knowledge
- Linking of theory and practice in a natural way
- Deeper understanding of the intricacies of heat treating
- A more satisfied and responsible heat-treat operator
- Individuals who see their company investing in them for the future
- A more confident management team due to the investment in personnel training
- More confident employees that are more willing to take on new responsibilities and offer positive suggestions for continuous improvement
- Customers who gain confidence in the company's ability to provide a quality product or a solid professional service
- The satisfaction and, more importantly, the motivation gained by knowing that things are done correctly

The Need for Continual Reinforcement

It is essential that *both* initial and annual reinforcement training be done. Too often companies train individuals and feel their obligation is complete. It is well known in education circles that a student must be exposed to the same subject matter (in slightly different forms) a minimum of 4-6 times before they gain a thorough understanding of the subject. An instructor knows that they must present a subject at least three times themselves before he or she fully understands it! So, how can a student only be exposed once? New hires must also go through both basic and advanced training in order to be successful.

Finally, classroom training alone cannot prepare an employee for what it is like in the heat-treatment shop. On-the-job training, done right, is a vital component in the overall training methodology. There are many different types of controls, equipment and products involved in heat treatment. OJT allows the operator to learn the specifics of what they are assigned to do. It also allows the operator to become familiar with the different products that are being heat treated at his/her facility. In aluminum heat treatment, for example, air space, racking/fixturing and distortion are all factors that must be considered while on the job. Just understanding that these are important is not enough without the experience of day-to-day activities.

Training Course Outline for Atmosphere Heat Treatment

Atmosphere heat treatment provides one example of a basic in-house training program that combines classroom instruction with training on the shop floor. Subjects include (in broad terms): fundamentals of heat treatment; principles of furnace atmospheres; and atmosphere process, equipment and applications. Here's a typical course outline.

A. Introduction to Heat Treatment
1. Learning objectives
2. What is heat treating, and why do we do it?
3. Ferrous metallurgy
4. Common heat-treating processes
5. Understanding the language of heat treating
6. Interactive discussion (question-and-answer session)
7. What we have learned

B. Furnace Atmospheres and Atmosphere Safety
1. Learning objectives
2. Introduction to furnace atmospheres
    a. Role and purpose of furnace atmospheres
    b. Common furnace atmospheres
    c. Volume requirements
    d. Classification of generated atmospheres
3. Control of furnace atmospheres
    a. Methods of control (dew point, infrared three-gas analysis, oxygen probe)
    b. Verification of control (tests and correlation)
4. Understanding gas generators
    a. Key components and their function
    b. Operational issues
    c. Control issues
    d. Maintenance
5. Applications of endothermic gas
    a. Batch-style furnaces
    b. Continuous-style furnaces
6. Understanding synthetic atmospheres
    a. Pure atmospheres
    b. Gas blends
    c. Nitrogen/methanol
7. Atmosphere safety
    a. Gas characteristics
    b. Rules and regulations
    c. Operator and maintenance considerations

8. Interactive discussion (question-and-answer session)
9. What we have learned

C. Process Applications
1. Learning objectives
2. Importance of cleaning
3. Steel heat treatment
    a. Hardening
    b. Case hardening
    c. Tempering
    d. Special processes
4. Quenching and managing distortion
    a. Quenchants
    b. Types and design of quench tanks
5. Interactive discussion (question-and-answer session)
6. What we have learned

D. Equipment Design Considerations
1. Learning objectives
2. Key features of various types of atmosphere furnaces
    a. Heating chambers
    b. Quench chambers
    c. Transfer mechanisms (if applicable, internal and external)
3. Interactive discussion (question-and-answer session)
4. What we have learned

E. Maintenance Considerations – Atmosphere Equipment
1. Learning objectives
2. Setting up a planned preventive-maintenance program
3. Critical components
    a. Establishing MTBF (mean time between failure)
4. Cause-and-effect relationship
5. Internal components
6. External components
7. Controls
8. Spare parts
    a. Original OEM vs. alternative sources
9. Documentation
10. Interactive discussion (question-and-answer session)
11. What we have learned

F. Miscellaneous
1. Industry standards and specifications (Nadcap, CQI-9, AMS, ASTM, etc.)
2. Interactive discussion (question-and-answer session)
3. What we have learned

G. Safety
1. Safety in the heat-treat shop
2. Atmosphere safety
3. Confined-entry spaces
4. Interactive discussion (question-and-answer session)
5. What we have learned

H. Management Overview
1. Management responsibilities
2. Lean heat treatment
3. The challenges ahead
4. Interactive discussion (question-and-answer session)
5. What we have learned

Training Course Outline for Aluminum Heat Treatment

Aluminum heat treatment is another common in-house training program that combines classroom instruction with training on the shop floor. Subjects include (in broad terms): heat-treatment fundamentals; understanding the properties and characteristics of aluminum (and other lightweight materials); and aluminum equipment, processing and specification compliance. Here's a typical course outline.

A. Fundamentals of Aluminum Heat Treatment
1. Materials science (an introduction)
2. What is heat treating, and why do we do it?
3. Introduction to nonferrous heat treatment
4. How parts heat
5. Interactive discussion (question-and-answer session)
6. What we have learned

B. Aluminum Metallurgy
1. Introduction to aluminum and other lightweight alloys
2. Strengthening mechanisms of aluminum
3. Wrought and cast aluminum alloys

4. Interactive discussion (question-and-answer session)
5. What we have learned

C. Heat Treatment of Aluminum and Aluminum Alloys
1. Aluminum heat-treatment overview
2. The heat treatment of cast-aluminum alloys
3. The heat treatment of wrought-aluminum alloys
4. Problems in the heat treatment of aluminum and aluminum alloys
5. Interactive discussion (question-and-answer session)
6. What we have learned

D. Aluminum Processes and Applications
1. Homogenizing
2. Annealing
3. Solution heat treatment
4. Quenching
5. Aging
6. Interactive discussion (question-and-answer session)
7. What we have learned

E. Equipment Overview
1. Furnace and oven design
2. Types of heat-treatment equipment for processing aluminum
3. Interactive discussion (question-and-answer session)
4. What we have learned

F. Quenching and Managing Distortion
1. Types of quenchants
2. Water and boiling-water quenching applications
3. Polymer quenching applications
4. Design and maintenance of quench baths and quench tanks
5. Interactive discussion (question-and-answer session)
6. What we have learned

G. Testing
1. Principles of mechanical testing
2. Rockwell and Rockwell-superficial hardness testing
3. Conductivity testing
4. Questions related to heat-treatment testing specifications for aluminum alloys

5. Interactive discussion (question-and-answer session)
6. What we have learned

H. Specification Compliance
1. The importance of specification compliance
2. A review of AMS 2770 (Heat Treatment of Wrought Aluminum Alloy Parts)
3. A review of AMS 2771 (Heat Treatment of Aluminum Alloy Castings)
4. A review of AMS 2750 (Pyrometry)
5. Other pertinent specifications (AMS, ASTM, ISO, etc.)
6. Interactive discussion (question-and-answer session)
7. What we have learned

I. Maintenance and Troubleshooting
1. Maintenance practices, procedures and tips
2. Maintenance of polymer quench tanks
3. The importance of airflow in ovens
4. Energy conservation
5. Interactive discussion (question-and-answer session)
6. What we have learned

J. Safety
1. Safety in the heat-treat shop
2. Confined-entry spaces
3. Interactive discussion (question-and-answer session)
4. What we have learned

Why In-House Training?

Valid and compelling arguments can be made for online training (or so-called distance learning), home-study courses and off-site training in a formal classroom environment. Each of these methods is important and necessary, but they should not take the place of in-house education that includes a hands-on element. Home-field advantage should not be underestimated. Individuals to be trained need to see that what they have learned can help them right away, on a day-in and day-out basis. Too often more formalized training materials sit on shelves and are sadly never looked at again.

In most organizations, OJT is the default training method, used in lieu of more formalized training. Traditional OJT has been found to be somewhat erratic in that it can produce inconsistent or unequal results from individual to individual. In the author's experience, one year of OJT is only equivalent to approximately 20 hours of classroom training.

OJT is limited by the skill and experience of the instructor and has been found to take up to twice as long as anticipated. It has been reported[4] that traditional OJT training could be replaced by a structured training approach. As a result, training time reduced by 72%, problem-solving ability increased by 130% and wasted time and effort reduced by 76%.

Summary

Training based on a "read it, see it, do it" philosophy combines sound scientific and engineering principles with practical real-world experience, and it should include plenty of practical examples. Individuals who follow this approach not only learn the subject but also understand how to apply it in their everyday jobs.

REFERENCES

1. Herring, Daniel H., "ARP-1962: The Need for In-Plant Training and Education Programs," *Industrial Heating*, November 2010
2. Herring, Daniel H., *Atmosphere Heat Treatment, Volume I*, BNP Media, 2014
3. SAE International (sae.org)
4. Sisson, Gary R., "Hands-On Training: A Simple and Effective Method for On-the-Job Training," 2001
5. Pye, David, "Heat Treatment and Metallurgical Training," The Experts Speak, *Industrial Heating* website, June 18, 2010
6. Hands-On-Training (handsontraining.org)
7. Nick Fajerson, Denison Industries, technical contributions and private correspondence

11.2

AUDITS

Audits of the heat-treat department, whether in a captive or commercial shop, are a critical part of any good quality program, either as part of a self-assessment (to determine where you are and where you need to be), an ISO program (to maintain or establish a total quality program), for compliance with an industry requirement (e.g., Nadcap, CQI-9, TS16949, AQ9100) or to qualify the department to another industry-established standard. Audits are also a vital tool in continuous-improvement (CI) programs.

A heat-treat audit can be looked at as an examination of a supplier and their actions or as a critical look at internal operations. Audits are a very big part of any heat-treat facility. They can be specific to a part, customer, process or facility. The actual audit can consist of a single form that may take only an hour to complete or multiple checklists that could take up to several days or a week to complete.

Why audit?

The goal of a heat-treat audit is to find variability. This can be equipment-induced or process-induced. Audits can also help one understand the current balance of productivity and quality in the department or shop (Fig. 11.2.1). One of the many objectives of such an audit is to ensure that the parts are processed in accordance with the applicable customer specifications. The fear most people or organizations have is that they will interrupt current production and that the implementation of audit findings will further interrupt workflow. Therefore, the department will not be able to produce the required production volumes with rigid quality standards in place. This fear is unfounded, especially in today's product-recall-sensitive world.

FIGURE 11.2.1 | Balance between productivity and quality

What are the objectives of a heat-treat audit?
The audit process is designed to ask and answer basic questions such as:
- ❖ Who is performing the heat treatment, and are they competent?
- ❖ What procedures are being used to carry out the heat-treating operation?
- ❖ Is the audit adequate to ensure proper quality?
- ❖ Where is the work being done, and is the shop itself capable of performing the required task(s)?
- ❖ When was the last assessment, and was it representative of current practice?
- ❖ Why is an assessment required now? In other words, did a quality issue or problem trigger the event, and, if so, will the audit help solve it?
- ❖ How will the audit be performed? Are the right personnel in place to reach meaningful conclusions?

What constitutes a good audit?
Surprisingly, most audits that fail do so because they do not reveal the true nature of what is happening within the heat-treatment shop or department. By contrast, successful audits find compliance with respect to:
- ❖ Quality (forms, instructions, records)
- ❖ Performance (process or equipment variability)
- ❖ Testing (methods and procedures)
- ❖ Specifications (AMS 2750, etc.)
- ❖ Training (SAE-ARP-1962, etc.)

WHAT DOES A HEAT-TREAT AUDIT LOOK FOR?
In general, heat-treat audits look at people, methods, equipment and testing procedures to measure the overall success of a department's quality. Specific

areas of focus include:
- General facts (company/department profile)
- Capabilities (ability to meet customer requirements)
- Instructions (recipes, records)
- Sampling guidelines (testing methods and requirements)
- Continuous-improvement programs (plans, progress)
- Specific compliance (audit questions/answers)
- Nonconformance (major and minor deviations)
- Corrective actions (root-cause determination)

Audits can be stressful. They tie up resources, cost the company time and money (in some cases) and often bring to light issues that otherwise would remain hidden. A successful audit can boost morale and invigorate a company by opening doors to better opportunities.

Audit preparation is vital to success. You must make sure all your drawings and purchase orders are up to date, all specifications are up to date and available, any and all open corrective actions (both internally and externally) have been answered, and any approvals you may have from customers are available.

A self-audit is invaluable in preparing for a third-party audit because you will become aware of what the auditor is looking for and you will have the appropriate documentation available to present. Assume the auditor has no idea of what you do or how your company performs heat treatment. This will assist in preparing the necessary information. Make sure you have a clean, quiet workspace that will allow any and all people that may be involved with the audit to attend. Prepare to spend the allotted time with the auditor uninterrupted. Have access to all prints, forms, specifications, training records and maintenance records you may need for the auditor.

WHAT DOES AN AUDIT ACHIEVE?

Too often we are only concerned with the heat-treat department's capabilities either in terms of overall capability, services we can offer to our customers or the size/quantity of equipment in our shops. Instead, we need to focus on our limitations. We need to know what we can't do and set our sights on overcoming these limitations. This is precisely what a good audit achieves.

Most audits should be done in two distinct phases: a general-condition assessment and an equipment/process capabilities specific assessment. Each phase performs a different role.

Phase I: A general condition assessment asks questions such as:
- Where are we today?
- What is the true condition of our equipment?
- Are we monitoring and controlling the right variables?
- Are our maintenance practices adequate?
- Do our methods and control systems measure up?

Phase II: A process/equipment capabilities assessment looks into:
- Process recipes, cycle times and results
- Historical records (e.g., defective parts and corrective actions, types of product failures being experienced and root-cause determination, etc.)
- Safety and environmental conditions
- Customer-complaint resolution

What is the outcome of a heat-treat audit?

Simply stated, the end result of any heat-treat audit is to help establish a world-class heat-treatment department (Fig. 11.2.2), perhaps not in terms of all new equipment but in terms of producing the highest possible quality at the lowest possible cost.

FIGURE 11.2.2 | Example of a world-class heat-treat department (courtesy of Ipsen)

The audit process forces companies to examine their weaknesses and helps to build a better system. Companies can save time and money by using the audit process as a tool to reduce rework and spending and also to streamline processes to increase productivity and efficiency. By following an audit and digging deep into root-cause corrective actions (RCCA), companies can bring to light: deficiencies that may hide in the flow-down process from a customer to the shop floor, holes in the planning side of operations, tribal knowledge, issues in record keeping, and/or quality problems that are not apparent in day-to-day operations. All of these issues and others are faced by heat treaters and must be examined during the audit process.

11 | HEAT-TREATMENT SPECIFICATIONS, COMPLIANCE

What are the audit steps?

A typical audit looks to address the following issues (in sequential order) each time it is to be conducted.

- Schedule a kickoff meeting to discuss the audit objectives, timing, and report format and distribution. Be sure to include all necessary disciplines and assign an overall audit coordinator.
- Assess the readiness of the heat-treat operation, controls and systems in place for an audit.
- Test these systems to ensure proper operation.
- Discuss with management all preliminary observations.
- Draft realistic and achievable short-term and long-term corrective actions.
- Discuss with management the draft audit report and responses, if available, prior to release of the final audit report.
- Follow up on critical issues raised in audit reports to determine if they have been successfully resolved.

INDIVIDUAL JOB AUDITS

One of the best ways to understand if a general audit is necessary is to perform individual job audits. A job audit is a step-by-step review of all processes conducted on a given batch of parts to evaluate if they fully meet customer requirements. A good heat-treat job audit will cover items such as:

- Incoming part inspection and material verification
- Material specification review
- Component part drawing/heat-treating specification reviews
- Part fixturing/racking method
- Procedure for recipe creation and overall process cycle review
- Equipment selection/capability
- Process review (including atmosphere requirements)
- Testing requirements/methods/capabilities/certifications
- Shop paperwork (forms and record keeping)
- Operator and quality sampling/testing and acceptance criteria

There is also great value in conducting a thorough self-audit. To do so, one must be open-minded and honest. Every question must be examined and understood by the auditor. As the auditor conducts the self-audit, they must look for answers to each question and make sure there is documentation to prove each answer to an outside auditor. The most successful self-audit is often the one where knowledge of the audit process is known by the auditor but not the subject being audited. The auditor must then be educated on what is being done by answers and documents.

This brings discrepancies to light prior to the actual audit and allows time to fix the findings before an outside audit is performed. During a self-audit, it is often best to see the question from the viewpoint of floor personnel to the customer. If the floor personnel can access and answer all necessary information that a customer would ask, then you are off to a great start.

PRE-AUDITS

The value of a pre-audit should not be ignored. Whether it is done by an outside individual or agency or as part of a self-assessment, going through the audit process is an invaluable exercise. Be sure you are in conformance to written procedures.

Look at the operation. Just because something has been done the same way for years, do not assume it is correct. Remember, if it is called out in a specification – even if it may not apply to the particular situation – it must be documented. In other words, if it is stated in a document you are working on, it must be mentioned in your audit procedure, whether it applies or not.

Pre-audits are worth their weight in gold simply because they force everyone in the organization to think about their responses, question their procedures and look inward at the problems they face every day.

SAMPLE AUDIT REQUIREMENTS

Areas such as data management and operator/maintenance personnel training in accordance with SAE-ARP-1962 (Section 11.1) should not be overlooked. Critical questions need to be asked. Does the heat-treat equipment have the ability to display, achieve, store and retrieve critical process data? Does the company have a defined training program? Is reinforcement training done on an annual basis?

Finally, companies should consider independent verification of their auditing practices. Personnel conducting such audits would observe processes being run; review equipment and methods in use; talk with supervisors, managers and engineers about policies and procedures; collect evidence that activities are as claimed; and verify operations (including outputs) at the various steps.

Internal audits have broad-based objectives, including verification of items such as:
- Special procedures or tests
- Product characteristics and performance
- Implementation of procedures to match the requirements
- Defects or nonconformities (and if are they being addressed)
- Training, equipment capabilities and process settings

By comparison, external audits look to verify:
- ❖ All appropriate requirements are being followed, including supplier processes and calibration procedures, heat-treatment methods and maintenance activities
- ❖ Supplier product/service characteristics or performance measures
- ❖ Conformance to contract requirements
- ❖ All nonconformities have been addressed
- ❖ Material is as specified and all sources are traceable

AUDIT DETAILS

To provide maximum value, heat-treat audits need to ask realistic questions, not just be forms in which the auditor fills in the blanks. In other words, the true status of a heat-treat department or shop is revealed only in the details. Many departments or shops that are audited are hoping the auditor won't find what they know is wrong or won't look where they don't want them to look. The belief of many heat treaters is that they will correct the problems on their own, but it never seems like we find the time to do what we know we must. This is why we need the audit process.

Heat treaters should not be afraid to point out areas of deficiency and ensure that these are included rather than excluded from the audit. In other words, don't fear a major or minor nonconformance; embrace the need to fix what is not right.

It is critical that audits drill down to the level that the physical work is being done. A good heat-treat audit spends less time in the office than on the shop floor. If the metallurgical laboratory is to be included in the audit, it should not be the last area audited or only superficially looked at due to time constraints. Finally, auditors must reward well-run operations and not hesitate to give them top scores when it is deserved.

Here's a look at some of the critical information necessary.

A. **General (company/department profile)**
   1. Audit date, supplier's name and plant location
   2. Key contact information, including corporate contacts (if appropriate), plant manager, quality manager, metallurgist and shift supervisors
   3. Financial viability (to the extent appropriate to the audit goals)

B. **Capabilities (general requirements)**
   1. List all part numbers, cross-indexing them to their corresponding engineering drawings, specifications (including all testing requirements), and special needs (e.g., distortion and handling concerns, dimensional tolerances, etc.).

2. List the types of materials that the heat treater is qualified to run.
3. List the heat-treatment processes capable of being run. Be sure to tie each heat-treat process with the specific equipment involved by part number. Be sure to check that each process is under control as evidenced by quality records.
4. List each heat-treat cycle, including the type of quench. Be sure to identify all process and equipment variables involved.

C. **Instructions (for auditors)**
1. Clearly define what will be required in a heat-treat audit and communicate this information to the intended parties well in advance of the physical audit so that the necessary information can be gathered ahead of time.
2. Create a consistent and fair rating guideline, and adhere to the categories and questions selected.
3. Be sure that corrective actions and completion dates are agreed to by both parties and responsibilities are clearly delineated.
4. Follow up personally within the specified time frame.

D. **Sample rating guidelines (for audit questions)**
1. If a required activity is not being performed (rating = 0)
2. If there is only rudimentary activity (rating = 1) or if the activity is being performed and documented but has (minor) deficiencies (rating = 2)
3. If the activity is inadequate for the task required (rating = 3) or if the activity is properly documented but not properly performed (rating = 4)
4. If the activity is being adequately performed and documented (rating = 5); in addition, if there is evidence that the activity achieves the task(s) required (rating = 6)
5. If the activity is being well performed and well documented (rating = 7); in addition, if continuous improvement is evident (rating = 8)
6. If the activity is being performed and documented beyond expectation (rating = 9); in addition, if continuous improvement is realistically being achieved (rating = 10)

E. **Continuous improvement program (areas to review for evidence of)**
1. Good-better-best practices related to heat treatment and testing.
2. Process parameter variability is being controlled.
3. Equipment variability is being controlled.

11 | HEAT-TREATMENT SPECIFICATIONS, COMPLIANCE

   4. Laboratory best practices are being used.
   5. Scrap-reject-rework plans and procedures are being used.
   6. Documented planned preventive maintenance exists.

F. **Audit questions (typical)**
   1. Are part handling, processing and storage adequate to preserve product integrity and quality?
   2. Are adequate controls employed to ensure that the processing and inspection status of the product are known throughout the heat-treating operation, and are process/product monitoring and control functions (and responsibilities) clearly defined?
   3. Is the responsibility for and practice of heat-treat process (recipe) development, testing methods and quality planning clearly defined?
   4. Does the heat treater use a procedure for reviewing part design and specifications in relation to method of loading as well as heat-treat process parameters and equipment selection?
   5. Are process verification and/or capability studies conducted on all new part numbers?
   6. Are control plans and processes (FMEAs) used as a basis for establishing quality programs for heat-treat processes?
   7. What procedures are in place and how does the heat treater react to customer concerns (internal or external indicators)?
   8. Are controls in place and being used on the shop floor to effectively monitor the process?
   9. If necessary, are statistical process control (SPC) methods utilized for key product parameters?
   10. Are written procedures/work instructions defining heat-treat and quality functions available and in use on the shop floor? Is the quality manual a living document?
   11. Are adequate in-process monitoring and inspections/tests performed, and are there adequate records?
   12. If on-site, does the testing or metallurgical laboratory have the tools, procedures and expertise to accurately determine part quality? If off-site, is the testing laboratory properly accredited?
   13. If part testing and/or PPAPs are performed, are records available with supporting documentation for the relevant heat-treated products?
   14. Are documented and verifiable heat-treating reject, reprocessing and/or scrap records available?

15. Is there an effective preventive-maintenance program in place for both the heat-treating and process-monitoring equipment?
16. Does the heat treater have an effective system for ensuring the quality from his suppliers and service providers (instrumentation calibrations, quench-oil checks, etc.)?
17. Are plant cleanliness, housekeeping and environmental and working conditions conducive to a safe, efficient operation in which continuous improvement can take place?

G. Nonconformance (document in detail)
   1. Major and minor nonconformances
   2. Pertinent general and specific observations

H. Corrective actions (for each supplier location)
   1. Detailed statement of findings
   2. Corrective action(s) required
   3. Responsibility
   4. Implementation date
   5. Root cause found
   6. Follow-up plan (actions and dates)

AUDIT FOLLOW-UP

Don't forget to conduct a follow-up audit, typically within six months of the initial (internal or external) audit. This should be scheduled after the initial audit takes place. The purpose of this follow-up is to critically evaluate if corrective action has been taken on the issues in the original report.

Summary

"We're too busy" is the most common excuse for not performing an audit. However, it always seems there is enough time to do the task over again if it was not correct in the first place. Let's do it right the first time and every time. Perform an audit, find areas where improvement is needed and take the necessary corrective action(s). You'll be happy you did.

REFERENCES

1. Herring, Daniel H., "How to Conduct a Heat Treat Audit," *Industrial Heating*, September 2006
2. Hafeez, Arshad, "Demystifying Nadcap Heat-Treat Audits," *Industrial Heating*, April 2010
3. Russell, J.P., *Verification Auditing*, J.P. Russell & Associates, 2008
4. Nick Fajerson, Denison Industries, technical contributions and private correspondence

Specialized Heat Treating Since 1962

10008 Miller Way South Gate CA 90280
www.accuratesteeltreating.com

11.3

ACCREDITATION

Nadcap and CQI-9 are two of the most recognized accreditation programs in the heat-treatment industry. Nadcap is a global cooperative accreditation program for aerospace engineering, defense and related industries. CQI-9 addresses the automotive and associated industries. While not necessarily exclusive, most customers require one or both of these accreditations prior to conducting their own audit or issuing purchase orders for heat-treatment services.

Both accreditation processes involve different types of audits, but what they share in common is that they are very thorough. These audits are intended to improve heat-treatment operations by looking to analyze and find the root cause of any nonconformance. An experienced team member must handle these audits carefully because failing a Nadcap or CQI-9 audit can cause many issues with your customers. If successful, these audits can bring in additional clients.

Nadcap Accreditation

The goal of Nadcap is to help create an industry-managed, standardized approach to conformity assessment of so-called special processes (including heat treatment and testing) that brings together technical experts from prime contractors, suppliers and representatives from government to establish requirements for supplier approval. Unlike traditional third-party programs, Nadcap approval is granted based on industry consensus.

HOW DOES THE NADCAP PROCESS WORK?

The Performance Review Institute (PRI) administers the Nadcap program. PRI will schedule an audit and assign an approved auditor who will conduct the audit against an industry-agreed standard using an industry-agreed checklist. At the end of the audit, any nonconformity issues will be raised and nonconformance reports issued. PRI will administer closeout of these reports. Upon completion, PRI will present the completed audit package to a special-process task group made up of members from industry that will review it and vote on its acceptability for approval. Accreditation is granted when all nonconformance issues are closed.

WHAT IS A NADCAP TASK GROUP?

Nadcap task groups are composed of personnel (prime contractors, government representatives, suppliers) with expertise in the particular product, process or service for which suppliers are to be accredited (e.g., NDT, heat treating). The task group has responsibilities for the accreditation program that include deploying the accreditation process, recommendations for auditor hiring and training needs, standard and checklist development, metrics and continuous improvement.

WHAT WILL NADCAP DO FOR ME, AND WHAT ADDED VALUE DOES IT SERVE?

The Nadcap program offers the opportunity for suppliers to gain tighter controls of their special processes. Nadcap provides a forum for changing audit requirements and networking.

WHAT ARE THE QUALIFICATIONS TO BE AN AUDITOR?

Nadcap has prescribed requirements (as defined in PRI/Nadcap Procedure NIP-002 "Auditor Selection, Approval and Training") for auditors for general education, experience, professional accomplishments and technical specific requirements. Each task group defines their specific criteria for approval. The Nadcap auditor base averages over 30 years of experience in the aerospace industry with a high degree of technical expertise in areas such as heat treatment, testing laboratories, welding, etc.

NADCAP PROCESS DETAILS

Audit checklists from PRI to the heat treater for preparation simplify the actual Nadcap process. PRI, however, recommends that a thorough self-audit using the audit checklists be performed to allow for adequate preparation time prior to scheduling of the audit. Nadcap audits follow general auditing protocol with an in-briefing meeting and an exit or out-briefing meeting. Auditors identify any/all nonconformance issues on a daily basis. The auditor will contact the supplier prior to the audit to determine the audit plan so as to disrupt operations as little as possible. The audit itself is quite detailed and generally takes two to five days to conduct.

Nadcap requires a written response to nonconformities to include: root cause, immediate corrective actions, action taken to prevent recurrence, product impact evaluation (if applicable) and submittal of objective evidence. Nadcap approves the processor's corrective action and requires objective evidence of implementation for all findings (major and minor) for initial audits. For sub-

11. HEAT-TREATMENT SPECIFICATIONS, COMPLIANCE

sequent reaccreditation audits, objective evidence is required for all major findings. All corrective actions are verified as part of subsequent audits (and/or as part of a follow-up audit, if required).

Nadcap task groups generally require follow-up audits for two reasons: when several nonconformance issues are written indicating systemic problems that result in numerous corrective actions (a follow-up audit will be conducted to test the new systems put into place) and when major findings are identified that indicate product-impact issues. Nadcap has an advisory process in place to identify potential product-impact issues immediately to the prime contractor/government representatives through the Supplier Advisory notification system.

Nadcap audits check for job compliance. This involves the tracing of a job from arrival to completion. Through this method, Nadcap verifies that the heat treater is meeting customer requirements through the flow-down of the process. If the customer requirement is for a "process-implementing procedure," then Nadcap will audit this process.

Technologies Covered

Nadcap defines "special processes" to include heat treatment, welding, coating, chemical processing, nondestructive testing and materials testing laboratories covering the following areas.

- ❖ Heat treating: stress relief, annealing, carburizing, nitriding, carbonitriding, ferritic nitrocarburizing, ion nitriding, vacuum heat treating, vacuum oil quenching, hardening, induction hardening, furnace brazing, dip brazing, induction brazing, vacuum furnace brazing, flame hardening, cryogenic treatments, hot forming/hot sizing, die quenching, hipping, hardness and metallography.
- ❖ Welding: torch/induction manual brazing, flash welding, electron-beam welding, fusion welding, laser welding, resistance welding, friction/inertia welding, diffusion welding and percussion stud welding.
- ❖ Coating: thermal spray, vapor deposition, cementation, stripping, coating evaluations, plating of coated parts and heat treatment of coatings.
- ❖ Chemical processing: plating, anodizing, conversion/phosphate coatings, paint and dry-film lubricants and temper etch.
- ❖ Nondestructive testing: liquid penetrant, magnetic particle, ultrasonic and X-ray.
- ❖ Materials testing laboratories: chemical, mechanical, metallography and microhardness, hardness, corrosion, mechanical test specimen preparation, differential thermal analysis, X-ray diffraction, coating evaluations and fastener testing.

The frequency of a Nadcap audit can vary for each special process. Nadcap policy is that the first subsequent audit (first reaccreditation audit) is to be conducted within 12 months of the initial audit. Subsequent reaccreditation audits may be conducted per the Supplier Merit Program on 18- or 24-month intervals based on audit performance.

Opinions on Nadcap accreditation vary throughout the industry, but here is a general consensus of those who have gone through the process:

- Nadcap accreditation can be a *very* challenging process and quite frustrating for initial audits. The level of detail involved is surprising – it is the most detailed and comprehensive audit process most people participate in.
- While not guaranteed, obtaining Nadcap accreditation should significantly decrease/eliminate prime customer audits and result in a considerable drop-off in audits from individual customers.
- Nadcap accreditation seems more and more like an expectation/prerequisite to do business in the aerospace industry.
- Audits are costly (a week-long audit costs thousands of dollars). There are also financial penalties for rescheduling an audit.
- Participation in Nadcap meetings is a good way to learn about the program, interact with prime customer representatives and PRI personnel, obtain clarification on specific questions and learn about upcoming changes/revisions. Suppliers are also able to serve on task-group initiatives and have an opportunity to provide their input regarding technical matters.
- Comprehensive online resources through the PRI website assist with the accreditation process, particularly the Heat Treat Pyrometry Guide (which helps provide clarification/interpretation of a variety of AMS 2750 details), a list of the most common audit findings, audit checklists, meeting notes, training guides, auditor handbooks, audit advisories and operating procedures (NOPs – includes audit-preparation requirements, criteria for audit failure, etc.).
- Suppliers new to the process may want to consider attending training courses offered by PRI, most notably Nadcap audit preparation courses specific to a discipline (e.g., heat treatment) and root-cause and corrective-action (RCCA) training. These are offered through eQuaLearn in cooperation with PRI and are *highly* recommended.
- It is not uncommon for initial RCCA responses to be rejected, requiring additional investigation into root cause and/or providing objective evidence.
- In addition to the core checklists for various disciplines, there is also a

11 HEAT-TREATMENT SPECIFICATIONS, COMPLIANCE

supplemental checklist with requirements specific to a prime customer. If you have identified that you do business with any of these customers as part of the defined scope of your audit (done in advance of audit), then you must also complete all questions on this checklist applicable to those customers.

- A self-audit is a requirement, not a recommendation. Failure to perform a self-audit prior to the arrival of the PRI auditor will result in a major finding and could jeopardize the entire audit. The self-audit should be done 90 days or sooner prior to a scheduled audit. This allows for any findings to be addressed prior to the PRI audit. (Note: Objective evidence of RCCA efforts are still required and should be available as part of the supplier's internal quality-assurance and corrective-action system.)
- If practical, it is good to have more than one set of eyes available during the self-audit. It is often the case that those closest to the process make assumptions and miss some of the more important details in terms of checklist requirements. An independent, properly trained auditor can avoid these oversights.
- PRI has worked very hard (and continues to work hard) to improve the consistency of its audits. This has been a rather significant complaint from suppliers for some time (and still is), but PRI solicits feedback from suppliers and performs analytical studies on audit results.
- It is interesting to note that different auditors bring different strengths/experience to the audit based on their background.
- As part of a self-audit for reaccreditation audits, it is important that corrective actions from previous audit findings are verified as still being effective. Repeat findings are automatically a major finding.
- Every audit looks *very* closely at pyrometry-related equipment: calibrations, certifications, PM and periodic test results (SAT, TUS, vacuum leak rates, dew-point measurements, etc.). Auditors tend to look very closely at instrument calibrations (are they done at the correct points in the applicable range, per manufacturer requirements?). They also look very closely at all thermocouple certifications and calibration points (process T/C, SAT T/C, TUS T/C, quench-oil T/C, etc.).
- While it may not seem clear from the checklist, when it states "are procedures available to…," you *must* have a controlled document that covers the applicable requirement. If a requirement is listed in an industry standard, then your procedures must comply.
- "Say what you do and do what you say." This phrase best describes the Nadcap process.

- Documented training/competency testing and accompanying procedures are required for heat-treat audits along with support staff (hardness test inspectors, metallurgical lab personnel, TUS/SAT checkers if applicable). A heat-treat training program compliant to SAE-ARP-1962 is mandatory.
- The supplier is ultimately responsible for all aspects of compliance, even if certain functions are performed by outside companies. Any documents certifying compliance (e.g., instrument calibrations, thermocouple certifications, quench-oil test results, etc.) must be reviewed and signed by the recognized technical authority (e.g., metallurgist, quality engineer, technician, supervisor, etc.). The supplier is responsible for making sure any outside party has provided accurate and compliant paperwork.
- Nadcap accreditation requires a significant level of commitment throughout the year. In order to maintain compliance, a supplier must keep current on documentation, revision changes (PRI checklists/requirements and aerospace industry standards), internal procedures, periodic testing, preventive maintenance and record control/verification.
- Quality-system checklist requirements are minimized if a supplier is AS9100 accredited (or equivalent).
- Be careful to make sure any references to temperatures, times, rates, carbon potential and similar data have appropriate tolerances clearly stated in internal procedures. Where appropriate, these should be consistent with industry and/or customer-specific requirements (e.g., Class 3 furnace for carburizing: +/-15°F, soak time -0/+5 minutes, carbon potential $1.0 \pm 0.05\%$, etc.). Even if a tolerance is not specified in an industry standard or customer specification, suppliers should specify their own tolerance bands. Failure to do so will result in nonconformance findings during job audits because it is likely the actual value will fluctuate to some extent during the process.
- Be prepared to have representative jobs (by customer, material, process, configuration, etc.) available to cover all aspects of the identified scope of your audit.
  - For long job audits (typically two needed) the auditor will usually select one that starts being processed on the first day of the audit, and he monitors all aspects of this job through his time on-site. The other long job audit is one that has already been completed within the past few months (selected randomly).
  - For short job audits (quantity depends on scope of audit – a review of eight jobs is not uncommon) typically half are jobs the auditor wants to see in-process while on-site, and the others can be recently completed jobs. If you have multiple customers, the audi-

tor will select jobs that touch on as many of these as possible. If your scope includes carburizing, you *must* have a carburizing job ready to be processed or select one that has recently been completed if there is nothing in queue. Collecting all process paperwork (purchase orders, prints, routings, records, product test results, equipment test results) can take some time. It is good to have resources available to assist with collection of this information if it is not readily available.

CQI-9 Accreditation

CQI-9 is a stand-alone heat-treat assessment program intended to promote good practices throughout the industry. It applies to many of the topics discussed in this book. The specification is focused on the concept of continuous improvement both with respect to atmosphere control and pyrometry. In fact, the fundamentals upon which many of the process tables are created are to ensure good process control in atmosphere heat treatment.

Continuous control and data acquisition are required for furnace atmospheres such as endothermic gas. Common control parameters for these types of atmospheres include dew-point, infrared and oxygen-probe sensors and controls (Section 9.7). This control is critical to ensuring that the prepared atmosphere used is the correct composition so that accurate furnace control can be achieved. On a daily basis, for example, heat treaters are required to provide an alternate method of verification to ensure the control is operating properly. It is common to use infrared analysis or portable dew-point analyzers for this verification.

Continuous controls and daily checks are also required for process parameters such as carburizing. Most atmosphere furnaces have in-situ monitoring or continuous sampling for control parameters. The type of furnace usually determines the best application for continuous control. In continuous rotary-retort furnaces, for example, sampling of the atmosphere for control is often preferred over the use of oxygen probes.

The goal of CQI-9 is to ensure that controllable parameters can be achieved continuously and are properly recorded. Alternate methods for daily checks and verification such as dew point, three-gas infrared analysis, shim stock and resistance wire (Section 13.6) are highlighted for specific processes to provide confidence in the continuous controls and ensure that parameters are met properly

The purpose in what follows is to provide the reader with a deeper appreciation for the more important aspects involved with CQI-9 accreditation. Again, the goal of CQI-9 is to help develop a heat-treat management system that provides for continual improvement while emphasizing defect prevention and the reduction of variation and waste in the (automotive) supply chain. Application of CQI-9 is intended to allow the heat treater to meet regulatory requirements and identify areas of improvement while enhancing customer satisfaction.

REFERENCE DOCUMENTS

The most common reference documents for the CQI-9 standard are:
- TS16949 – quality management system
- ISO9001 – quality management system
- ISO17025 – calibration laboratory standard
- IEC60584-1 – thermocouple reference tables
- ASTM E230 – thermocouple specifications

HEAT-TREAT SYSTEM ASSESSMENT

The heat-treat system assessment can be broken down into the following general categories.
- Management responsibility and quality planning – questions involving the management oversight of the audit process as well as ensuring the process-planning documents have been completed and reviewed for each process.
- Floor and material-handling responsibility – questions involving product, creating paperwork and tracking the product throughout the entire process (including inspection requirements).
- Equipment – questions involving equipment checks.

The first question heat treaters often ask is, "What should I be checking?" In simplest terms, it is the following.
- Furnace atmosphere – verify gas composition or dew point
  - Furnace atmosphere carbon potential and dew point must be continuously monitored, automatically controlled and documented.
  - The primary atmosphere-control method must be supported by a backup method (e.g., dew point verified by infrared).
- Temperature – thermocouple and instrument accuracy and uniformity
  - Continuous strip charts and/or data loggers are required for temperature and carbon monitoring. Electronic records are write-once, read-only.
  - Guaranteed soak times and data logging
- Carbon potential – verification of the oxygen (carbon) probe readings
  - Carbon potential must be within 0.05% of setpoint, meaning the configuration of the controller and stability of the atmosphere must be correct.

- ❖ Generators – monitoring of control systems
- ❖ Data logging (discrete and continuous) – load tracking (software, documentation) for traceability, load identification and processing parameters (e.g., belt speeds, guaranteed soak times, load size, quench-delay time, testing results, etc.)

THERMOCOUPLES

Guidelines and requirements for calibration of thermocouples as well as accuracy and replacement requirements for control, monitoring and recording thermocouples, load thermocouples, and SAT and TUS thermocouples are provided within the specification.

INSTRUMENT CALIBRATION

Calibration is performed to ensure the process instruments are providing a true reading (within limits) of the actual temperature being measured. A field test instrument calibrated to a higher level of accuracy is used to inject signals into the process instrument to see if any deviation is observed. This is usually performed at the minimum, midpoint and maximum temperature of the operating range for process instruments.

SYSTEM ACCURACY TESTS

System accuracy tests (SATs; Section 11.4) are performed to assess the accuracy of the complete measurement system by using an independent device. By placing a test or alternative thermocouple in close proximity to the zone thermocouple, the differential provides a good indication of the ongoing accuracy of the thermocouple and instrument setup.

TEMPERATURE UNIFORMITY SURVEYS

Temperature uniformity surveys (TUSs; Section 11.4) are performed to assess the temperature variation within the stated work zone. Typical configurations for thermocouple positions can be rectangular or cylindrical, depending on the working zone of the furnace.

PROCESS TABLES

The process tables found in CQI-9 call out the tolerance and frequency for checking process-control parameters and component parts (Table 11.3.1). The general categories include process and test equipment, pyrometry and process-monitoring frequencies, in-process and final test frequencies, and quenchant/solution test frequencies.

TABLE 11.3.1 | Overview of process tables[2,3]

Process table	Process descriptors	Process and test equipment	Pyrometry	Process-monitoring frequencies
A	Carburizing, carbonitriding, carbon restoration or correction, neutral hardening (quench and temper), austempering, martempering, tempering, precipitation hardening (aka aging)	[a]	[b]	[c]
B	Nitriding (gas), ferritic nitrocarburizing (gas or salt)	[d]	[e]	[f]
C	Aluminum heat treating	[g]	[h]	[i]
D	Ferrous – induction heat treating	[j]	[k]	[l]
E	Annealing, normalizing and stress relieving	[m]	[n]	[o]
F	Low-pressure carburizing	[p]	[q]	[r]
G	Sinter hardening	[s]	[t]	[u]
H	Ion nitriding	[v]	[w]	[x]

Description of Notes for Table 11.3.1

Examples of process and test equipment [a] requirements are:

- ❖ All furnaces, generators and quench systems shall have temperature-indicating instruments.
- ❖ Continuous strip charts and/or data loggers are required for temperature and carbon monitoring.
- ❖ A program for furnace and generator burnout is required (carbon-bearing atmosphere).
- ❖ Furnace weigh scales shall be verified quarterly and calibrated annually.
- ❖ Dew pointers, three-gas IR and spectrometers to verify carbon potential shall be calibrated annually.
- ❖ Verification of calibration of spectrometers and carbon IR shall be checked daily prior to use.
- ❖ Verification of calibration of three-gas IR, with zero gas and span, shall be performed weekly.
- ❖ Oxygen-probe controllers shall be calibrated quarterly (single/multi-point) or every six months (multi-point).
- ❖ All hardness-testing equipment shall be calibrated annually and verified daily per ASTM standard or equivalent.
- ❖ Files for testing hardness shall be verified per the customer requirement.

❖ Refractometers shall be verified prior to use and calibrated annually (quenchants, water solutions).

Examples of pyrometry [b] requirements are:
❖ Thermocouples and calibration shall conform to stated accuracy requirements.
❖ Instrumentation shall conform to stated accuracy requirements.
❖ SAT checks of the control thermocouple.
❖ Annual TUS (or after major rebuild) with hardening process, ±14°C (±25°F), and tempering process, ±10°C (±20°F).
❖ Recorded temperatures for austenitizing controlled within ±10°C (±20°F). Furnace temperature shall be controlled, with soak times starting at the lower tolerance limit.
❖ Recorded temperatures for tempering controlled within ±5.5°C (±10°F). Furnace temperature shall be controlled, with soak times starting at the lower tolerance limit.
❖ Infrared pyrometers shall be calibrated annually using approved methods.

Examples of process-monitoring frequencies [c] are:
❖ Primary temperature-control instrument
  • Continuous recording with sign-off every two hours or each batch (if under two hours). Alarm systems can satisfy the sign-off.
  • Sign-off is required for each shift for generators.
❖ Monitor atmosphere generation
  • Continuously monitored and alarmed; other systems (e.g., nitrogen-methanol) can be continuously monitored and alarmed or signed off every two hours.
❖ Monitor primary atmosphere controls
  • Continuous recording with sign-off every two hours or each batch (if under two hours). Alarm systems can satisfy the sign-off.
❖ Verify primary furnace atmosphere control method by backup method daily.
❖ Check salt chemistry or part decarburization from austenitizing salt baths daily.
❖ Quench-media process parameters
  • Temperature – continuous recording with sign-off every two hours or each batch (if under two hours). Alarm systems can satisfy the sign-off.
  • Quench level – continuous monitor with alarm or daily verification.
  • Agitation – daily visual checks or monitor the agitation during quenching with alarm systems set with limits.

Examples of process and test equipment, pyrometry and process-monitoring frequencies [d-f] should refer to the latest revision of CQI-9.

Examples of process and test equipment [g] requirements are:
- All furnaces, generators and quench systems shall have temperature-indicating instruments.
- Continuous strip charts and/or data loggers are required for temperature and carbon monitoring.
- Furnace weigh scales shall be verified quarterly and calibrated annually.
- All hardness-testing equipment shall be calibrated annually and verified daily per ASTM standard or equivalent.
- Refractometers shall be verified prior to use and calibrated annually (quenchants, water solutions).

Examples of pyrometry [h] requirements are:
- Thermocouples and calibration shall conform to stated accuracy requirements.
- Instrumentation shall conform to stated accuracy requirements.
- SAT check of the control thermocouple.
- Annual TUS (or after major rebuild) – minimum and maximum temperature ranges shall be defined. If operating range is less than ±85°C (153°F), then only one temperature needs to be tested.
- Solution treating and aging – recorded temperatures controlled within ±5.5°C (±10°F). Furnace temperature shall be controlled, with soak times starting at the lower tolerance limit.
- Annealing – recorded temperatures controlled within ±10°C (±20°F). Furnace temperature shall be controlled, with soak times starting at the lower tolerance limit.
- Infrared pyrometers shall be calibrated annually using approved methods.

Examples of process-monitoring frequencies [i] are:
- Primary temperature-control instrument
  - Continuous recording with sign-off every two hours or each batch (if under two hours). Alarm systems can satisfy the sign-off.
- Quench-media process parameters
- Temperature
  - Continuous recording with sign-off every two hours or each batch (if under two hours). Alarm systems can satisfy the sign-off.
- Quench level
  - Daily verification

- ❖ Agitation
  - Daily visual check or monitor the agitation during quenching with alarm systems set with limits.
- ❖ Process-monitor frequencies
  - Monitor process cycle time or belt speed (each batch or twice/shift or speed change for continuous furnace).
  - Monitor load size of fixturing or loading rate (each batch or twice/shift or loading-rate change for continuous furnace).
  - Quench-delay time – door starts to open at the time the load is at the bottom of the quench tank (each batch or basket for pusher furnace or roller-hearth continuous furnace).
- ❖ In-process/final test frequencies
  - Hardness or tensile testing (each batch, every four hours for continuous).
- ❖ Quenchant and solution test frequencies
  - Polymer quench-media concentration (daily), cooling curve (semi-annual)
  - Water-quench suspended solids (semi-annual)
  - Washer concentration (daily), temperature of solution if above ambient (each shift)

Examples of process and test equipment, pyrometry and process monitoring frequencies [j-x] should refer to the latest revision of CQI-9.

Finally, it is important to note that any process that doesn't affect the final characteristics of the part is exempt from conforming to CQI-9.

Summary

Accreditation is an important goal of the heat-treatment industry. It is intended to validate the work we do and the equipment in which it is performed. The organizations highlighted here and others help convince the customers we serve of our commitment to quality and value.

REFERENCES
1. The Boeing Company (www.boeing.com)
2. CQI-9, 3rd Edition (www.aiag.org)
3. Sherwin, Peter, "Understanding CQI-9 Version 3, Intermediate (Heat Treatment," SECO/WARWICK Seminar on Aluminum Heat Treatment, June 2014
4. Steve Carey, New Hampshire Ball Bearing, technical contributions and private correspondence
5. Nick Fajerson, Denison Industries, technical contributions and private correspondence
6. Peter Sherwin, Eurotherm by Schneider Electric, technical contributions and private correspondence

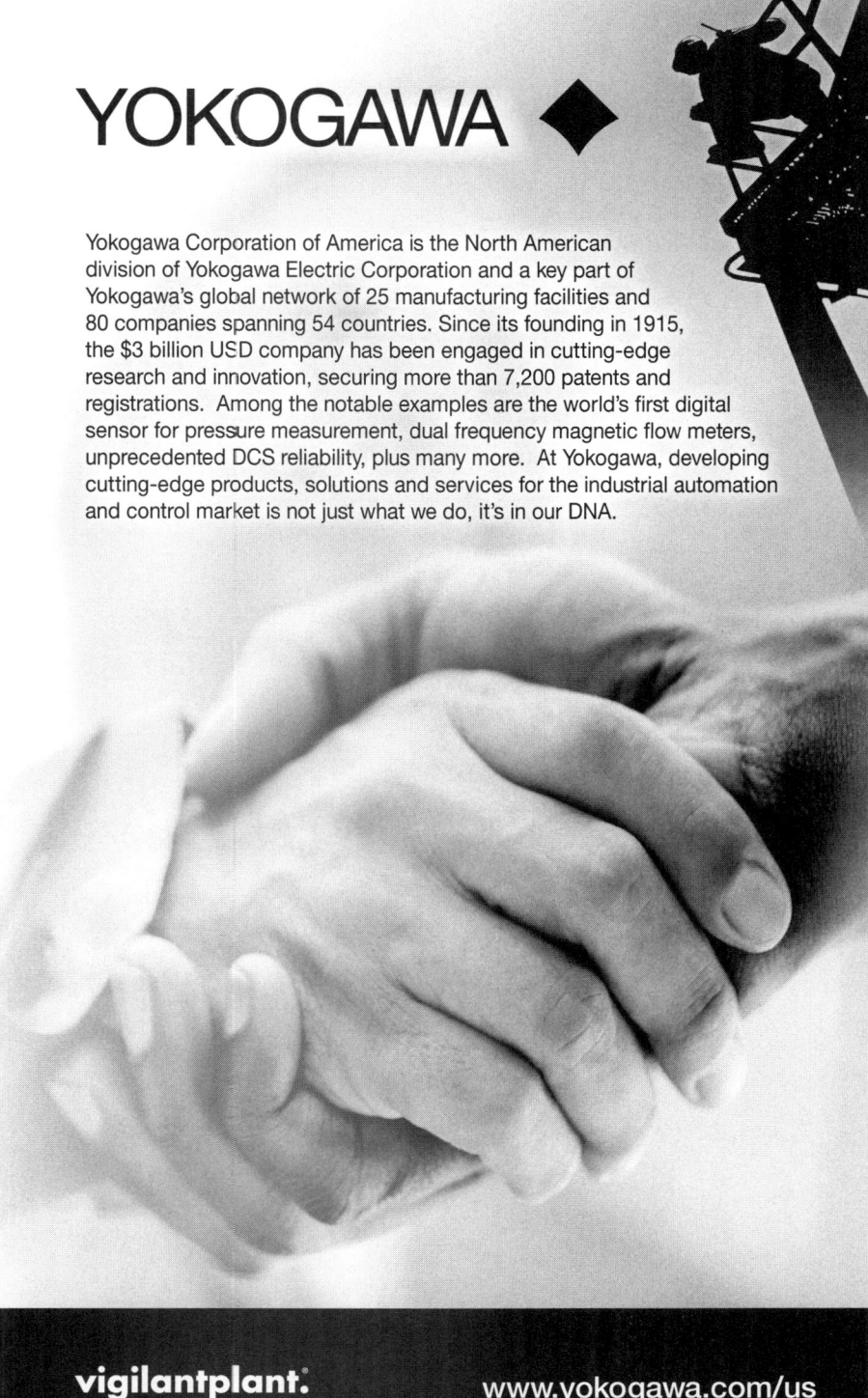

50 years of groundbreaking firsts and innovative thinking

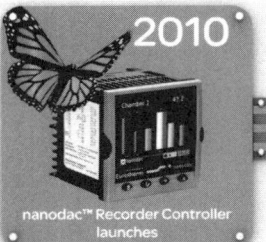  

www.eurotherm.com

11.4

SYSTEM ACCURACY TESTS, TEMPERATURE UNIFORMITY SURVEYS

Two of the more recognizable tests to ensure that the temperature of our heat-treatment equipment is accurate and under control are the system accuracy test (SAT) and the temperature uniformity survey (TUS). Specifications such as AMS 2750 (pyrometry) provide greater detail on each.

As heat treaters, we know (Section 2.4) that changes in temperature have the most profound effect on our heat-treatment processes and that controlling this most influential of process variables is a major goal. These techniques apply to atmosphere and vacuum furnaces as well as ovens.

System Accuracy Test (SAT)

SATs check the accuracy of our control thermocouples. It is performed to ensure that the furnace or oven being used is properly controlled and that the recording system in each control zone is providing accurate information about the process being run.

SATs are normally performed on a preset basis (e.g., weekly) in order to confirm that the furnace or oven thermocouples have not drifted due to deterioration from temperature, atmosphere or mechanical damage. SAT is an on-site comparison of the instrument/lead wire/sensor readings or values to readings or values of a calibrated test instrument/lead wire/sensor to determine if the measured temperature deviation is within certain applicable limits (Table 11.4.1).

Sensors (monitoring or resident) in each control zone can be Type B, N, R or S (Section 16.4) and shall be nonexpendable if exposed to temperatures above 538°C (1000°F).

Thermocouples for control, monitoring or recording must be replaced if they are damaged, if an SAT fails or at a prescribed interval. When complying with CQI-9, for example, base-metal thermocouples (K, N, J and E) must be calibrated before first use and replaced annually if used at or above 760°C (1400°F) or every two years if used below this temperature. Noble-metal types (B, R and S) must be replaced every two years regardless of use temperature.

Two SAT methods are permissible: the probe method and the comparative method. The maximum permitted uses are defined for the various thermocouple types. In the absence of documentation of the number of uses for nonexpendable base-metal thermocouples, the thermocouples must be replaced after six months.

TABLE 11.4.1 | SAT differences by furnace class

Furnace class	Maximum overall SAT difference, °C (°F)	Maximum permitted SAT adjustment (offset)	
		°C (°F)	%
1	± 1.1 (± 2)	± 1.5 (± 2.5)	
2	± 1.7 (± 3)	± 3 (± 5)	
3	± 2.2 (± 4)	± 5 (± 8)	± 0.38
4	± 2.2 (± 4)	± 6 (± 10)	± 0.38
5	± 3 (± 5)	± 7 (± 13)	± 0.38
6	± 6 (± 10)		± 0.75
Refrigeration and quench	± 3 (± 5)	± 6 (± 10)	

TIPS FOR CONDUCTING A SUCCESSFUL SAT

When conducting an SAT, it is important to keep the following in mind.
1. System accuracy testing shall be performed on the temperature control and recording systems for each piece of thermal-processing equipment.
2. An SAT is not required for sensors whose only function is over-temperature (i.e., excess temperature) control.
3. While it may be possible to free up a thermocouple port for the SAT probe by using a dual thermocouple for control and over-temperature indication, be sure that the spacing between the SAT probe and the other thermocouples is within allowable specification limits – 50 mm (2 inches) or 75 mm (3 inches), depending on the controlling document.
4. SATs shall be performed upon installation and periodically thereafter at a frequency in accordance with the furnace class. For example,

a Class-2 furnace (±5.5°C/±10°F) with type-D instrumentation shall have an SAT performed on a weekly basis. In another example, a cryogenic unit shall have an SAT performed on a monthly basis using a type-T test sensor.

5. A new SAT shall be performed after any maintenance that could affect SAT accuracy. A quality-assurance representative (as opposed to maintenance or heat-treat personnel) should be consulted for direction on whether specific maintenance requires a new SAT.
6. A resident SAT sensor may be used subject to the following limitations. It shall be restricted to a type B, N, R or S thermocouple and shall be a different type than the sensor being tested. Resident SAT sensors shall be subject to replacement or recalibration based on the specification requirement.
7. The difference between the uncorrected reading of the sensor system being tested (sensor, lead wire and instrument) and the corrected reading of the test-sensor system (after test sensor and test instrument correction factors are applied) shall be recorded as the system accuracy difference. Applicable correction factors shall be applied algebraically.
8. Once the furnace has stabilized at the test temperature, the temperatures of the test sensor and the work instrument shall be recorded on a permanent record form.
9. Any SAT performed using expendable thermocouple wire shall be limited to a single use.
10. The tip of the test sensor shall be positioned in the furnace such that it is within 50 mm (2 inches) or 75 mm (3 inches) of the tip of the process sensor as required by specification. Subsequent SATs shall utilize test sensors placed in the same location (position and depth) as the initial SAT. Technicians performing an SAT shall verify the position of the test sensor relative to the working sensor for each SAT. The technician shall acknowledge the position of the test sensor on the permanent record form.
11. The following additional details shall also be included on an applicable form created for recording SAT results:
    a. Furnace tested
    b. Date and time of test
    c. Initials of technician performing the test
    d. Serial number of the test instrument being used (e.g., Fluke)
    e Initials of technician acknowledging test sensor within specified distance of the work sensor
    f. Identification of the sensor being tested (serial number)
    g. Identification of the test sensor (serial number)

      h. Test-sensor type (resident or expendable thermocouple)
      i. Due date for resident test-sensor replacement/recalibration (if applicable)
      j. Observed control or recording instrument reading
      k. Observed test instrument reading
      l. Test sensor and test instrument correction factors
      m. Corrected test instrument reading
      n. Calculated system accuracy difference
      o. Indication of test acceptance or failure
      p. Authorized quality-assurance representative sign-off

12. The difference of any SAT result must not exceed the allowable difference specified. For example, the maximum allowable system accuracy difference for a Class-2 furnace shall be ±3°C (±5°F) or ±0.3% of the reading in °F, whichever is greater. In another example, the maximum allowable system accuracy difference for the cryogenic unit shall be ±3°C (±5°F). The failure shall be documented, the cause of the difference determined and corrective action taken before commencing additional thermal processing. Test failures shall be immediately brought to the attention of the authorized quality-assurance representative. Corrective action may include, but is not limited to, any of the following:

      a. Replacement of the out-of-tolerance sensor and/or lead wire
      b. Recalibration of the out-of-tolerance instrument
      c. If the cause is wholly or partially the result of movement of the sensor being tested from its documented position, the sensor shall be returned to its documented location and the SAT repeated.
      d. Adjustment of the control or recording instrument calibration is permitted within the maximum adjustment limitation of the appropriate table in AMS 2750. For a Class-2 furnace used for raw material, the maximum adjustment limitation is ±3°C (±5°F). The effect of this adjustment over the entire operating temperature range shall be evaluated. Any offset shall be documented on the permanent record form.

13. If any furnace maintenance is performed that could affect heating characteristics (e.g., replacement of test sensor, repositioning of work sensor, significant furnace repair), an SAT shall be performed on the temperature control and recording systems.

Temperature Uniformity Survey (TUS)

As stated earlier, meeting the requirements of AMS 2750 (pyrometry) for temperature uniformity ensures that one of the most important process variables is under control. In addition, it is an excellent place to start when defining statistical capability for heat-treatment furnaces and ovens. Meeting this type of specification gives the heat treater a greater level of confidence that the equipment will be able to perform to the required temperature tolerances.

When working to AMS 2750, it is important to avoid failing a survey. If a survey fails, all jobs run in that particular furnace from the time of the last successful survey until the time when the failure is observed must be reviewed for effect and disposition. This can be a daunting task in a job-shop environment. Using statistical-analysis techniques to monitor the temperature uniformity can help eliminate such discrepancies.

One of the more important decisions that must be made initially is to decide if a survey will be conducted with or without a load. This has been the subject of serious debate by proponents of each approach for a very long time. Some feel surveys with an empty load represent a worst-case scenario for uniformity. There is little thermal mass to absorb the heat with an empty load (a test rack alone that extends to the extreme locations of the work zone), so there is little conduction – just convection and/or radiation from the elements.

Commercial heat treaters often prefer this approach because they cannot predict what a typical load size will be. For captive shops that heat treat the same part types and same load sizes every day, it makes sense to run the survey with a load. The feeling is that it is more representative of production conditions and that more accurate temperature uniformity results can be achieved. Also, the heat-up data can be used to determine how long it takes a load to reach temperature in order to predict soak times during production.

Temperature profiling is the process by which we record and then interpret the effects of temperature on our heat-treated products either in a continuous or batch process. The collected numeric data is converted, often by temperature-analysis software, into meaningful information and usually displayed as a thermal-profile graph. This information tells us the temperature the product will reach, for how long and at what point of the process. Heat-treat engineers know what the ideal profile for their product should be and look for variations to indicate a potential quality problem.

The necessary components of an effective temperature-profiling system include thermocouple sensors to gather temperature information, data-acquisition loggers to collect the data, thermal barriers to protect the data loggers, and temperature-profiling software for the analysis and archiving of all temperature data.

The most obvious benefit of conducting a TUS is that it will identify hot/cold areas inside the furnace. It can also verify mathematical models; help us better understand brazing windows for critical processes such as CAB; and provide validation and proof of process-control compliance for Nadcap, CQI-9 and other ISO or quality standards.

TIPS FOR CONDUCTING A SUCCESSFUL TUS

Here is some advice gathered from a number of industry experts on what you should watch out for when conducting a TUS. It is important to keep the following in mind.

1. The advertised "working zone" must be the same size as any racks or setup used (Fig 11.4.1). It is all too common to find survey racks smaller than the actual working zone of the oven. Auditors routinely measure the rack and compare it against the advertised working zone. If the advertised working zone is bigger than the rack, it will be a major nonconformance.

FIGURE 11.4.1 | Typical survey rack in an annealing furnace (courtesy of Nabertherm)

2. Make sure the thermocouples used are correct to the certification of calibration. Correction factors can be as much as 7°C (13°F) or 0.75% of the reading, whichever is greater, depending on furnace class. If the wrong correction factors are used, it will be a major nonconformance.
3. Make sure the location of the sensors is correct to the record. If you make adjustments to bring the working zone into a better uniformity and the sensor locations are incorrectly logged, you may be adjusting it incorrectly, making it worse and wasting time.
4. When crunching the numbers afterward, make sure the approach to temperature is included in the "high" reading calculation for the test sensors and any other recording sensor(s). A sensor could overshoot a lit-

tle, and if you did not analyze the approach and include it in the report, you are not certifying the furnace accurately.

5. Make sure the thermocouple wires are not bent sharply or stretched too tightly when routing them to their respective locations. The wires themselves go through phases of expansion and contraction, and they can become shorted during the process if they're too tight. A short reading will most certainly skew the results or cause a survey to be aborted.

6. Make sure to include the known deviation of the portable TUS recorder into the correction of the readings. An instrument may be off as much as 0.6°C (1°F) or 0.1% of the reading, whichever is greater. This could add a deviation of 1.1°C (2°F) at 1095°C (2000°F). If not compensated for, this can seriously affect the results of the TUS, and it could be a major finding.

7. Do not "step" the ramp for a TUS to prevent overshoot; that is, do not ramp at one rate up until, for example, 10°C (50°F) below test temperature and then drop to a slower ramp rate into the setpoint of the TUS. Also, it is imperative that production ramp rates are used. An auditor knows this can be a bad habit and will typically look for it, causing a major finding in an audit.

8. Always measure the insertion of the controlling thermocouple before each survey. It must always be the same because this is critical to setpoint-versus-uniformity distribution. If a thermocouple is replaced and inserted to a different depth, the uniformity from setpoint may change.

9. Thermocouple wire and connections are an insidious source of EMF error. If using a thermocouple jack panel, be sure the panel is absolutely clean between the legs as any slight metal film could produce errors in the readings. Always keep dummy plugs in unused jack locations. Remember that fiber insulation attracts and holds contamination and water vapor, which can bridge the gap between the wires and cause a slight EMF error. It is important to keep in mind that even on the more-forgiving type-K thermocouple, the difference between 300°C (572.0°F) and 301°C (573.8°F) is a mere 0.041 millivolts. It doesn't take much contamination to cause an error in the difference between the hot and cold junction of a thermoelectric circuit.

10. Be absolutely certain that the instrument used to calibrate the portable TUS recorder meets the accuracy requirements of a "secondary standard," which is ±0.2°C (±0.3°F) or ±0.05%, whichever is greater. Many common field test instruments do NOT meet this accuracy requirement. If you outsource the calibration, you should always have a current copy of the calibration from the field test instrument used and make sure it meets this requirement. In addition, the calibration should be performed as the

instrument is used. For example, if the field test instrument is used as a thermocouple simulation source (sending the temperature signal to the device being calibrated), the appropriate thermocouple types should be checked while the instrument is in the source mode. Having a calibration certificate that shows how accurate an instrument is in read mode does not always qualify it in the source mode. Communicating these kinds of requirements to your supplier is very important to get the service you need in order to pass your audit without findings.

11. Check the element resistance to ground before conducting a TUS.
12. With a calibrated run-up box (thermocouple generator source), send a signal through every channel to ensure correct readings back to the official recording system.
13. Confirm that the reference thermocouple is within 50 mm (2 inches) or 75 mm (3 inches) of the control thermocouple, depending on the specification requirement. Note: This may require the furnace to be cold and inspected internally before commencing the survey.
14. Ensure load thermocouples are new and preferably (not required) from the same lot of source wire for ease of tracking deviation factors.
15. Make sure that your load thermocouples maintain correction factors that are less than the survey spread allowance. For example, surveying for a spread of ±5.5°C (±10°F) and utilizing load thermocouples with correction factors of ±6.1°C (±11°F) or higher is not allowed.
16. Extrapolating setpoint temperatures on load thermocouples in steps greater than 120°C (250°F) is not allowed. Load thermocouples for AMS 2750 are supplied with certification sheets from 538°C (1000°F) up to 1200°C (2200°F) in 120°C (250°F) steps. Therefore, no extrapolation is required.
17. Confirm the furnace has recently passed the SAT.
18. Ensure all elements are functional and operating within established limits.
19. Confirm that the type of sheathed load thermocouples (fiber or metal) are allowed by the specification you are conforming to.
20. Ensure load thermocouples are ungrounded to prevent transient noise from reaching the control system.
21. Always utilize a temperature survey fixture with appropriately sized heat sinks unless otherwise specified by a particular customer-driven internal specification.
22. Always center the temperature survey fixture front to back and left to right for consistencies in readings and repeatability in future surveys.
23. Utilize a load survey fixture, heating rate and atmosphere (if other than air) that best emulates your process requirements as mandated (vaguely) in AMS 2750.

24. Heat to first test temperature at or around 538°C (1000°F) and soak for typically one hour. Allow the furnace ample time to reach steady state.
25. Each progressive setpoint will take less time to soak out due to the fact that radiation effects are increasing over the black-heat area of 538°C (1000°F) range.
26. Follow in-house or AMS 2750 documents for sampling rates, time at temperature and stabilization periods.

Surveying Atmosphere Furnaces

Atmosphere furnaces come in all shapes and sizes. Surveys can be particularly challenging, depending on the amount of manual intervention needed to feed the thermocouple wires through doors (Fig. 11.4.2). For this reason, many companies now offer devices that can be placed on a grid or in a basket before being run through the furnace (Fig 11.4.3). Be aware that there are certain restrictions (e.g., time at temperature) that apply to these devices (Section 13.6).

FIGURE 11.4.2 | Manual survey method for an integral-quench furnace (courtesy of Super Systems Inc.)

FIGURE 11.4.3 | In-process thermal-profiling tools (courtesy of Datapaq Ltd.)

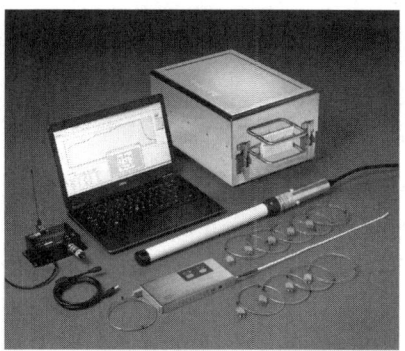

SURVEY TIPS

When conducting a survey, it is important to keep the following in mind.

1. Make sure the proper classification is given to each furnace (Table 11.4.2, p. 273). One furnace can be classified and used in more than one way. You cannot use a furnace for a process in which it has not met temperature uniformity.
2. Take the proper number of measurements. Surveys require data points in two-minute intervals for each thermocouple, including clear evidence of ramp-up and clear evidence of soak (Fig. 11.4.4).

FIGURE 11.4.4 | Typical survey chart (courtesy of Super Systems Inc.)

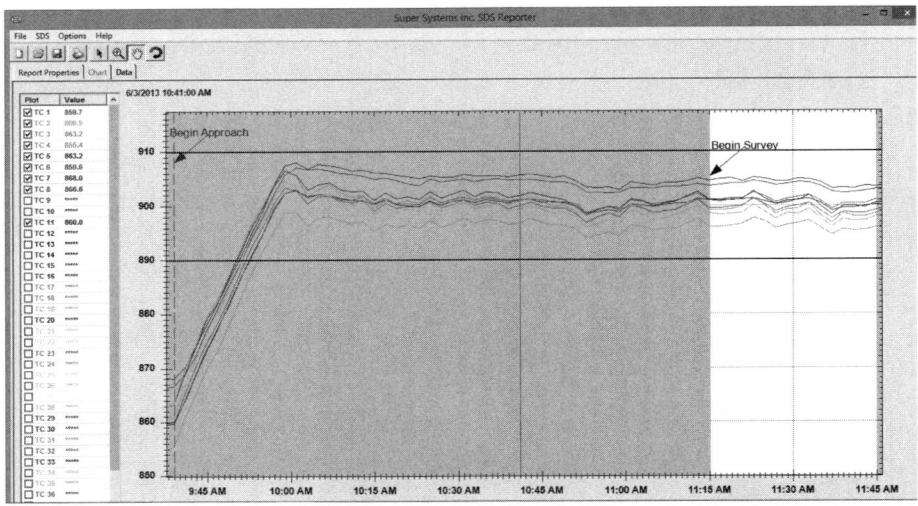

3. Be thorough. Make sure your records show all required details.
4. Make sure any resident SAT probes are not of the same thermocouple type as the control thermocouple. Also, the SAT thermocouple must be located within 50 mm (2 inches) or 75 mm (3 inches) of the control thermocouple to be valid, depending on specification. Have a calibration report for both the SAT probes and thermocouples. You now need to periodically perform SAT tests on freezers and quench oils (monthly is acceptable). The frequency of SAT depends on instrumentation type (as does TUS).
5. Instrumentation calibration is mandatory. Make sure the company meets ALL calibration requirements for instrumentation. The certification must clearly state each is met, along with objective evidence. Each instrument parameter should be calibrated in at least three points. Make sure the points calibrated cover the entire operating range.

6. Thermocouple wire traceability is required. Calibration records are needed with clearly stated correction OR error factors (these must be properly applied algebraically). Make sure any rolls of wire greater than 75 m (1,000 feet) are calibrated at both the beginning and end of the purchased roll (use average between beginning and end in TUS), and make sure the beginning and end values do not differ by more than ±1.1°C (±2°F).
7. Use type N for survey thermocouples, if possible. In many cases, they are now preferred over type K from an initial cost and reuse basis.
8. Selection and placement of thermocouples are critical. Attention to every detail demanded by the specification is mandatory.
9. Be sure to maximize the contact between the thermocouples and the rack, test slug or part. If the integrity of the contact between these two surfaces is not maintained, you might be logging an extraneous temperature instead of a process temperature.
10. Do not weld heat sinks to the survey frame since the contact area makes the heat sink larger than it should be. Heat sinks have been found to make a difference with respect to overshooting and also the degree of overshoot.
11. Understand the impact of the thermocouple's thermal mass. Every thermocouple has an effect on the temperature or heating characteristics of the product being measured. It is important to match the thermocouple to the job required.
12. Understand the compromise that must be made between thermocouple life and maneuverability. For example, while 3.2-mm-diameter (1/8-inch-diameter) thermocouples are common, they are more rigid than 1.6-mm-diameter (1/16-inch-diameter) thermocouples.
13. Know and record the position of the control thermocouple in relationship to the survey thermocouples. Control thermocouples located too close to the walls or too far into the chamber can cause large temperature variations and different response times. Also, the survey must be redone if the control thermocouple was replaced with a different type, different gauge-thickness wire, different protection tube or placed in a different location after surveying.
14. Clearly define the work envelope. Workloads or parts that fall outside survey dimensions must be considered having been run in a nonconforming piece of equipment.
15. Clearly document procedures for performing a TUS. Make sure that procedures for the TUS activity are readily available during an audit and that the in-house technicians who perform the surveys know the procedures

and where to access them. If a third party performs the survey, make sure that the quality department reviews and signs off on the paperwork prior to filing it.

16. Make sure the equipment being used meets the specified accuracy and readability requirements. Tables in AMS 2750 define a TUS device as a field test instrument. Calibration accuracy for the device is ±0.6°C (±1°F) or ±0.1% reading in °F, whichever is greater.

17. Have documented procedures on how correction factors are used. Make this procedure as simple as possible and provide examples so that pyrometry personnel and auditors can easily evaluate survey results and documentation on corrections used.

18. Make reporting easy. Since surveys take time to perform, make the reporting process simple, consistent and repeatable so that survey results can be documented, signed off and technicians can focus on spending their time getting equipment ready for production.

19. Don't try to qualify a furnace for uniformity that is not required. If you are only running parts that require a Class-5 furnace (e.g., AMS 2750, ±14°C/±25°F), don't try to achieve a uniformity that is tighter than that.

20. Design and fabricate a proper test rack and be sure it is suitable for use in the type of furnace, atmosphere or vacuum (Fig. 11.4.5) being surveyed as well as the atmosphere being used. This is especially important when running a survey without a load. Be sure to place the thermocouples in the same location each time, and run each TUS in as similar a fashion as possible.

FIGURE 11.4.5 | TUS rack of carbon/carbon-fiber material (courtesy of Schunk Graphite Technology LLC)

Surveying Ovens

The key to good temperature uniformity in an industrial oven is airflow. Temperature uniformity in ovens should never be assumed, which is why it needs to be checked on a routine basis. One of the greatest challenges faced by manufacturing is the number of older ovens (some dating back over 60 years) limping along and/or ovens in terrible shape still being used in critical applications. Many times ovens are overloaded or restrictions on uniform temperature areas are ignored. Ovens can also be subjected to mechanical damage due to the method of loading (e.g., forklift trucks), which damages internal airflow louvers. Should this happen, the oven must be resurveyed.

It is important to know what the proper airflow should be and what the temperature really is in all locations of the oven. Remember, ovens maintain heating rates and uniformities if they are initially well-designed, used for the intended purpose (this may present a significant challenge over their lifetime), regularly and properly maintained, and have good loading practices (Fig. 11.4.6). A well-documented TUS will establish actual heating rate, uniformity and become an important problem-solving tool for later.

Ovens are simple machines. Their fans need to be maintained, balanced and operating within design parameters; louvers need to be properly adjusted (or readjusted); workloads need to be adequately placed within them, allowing good airflow for convective heating; and the heating system needs to be properly adjusted. Paying attention to these details will result in your ovens easily passing TUS and conforming to the requirements of AMS 2750.

FIGURE 11.4.6 | Well-spaced and supported oven load of aluminum coils (courtesy of SECO/WARWICK Corp.)

SURVEY TIPS[9]

Years of working to enhance oven performance and struggling to maintain proper oven operation has provided insights into what might go wrong and what one needs to be aware of.

1. **Deliver the air where it is needed.** So-called combination airflow (Fig. 11.4.7) ensures that heat is uniformly distributed along the length of the chamber. In this flow pattern, air is delivered along the full length of the work chamber via supply ducts located on the sidewalls. A return duct normally located in the roof helps to evenly remove the air after it passes over the load. Both the supply and return ducts should have adjustable louvers in order to achieve the best uniformity. Horizontal airflow is also common. Here, all the air is delivered to one side of the heating chamber, flows across the parts and returns on the opposite sidewall. This design is used for multi-layer loads (e.g., shelf or rack ovens) that would prevent air from passing vertically through the load.

FIGURE 11.4.7 | Combination airflow oven design[10]

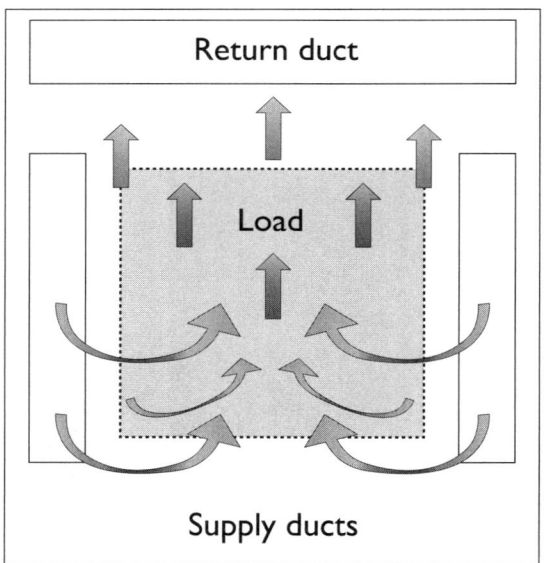

2. **Maintain a positive pressure inside the oven.** To operate efficiently and achieve good temperature uniformity, the pressure inside the oven must be neutral or slightly positive. The proper balance prevents cold air from being drawn in. To check your oven's pressure balance, use a pressure gauge calibrated in inches of water column (inches w.c.) or Pascals (Pa) mounted through the oven wall. A well-balanced oven will show a positive pressure of 0-50 Pa (0-0.2 inch w.c.) in relationship to the ambient conditions.

A simple method of checking to see if you have proper airflow throughout the oven work chamber at room temperature is to tie ribbons or hang pieces of yarn onto parts and fixtures. With the oven at room temperature and the door blocked by a Plexiglas sheet (so you can see through it), turn on the fan and observe the pattern of the streamers. Adjust louvers as necessary to improve the flow pattern, then remove the test rig and perform a TUS before returning the oven to service.

Leakage around the door can also be checked using streamers. If the streamer hangs limp or is blown slightly outward by air escaping from below the door, the oven balance is usually acceptable. If the streamer is being sucked in, the oven has a negative pressure, which is undesirable. The balance can be corrected by adjusting the exhaust and air-inlet dampers. If the oven pressure is negative, open the inlet damper or close the exhaust damper until balance is achieved.

3. **Work with your supplier to ensure adequate airflow is inherent in the design.** An oven with insufficient air will not achieve proper temperature uniformity and be a poor performer overall. Generally, the greater the volume of air circulated, the tighter the temperature uniformity. Work with your selected oven vendors to understand how they calculate airflow and what exit velocity can be expected coming out of the ducts so that you can measure it over time once the unit is in the field. Increasing the blower RPM can also improve the uniformity (this is also true for existing ovens). Finally, pay attention to the style of adjustable duct louver and the ease of its adjustment – these are key to long-term temperature uniformity.

4. **Set the intake and exhaust dampers properly.** One of the most overlooked items when running ovens day in and day out is damper adjustment (both intake and exhaust dampers). It is not uncommon to find that these settings have never been adjusted and that no one can recall how they were originally. Static pressure inside the oven and velocity in the ductwork can be negatively affected.

5. **Pay attention to exhausters.** Ovens, especially those running workloads that involve potentially combustible/flammable gases or other hazardous materials, are normally equipped with exhausters designed to dilute the mixture to non-explosive limits. Exhausters pull a tremendous amount of air through the oven, which can have a dramatic influence on heating and temperature uniformity. Remember that ovens should be tested under the conditions in which they will be used.

6. **Size ovens properly.** It is too often the case that large loads are placed in oven chambers that are too small for them, or small loads

are placed in large ovens. In either case, there is a danger that the loads will not heat uniformly. The oven should be sized properly for the load being processed.

7. **Perform regular TUSs.** It is important to understand that operating conditions change over time, and just because they operate at fairly low temperatures, ovens are no exception. As time passes, fans go out of balance, belts stretch and wear, heating elements burn out, door seals become frayed and worn, louver alignments need correction, louvers become damaged and air passages can become restricted. Separately or in combination, these factors lead to a loss of temperature uniformity. This is why performing TUS checks (Fig. 11.4.8), perhaps even more frequently than specification requirements, is very necessary.

FIGURE 11.4.8 | Oven temperature uniformity survey (courtesy of Super Systems Inc.)

8. **Understand temperature rating.** Ovens come in a variety of temperature ranges. It is not uncommon to purchase ovens designed for a limited temperature. Heat input is often limited as is the amount of insulation used. Most ovens operating up to 260°C (500°F) will have at least 100 mm (4 inches) of insulation, typically mineral wool. Ovens operating in the range of 260-538°C (500-1000°F) require a minimum of 150 mm (6 inches) of insulation. Ovens can be rated as high as 760°C (1400°F), so insulation type and thickness are important considerations for good temperature uniformity. If purchasing a new oven, remember that adding an additional 25-50 mm (1-2 inches) of insulation is an inexpensive way to improve uniformity. In older ovens, hot spots (often designated by peeling or discolored paint) are signs of areas where insulation has deteriorated or compressed and are often indicators of temperature uniformity problems.

9. **Properly arrange the load in the oven.** To maximize even heating, parts must be arranged in a manner that allows the heated air to reach all areas.

If racks or shelves are used, be sure they elevate the load off the floor and do not block the air. There must also be sufficient space between each layer for air to circulate and transfer the heat properly. This sounds obvious, but overloading is frequently a hidden cause of heating and uniformity problems. In addition, the cart or racks should be constructed using lightweight materials to minimize the energy required to heat them.

10. **One size does not necessarily fit all.** The oven chamber should only be slightly larger than the load being heated to ensure good temperature uniformity and even heating. In a small or large oven, for example, air may pass around rather than through the load, resulting in poor uniformity between the outside and the center of the parts being treated.

TABLE 11.4.2 | Furnace class and uniformity requirements (AMS 2750)

Furnace class	Temperature uniformity range, °C (°F)
1	±3 (±5)
2	±6 (±10)
3	±8 (±15)
4	±10 (±20)
5	±14 (±25)
6	±28 (±50)

Summary

Putting forth the effort and taking the time to do SATs and TUSs properly ensures that we have both accurate temperature measurement and temperature control in place. This eliminates concern over one of the most (if not the most) influential process variables in our heat-treatment operation.

REFERENCES

1. Herring, Daniel H., *Atmosphere Heat Treatment, Volume I*, BNP Media, 2014
2. "Instrument System Accuracy Tests and Eurotherm Products," Technical Note, Eurotherm by Snyder Electric
3. Houghton, Richard, "SPC in Furnace Temperature Surveys," *Heat Treat Progress*, June 1999
4. Grande, Mike, "7 Ways to Improve Temperature Uniformity," *Process Heating*, January 2005
5. Herring, Daniel H., "Temperature Uniformity Survey Tips Part One: Atmosphere Furnaces," *Industrial Heating*, May 2009
6. Herring, Daniel H., "Temperature Uniformity Survey Tips Part Two: Vacuum Furnaces," *Industrial Heating*, June 2009
7. Herring, Daniel H., "Temperature Uniformity Survey Tips Part Three: Ovens," *Industrial Heating*, July 2009
8. James Oakes, Super Systems Inc., technical and editorial review, private correspondence
9. Elie El Choueiry, Datapaq Ltd., technical contributions and private correspondence
10. David Plester, Datapaq Inc., technical contributions and private correspondence
11. Mike Moyer, Solar Atmospheres, technical contributions and private correspondence
12. Richard Houghton, Spectrum Thermal Processing, private correspondence
13. Steve Carey, New Hampshire Ball Bearing, technical contributions and private correspondence
14. James E. Grann, Ipsen, technical contributions and private correspondence

11 | HEAT-TREATMENT SPECIFICATIONS, COMPLIANCE

High Quality Work Begins with Exceptional Furnace Control!

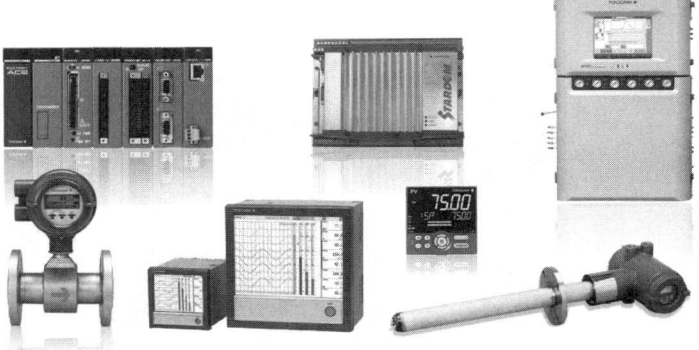

Yokogawa offers multiple automation platforms, control strategies and solutions for every application size and complexity. Our turnkey solutions and factory level integration gives you the assurance that our custom and pre-engineered solutions will work right the first time. Yokogawa's products deliver unsurpassed performance. Not only do we have vast experience in the areas of instrumentation and field devices, but also in all areas of furnace controls and operations. High quality measurement and control with Yokogawa's field-proven sensors, controllers and data acquisition systems ensure efficient, reliable and economical plant operations. Contact Yokogawa to see how our solutions will exceed your expectations.

vigilantplant.
The clear path to operational excellence

YOKOGAWA ◆
www.yokogawa.com/us

ATMOSPHERE HEAT TREATMENT

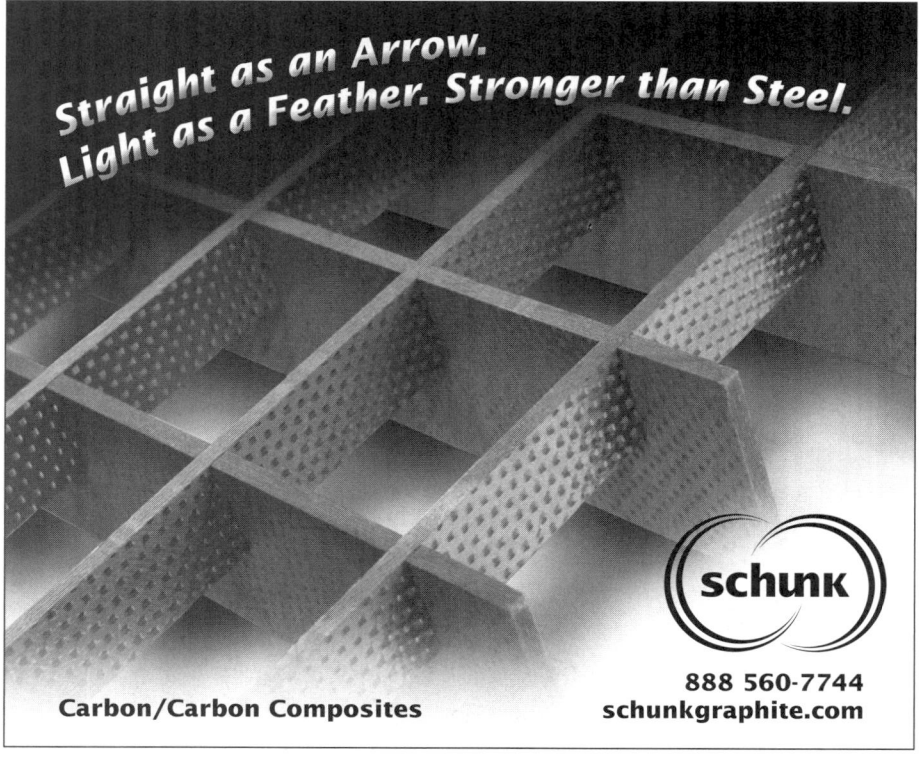

11.5

INSTRUCTIONS AND BLUEPRINT REQUIREMENTS

One of the most significant challenges faced by the heat-treatment community is consistency in instructions and blueprint callouts. In addition, we must be sure that we all use correct definitions of terms relating to heat-treatment processes (Section 16.1) and call out current revisions of all required specifications.

The Four Basic Types of Heat-Treatment Processes

Heat-treatment processes are performed to change certain characteristics of metals and alloys in order to make them more suitable for a given service application. The large number of alloys and the vast number of service requirements make for a considerable number of heat-treatment processes. In general, heat treatments can be classified as:

- Softening operations – annealing, normalizing, stress relief (e.g., cold-worked material) and tempering
- Hardening operations – through-hardening, development of properties in solid-solution alloys
- Case-hardening operations – boronizing, carbonitriding, carburizing, nitriding, nitrocarburizing and hardening by applied energy (e.g., induction, flame, laser)
- Special operations – those designed to achieve special properties or characteristics (e.g., austempering, martempering and cryogenic treatments)

Setting Up an Internal Heat-Treat Standard

Internal heat-treatment standards and work instructions should always be created after careful consultation between the engineering and manufacturing departments. This process should involve the metallurgist (Section 15.4), who must be an integral part of the team, not only at the design stage but also throughout manufacture, heat treatment and quality inspection.

Some of the items in a good specification include but are not necessarily limited to:

- Part number(s)
- Material type(s)
- Material (chemical) requirements/specifications (e.g., chemistry restrictions, grain size, microstructure, allowable grain-refining elements, mill treatments, mill testing, hardenability/DI range, maximum partial and total decarburization, etc.)
- Material certification – ISO, AMS, ASTM or other applicable standards
- Incoming material inspection requirements (e.g., samples for in-process hardness checks, samples for chemistry check and samples for microstructural checks)
- Dimensional requirements – pre- and post-heat treatment
- Finish requirements
- Identification/packaging – color-coding, tags, heat number, manufacturer's name or trademark, designation of the steel grade
- Prohibited practices (e.g., thermal treatments for carbon correction, grinding of defects, etc.)
- Heat-treatment/process specifications – hardness and microstructure (acceptable and unacceptable criteria)
- Ancillary processes (e.g., shot blast or shot peen)
- Inspection criteria (Table 11.5.1)
- Test procedures – including sample size and sample location
- Report requirements – summary of internal and/or external testing (if applicable)
- Definitions and examples (e.g., core, effective case depth, surface decarburization)
- PFMEAs and control plans
- PPAP requirements and procedures (e.g., first-time production runs, subsequent changes to the heat-treat process, approvals)
- Customer specifications
- Originator information – person and date
- Revision level – person, date, applicable section or paragraph, change made

11 | HEAT-TREATMENT SPECIFICATIONS, COMPLIANCE

TABLE 11.5.1 | Example of inspection criteria for carburized, quench-and-tempered steel

Part number	Surface hardness, HRC	Effective case depth @ 50 HRC		Core hardness, HRC
		mm	inches	
A	50-56	0.50-0.75	0.020-0.030	40-44
B	58-60	0.50-0.75	0.020-0.030	40-44
C	58-60	0.50-0.75	0.020-0.030	38-42
D	58-60	0.58-0.84	0.040-0.060	30-34
E	50-56	0.58-0.84	0.040-0.060	32-30
F	50-56	0.58-0.84	0.040-0.060	30-28
G	50-56	0.58-0.84	0.040-0.060	26-28

Setting up an External Heat-Treat Standard

Commercial heat treaters spend much of their time trying to clarify what their customer's expectations are on each order they process. Clearly defining the nature of the materials and the heat-treatment requirements is the first step in making sure the heat treater will be able to meet or exceed customer expectations. If possible, the requirements of each job sent to the heat treater should be discussed *before* the start of manufacturing, not after they are final-machined. It is also important to provide written documentation to the heat treater, including a duplicate copy sent with the materials/components, that represents the final agreed-upon treatments and specifications.

MATERIAL GRADE

The first piece of information that must be supplied is the material grade, typically the SAE/AISI number. Examples of these designations include 1095, 1144, 4140, 52100, H11, M4, etc. Other common material specification systems, such as DIN or JIS, are also used in today's global marketplace. Cross reference to SAE/AISI grades should be provided, if possible. To avoid confusion, listing trade names in place of these designations should be avoided. If these are provided, the specific manufacturer's material data sheets should be included in the paperwork package.

The second most important document you can offer is the material certification sheet (Section 3.1). This is provided by the material supplier and details the exact chemistry of the material being heat treated (including trace elements), grain size, cleanliness of the steel, prior processing and hardenability. These are invaluable aids for the heat treater when it comes to understanding how to correctly process the material and evaluating if additional steps may be required.

CONDITION SUPPLIED

The mill supplies most steels in the annealed condition. However, the material can also be supplied in conditions such as normalized, normalized and tempered, or hardened. In addition, the material may have come to you from the mill as a sheet, bar, rod, forging or casting. It is critical that you notify the heat treater of the condition of the steel you are sending to them. If not, it is possible that the parts may not respond properly to the heat-treating process. For example, if you send parts made from 17-4PH to be processed to condition H-900 and the material was purchased in condition H-1150, it will not respond properly to the standard H-900 process. Your material supplier should notify you of the as-supplied condition in their material certification sheet.

INSTRUCTIONS AND SPECIFICATIONS

There are three types of instructions (purchase orders) heat treaters typically receive from their customers: commercial practices, customer specifications and industry specifications. Examples of each of these are as follows.

- Commercial: 440C material, heat treat to 58-60 HRC
- Customer: 440C material, process per BPS-4602 (Bell Process Specification)
- Industry: 440C material, process per AMS 2759/5 (Aerospace Material Specification)

For a commercial purchase order, the heat treater is able to choose their own process (process temperatures, equipment type, atmosphere, soak durations, etc.). This often gives them the ability to combine orders and reduce processing time and cost. It is fair to argue that this type of purchase order gives the heat treater too much freedom. Purchase orders that call out specific customer or industry specifications are usually much more stringent. They may also bring other specifications and requirements into play.

It is important for customers to realize that calling out a customer or industry specification on the purchase order may significantly increase the heat-treatment cost and lead time. On a commercial order, for example, 440C may only require two processes (hardening and tempering) in order to meet a given hardness requirement. If the same 440C order is required to be processed per a customer or industry specification, it may require that the parts be subjected to multiple tempers and deep-freeze operations. It may also bring additional requirements into play, such as dew point, furnace atmosphere, furnace pyrometry, instrument calibration, intergranular attack limits, decarburization callouts, surface contamination tests, special documentation, additional destructive testing, larger hardness-testing sampling sizes, training, cooling rates, process temperature setpoints and soak

durations. There are usually very good reasons for these special requirements, but it may be worthwhile to investigate them to make sure they are all truly required and not forgotten holdovers from previous generations of the specification.

PART DRAWINGS (PRINTS)

It is extremely useful for your heat treater to have a copy of the current part drawing so that he/she can verify dimensions; note critical dimensions; and understand the geometry of the part with respect to radii, sharp corners, ruling (thickest) section, the location of thin sections and surface-finish requirements. Part drawings should clearly call out the heat treatment and resultant property requirements or reference documents/standards where this information can be found.

Many customers also provide copies of their routers so that the heat treater can see how his processing fits into the overall scheme of the part manufacturing process. It is not uncommon for them to raise questions based on their experience and what they see on the print. Remember, your heat treater has seen and dealt with literally thousands of shapes and sizes, and his/her opinions are invaluable.

REQUIRED CERTIFICATIONS

Customers often require that their work be sent to heat treaters that have certain approvals in place. These approvals may include: ISO9001, AS9100, TS16949, CQI-9, etc. In order to gain these types of approvals, the heat treater must have an effective quality system in place, and it must be verified by an outside service. ISO9001, AS9100 and TS16949 quality approvals are usually general in nature (quality-management system, management responsibility, product realization, etc.) and are not heat-treating process-specific.

Nadcap (Section 11.3)

This aerospace approval is the highest industry-wide accreditation a heat treater can achieve. Not only does it cover the quality systems described, it also includes audits on specific heat-treating processes. A portion of the week-long audit also includes the witnessing of actual heat-treat runs. The process-related portion of the audit confirms compliance with aerospace and aerospace-prime specifications. PRI (www.pri-network.org) is the accrediting body for this program.

Prime Approval

Even if a heat treater is approved and accredited, they still may not be able to process work for certain aerospace primes (e.g., Boeing, Pratt & Whitney, Airbus). These companies also require that your heat treater pass their site audit. If the parts being submitted are related to such companies, they can only be processed at heat-treating facilities approved by the prime. The nuclear industry is another example of where prime approval is an additional requirement.

TESTING REQUIREMENTS

The purchase order or request for quotation should clearly state the testing requirements. If you are asking the heat treater to process the order to a certain AMS specification, these requirements may already be defined. Otherwise, you need to communicate what your requirements are or what exceptions you will allow. A few examples of typical testing requirements include surface hardness, microhardness, tensile strength, surface carbon content, IGO/IGA and microstructure. Be aware that test coupons may be required to facilitate these types of tests. Also, keep in mind the consequences of the tests. If an order is to be 100% Rockwell hardness tested, for example, all of your parts are going to be returned with hardness indentations on them.

DIMENSIONAL REQUIREMENTS

To some degree, all materials will change size and shape during heat treatment. You need to plan your manufacturing process to accommodate these changes. Stating "keep flat" or "keep straight" on purchase orders is not realistic. However, there are materials and processes that can be chosen to minimize these changes. Putting a flatness or straightness callout on your purchase order is good practice only if it is realistic and achievable. Involving your heat treater *early* in the project can help in this regard. The heat treater can assist in specifying the most suitable material and process sequence that will achieve results that meet your expectations.

Cosmetic Requirements

If your parts require special handling, please inform the heat treater. Keep in mind that individual handling and racking of parts can add significant cost. In most cases, heat treaters will handle your parts with care. If you require that a certain surface remain free of hardness indentations, however, please note that instruction on the purchase order. Other cosmetic requirements may include glass-bead cleaning or peening after heat treatment, processing to keep parts clean, and keeping edges and corners of parts free from nicks and dings.

Documentation Requirements

Several levels of documentation can be supplied by the heat treater:
- ❖ Shipping ticket only
- ❖ Product certifications (hardness, microstructure, mechanical properties, etc.)

- ❖ Process certifications
- ❖ Furnace chart recorder or electronic data

The shipping ticket is signed by the customer and used as the proof-of-delivery document. Product certifications typically show the number of parts tested and their range of values. It should also state the specification with which the testing complied. The process certification usually shows the process that was run (soak times, temperature setpoints, quenchants used, etc.). It will also state the specifications with which the process was in compliance.

For commercial work, only a shipping ticket and product certification are usually supplied. For aerospace work, process certifications are also provided. The furnace chart-recorder data is usually kept on file by the heat treater for a predetermined number of years. It can be made available to the customer on request. If the heat treater is running a proprietary process, however, he may be unwilling to provide this intellectual property to the customer.

A true partnership needs to exist between the customer and heat treater in order to optimize the performance of the end-product. Heat treaters have their customer's best interest at heart, but they are not mind readers and have limitations that are best overcome by mutual information flow and good planning combined with knowledge and experience.

Industry Best Practice

The industry continues to struggle to define industry best practice in heat treatment. It is no longer acceptable for heat treaters to avoid significant responsibility for the outcome of their heat-treatment operations. Such acceptance can introduce considerable complexity and cost, but for the most part, sound engineering by all parties in the manufacturing chain can minimize these issues. Heat treaters must essentially become partners in the design and manufacturing process and be in a position to offer experience-based opinions on what might go wrong during the heat-treatment portion of the process chain.

In conducting a design/material/manufacturing analysis, the basic questions to answer first are:

1. **What must this component endure during service (i.e., what are the product requirements)?** What are the rigors of the application, and what is the design life? Must it provide premier service or is there an adequate design life that is involved? What other factors will end its service long before its useful life is expended? What loading, lubricants, temperature and contaminants are involved? Other service/performance aspects specific to a particular product must also be factored into the selection process.

2. **How is this entity to be made (i.e., what are the process requirements)?** How will its basic form be generated, and how will it be heat treated (if at all)? Will you try to introduce particular mechanical properties? If so, will you do so by heat treatment or mechanical means? Is geometry or surface finish important? Will special coatings be used? Is dimensional control (stability or stability at temperature) an issue? Other processing aspects specific to a particular product must also be considered.

Obviously, process development and product development are not separate entities. They are highly interdependent, and both must be considered. However, one must also recognize that today's cost demands often require some compromises in material selection (Table 11.5.2) to meet logistical, supply-chain and inventory requirements. Fortunately, that does not mean that material selection needs to be minimized. If done correctly, the needs of all parties can usually be met with excellent success while maintaining realistic economic, manufacturing and performance goals.

TABLE 11.5.2 | Example of process variables in material selection

Variables			
Raw material	**Design**	**Process**	**Application**
Cleanliness	Innate validity	Heat treatment (type, properties)	Lubricant type (base, viscosity, temperature, oxidation stability)
Alloy selection	Loading	IGO/IGA control	Contamination (chemical/mechanical)
Alloy design	Applied stresses	NMTP control	Lambda ratio
Formability	Residual stresses	Distortion control	
Machinability	Process selection	Residual stress (heat treat, post processes)	
Chemistry/hardenability control		Geometry/contour control	
Strength/toughness		Surface finish (grinding, polishing)	
Fatigue strength		Surface treatment/coatings	
Inclusions			
Surface quality			
Dimensional quality			
Price/delivery			

11 | HEAT-TREATMENT SPECIFICATIONS, COMPLIANCE

Organizations such as the Metal Treating Institute (www.heattreat.net) have developed guidelines in the form of an industry statement for heat treaters. For instance, it recognizes that inherent problems in material chemistry, part design and manufacturing methods can make heat treatment very difficult or, in some cases, impossible. Without knowing specific details and the ultimate end-use requirements of the part, it could be risky to proceed with heat treatment. Several illustrative examples are provided to show a few typical design issues. Equally important are the proper selection of the basic manufacturing method and a material that will meet all processing and product requirements.

EXAMPLES OF PROPER AND IMPROPER PART DESIGN[3]

A die (Fig. 11.5.1a) prone to cracking due to uneven section thickness can be modified to better balance section thickness (Fig. 11.5.1b), thus improving the uniformity of heating and cooling.

FIGURE 11.5.1 | Example of balancing section size to minimize cracking

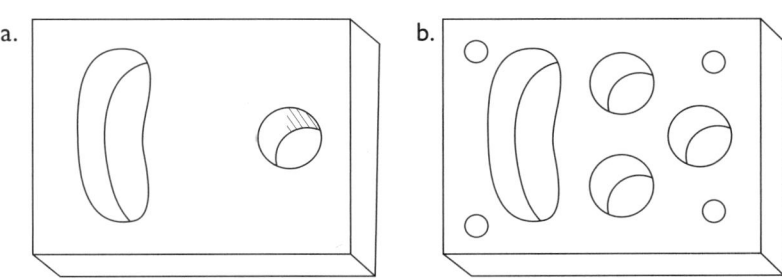

Sharp or re-entrant angles (Fig. 11.5.2a) act as stress-concentration centers and should be avoided in part design if possible. Cracks are very likely to develop in these corners during quenching, and a fillet in the design (Fig. 11.5.2b) will help to minimize this risk.

FIGURE 11.5.2 | Example of reducing stress risers

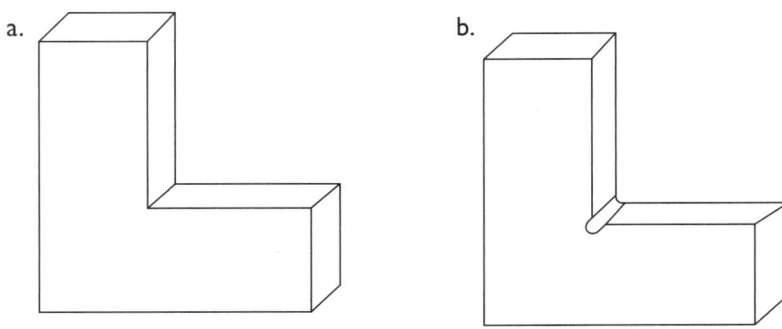

Sometimes it is nearly impossible to design a part without incorporating adjoining thin (light) and thick (heavy) sections (Fig. 11.5.3a). An improved design (Fig. 11.5.3b) might still be at risk for cracking where the shaft meets the base. Also, the cylinder could warp during quenching because of the two unbalanced masses. Before committing an unstable part design to manufacturing, metallurgical engineering and the heat treater should be consulted.

FIGURE 11.5.3 | Thin and thick sections on the same component part

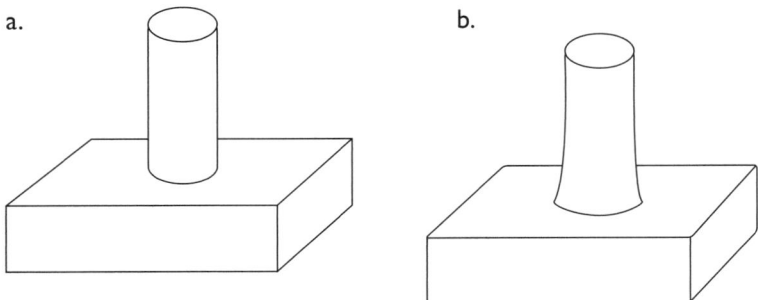

Keyways in shafts (Fig. 11.5.4a) often present stress risers that could result in cracking. The corners of the keyways should be filleted (Fig. 11.5.4b) to prevent cracks, and the keyways should be cut opposite, at 90 degrees to each other, to balance the piece so it will stay round when quenched.

FIGURE 11.5.4 | Keyways in shafts

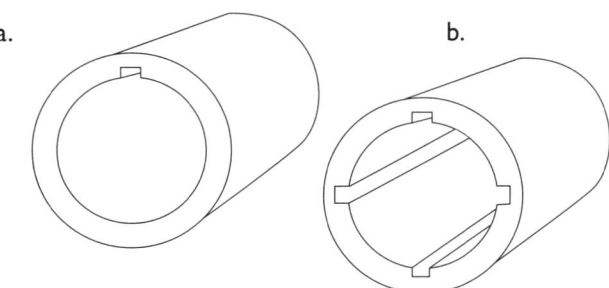

Improper design in a double-ended side-mill or spot-facer (Fig. 11.5.5a) found each side of the tool with three teeth placed opposite each other. This is an unbalanced design in the cross section of the piece, and it is made more serious by a sharp corner at the base of the teeth. This tool design is most likely to crack at the junction between the light and heavy sections. This condition may be corrected by staggering the teeth on opposite sides of the tool (Fig. 11.5.5b) and eliminating the sharp corner at the base of each tooth.

FIGURE 11.5.5 | Milling cutter at risk of distortion and cracking

a.
b.

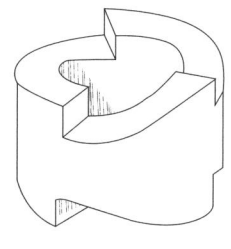
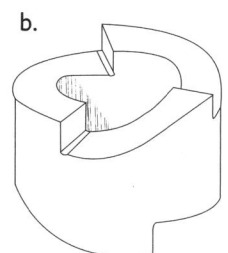

When cutting tools are designed, it is best to keep the hub and cutting edge as near to the same thickness as possible. The heavy mass in the hub will tend to warp and buckle, if not crack, the lighter section.

Gear teeth (Fig. 11.5.6a) can be prone to cracking if a radius is not provided to avoid a stress-concentration point. The design should be able to be corrected by leaving a generous radius in the root (Fig. 11.5.6b).

FIGURE 11.5.6 | Gear teeth with stress-concentration points

a.
b.

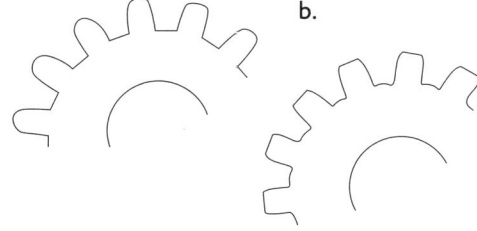

The design of the blanking die (Fig. 11.5.7a) is not correct since the setscrew holes are in direct line with the sharp angles of the blanking section. In all likelihood, this will lead to a crack. This condition also reduces the "land" in that area, which makes the die unbalanced. A better design (Fig. 11.5.7b) moves the setscrew holes to another area.

FIGURE 11.5.7 | Improved design of a blanking die to avoid cracking

a.

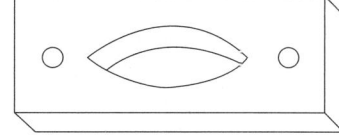

b.

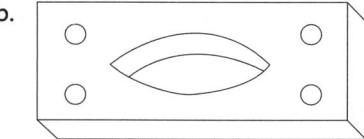

Longitudinal sections of cold-drawing dies can have stress set up in heat treatment and in service at the sharp corners of the small opening (Fig. 11.5.8a), which can easily cause spalling and flaking. The use of a fillet in this area (Fig. 11.5.8b) avoids this problem.

FIGURE 11.5.8 | Cold-drawing dies with sharp corners

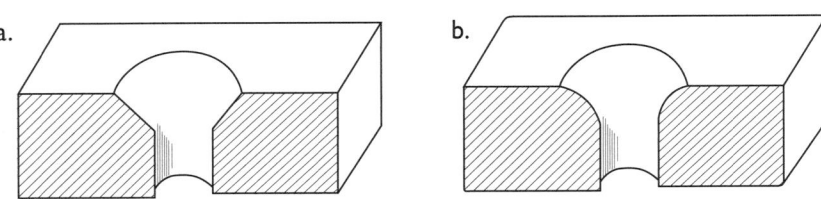

Thousands of dollars are spent each year on straightening of long shafts with keyways (Fig. 11.5.9a), which have a tendency to warp when quenched, whereas a shaft with a balanced section (Fig. 11.5.9b) will stay more straight. If the keyways are impossible to add to heat-treated shafts, the parts should be processed in furnaces so that they can be held in the vertical position.

FIGURE 11.5.9 | Long shafts with keyways

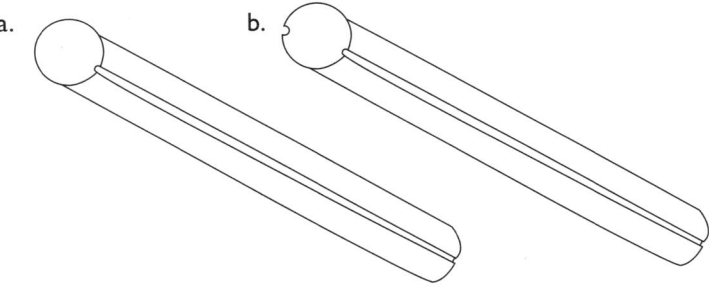

The preventive step needed in all of the previous examples would be for the part designer to meet with an experienced metallurgical engineer and the heat treater prior to committing the design to production. This can save both time and money and prevent parts from failing.

UNDERSTANDING THE STRESS STATE

Stress can occur from a variety of causes:
- ❖ Stress originating in the raw material received from the mill
- ❖ Stress created during the manufacturing process
- ❖ Transformational stresses created during heat treatment

For example, simple cooling may cause a component part to shrink, but the structure change that developed hardness could result in an increase in volume. As long as these changes occur uniformly throughout a piece, the resulting stress effect should be minor. If, however, the shape of a piece is such that a thin area cools faster and becomes hard while a thicker area is still cooling and becoming hard, the resulting stress can be great enough to cause cracking or unanticipated distortion.

Summary

Whether creating internal specifications for use by your heat-treat department or sending work out to commercial heat-treatment shops, it is imperative that designers and manufacturers clearly define their requirements, understand the challenges that face heat treaters and mitigate potential problems.

REFERENCES
1. Kern, Roy, "Achievable Carburizing Specifications," *Gear Technology*, January/February 1990
2. McKenna, Patrick, Daniel H. Herring and Craig Darragh, "How to Best Communicate Your Needs to the Heat Treater," *Industrial Heating*, April 2010
3. Peter Hushek, Phoenix Heat Treating, Inc. (www.phoenix-heat-treating.com), technical contributions and private correspondence
4. Metal Treating Institute (www.heattreat.net)
5. Craig Darragh, AgFox LLC, technical review and private correspondence

Manufacturers of the Finest Automated Blast Finishing and Shot Peening Equipment

ENGINEERED ABRASIVES®

ISO/TS 16949
ISO 14001
Ford Q1
Certified
Job Services

Complete Stock of Parts
ENGINEERING SERVICE
- Let us automate your operations
- Benefit from our 75 years experience in commercial blast finishing and shop peening

Manufacturers of the Finest Automated Blast Finishing, Air Peening and Deburr Equipment
SALES
- Custom automated equipment for grit & shot
- Shot flow controllers & monitors
- CNC machines
- Robot load/unload
- Pick & place load/unload

One stop source for all your shot peening and blast finishing needs

Michael J. Wern • President
Phone: 708.389.9700 or 773.468.0440
Fax: 708.389.4149 • Cell: 312.403.2462
Email: mwern@engineeredabrasives.com

11631 S. Austin • Alsip, Illinois 60803
www.engineeredabrasives.com

12 | TESTING

12.1
RATIONALE

Testing provides the heat treater and quality-control engineer with critical information that allows informed decision making, especially about the effectiveness of the heat-treatment process performed.

Proper testing will indicate the success or failure of a particular heat-treatment process; provide data and feedback to the heat treater, quality-assurance team and management; contribute to continuous quality improvement; and enhance our knowledge base and promote future success.

Furthermore, testing provides a comparison to baseline data or statistical data (Figs. 12.1.1, 12.1.2) so that valid conclusions can be drawn. Testing also:

- ❖ Focuses our attention on how the process may be changing over time (i.e., if equipment and process variability in the heat-treatment department is under control or if changes or maintenance are needed)
- ❖ Confirms the effectiveness of our heat treatment
- ❖ Provides a certain measure of assurance that the component part will perform in service
- ❖ Provides insights into how the heat treatment will affect the performance of a component part in service
- ❖ Allows us to learn what to do and what not to do for a given material or process

FIGURE 12.1.1 | Example of the use of statistical data to evaluate an in-control process

a.

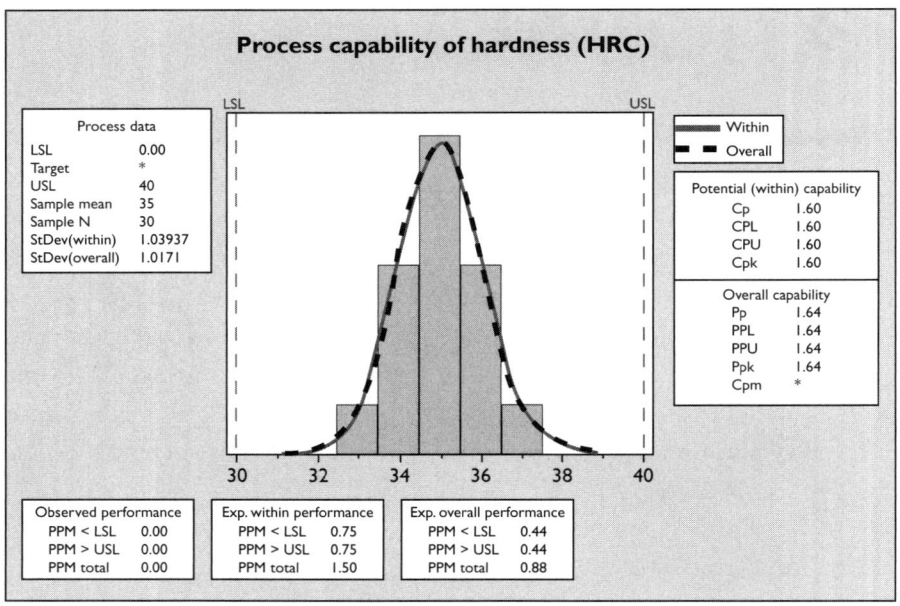

b.

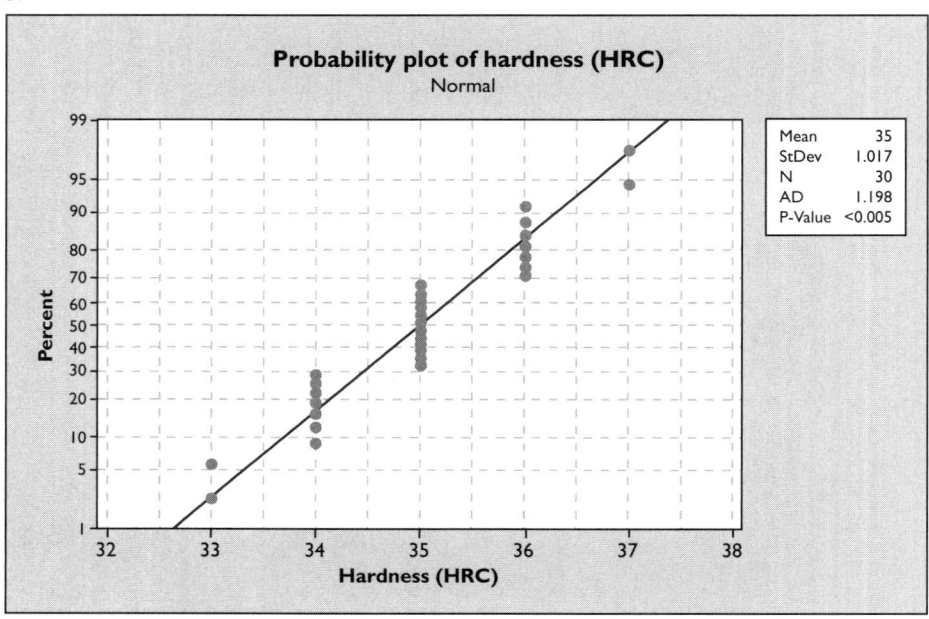

a. Process-capability histogram – data is centered within the specified limits with very good capability (Cpk = 1.60)

b. Probability plot showing data is normally distributed

12 | TESTING

FIGURE 12.1.2 | Example of the use of statistical data to evaluate an out-of-control process

a.

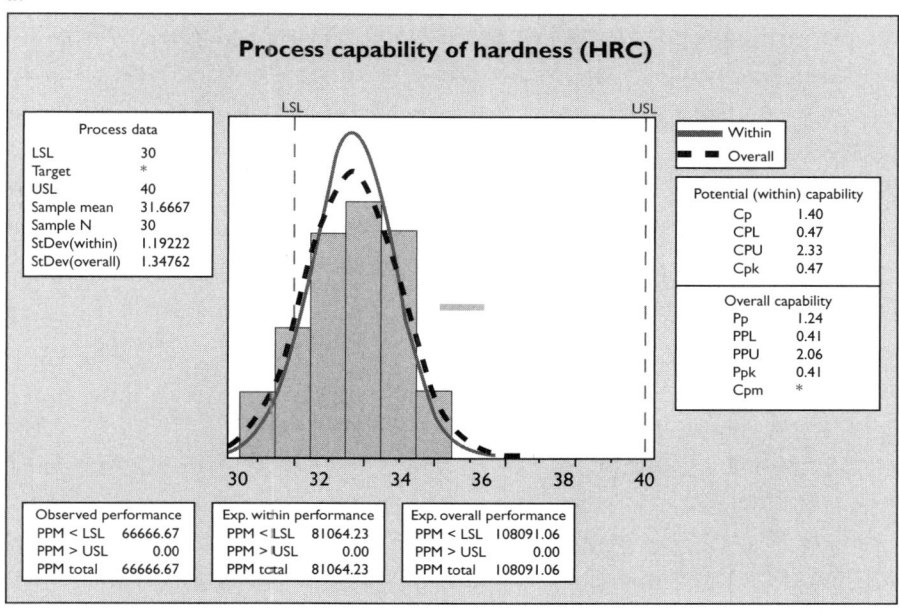

b.

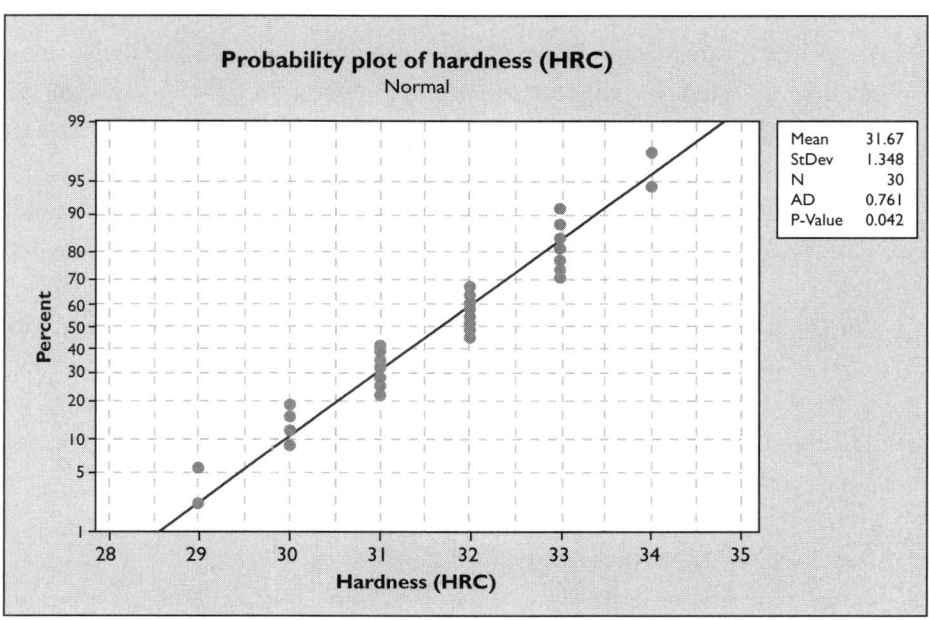

a. Process-capability histogram – data is normally distributed but skewed to the low side of the specification with data points outside the lower limits; undesirable situation with poor capability (Cpk = 0.47)

b. Probability plot showing data is normally distributed

From a purely philosophical standpoint, we gain knowledge about the quality of our heat-treatment operations by systematic planning, processing, observation and verification of results – the latter being what we call testing. There are many things we could desire to know about a component part. Most often we simply want to determine if it will actually perform in service as we intended it to perform.

From a shop-floor perspective, testing focuses our attention on process or equipment problems; prevents the shipment of bad product; documents discrepancies, changes and maintenance needs; supports traceability; and, most importantly, prevents expensive product recalls.

Role of Quality

In the classic definition, quality control is the process of verifying predetermined requirements for quality. On the other hand, quality assurance is more about providing the continuous and consistent improvement and maintenance of the process that allows quality control to do its job. The purpose of these functions is to ensure that the testing being done is consistent, repeatable and representative of the desired performance requirements as outlined in specifications or on part drawings.

In addition, each component part or representative sample population must be adequately tested, and the test results must be reported in a clear and concise manner so that management can make informed decisions on the success or failure of the process being performed.

Nowhere is testing more critical than in heat treatment. Changes to a part as a result of the heat-treatment process happen inside the component (Section 1.1) and, as such, may or may not be readily observable with the naked eye.

On the shop floor, we must be sure that we have selected a representative number of parts to test and that they are from locations that adequately reflect all areas of the load. We must also make sure that any testing on the shop floor is performed with properly calibrated equipment and that we are not testing in such a way as to simply "pass" the part.

Which tests should I perform?

The choice of tests to perform on a component part or representative sample thereof is highly dependent on the material, required properties, specifications and the heat-treatment process(es) performed. Testing is a critical part of the heat-treatment process. It is not something to be manipulated, ignored or considered the responsibility of someone else. The cost of a product recall warrants comprehensive testing programs.

12 | TESTING

THE VALUE OF HARDNESS TESTING

After visual inspection, hardness testing (Section 12.2) is undoubtedly the most common test performed on the shop floor. It is used extensively by individuals involved in quality control, making it very useful in manufacturing and R&D applications.

What is critically important to remember is that hardness testing is an indication (only) and not an absolute guarantee of the success or failure of a particular heat-treatment process. While hardness is used extensively to characterize materials, the hardness test should not be the only test to determine if a material is suitable for its intended use. Hardness testing is popular for a variety of reasons, including:

- Easy to set up and perform
- Quick (1-30 seconds)
- Simple to understand and apply
- Generally considered nondestructive
- Flexible (virtually any size, shape and weight can be tested)
- Practical for both incoming or outgoing product
- Relatively inexpensive

It is important to remember that hardness is not a fundamental material property. Furthermore, the hardness of a material is test-method dependent (i.e., different hardness values can be obtained on the same material by different test methods). Factors that must be considered in order to determine the correct hardness test method for a given application include:

- Material type and material characteristics (e.g., grain size)
- Approximate hardness
- Physical dimensions (e.g., shape, thickness, size)
- Heat-treatment condition
- Production/sampling requirements

While many product designers and heat treaters look to establish a correlation between the hardness result and the required/desired material properties, this is not always possible.

THE VALUE OF MECHANICAL TESTING

The goal of mechanical testing (Section 12.3) is to evaluate the behavior of a material when subjected to a mechanical load. Mechanical testing often measures properties such as strength, hardness, ductility, impact resistance and fracture toughness of materials at various temperatures and load conditions. Examples include tension and compression, torsion, bending (flexure), fatigue, creep and residual stress to name a few. Most, but not all, mechanical tests are performed in the laboratory. Simple mechanical tests that can be (and often are) performed on the shop floor include bend tests and contact (impact) tests.

Mechanical testing (Table 12.1.1) reveals the elastic and inelastic behavior of a material when a force is applied. A mechanical test shows whether a material or part is suitable for its intended mechanical applications by measuring elasticity, tensile strength, elongation, hardness, fracture toughness, impact resistance, stress rupture and fatigue limit.[2]

TABLE 12.1.1 | Typical mechanical tests[2]

Mechanical test	Purpose
Bend	To determine the ductility or the strength of a material by bending it through a given radius
Charpy impact	To determine the amount of energy absorbed by a material during fracture
Fatigue	To simulate the progressive and localized structural damage that occurs when a material is subjected to cyclic loading
Hardness (macro and micro)	To measure how resistant a material is to various types of shape change (usually permanent) when a force is applied
Other	To determine specific mechanical properties by customized test methods; examples include mechanical durability, flow, stress, pressure cycling, structural integrity
Proof load	Used interchangeably with yield strength, it refers to the tension-applied load that a test sample must support without evidence of deformation.
Shear	To measure a material's response to shear loading, a force that tends to produce a sliding failure on a material along a plane that is parallel to the direction of the force
Tensile	To determine the maximum (uniaxial) tension a material can resist before failure
Torque	To determine the maximum resistance of a material when a force is applied to create rotation about an axis

Mechanical properties are governed by the basic concepts of elasticity, plasticity and toughness. Elasticity and plasticity, for example, can be determined from tensile test results and hardness measurements. Toughness is often obtained from impact bending tests. In simplest terms:
- ❖ Elasticity is the capacity of a metal to undergo temporary deformation. As soon as the load that caused this deformation is removed, the metal returns to its original shape.

12 | TESTING

- Plasticity, or ductility, is the ability to undergo permanent deformation without failure. This property is used in metal forming in order to permanently modify the shape.
- Toughness represents the ability to absorb shocks without suddenly failing and can be regarded as the opposite of brittleness.

Mechanical testing is a part of the broader category called materials testing, which encompasses mechanical testing; testing for thermal properties; testing for electrical properties; testing for resistance to corrosion, radiation or biological deterioration; and nondestructive testing (Section 12.6). ASTM International (www.astm.org), NIST (www.nist.gov) and international organizations such as ISO (www.iso.org) are some of the groups providing standards for testing.

THE VALUE OF METALLOGRAPHIC TESTING

Metallurgists know the value of microstructural evaluation (Section 12.4) for providing definitive information in our quest to determine the success or failure of our heat-treatment efforts. Heat treaters strive to determine if a particular component part is good or bad. They often do so by relying on hardness testing alone, which is only a partial indication that all is well. There is really only one place to look to be positive, and that is the microstructure. This is not to say that there is no value in hardness testing, mechanical testing or other analysis techniques, but (as all metallurgists know) the microstructure doesn't lie!

Most informative technical articles and scholarly works rely on microstructures to provide evidence, confirm conclusions or prove certain facts. The key, however, is for the observer to understand the significance of what he is seeing. Therefore, metallographic interpretation becomes an invaluable skill.

Metallurgists secretly enjoy destroying things. If we had our way we would cut up every part in a load just to confirm they were all good. The problem, of course, is that there would be no parts to ship to the customer – not our problem, as metallurgists like to say! We make the argument that other types of testing only give an indication that all is well but cannot make a definitive statement one way or the other. We must also be careful to draw conclusions only from a truly representative sample. There is no greater sin than spending considerable time and effort only to reach a false conclusion – a red herring, if you will pardon the pun.

Metallography Secrets

Any analysis starts with a good visual or low-magnification (5-50X) stereographic inspection of a component part (Fig. 12.1.3). This helps to set the direction for further investigation and often provides a clue as to the best starting point for the investigation, which may be missed if we begin the analysis at the microscopic level.

Metallography, whether it is via optical microscopy (Figs. 12.1.4-12.1.7) or through the use of more advanced tools such as a scanning electron microscope (Fig. 12.1.8), is often the easiest, fastest, most direct and most reliable way to determine the acceptability of a component part, solve a problem or determine the root cause of a failure. In the hands of a skilled metallographer, answers come quickly and positive conclusions can be drawn.

One of the secrets to success in metallographic interpretation is the ability to spot something in the microstructure that is abnormal. Management often doesn't understand why metallographic equipment costs so much, why more and more sophisticated tools are necessary or why it takes so long to reach a conclusion. The metallographer is akin to a detective, constantly piecing together clues until the crime is solved. To the uninformed, however, getting answers is an exasperating experience because metallurgists never seem to have enough evidence to conclude with absolute certainty.

The Role of Comparative Metallography

It is a practical reality that many companies find themselves without a metallurgist or with fewer metallurgists having less time to support production on the shop floor. As such, technicians or those less skilled in the art are called upon to make daily decisions. This is where comparative metallography comes in.

Having examples of acceptable and unacceptable microstructures allows non-metallurgists to make decisions. The key is to have these comparisons prepared by a metallurgical laboratory that knows and understands what is needed. For example, there is no need for color metallographic images if the people doing the work day in and day out do not use this technique. Photomicrographs should be provided at a magnification used for routine evaluation.

Metallurgical Reports

A comment about reports is also in order. If you request an analysis from an outside metallurgical laboratory, recognize that their reports are most often written by metallurgists and intended for other metallurgists, so interpretation by those with a different skill set is sometimes difficult (if not impossible). Reports from outside testing laboratories often do not attempt to interpret the facts, which is understandable since they do not have enough background information to understand the full scope of the problem under investigation. Asking the right questions of them and, if necessary, bringing in third parties for help in deciphering the results is of critical importance.

12 | TESTING

Is metallographic interpretation a lost art?

Universities, technical societies, suppliers of metallographic equipment and independent third-party consultants offer courses in metallographic interpretation. These are important and necessary to teach the basic skills, but they are no substitute for hands-on application of the lessons learned. Metallography is a practical science gained by doing and doing over and over again. While one etchant or method may reveal the desired answer, experimenting with other techniques – perhaps a different edge-retention method, polishing cloth or etchant – may produce new insights. Don't be afraid to try new things. You will like what you find!

FIGURE 12.1.3 | 440C stainless steel (composite topographical image, 30X) – example of thread damage from machining (courtesy of Aston Metallurgical Services Co. Inc.)

FIGURE 12.1.4 | A356.0 aluminum casting (unetched, 125X) – blistering due to hydrogen embrittlement

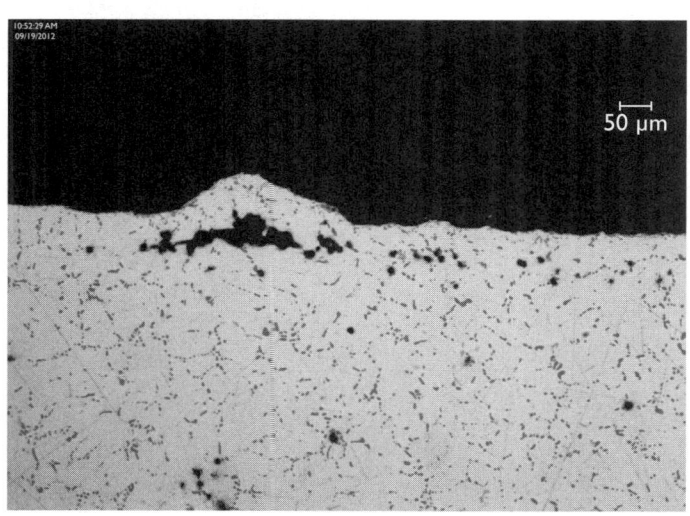

FIGURE 12.1.5 | Inconel 600 retort failure on the raw ammonia (inlet side) of an ammonia dissociator (Kallings, DIC, 25X) – effect of gaseous nitriding (inside to outside) after seven years of service (courtesy of Aston Metallurgical Services Co. Inc.)

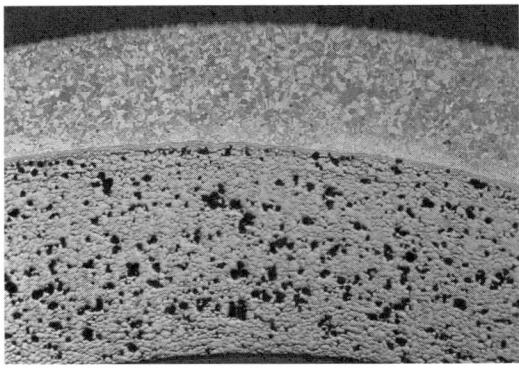

FIGURE 12.1.6 | SAE 8620 (1250X, 2% nital) – pearlite and ferrite formed in the core of a carburized 8620 gear tooth, indicating improper quenching (courtesy of Aston Metallurgical Services Co. Inc.)

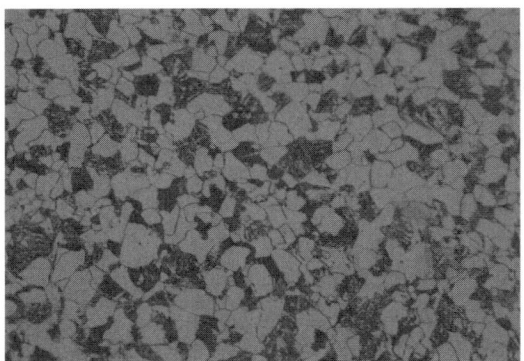

FIGURE 12.1.7 | 330 stainless steel alloy illustrating catastrophic carburization (Kallings, 1000X) – radiant-tube failure in the area where the tube passed through the insulation (courtesy of Aston Metallurgical Services Co. Inc.)

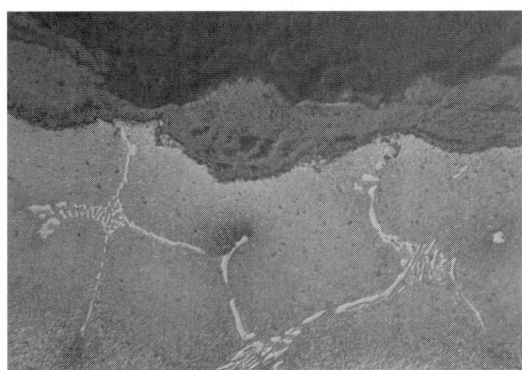

**FIGURE 12.1.8 | 52100 bearing race (3680X, 2% nital) – SEM reveals an acceptable microstructure of fine carbide distributed in a matrix of tempered martensite

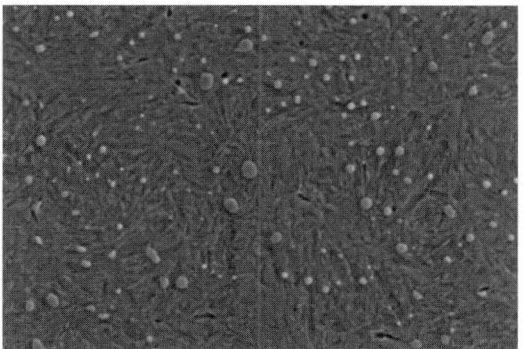

Summary

Testing must be an integral part of what we do. Our job as heat treaters isn't finished until a representative number of heat-treated parts or test samples have been checked and found to be in compliance to specification. Then, and only then, the paperwork can be completed.

Remember, testing is not intended to be a condemnation of our efforts but rather one of the last lines of defense to ensure that the parts in question have been properly processed. Testing is confirmation that we have indeed produced the finest quality component that any technological process can produce.

REFERENCES

1. Instron (www.instron.us)
2. Element Materials Technology (www.element.com)
3. Encyclopedia Brittanica (www.britannica.com)
4. Herring, Daniel H., "The Value of Metallographic Interpretation," *Industrial Heating*, March 2013
5. Alan Stone, Aston Metallurgical Services Co. Inc. (www.astonmet.com), technical and editorial review, private correspondence
6. Steve Carey, New Hampshire Ball Bearing, technical contributions and private correspondence
7. George Vander Voort, Vander Voort Consulting, private correspondence

12.2 HARDNESS AND MICROHARDNESS TESTING

Hardness testing is one of the most important and seemingly easiest tests to perform on the shop floor, in the quality-control department or in the metallurgical laboratory. However, it's arguably one of the hardest tests to perform properly. It is often, though incorrectly so, used as the sole determination of the success or failure of a particular heat-treatment operation.

What is hardness?

A common definition is the measure of the resistance of a material to an applied force involving the use of an indenter of fixed geometry under static load. However, hardness can also refer to stiffness or temper resistance or resistance to scratching, abrasion or cutting. It is the property of a metal that gives it the ability to resist being permanently deformed (i.e., bent, broken or have its shape changed) when a load is applied.

The greater the hardness of the metal, the greater resistance it has to deformation. The ability of the material to resist plastic deformation depends on its microstructure. Therefore, the same material can have different hardness values depending on its microstructure, which is influenced by the heat-treatment process.

Hardness is not a fundamental material property but rather a composite value. It is of such great interest because hardness can be directly related to the expected strength of the material, which would otherwise require destructive testing to measure. The location of the hardness indentations away from working surfaces is often an important consideration if hardness testing is to be considered a nondestructive test.

Hardness measuring methods can be sorted into three general categories depending on the manner in which the tests are conducted: scratch hardness (Fig. 12.2.1), indentation hardness (Fig. 12.2.2) and rebound (i.e., dynamic) hardness (Fig. 12.2.3).

FIGURE 12.2.1 | Typical scratch (file) hardness testing arrangement (courtesy of Fred V. Fowler Company)

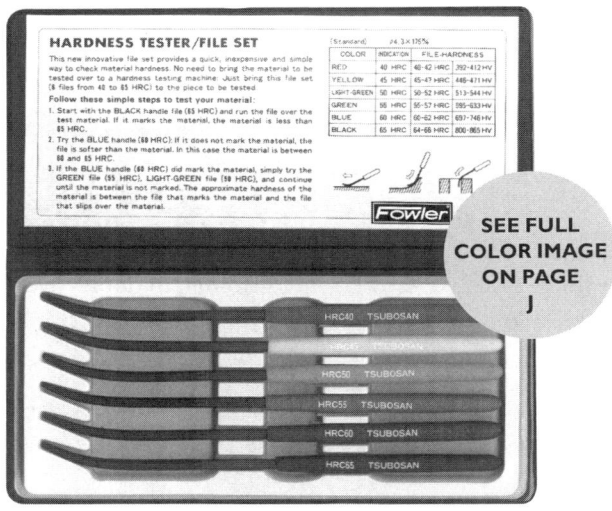

FIGURE 12.2.2 | Typical indentation hardness testing arrangement (courtesy of Buehler, an ITW Company)

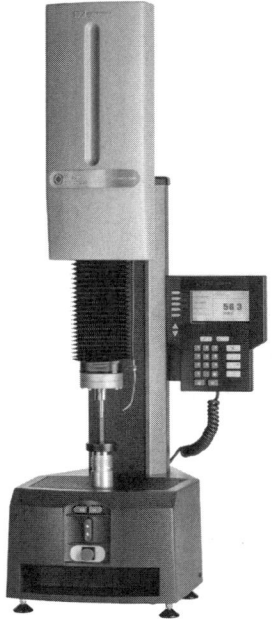

FIGURE 12.2.3 | Typical rebound (dynamic) hardness testing arrangement (courtesy of Ray Company)

a.

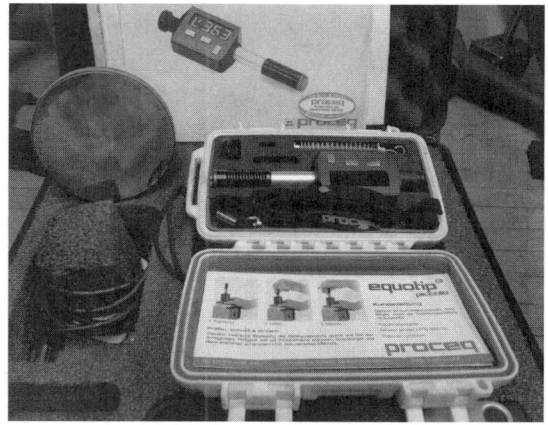

b.

a. Instrument

b. Probe close-up

Of all the mechanical testing equipment, heat treaters are probably the most familiar with hardness testers. Almost every shop has either a Rockwell or Rockwell superficial tester since these machines are most often used for ferrous and nonferrous materials. Many shops also have Brinell hardness testers and/or microhardness (microindentation) hardness-testing capability.

Testing Basics

A part can be, and often is, hardness tested before and after heat treatment. One begins by selecting the proper hardness tester, test method and hardness scale based on information about the anticipated condition of the sample, making sure that the hardness tester selected has been properly calibrated. This is done by choosing a test block of known hardness in a hardness range representative of the parts to be tested.

Next, one selects the indenter (Fig. 12.2.4) and the load (weight) appropriate for the particular part under evaluation. The material, specification requirement and/or thermal treatment(s) that the component has undergone are important factors in this decision-making process. Finally, placing the sample on the correct anvil (Fig. 12.2.5) and, if necessary, supporting it to prevent movement is part of the process. Be aware that if the anvil or indenter is changed prior to or during testing, the tester must be recalibrated using the appropriate test block.

FIGURE 12.2.4 | Various styles of Rockwell hardness indenters (courtesy of Gilmore Diamond Tools Inc.)

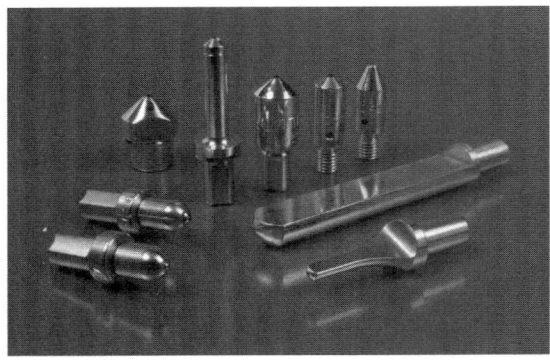

FIGURE 12.2.5 | Various styles of support anvils

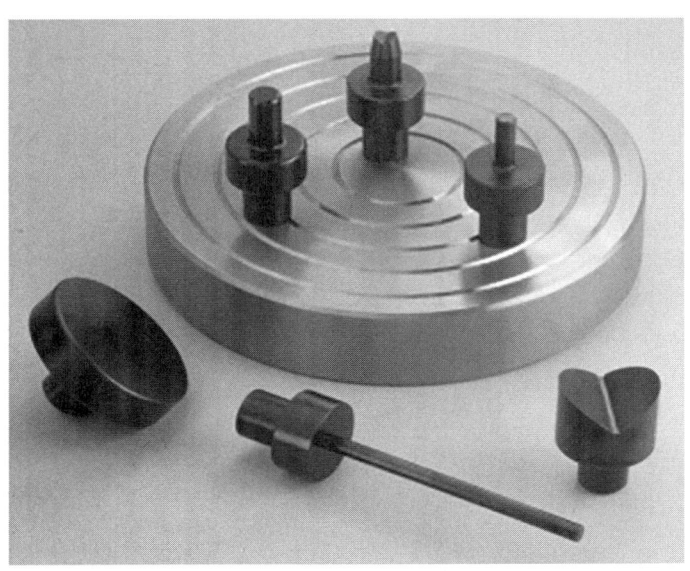

12 | TESTING

The hardness value obtained is a function of either the size of the indent (Brinell and Vickers/Knoop) or the depth of penetration by the indenter (Rockwell). The penetration is measured by either depth or area and translated into a hardness number. It is important to recognize that all conventional hardness-testing methods involve sampling some volume of material. The amount of material actually sampled is a function of the indenter selected, applied load and material properties. If the sampled volume is limited by the physical size of the piece to be tested, then you may actually be sampling the underlying anvil or pushing out beyond the edge of the sample.

Types of Hardness Tests

The most commonly used indentation hardness tests are the Brinell, Rockwell, Rockwell superficial and microhardness methods.

BRINELL HARDNESS TESTING[3]

Brinell testing dates back well over 100 years and is particularly well suited for forgings and castings, inhomogeneous material (e.g., cast irons) or variable microstructures (e.g., coarse-grained material). Brinell testing uses the largest size indentation of all the test methods and, as such, offers the best indication of bulk hardness.

Brinell test hardware is often configured to suit the testing needs of a specific part or manufacturing process. Portable Brinell testing gives the option of bringing the hardness test to the part rather than the part to the hardness tester. The use of light-load Brinell testing is less common in North America than in Europe and the rest of the world.

A Brinell tester has two basic components: the indentation system and the microscope.

Brinell Indenter

The indentation system consists of the indenter, the method of loading and the support frame. The indenter itself is made from tungsten carbide (with a specific chemistry and hardness). The standard indenter for iron and steel is 10 mm (0.4 inch) diameter. However, smaller diameters (such as 1, 2.5 or 5 mm) may be used for certain applications.

As a rule, the indentation should be at least 24% of the ball diameter to avoid damaging the indenter and to avoid problems accurately measuring the indentation. The test indentation should be no more than 60% of the diameter of the indenter to avoid excessive plasticity around the indentation and reduced sensitivity of the measurement. The indenter surface must be free of scratches and defects. Steel-ball indenters were previously used, but they are no longer allowed.

Brinell Scopes

Brinell scopes can vary from simple viewers (20X) to a fully automated system that reads the indentation using imaging software. Some systems also measure the depth of the indentation and convert the depth reading into the expected indentation diameter.

Brinell Measurements

The following steps describe the Brinell test:
1. The test sample is prepared, typically by grinding, to provide a smooth and clean surface.
2. The indenter is forced into the test sample using the desired load for the prescribed dwell time.
3. The load and the indenter are removed, leaving a concave-shaped indentation.
4. The diameter of the indentation is measured in two perpendicular locations using the calibrated eyepiece in the Brinell scope.
5. The diameter is converted into a Brinell hardness using tables or by calculation.

Measuring the diameter of the indentation provides a reading in mm. If a 10-mm ball was used in conjunction with a test force of 3,000, 1,500 or 500 kgf, then ASTM E-10 provides a conversion table to Brinell hardness. If a non-standard ball or force is used, however, the reading can be calculated (Equation 12.2.1).

12.2.1) $$HB = 0.102 \cdot \frac{2F}{\pi D(D-\sqrt{D^2-d^2})}$$

where:
D = ball diameter (mm)
F = force (N)
d = mean diameter of the indentation (mm)
π = Pi, a constant having a value of 3.1416

The symbol for Brinell hardness is HB, and the test number is reported to three significant digits. For example, correct test numbers would be HB 201 or HB 99.5. This is the measured value converted to a hardness number using the Brinell hardness number table. Next, the ball diameter is reported in millimeters (mm), the test force in kilograms force (kgf) and the test duration in seconds. For example, 555 HB 10/3000/15 means that a Brinell hardness of 555 was obtained using a 10-mm-diameter tungsten-carbide ball with a 3,000 kgf test force for a period of 15 seconds. The former designations of HB_W and HB_S are now obsolete.

ROCKWELL AND ROCKWELL SUPERFICIAL HARDNESS TESTING
Hardness File Method

Testing of a metal surface by the use of a file of known hardness (Table 12.2.1) is a well-established technique that is simple to perform and surprisingly accurate. Starting with the highest hardness file (e.g., 65 HRC) and using light to moderate pressure, move the file across the metal surface in one direction only one time. If the resistance of the metal to the file forces the file to "drag" along the surface (i.e., marks the metal's surface), the material is less than 65 HRC.

Repeat the process with the next softer file (e.g., 60 HRC) and so on and so forth until no marking of the surface (i.e., resistance to the movement of the file) is felt. The hardness of the material will be between the last number that marked the metal and the first number that did not. Keep in mind this test might be considered destructive testing.

TABLE 12.2.1 | Hardness ranges for testing files[5]

Hardness range, HRC (HV)	File color	Indication, HRC
40-42 (392-412)	Red	40
45-47 (446-471)	Yellow	45
50-52 (513-544)	Light green	50
55-57 (595-633)	Dark green	55
60-62 (697-746)	Blue	60
64-66 (800-865)	Black	65

Rockwell Method

The Rockwell hardness test and tester is unquestionably the most widely used and versatile of the hardness tests for materials in their heat-treated condition. Sheet material in heavier gauges and cemented carbides can also be tested. The expected hardness of the material, along with the type of indenter and test load, determines the correct scale and weight to use in the test (Table 12.2.2).

In the Rockwell method of hardness testing, the depth of penetration of an indenter under certain arbitrary test conditions is determined. The indenter may either be a steel ball of some specified diameter or a spherical, diamond-tipped cone of 120-degree angle and 0.2-mm tip radius called a Brale®. The hardness numbers themselves have no units. The higher the number in each of the scales, the harder the material.

A Rockwell hardness measurement can be defined as macro-, micro- (or even nano-) scale according to the forces applied and displacements obtained. Typical indenters used during Rockwell tests include diamonds and carbide or steel balls. The type of indenter and the test load determine the hardness scale. Rockwell B and C scales are the most common, although many other scales can be used for specific test purposes.

Rockwell hardness measurements are actually a conversion of the depth of the indentation left by the indenter. Each Rockwell point (in the English system) represents 0.00008 inches of depth. For a reading of 60 HRC, subtract the major and minor loads – 150-10 x 0.00008 = 0.0112 inches. So, the depth of a 60 HRC reading is 0.0072 inches.

In some cases, the part to be tested may be particularly large or heavy (Fig. 12.2.6), which necessitates the proper choice of anvil and support system. Some hardness testers are equipped with Steady-Rest® or similar devices to support the load. Any movement of the sample being tested voids the reading observed.

FIGURE 12.2.6 | Typical Rockwell test

TABLE 12.2.2 | Typical Rockwell hardness scales by applications

Rockwell scale	Typical application of the scale
A	Cemented carbides, thin steel and shallow case-hardened steel
B	Copper alloys, soft steels, aluminum alloys and malleable irons
C	Steel, hard cast irons, pearlitic malleable iron, titanium, deep case-hardened steel and other materials harder than HRB 100
D	Thin steel, medium case-hardened steel and pearlitic malleable iron
E	Cast iron, aluminum and magnesium alloys, bearing materials
F	Annealed copper alloys and thin soft sheet metals
G	Phosphorous bronze, beryllium copper, malleable irons and materials softer than HRG 92
H	Aluminum, lead and zinc
K, L, M, P, R, S, V	Bearing materials and other very soft or thin materials. Use the smallest ball and heaviest load that does not give anvil effect.
N	Applications similar to the A, C and D scales but of thinner gauge or shallower case depth
T	Applications similar to the B, F and G scales but of thinner gauge
W, X, Y	Very soft materials

The Rockwell test uses two loads, one applied directly after the other. The first load, known as the "minor" load of 10 kg, is applied to the specimen to help seat the indenter and remove the effects of any surface irregularities. In essence, the minor load creates a uniformly shaped surface for the major load to be applied to.

The Rockwell hardness is calculated (Equation 12.2.2) from the difference between the depth of the indentation after the application of the total force and the initial depth under the application of the preliminary (minor load) force. This difference is h, measured in millimeters.

12.2.2)   Rockwell hardness value = N − (h/S)

where:

N = numbers specific to the Rockwell hardness scale: 100 for scales A, C, D, 15N, 30N, 45N, 15T, 30T and 45T; and 130 for scales B, E, F, G, H and K.

h = permanent depth of indentation, in mm

S = scale unit specific to the Rockwell hardness scale: 0.002 mm for scales A, B, C, D, E, F, G, H and K; and 0.001 mm for scales 15N, 30N, 45N, 15T, 30T and 45T.

There are many Rockwell scales other than the B and C scales (often called the common scales). These other scales also use a letter for the scale symbol prefix, and many use a different-sized steel-ball indenter. A properly used Rockwell designation will have the hardness number followed by "HR" (hardness Rockwell), which will be followed by another letter that indicates the specific Rockwell scale.

An example is 60 HRB, which indicates that the specimen has a hardness reading of 60 on the B scale.

For soft materials such as aluminum alloys, a 16-mm-diameter (1/16-inch-diameter) steel ball is used with a 100-kg load. The hardness is read on the B scale. In testing harder materials such as steel alloys, a 120-degree diamond cone is used with up to a 150-kg load, and the hardness is read on the C scale.

ASTM E18 defines how readings are to be reported, which is different from common industry practice in some instances.

Rockwell Superficial Hardness Testing

Rockwell superficial is used where lighter loads are required for testing thin case-hardened surfaces, decarburized layers and sheet material in thin gauges. These testers can be dedicated machines or combination (Rockwell and Rockwell superficial) units.

Rockwell superficial hardness testers work similarly to the standard Rockwell tester but use a reduced minor load (just 3 kg) and have the major load reduced to either 15, 30 or 45 kg, depending on the indenter. Using the 16-mm-diameter tungsten ball indenter, a T is added (designating thin-sheet testing) to the superficial hardness designation. An example of a superficial Rockwell hardness is 22 HR15T, which indicates the superficial hardness as 22, with a load of 15 kg using the tungsten ball. If the 120-degree diamond cone is used instead, the scale designation is N rather than T. The use of steel balls is now obsolete.

MICROHARDNESS TESTING

Microhardness tests are typically used for very small, intricate shapes, thin parts and case-depth determination. Microhardness testing (Fig. 12.2.7) requires care in order to achieve accurate and repeatable results. The testing requires more-involved sample preparation and operator skill as well as separate testing equipment. Unlike Rockwell scales, the respective applied load and the location in the microstructure where the hardness is measured affect microhardness values.

FIGURE 12.2.7 | Microhardness testing (courtesy of Struers Inc.)

12 | TESTING

Microhardness testing is performed to measure the hardness of constituent phases, segregation effects and the carburizing or nitriding gradients in a metal. Indenters for microhardness testing (Fig. 12.2.8) include Knoop and Vickers with applied loads ranging from 1 gram to 1,000 grams. The test is destructive for both methods since careful preparation of the metallographic sample is necessary. The indentations are typically measured optically at magnifications of 400-1,000X. High-quality optics are critical, especially when using magnifications of 500X and greater. Accurate measurement of the indentation boundaries is mandatory, especially when using lighter loads.

Microhardness testing is ideal for precisely sampling a small region, a very thin or soft sample, or a material with either hard or soft particles you wish to include or exclude from the field of measurement (Fig. 12.2.9).

FIGURE 12.2.8 | Typical microhardness indenters[1]

a.

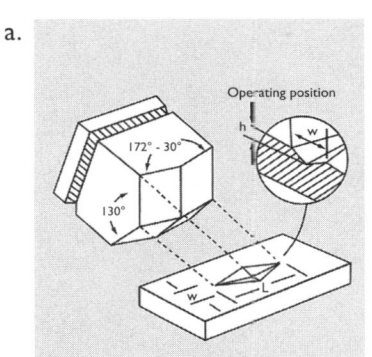

b.

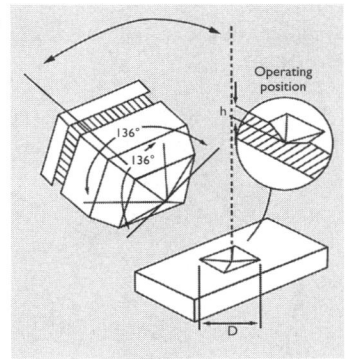

c.

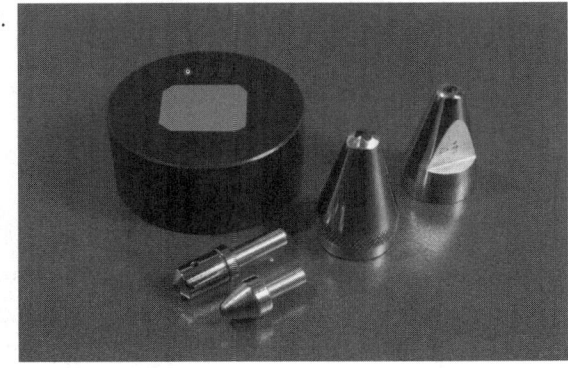

a. Dimensions for Knoop indenter

b. Dimensions for Vickers indenter

c. Knoop and Vickers indenters and test block (courtesy of Gilmore Diamond Tools)

FIGURE 12.2.9 | Minimum thickness charts[2]

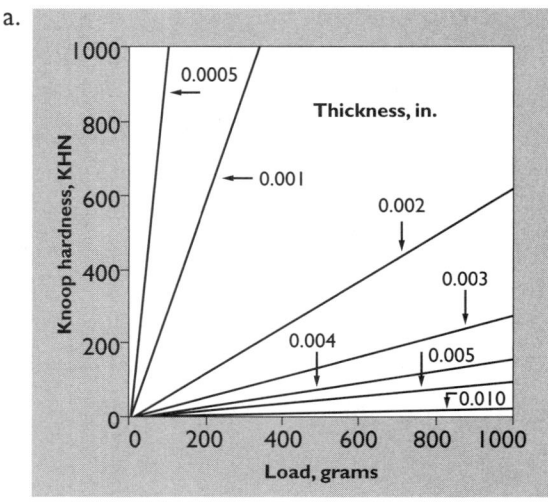

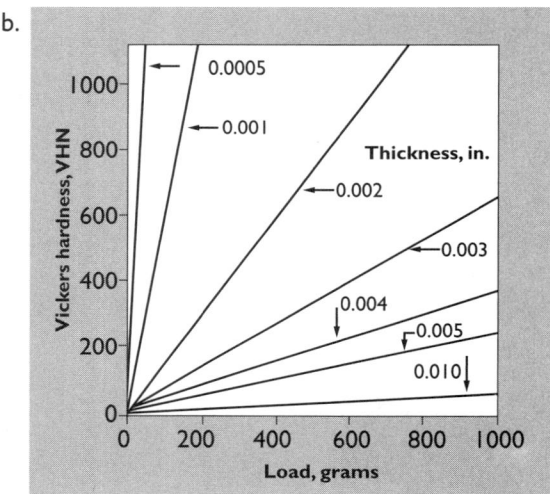

a. Knoop

b. Vickers

Microhardness testing is sensitive to a number of variables, such as indentation spacing, material segregation, inclusions, edge effects and cold working. For example, a Knoop hardness value of 100 taken with a 500-gram load (100 HK 0.5) is not equivalent to a Knoop hardness value of 100 taken with a 50-gram load (100 HK 0.05). Hence, it is crucial to report the applied load along with the test result. In addition, ASTM E140 only recognizes microhardness conversions to the Rockwell C scale with minimum loading of 500 grams. Therefore, conversions from other Rockwell scales produce errant values at increasingly lighter loads (Fig. 12.2.10).

FIGURE 12.2.10 | Microhardness values as a function of test load[2]

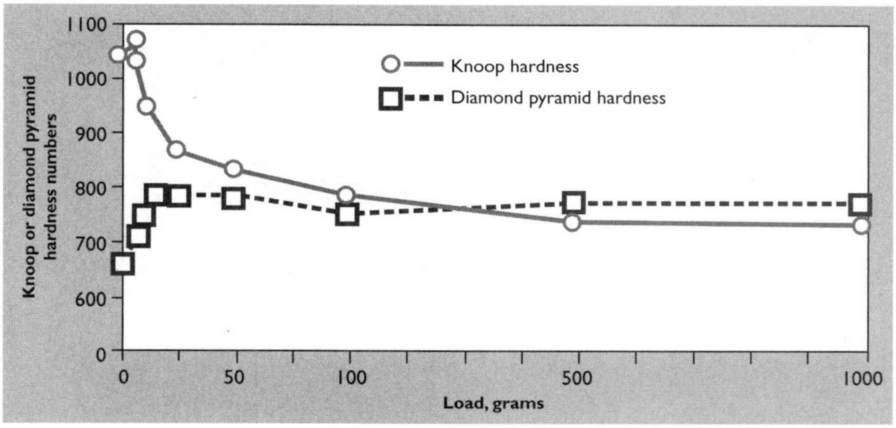

Practical Considerations

To the layman, hardness testing sounds pretty simple … until you factor in real-world considerations. What happens if there is a hard case (e.g., nitriding, carburizing or plating) over a soft core? How do you evaluate a soft layer (e.g., decarburization) over a hard core? What impact do soft inclusions or hard particles such as carbides have on machinability? What if the shape is complex and you don't have flat parallel surfaces? What if you can't get your indenter to the region of interest? These and many other questions make the task of taking a simple hardness measurement far more complex.

One of the most common problems we see is that the material is too thin, soft or irregularly shaped to allow for Rockwell testing. Another problem is that the individual or method fails to take into account the three-dimensional aspects of the part, or the selected test method does not truly sample the appropriate volume of material.

For example, imagine a pen being pushed into a small cubic chunk of clay and then retracted (Fig. 12.2.11). You are left with a hole whose depth is dependent on the force of your applied load and the resistance of that clay to the indentation. A different sized or shaped pen (Fig. 12.2.12) produces a different type of indentation. The heavy loads produce deeper indentations and sample more material volume. The higher the hardness of the material, the more resistant it is to penetration and the shallower the resultant indentation.

Hardness testing is not immune to a variety of issues, depending on the type of test being performed.

FIGURE 12.2.11 | Large pen indenter applying load into clay cube[3]

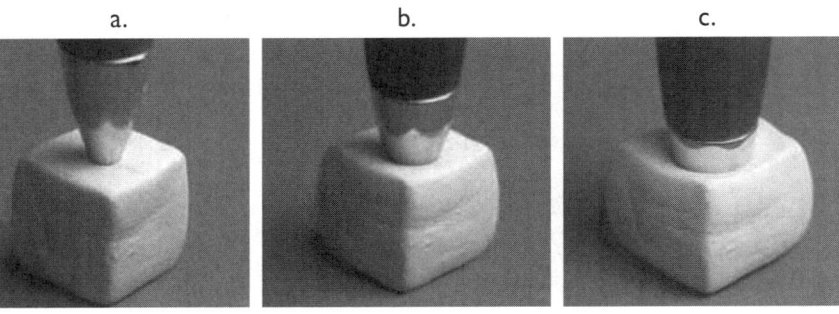

a. Initial force

b. Intermediate force

c. Excessive force (bulging or side distortion in evidence)

FIGURE 12.2.12 | Small pen indenter applying load to clay cube with minimal bulging (side distortion)[3]

COMMON PROBLEMS IN BRINELL HARDNESS TESTING

Here are the most common problems encountered in Brinell testing and possible solutions for them (in no particular order).

1. The sample is too hard.
    - Brinell testing should be limited to metallic materials with hardness values not exceeding 650 HB.
2. The sample is too thin.
    - The test specimen should be five times thicker than the depth of hardness indentation.
3. An incorrect measurement of the impression has taken place.
    - The impression should be essentially round and measured in two positions approximately 90 degrees apart.
    - The reported value is the sum of these measurements divided by two.

4. An incorrect eyepiece resolution has been used.
   - ❖ The micrometer eyepiece in the microscope should have divisions of 0.1 mm and be able to estimate to 0.05 mm.
5. The sample surface is dirty.
   - ❖ Parts should be clean and free of oils or contamination.
   - ❖ Parts can be ground or polished if necessary.
6. The curvature of the surface is wrong.
   - ❖ Radius of the curvature should be no more than 2.5 times the diameter of the indenter ball.
7. The distance from the edge is too close.
   - ❖ Indentions should be done 2.5 indenter diameters from the part's edge.
8. The distance between indentations is too close.
   - ❖ Indentions should be done 2.5 indenter diameters from the next indentation.
9. The indenter is not properly aligned with the part.
   - ❖ The test force should be applied at a 90-degree angle to the part, plus or minus only two degrees.
10. There is excessive plasticity in the sample.
    - ❖ If the indenter pushes into the sample with more than 60% of the diameter penetrating into the surface, excessive material will dimple up around the indentation. This makes the reading less accurate and more difficult to read. Use a larger indenter or a smaller test piece.
11. Hardness conversions are inaccurate.
    - ❖ Due to the nature of Brinell testing, conversion to the other hardness scales may be inaccurate.
12. The test duration is too long or too short.
    - ❖ The typical test time for steel and iron is 10-15 seconds. Other materials, such as aluminum, may require a 30-second dwell time. The test duration should be reported as part of the reading. Shorter test durations are acceptable if this has been proven to not influence a correct result.

COMMON PROBLEMS IN ROCKWELL HARDNESS TESTING

Here are some of the most common problems encountered in Rockwell hardness testing and possible solutions for them (in no particular order).

1. Cleanliness of the part and tester is paramount.
   - ❖ Remove any scale, debris, dirt and oil on the part or the machine before testing. A small amount of debris can alter the reading by as much as several Rockwell points.

- ❖ Remove and clean the indenter and anvil prior to operation and at shift change.
- ❖ Light sanding of both the bottom and top surfaces may be necessary before hardness testing. Wipe or blow off any debris left from sanding operations.
2. Non-flat surfaces can alter readings.
    - ❖ Extremely rough or textured surfaces (e.g., machining marks) may give inconsistent readings. Light sanding of both the bottom and top surfaces may be necessary before hardness testing.
3. Take into account the curvature of the surface.
    - ❖ A correction factor must be added to the hardness reading of small-diameter shapes for Rockwell A, C and D scales and varies with the apparent hardness and part diameter. The correction factor to be added is shown in the appropriate ASTM E18 tables.
    - ❖ Surfaces not perpendicular to the indenter will give false readings. Surfaces should be flat within 2 degrees. Be careful when taking readings on mounted samples. They must be flat, thick and not flex under load. A microhardness test may be more appropriate.
    - ❖ Readings taken too close to the sample edge will both damage the indenter and give false readings. Indentations should be no closer than 2.5 times the indenter diameter from the edge. If the metal buckles outward, the indenter is too close to the edge and the reading is invalid.
    - ❖ Readings taken too close together will give false (higher) hardness readings. Indentations should be three diameters apart.
    - ❖ A damaged indenter (chipped or cracked diamonds or flattened balls) will produce false readings. Periodically remove the indenter from the hardness tester and inspect the tip using a low-power magnification (10-50X) such as a stereomicroscope or loop to check for damage. Flattened balls are sometimes difficult to detect unless you inspect all surfaces – often at an angle.
    - ❖ Parts that are not properly supported will give false readings. Large and irregularly shaped parts need to be well supported. Parts that move, even slightly, during the test produce a false reading – even if that reading falls within the desired hardness range.
    - ❖ Change the anvil to one that keeps the part stationary using the variety that should be available with your tester. Additional outside support (such as a Steady-Rest®) may also be required.
    - ❖ A sample that is too thin will yield false readings. The material being tested should have a thickness at least 10 times the depth of

the indentation. Minimum acceptable thicknesses can be found in ASTM E18 tables. Special (pre-hardened) anvils can be used when hardness testing thin sheet or foil material.

COMMON PROBLEMS IN MICROHARDNESS TESTING

Here are some of the most common problems encountered in microhardness testing and possible solutions for them (in no particular order).

1. The optical measurement of the indent is the most common source of errors.
   - Make sure the magnification is correct. The image should fill from 50-75% of the total field of view. Practice by reading the calibration indents on certified test blocks.
   - Adjust the optics for maximum contrast.
   - Avoid vibration.
2. The sample surface preparation is important.
   - The smaller the indent, the better the required surface finish.
   - Make sure that there is no dirt or other foreign material on the top or bottom surface of the sample or the indenter.
   - Make certain that the surface preparation does not affect the hardness of the test location.
3. Sample mounting is important.
   - The sample must be mounted so that the test point is perpendicular to the indenter.
   - The sample must not move or deflect during the test-force application. Avoid excessively large loads too close to the edge of the sample (springboard effect).
   - If measuring near the edge of the sample, the type of mounting material may influence the readings.
4. Verify the tester calibration frequently using traceable, certified reference material.
5. Make certain that the test-force application and dwell times are correct for the samples you are testing.
   - Materials that exhibit significant cold-flow characteristics may require a longer dwell time. Check the test specification.
   - Location of the indentation with respect to the edge of the sample and proper spacing of the indentations must follow the requirements specified in ASTM E384.
6. A microhardness tester is a sensitive instrument.
   - Make sure that there is no vibration or other environmental conditions present that could affect the results.

7. Make certain that you understand the test results.
   - ❖ Knoop and Vickers test results are considered test-force-dependent. For test forces below 500 g, the results can vary significantly. For example, compared to 500 g results, at 25 g the Knoop hardness can go up by 200 points, while the Vickers value could go down by 50 points on the same sample, due solely to the test-force difference.
8. Remember that conversion to Rockwell C values is only valid when using 500 gf loads. Refer to ASTM E140.
9. When testing thin parts, make sure that the thickness of the sample is at least 10 times the depth of the indent.
10. Always follow the appropriate ASTM requirement (typically ASTM E384, ASTM E92 or ISO 6507).
11. Sample orientation must be such that the sample is flat and the indenter is in contact with the surface evenly.
    - ❖ The sample surface must be tested perpendicular to the indenter.
    - ❖ Samples need to be mounted and polished using standard metallographic techniques.
12. A microhardness test is often used to show hardness gradients within multiphase materials.
    - ❖ A homogenous sample is optimal for general microhardness testing.
    - ❖ In the case of a multiphase alloy, it is often necessary to take multiple hardness measurements to obtain an adequate sampling of different phases (grains) within the alloy.
    - ❖ The sampling scheme should be determined on the basis of statistical calculation and the model that is chosen.

Test Standards

The more common test standards (and latest revisions) used in the heat-treatment industry include the following.

BRINELL

- ❖ ASTM E10 (Standard Test Method for Brinell Testing of Metallic Materials)
- ❖ ISO6506-1 (Metallic materials – Brinell hardness test – Part 1: Test Method)
- ❖ ASTM E103 (Standard Test Method for Rapid Indentation Hardness Testing of Metallic Materials)

ROCKWELL
- ASTM E18 (Standard Test Methods for Rockwell Hardness and Rockwell Superficial Hardness of Metallic Materials)

MICROHARDNESS (VICKERS OR KNOOP)
- ASTM E3 (Standard Practice for Preparation of Metallographic Specimens)
- ASTM E384 (Standard Test Method for Knoop and Vickers Hardness of Materials)

OTHER
- ASTM E140 (Standard Hardness Conversion Tables or Metals Relationship Among Brinell Hardness, Vickers Hardness, Rockwell Hardness, Superficial Hardness, Knoop Hardness, Scleroscope Hardness and Leeb Hardness)

In addition, wall charts and pocket tables showing the conversion between hardness numbers and values such as approximate tensile strength are available from a number of equipment manufacturers.

Summary

Most people need not be experts in all the intricate details of hardness testing. However, it is important that the user selects the appropriate hardness-testing method and scale, considers part geometry and test location, and is aware of equipment and testing limitations. Failure to do so can lead to improper interpretations of the true material condition, properties and hardness.

Should you find yourself in a dispute regarding hardness and hardness-testing methods, the first item to confirm is that the specified hardness is appropriate for that material. Next, investigate how the hardness was measured and if it was an appropriate method for that sample. While there can be shades of gray and varying levels of uncertainty between hardness-testing machines or laboratories, expect some level of consensus if the methods are correct. ASTM guidelines[6] say to report Rockwell results to the nearest integer. In one laboratory study, it was found that the uncertainty for the Rockwell C scale was only 1 HRC point over a 10-year period.

Everyone involved with hardness testing should be familiar with the appropriate ASTM specifications and ISO standards. These specifications address proper sample preparations, selection of loads and penetrators, sample geometry, minimum sample thickness considerations, roundness corrections, spacing and edge considerations, and conversions between scales.

REFERENCES

1. Lysaght, Vincent E. and Anthony DeBellis, *Hardness Testing Handbook*, American Chain and Cable Company, 1969
2. Stone, Alan and Daniel H. Herring, "Practical Considerations for Successful Hardness Testing," *Industrial Heating*, April 2006
3. Midea, S., "Brinell Testing," HOT TOPICS in Heat Treatment & Metallurgy, Vol. 2, No. 12, December 2004
4. Midea, S., "The Basics of Microindentation Hardness Testing," HOT TOPICS in Heat Treatment & Metallurgy, Vol. 1, No. 2, December 2003
5. Fred C. Fowler Company (www.fvfowler.com)
6. ASTM specifications (www.astm.org)
7. ISO specification (www.iso.org)
8. Instron Corporation (www.instron.us)
9. Alan Stone, Aston Metallurgical Services Co. Inc. (www.astonmet.com), technical and editorial review, private correspondence

12.3
MECHANICAL TESTING

Product reliability and fitness for purpose is ensured by testing. The purpose of mechanical testing is to help us predict how a given material and/or component (Fig. 12.3.1) will respond in its end-use environment – both in the short- and long-term – before its application in the field. Since heat treatment can have a profound influence on final properties, it is especially important to test the product at this point in the manufacturing process.

Types of Tests

The diversity of product types, sizes and end-use applications requires us to select from a variety of test methods (Table 12.3.1), from simple dimensional checks to rigorous mechanical testing. It is important to remember that the type of mechanical test method(s) used *must* be representative of the forces the component will experience in service. Some of the more common types of tests related to predicting mechanical failure of heat-treated components include:

- ❖ Hardness (Section 12.2)
- ❖ Tension (tensile strength and transverse rupture)
- ❖ Torsion
- ❖ Fatigue
- ❖ Creep and stress rupture
- ❖ Impact

Many other mechanical tests are used, however, and some are product-specific:

- ❖ Chemical analysis
- ❖ Corrosion
- ❖ Engagement
- ❖ Microstructural evaluation
- ❖ Pull-out/push-out

- ❖ Stress durability
- ❖ Vibration
- ❖ Wear
- ❖ Wedge tensile (10°)

Finally, it is important that testing be performed on the raw material, after heat treatment and on the finished product.

FIGURE 12.3.1 | Forged bolts (courtesy of Instron®)

In mechanical testing, parts are analyzed to determine their mechanical properties. Mechanical properties are those associated with elastic or inelastic behavior of a component when force is applied. It involves the relationship between stress and strain. A mechanical test shows whether the material is suitable for its intended application by measuring such aspects of performance as elasticity, tensile strength, elongation, hardness and fatigue limit.

TABLE 12.3.1 | Example of common test methods for fasteners[1]

Test	Method	Applicable Specifications
Chemical analysis	Optical emission spectrochemical analysis	ASTM E415, ASTM E1086
Mechanical and physical testing and inspection	Axial tensile strength of full-sized threaded fasteners	ASTM A370, ASTM F606, SAE J429, SAE J1216, ISO 6892
	Charpy impact (V-notch)	ASTM A370, ASTM E23
	Compression test of washers	ASME B18.21.1
	Cone-proof load of internally threaded fasteners (nuts)	ASTM F606
	Drive test	ANSI/ASME B18.6.4, SAE J933
	Embrittlement test of washers	ASME B18.21.1
	Hardness preparation	ASTM F606
	Microhardness of fasteners	ASTM E384
	Proof load of full-sized externally threaded fasteners	ASTM A370, ASTM F606, SAE J429, SAE J1216
	Proof load of internally threaded fasteners (nuts)	ASTM A370, ASTM F606, SAE J995
	Proof torque test	ASTM F912
	Rockwell hardness of fasteners	ASTM A370, ASTM E18, SAE J417, ISO 6508-1
	Rockwell superficial hardness of fasteners	ASTM A370, ASTM E18, SAE J417
	Tension testing of machined specimens for externally threaded fasteners	ASTM A370, ASTM F606, SAE J429, ASTM E8, ISO 6892
	Torsional strength test of thread rolling and self-drilling tapping screws	SAE J78, SAE J933, ANIS/ASME B18.6.4
	Total extension at fracture of externally threaded fasteners	ASTM F606, ISO 6892
	Twist test of lock washers	ASME B18.21.1
	Wedge tensile strength of full-sized threaded fasteners	ASTM A370, ASTM F606, SAE J429, SAE J1216, ISO 6892
Microstructural analysis	Decarburization and case measurement in fasteners	ASTM E1077, SAE J121, ASTM A574, ASTM E3, ASTM A490
	Macroscopic examination of fasteners by etching	ASTM E381, ASTM E340
	Surface discontinuities of externally threaded fasteners	ASTM F788/788M, ISO 6157-1, ISO 6157-3, SAE J123
	Surface discontinuities of internally threaded fasteners	ASTM F812, SAE J122, ISO 6157-2
Nondestructive inspection	Liquid penetrant inspection	ASTM E165, MIL-STD-271, MIL-STD-6866, SAE J426, ASTM E1417
	Magnetic-particle inspection	ASTM E709, MIL-STD-271, MIL-STD-1949, SAE J420, ASTM E1444, ASTM A275/A275M
	Ultrasonic inspection	ASTM A388/A388M, ASTM A577/A577M, ASTM 745/A745M, ASTM E114, ASTM E273, MIL-STD-271

TENSILE

The tensile test (Fig. 12.3.2) allows us to measure a material's response to loading and deformation. By measuring the force required to elongate a specimen to its breaking point, material properties can be determined that will allow engineering designers and quality managers to predict how their materials and products will behave in service.

Examples of products and industries that use tensile testing include the fastener industry (e.g., bolts, nuts, screws), automotive industry (e.g., seat-belt components) and the oil and gas industry (e.g., down-hole tube and pipe) to name a few. Tensile tests can predict pull-off force, peel and tear resistance and adhesion/bond strength.

In the test itself, one end of a specimen is typically clamped in a load frame while the other end is subjected to a controlled displacement or a controlled load. A transducer or servo-drive connected in series with the specimen provides information about the displacement (δ) as a function of the load (P) applied (or vice versa).

Tensile tests are designed to "pull" a specimen to failure. The destructive test measures the force and stretch of a material during testing. The test plots stress versus strain. Stress is the (applied) load divided by the cross-sectional area of the test specimen at its center. Strain is the change in a dimension divided by the original dimension and is measured over the central portion of the specimen length (where the cross section is constant).

As the specimen is stretched, the load required to induce each level of strain is measured. When the load is first applied, the tensile specimen stretches in proportion to the applied load. If the load is removed during this portion of the test, the specimen will return to its original length. This is called elastic deformation. The proportionality between the stress and strain is the elastic modulus (modulus of elasticity or Young's modulus). The binding forces of the atoms determine the modulus of elasticity. Since these forces cannot be changed without changing the basic nature of the material, it follows that this is one of the most structure-insensitive mechanical properties.

With continued loading and stretching the tensile specimen permanently deforms, exhibiting plastic deformation. The yield strength is the stress at which the specimen shifts from elastic (recoverable) stretching to plastic (permanent) deformation. By standard convention, the reported yield strength corresponds to a plastic strain of 0.2% (where observable deformation has taken place).

Tensile strength is the highest stress encountered in the tensile test. For many steels, this corresponds to the stress at fracture. For very ductile steels, the stress at fracture is lower than the tensile strength. For very brittle steels, the yield strength is the same as the tensile strength (and fracture strength).

A common measure of ductility is the elongation, which is given as total stretch to failure divided by the initial specimen-center test length, or gauge length. Elongation is a dimensionless number expressed as a percentage. Another measure of ductility is reduction in area, which is also dimensionless and expressed as a percentage. Reduction in area is a measure of the change in cross-sectional area at the point of failure (change in area divided by the original area).

The transverse rupture test is a strength test designed for low-ductility materials, including carbides and powder-metallurgy (PM) materials. This destructive test involves bending rather than pulling of the specimen. Maximum load, specimen dimensions and test time are used to calculate the stress needed to cause failure. A typical transverse rupture strength is 1.5-2 times the tensile strength.

Many manufacturers must not only test for ultimate tensile strength but also perform proof tests and ensure that no permanent deformation has occurred once the proof load is removed.

FIGURE 12.3.2 | Tensile-testing apparatus (courtesy of Instron®)

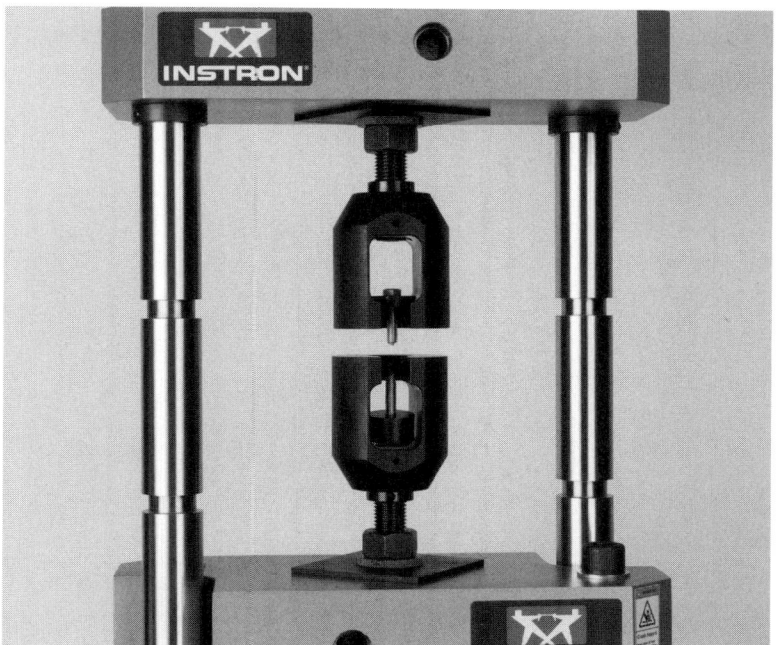

A Little Theory

The concept of stress, strain and strength of materials is at the core of the engineering discipline. Mechanical properties such as yield strength, tensile strength, ductility, toughness, impact resistance, creep resistance, fatigue resistance, stiffness and others all influence the design, fabrication and service life of equipment.

The engineering measures for stress (σ_e) and strain (ε_e) are determined from load and deflection readings using the original specimen cross-sectional area (A_0) and length (L_0). Formulas for these values are given in Equations 12.3.1 and 12.3.2:

12.3.1) $\quad \sigma_e = \dfrac{P}{A_0}$

12.3.2) $\quad \varepsilon_e = \dfrac{\delta}{L_0}$

STRESS-STRAIN CURVES

Once test data has been collected, stress-strain curves (Fig. 12.3.3) can be generated. They are then divided into "regions" that are descriptive of what is happening on the microscopic level, namely the:

❖ Elastic region
❖ Plastic region
  • Yielding
  • Strain hardening
  • Necking
❖ Failure (fracture)

FIGURE 12.3.3 | Various regions and points on the stress-strain curve[16]

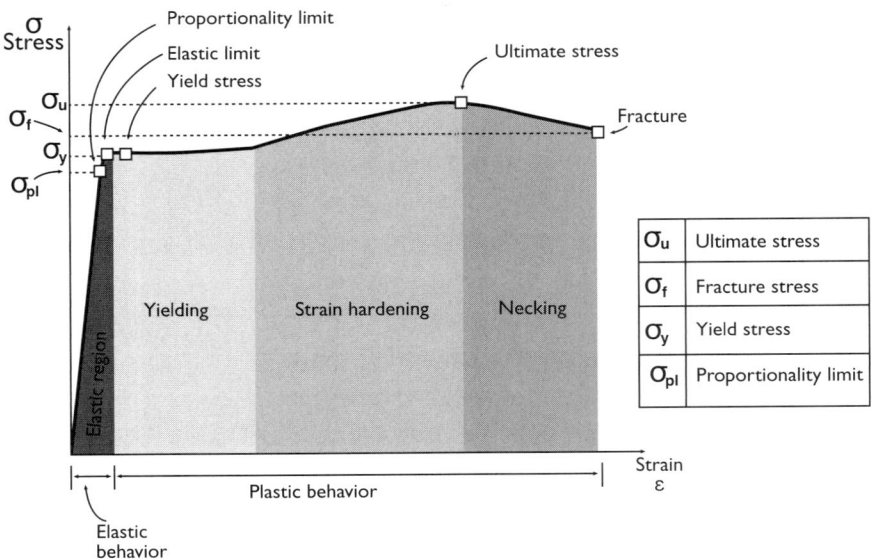

The shape and magnitude of the stress-strain curve of a metal (Fig. 12.3.4) will depend on its composition, heat treatment, prior history of plastic deformation, strain rate, temperature and state of stress imposed during the testing. The parameters, which are used to describe the stress-strain curve of a metal, are the tensile strength, yield strength or yield point, percent elongation and reduction of area. The first two are strength parameters, and the last two indicate ductility.

In the early (low-strain) portion of the curve, many materials obey Hooke's law, which states that the deformation is (within a reasonable approximation) linearly proportional to the stress. As a result, the stress is proportional to strain with the constant of proportionality being the modulus of elasticity (i.e., Young's modulus).

As strain is increased, many materials eventually deviate from this linear proportionality – the point of departure being known as the proportional limit. This nonlinearity is usually associated with so-called plastic deformation (flow) in the specimen. In this region, the material is undergoing rearrangement of its atoms (being moved to new equilibrium positions). The degree of plastic flow depends on the mobility mechanism, which in metals (i.e., crystalline materials) can arise from dislocation movement. Materials lacking this mobility – by having microstructural features that block dislocation motion, for instance – are usually brittle rather than ductile. The stress-strain curve for brittle materials is typically linear over the full range of strain, eventually fracturing without appreciable plastic flow.

As the plastic deformation continues above the yield strength, the engineering stress reaches a maximum point – the ultimate tensile strength (U.T.S.) of the material. The plastic deformation produces dislocations within the region in the curve between the yield strength and the tensile strength. The increasing dislocation density makes the plastic deformation harder. This phenomenon, known as strain hardening, is a key factor in shaping material by cold work.

The microstructural rearrangements associated with plastic flow are usually not reversed when the load is removed, so the proportional limit is often the same as or at least close to the material's elastic limit. Elasticity is the property of complete and immediate recovery from an imposed displacement on release of the load, and the elastic limit is the value of stress at which the material experiences a permanent residual strain that is not lost on unloading.

The strain at failure occurs after the specimen fractures. It is the interaction point of the plastic-recovery region on the strain axis. Thus, the ductility is the percent elongation at failure and indicates the general ability of the material to be plastically deformed.

FIGURE 12.3.4 | Typical stress-strain curves for various materials[6]

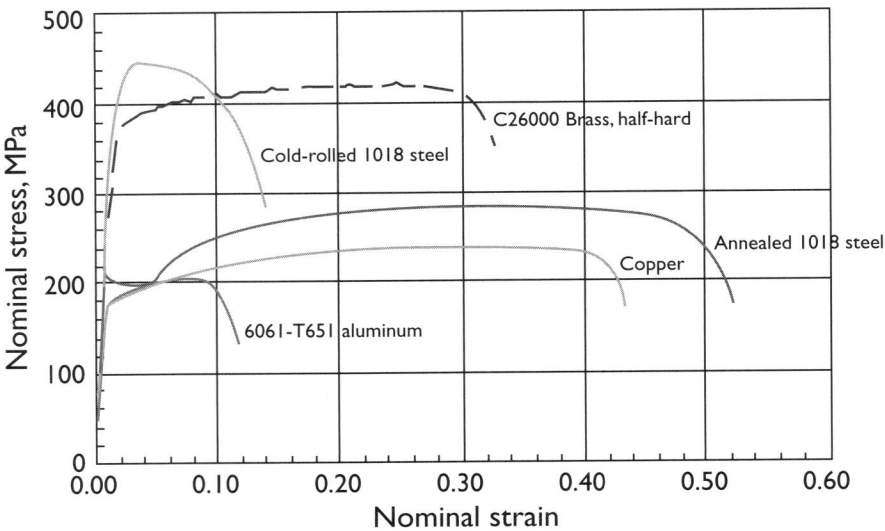

As most heat treaters know, hardness can be used to approximate the tensile strength of steel (Equations 12.3.3, 12.3.4), where HB is Brinell hardness (3,000-kgf load). This relationship is far from exact, however, and only a tensile test can reveal the true nature of the stress-strain relationship of a material.

12.3.3a) $\quad TS(MPa) = 3.55 \bullet HB(HB \leq 175)$

12.3.3b) $\quad TS(psi) = 515 \bullet HB(HB \leq 175)$

12.3.4a) $\quad TS(MPa) = 3.38 \bullet HB(HB > 175)$

12.3.4b) $\quad TS(psi) = 490 \bullet HB(HB > 175)$

Finally, recall that toughness measures the ability of a material to absorb energy and withstand shock up to fracture (Section 2.9); that is, the ability to absorb energy in the plastic range. In other words, toughness is the amount of energy per unit volume that a material can absorb before rupture and is represented by the area under the stress-strain curve.[2,3]

In Layman's Terms

The stress-strain curve is logically divided into two distinct deformation regions corresponding to elastic deformation and plastic deformation. The elastic deformation is temporary and fully recovered when the load is removed. By contrast, the plastic deformation is permanent and not recovered when the load is removed (even though a small portion of the elastic part in the deformation is recovered).

A simple analogy to explain these concepts is the rubber band. When a rubber band is stretched and returns to its original shape, we say that it is in the elastic region of the curve. The point at which the force we use is great enough to prevent the rubber band from returning to its original shape is the yield point, beyond which we are in the plastic-deformation region. Finally, when we stretch a rubber band beyond the material's ultimate tensile strength, it snaps and we reach the fracture point.

Tensile Testing Standards

There are a number of tensile testing standards, including:
- ASTM B913, D76, D1876, D3822, D412, D638, D828, E8
- BS 5G 178, BS EN 1895
- ISO 37, 527, 1924, 13934
- MIL-C-39029, MIL-T-7928

TORQUE/TORQUE-TENSION

Any part subject to torsional loading in service, such as shafts and axles, are torsion tested. Fasteners are another classic example because they are installed by applying a torsional force to the head or nut. This force (or seating torque) causes the fastener to stretch and effectively applies a preload to ensure a snug fit between components. One purpose of a torque-tension test is to determine the appropriate tightening torque required. This type of test will allow the nut factor (aka torque coefficient or k factor) to be determined as well as the overall coefficient of friction. Similarly, one often needs to determine the torque force at which the fastener will fail. By completing several similar tests, the variation in the torque-tension relationship due to frictional variation can be established for a given application.

Torque testing is used to determine properties such as modulus of elasticity in shear, torsional yield strength and modulus of rupture. Torsion testers (Fig. 12.3.5) consist of a twisting head with a chuck for gripping the specimen and applying the twisting moment. The weighing head grips the other end of the specimen and measures the twisting moment, or torque. Deformation of the specimen is measured by a twist-measuring device called a troptometer.

A torsion test specimen typically has a circular cross section (the simplest geometry for calculation of stress). Since the shear stress in the elastic range varies linearly from zero at the center of the specimen to a maximum at the surface, it is frequently desirable to test a thin-wall tubular specimen. This results in a nearly uniform shear stress over the cross section.

Torsion test results can be used to validate or expand on the information gleaned from tensile testing. Torsion testing provides a more fundamental measure of the plasticity of a material than does tension testing. It directly produces a shear-stress/shear-strain curve. In torsion, the critical shear stress for plastic flow is reached before the critical normal stress for fracture. In tension, meanwhile, the critical normal stress is reached before the critical shear stress.

FIGURE 12.3.5 | Torsion testing apparatus (courtesy of Instron®)

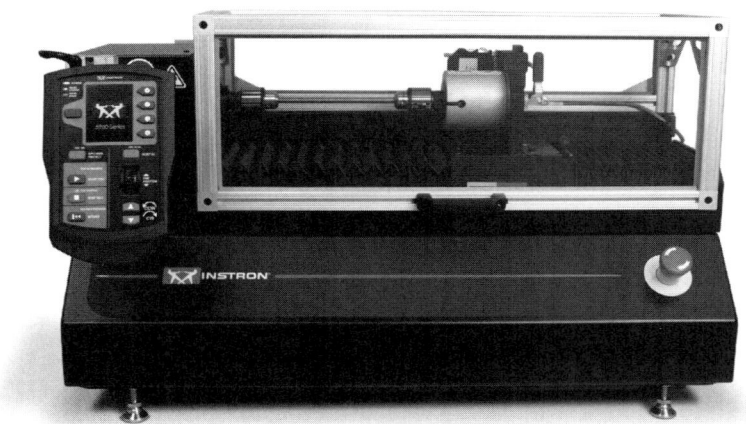

FATIGUE

Fatigue is a measure of the stress that a material can withstand repeatedly without failure. A fatigue failure is particularly catastrophic because it occurs without warning. Three basic factors are necessary to cause a fatigue failure: a maximum tensile stress of sufficiently high value, a large enough variation or fluctuation in the applied stress and a sufficiently large number of applied-stress cycles. Fatigue-life tests are performed on threaded fasteners, for example, by alternating loading and unloading of the part. Most testing is done at more severe strain than its designed service load but usually below the material yield strength.

The basic method of presenting engineering fatigue data is by means of the S-N curve (Fig. 12.3.6), which shows how the life of the specimen, expressed in number of cycles to failure (N), depends on the maximum applied stress (S). This particular curve shows that a linear reduction in applied stress results in an exponential improvement in fatigue life. Therefore, a 35% reduction in stress – from 760 MPa (110 ksi) to 485 MPa (70 ksi) – results in a 400% improvement in fatigue life (from 40,000 cycles to 160,000 cycles). Additional reductions in stress result in significantly more fatigue-life enhancement.

FIGURE 12.3.6 | Typical S-N curve[19]

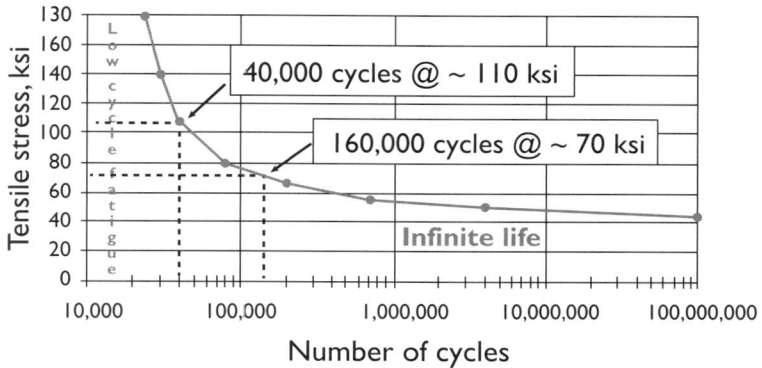

Fatigue testing equipment (Fig. 12.3.7) is usually designed to induce cyclic loading and unloading to a known (peak) stress before measuring the number of such cycles to failure of the specimen. Variants of the test include tensile, bending and rotating. The average stress at which a steel can withstand 10 million loading cycles without failure is reported as the fatigue strength (also called the endurance limit). As stress increases, the number of cycles to failure decreases.

FIGURE 12.3.7 | Fatigue testing equipment (courtesy of Instron®)

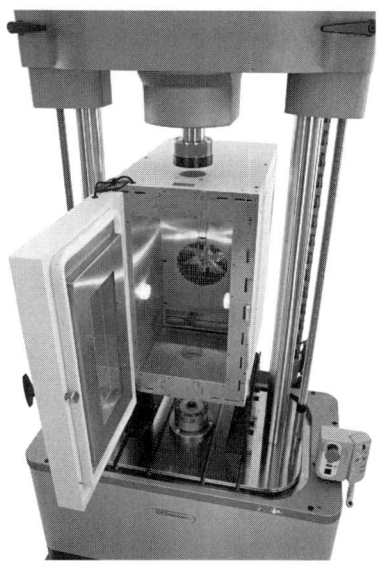

Fatigue failures are a combination of crack initiation and crack growth (propagation). Sophisticated design equations are available to predict component life under cyclic loading conditions. The fatigue strength is always less than the yield strength, and it is often 30-50% of the tensile strength.

Fatigue strength is significantly reduced by the introduction of a stress raiser (stress concentrator) such as a keyway, screw thread, fillet, press fit or hole, many of which are present in most structural parts. One of the best ways to minimize fatigue failures is to reduce the number of stress raisers as much as possible via careful design and proper heat treatment.

SHEAR STRENGTH

Shear strength is defined as the maximum load that can be supported prior to fracture when applied at a right angle to the component's axis. Simply stated, it is the force required to pull the base material in one direction and the surface material in the other direction until failure. Modes of failure include deformation and fracture. Bolted or riveted connections are those that are commonly subjected to shear stresses.

Unlike tensile testing, determining the ultimate shear load and detecting specimen failure can be difficult. The test system must be flexible enough to define different end-of-test criteria for each style of component. Most, if not all, shear testing for fasteners is done on the unthreaded portion of the fastener.

Single-shear testing applies a load in one plane and results in the component being cut into two pieces, while double shear (Fig. 12.3.8) produces three pieces. Single-shear values for fasteners are typically calculated based on the nominal body diameter or body shear area. There is a relationship between the tensile strength of a material and its shear strength. In alloy steel, for example, the shear strength is approximately 60% of its tensile strength. In corrosion-resistant steels (e.g., 300-series stainless steels), the tensile/shear relationship is usually only 50-55%.

FIGURE 12.3.8 | Double-shear testing apparatus (courtesy of Instron®)

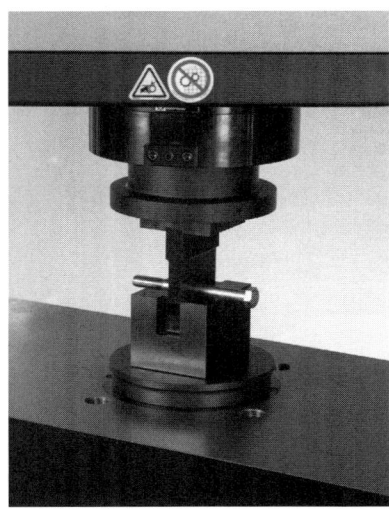

12 | TESTING

CREEP AND STRESS RUPTURE

It is often necessary to consider the effects of service temperature on the life of a heat-treated part. Since the mobility of atoms increases rapidly with temperature, diffusion-controlled processes can have a very significant effect on high-temperature mechanical properties. The effect of long-term exposure to these environments is a critical design consideration. Therefore, successful use of these parts requires knowledge of how their strength varies with time at high temperatures.

Creep and stress-rupture tests (Fig. 12.3.9) can be used to evaluate the performance of materials for elevated-temperature service. The creep test measures the dimensional changes that occur in a specimen during exposure to high temperature, while the stress-rupture test measures the effect of temperature on the specimen's longtime load-bearing characteristics. Other tests measure special properties such as thermal-shock resistance and stress relaxation.

Note that creep and stress-rupture data also play important roles in the heat-treatment industry. When selecting high-temperature alloys for furnace interior components or heat-treating fixtures, for example, a good measure of the reliability of a candidate alloy is the value of "1% creep in 10,000 hours" at the intended service temperature (Section 4.6).

Creep is time-dependent deformation of a material while under an applied load (below its yield point). Stress-rupture is the sudden and complete failure of a material held under a constant load for a given period of time at a specific temperature. These tests are used by a multitude of manufacturers to determine how their products will perform when subjected to constant loads at both ambient and elevated temperatures.

FIGURE 12.3.9 | Stress-rupture test (courtesy of Instron®)

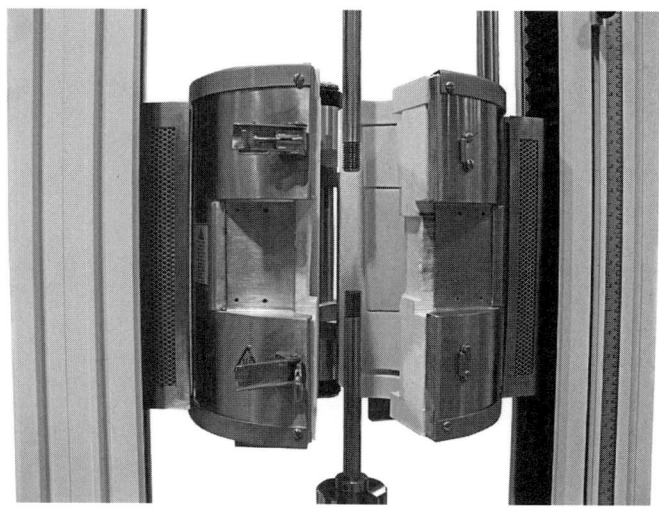

IMPACT

Brittle fracture is catastrophic, and factors that contribute to it include: triaxial stress, such as exists at a notch or stress raiser; low temperature; and a high strain rate or rapid rates of loading. All three do not have to be present at the same time.

Impact tests are designed to determine a toughness value, which indicates the susceptibility of a material to brittle fracture (tougher materials are less susceptible). Notched-bar impact tests are used most often. These tests detect differences between materials that are not revealed by a torsion test.

Notched Charpy and Izod specimens are considered the standards for impact testing. Charpy bars have a square cross section and a "V" or "keyhole" notch at the center of their length. The Charpy bar is struck at its center by the impact load to ensure failure, which results in a lower toughness value (reflecting a more conservative design parameter). Izod specimens have either circular or square cross sections and contain a "V" notch near one end. The specimen is clamped at the end with the notch and then struck by the impact load at the opposite end.

The most common test machine for measuring toughness (Section 2.9) or impact resistance involves a swinging pendulum with a test specimen positioned at the bottom of the pendulum swing (Fig. 12.3.10). A specimen of a tough material requires considerable energy for failure, consuming much of the pendulum's kinetic energy. Consequently, the pendulum will not swing very high after breaking the specimen. If the specimen has low toughness, the pendulum will end its swing considerably higher after fracture.

FIGURE 12.3.10 | Charpy impact test setup (courtesy of Instron®)

12 | TESTING

STRESS DURABILITY

Stress durability is used to test parts that have been subjected to any processing operation (e.g., electroplating) that may have an embrittlement effect. It requires loading the parts to a value higher than the expected service load and maintaining that load for a specified time, after which the load is removed and the part examined for the presence of cracks.

VIBRATION

Vibration tests are used to determine a component's life span and performance characteristics (e.g., self-loosening of fasteners) under vibratory conditions. A transverse vibration test machine (commonly called a Junker machine) is used to produce a preload-decay graph, an indication of resistance to self-loosening.

METALLURGICAL (STRUCTURE) ANALYSIS

Metallurgical testing (Section 12.4) can be performed to evaluate the component's microstructure. It can yield invaluable information on grain size, surface condition (e.g., carburization or decarburization) and heat-treatment response. The microstructure and grain size are most often influenced by heat treatment.

CHEMICAL ANALYSIS

The chemical composition of a steel heat is established at the mill and reported on a material certification sheet (Section 3.1). It is often desirable to know the exact chemistry of the component part in question – not only the principal elements, but the trace elements present. Grain size, prior processing (e.g., mill annealing) and hardenability are influenced by the elements present.

Hydrogen Embrittlement and Corrosion Testing

Components exposed to sources of hydrogen (e.g., electroplating operations) can fail prematurely at stress levels well below the material's yield strength. The effect is often a delayed one, meaning that it may occur in service. Higher-strength steels are more susceptible to hydrogen embrittlement than lower-strength steels. As a rule of thumb, steels below 30 HRC are considered to be far less susceptible. The problem can be controlled by careful selection of plating formulation, proper plating procedure and sufficient baking to drive off any residual hydrogen.

Many failures are due to corrosion, which can take place slowly over time or be surprisingly quick. Corrosive attack is usually defined as the amount of time before attack occurs and is measured in mm/year (inch/year). Another simple indicator is when white or red rust first appears on the surface of a steel or stainless steel component measured in terms of hours of resistance to a salt-spray (fog) test.

RESIDUAL STRESS

Heat treaters deal with issues of residual stress every day. These are the stresses that exist in a material when it is free from external forces. They are also sometimes referred to as internal or locked-in stresses. Residual stresses are responsible for warping and dimensional instability of heat-treated parts. They can also influence how a material reacts to externally applied stresses.

Residual stresses are produced whenever a material undergoes a non-uniform plastic deformation. In hardening, for example, quenching produces a temperature differential between the rapidly cooling surface and the slower-cooling core. This creates a mismatch of strain. The residual-stress pattern that results is a combination of the thermal volume changes and those resulting from the transformation of austenite to martensite.

Residual-stress analysis can be performed on component parts to better understand the true nature of the internal forces to which a part is exposed.

Statistical Methods

Engineers, heat treaters and quality-control personnel rely on statistical methods to benchmark the results of the heat-treatment processes. Control charts and linear-regression analyses are examples of useful everyday tools, while design of experiments and frequency-distribution techniques are tools helpful to the design or R&D engineer.

There are three principal reasons why a working knowledge of statistics is needed to interpret the results of mechanical testing.

1. Mechanical properties are structure-sensitive, so they frequently exhibit considerable variability or scatter. This makes statistical techniques useful, and often necessary, for determining the precision of the measurements and enabling valid conclusions to be drawn from test data.
2. Statistical methods can assist in designing experiments to provide the maximum amount of information at minimum cost.
3. Statistical methods based on probability theory can be used to help explain certain problems or phenomena such as the size effect in brittle fracture and fatigue.

Summary

Premature failure of a component in the field is unacceptable. Part testing, whether it is mechanical, electrical or highly specialized and product-specific, ensures this will not happen.

Determining the proper test methods for a particular application and executing them in such a way as to make sure that the actual testing does not introduce variability into the results are important parts of any good quality-control system. Practical shop tests have often been devised that produce highly valid results. They should not be discounted provided there is adequate historical and field data to support the validity of the tests.

A tremendous amount of time, money and resources are expended whenever a failure occurs, so avoiding failures must be our ultimate goal. The heat treater must embrace all types of informative testing and remember that disastrous consequences always result from loss of control or process/equipment variability. Mechanical testing is one component of a robust quality program, whether the parts will be used in a simple convenience item or in a life-critical application.

REFERENCES
1. Herring, Daniel H., "Basics of Mechanical Testing," *Heat Treating Progress*, March/April 2005
2. Herring, Daniel H., "Toughness," *Industrial Heating*, December 2010
3. Herring, Daniel H., "Toughness Revisited," *Industrial Heating*, March 2011
4. Instron (www.instron.com)
5. NIST (ts.nist.gov)
6. Collaborative Testing Services, Inc. (www.collaborativetesting.com)
7. Bolt Science (www.boltscience.com)
8. Westmoreland Mechanical Testing & Research, Inc. (www.wmtr.com)
9. Test Application Handbook Relating to Mechanical Fasteners, Division VI Aerospace Fasteners, 1995
10. Stork Materials Technology (www.storksmt.com).
11. Roylance, David, *Stress-Strain Curves*, MIT, 2001
12. Dolbow, John E., "The Stress-Strain Curve," Duke University
13. *Engineering Stress-strain Curve: Part One*, Key to Metals
14. "Stress-Strain Relationships," Material Testing Solutions (www.mts.com)
15. *Atlas of Stress-Strain Curves, 2nd Edition*, ASM International, 2002
16. Jaiswal, Praneet, Mechanical Engineering Blog (www.mechanicalengineeringstream.blogspot.com)
17. K-Street Studios (www.kstreetstudio.com)
18. Etomia, Molecular Simulations Software (www.etomia.org)
19. Penn State University, Department of Engineering Science and Mechanics
20. Herring, Daniel H., David J. Breuer and Gerald D. Lindell, "Selecting the Best Carburizing Method for the Heat Treatment of Gears," AGMA Fall Technical Conference, 2002
21. Alan Stone, Aston Metallurgical Services Co. Inc. (www.astonmet.com), technical and editorial review, private correspondence

12.4
METALLURGICAL EVALUATION

The success or failure of a heat-treatment process can best be revealed by a well-planned and carefully executed metallurgical evaluation (Fig. 12.4.1), which provides information about the microstructural and mechanical/physical properties of a component part or a representative sample thereof. The heat treater must resist the temptation of simply using hardness testing as the only criterion for success.

FIGURE 12.4.1 | Typical metallurgical laboratory setup (courtesy of Struers Inc.)

a.

b.

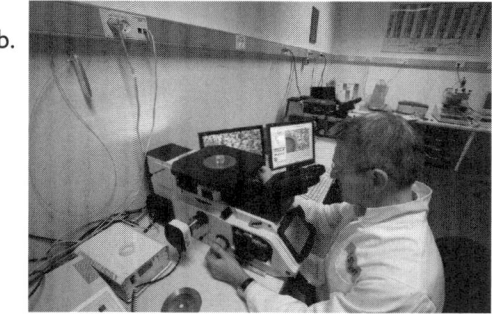

a. Sample preparation area

b. Sample evaluation area

The microstructure of a ferrous or nonferrous material, while complex, reveals the true nature of the material after heat treatment and is influenced by composition, homogeneity, processing procedures and section size. Sampling method, preparation, equipment and technique play important roles as well.

Role of Metallurgical Evaluation in Solving Problems

As heat treaters, we often seek answers to processing problems or component part failures from the metallurgical community without fully understanding (or appreciating) what metallurgists need in order to do a thorough job. We need to provide them with accurate background information and a representative set of samples (good and bad) for comparative analysis.

Even something as simple as protecting the surface of the component to be analyzed from damage is very important. We often don't communicate our expectations in precise terms. Therefore, we do not know what to expect from an analysis. Here are several insights to help us gain the most knowledge from whatever testing or analysis is performed. Having accurate information and reliable facts produce results upon which we can make informed decisions.

PHOTOGRAPHY

In the age of digital photography, a picture is indeed worth 1,000 words or more. Take photographs of everything, from multiple angles, and remember to use good lighting and high-resolution. Handle parts carefully so as not to induce damage, and note orientation and other salient features.

PROCESSING HISTORY/BACKGROUND INFORMATION

Don't assume that someone knows your process or product or its intended service application better than you do. Communicate the history of the part or process, separating assumptions from facts; provide necessary drawings, including mating components if appropriate, as well as required specifications; and explain the design requirements in detail. In other words, take the guesswork out of the analyst's job. Document anything of importance and give this information to the metallurgist, even if it means having a confidentiality agreement in place before you start.

If it turns out that the component is being returned from the field, extreme care must be taken to ensure a representative sample and avoid further damage. Even the sample mounts we prepare, or those supplied by others, and send to the lab should have their surfaces carefully protected. (Suppliers sell special plastic cups for this purpose. Tissue paper in multiple layers taped to the mount will suffice.)

Similarly, on failures, extreme care should be taken to prevent damaging the surfaces or allowing them to corrode. Even if you aren't planning on performing an analysis at that moment, make this a good habit and be rewarded if and when an analysis is required. Remember to resist the temptation to put mating surfaces of a failed part back together again!

INCOMING (RAW) MATERIAL ANALYSIS

Too often we are forced to begin an analysis by making assumptions about the raw material. Provide the laboratory with a copy of the mill's material certification sheets. In addition to the chemical constituents, the metallurgist will glean information based on the form of the raw material, its grain size, cleanliness and prior mill processing. Doing an actual chemical analysis is not necessarily redundant. For example, trace-element chemistry can play a significant role when investigating phenomenon such as temper embrittlement.

LABORATORY PROCEDURES

Discuss with your metallurgist or outside laboratory what type of tests will be conducted and in what order. Understand what will be achieved at each step in the analysis process so that you can ask questions or offer suggestions. This will also help explain the time and expense involved. This way you will be better able to interpret the final results.

Be aware that there is nothing more frustrating in the laboratory than to work hard on a job only to find out that it is not the right sample or that the damage observed was induced by extraneous factors. This translates into lost time and money. Be conscious of the fact that once the investigator has begun to work on your project, it is best that the analysis move forward uninterrupted. Be sure to define the scope of work and clarify the boundaries of what he or she is allowed to do once the investigation is under way. A not-to-exceed figure works well for this part of the investigation.

Selecting the proper tests may involve trade-offs due to cost or time, so be sure you understand the cost/benefit relationship of each test and what the expected outcome might be so that the right choices can be made. Insist on specificity to avoid open-ended analysis efforts. Here are some examples of what can be done in the laboratory.

- ❖ Stereomicroscopy
- ❖ Nondestructive testing
    - Eddy current
    - Ultrasonic

- Pressure testing (hydrostatic, pneumatic)
- Surface finish
❖ Macroetching
❖ Mechanical testing
  - Hardness/microhardness testing
  - Tensile testing
  - Impact testing (e.g., Charpy testing)
  - Fatigue testing
  - Torque/torque-tension
  - Shear and double-shear strength
  - Torsion testing
  - Creep
  - Stress-rupture and stress durability
  - Vibratory testing
❖ Sample preparation
  - Unetched part examination
  - Etched part examination
❖ Optical microscopy
  - Microstructural determination
  - Grain size
  - Micro-cleanliness
  - Intergranular attack
  - Inclusion characterization
  - Alpha case
❖ Image analysis
  - Plating depth (layer thickness)
  - Defect measurement
  - Grain size
❖ Scanning electron microscopy
  - Fractography
  - Feature/character recognition
❖ Energy-dispersive X-ray spectroscopy
  - Qualitative element analysis
  - Inclusion characterization
❖ Corrosion testing

SELECTING THE RIGHT LABORATORY

Not all laboratories are created equal, either in the talent of their researchers or in the tools available to do the job right. Talk to people you trust in the industry to help in your selection process. Be aware that many labs are better at some things than others and subcontract certain tasks to other labs. Be sure that you understand when and why this is being done and then determine if you are better off going direct.

COMPARATIVE ANALYSIS (GOOD vs. BAD)

If good parts exist, they can be invaluable aids in understanding why a bad part failed. Taking the seemingly extra step (and expense) of testing a good part along with a bad one will yield tremendous insight into the problem at hand. Do this whenever possible.

TIMING

In an effort to get answers, avoid the temptation to push the lab to the point where steps are skipped or time is not taken to investigate secondary factors that may prove to be major contributors. Ask for verbal reports at key milestones in the analysis work, but avoid taking up valuable analysis time by "checking in" too often. Meeting in person to begin a project is always beneficial.

LAB REPORTS

At times, lab reports can be difficult for the heat treater to understand. Be sure to ask the laboratory for clarification in these situations. There is a delicate balance between the facts and their interpretation, but this can often be handled by placing the interpretation in a discussion section of the report. Due to liability concerns, the trend is to simply report the facts and rely on the client to interpret them. If necessary, hire outside experts to put the information in the proper context in order for you to determine the right course of action. There is nothing worse than paying good money for a report you don't understand.

ROOT-CAUSE DETERMINATION

"What caused the problem and how can I avoid its reoccurrence?" This should be the objective of any process problem or failure analysis.

While it may or may not be possible to establish the root cause, it should always be the goal. There is only one root cause, which may be influenced by a combination of factors. The use of Ishikawa (fishbone) diagrams or other diagnostic methods to list all of the contributing variables to a successful outcome can be a big help. Sometimes it's the thought process itself and a discussion among various company departments that leads to the solution being implemented.

THE BOTTOM LINE: TO ANALYZE OR NOT TO ANALYZE

A cost/benefit analysis should be performed before and after any analysis or testing. Knowledge is strength and assumption is weakness, so do the metallurgical analysis when in doubt. It will amaze you what can be revealed.

Preserving the Sample for Analysis

In the real world, components fail. There are many reasons why. If a premature failure occurs, it is important to evaluate all possible causes and then isolate a root cause that can be corrected to ensure the problem will not reoccur.

Failures that happen during testing on the factory floor, in controlled laboratory conditions or during product manufacture can be classified as "regular" failures – ones in which a significant amount of resources (people and test equipment) can be brought to bear on the problem. By contrast, field failures are those in which an assembly has failed in its normal working environment, often remote to the manufacturer, and where resources for analysis can be limited and must be transported to the job site.

Field failures are particularly challenging. To help, here are 13 simple suggestions.

1. **Have a plan.** Failure analysis starts with documented procedures and good training. Anticipate your needs. Don't be surprised or confused with what has to be done. Do not attempt to investigate a failure in a haphazard way.
2. **Set up a team** composed of representatives of all disciplines (management, purchasing, engineering, service, metallurgy, manufacturing, quality and legal). Meet to discuss the problem and review the facts. This step is often overlooked, but it can be invaluable if someone is able to recall details of the circumstances surrounding how the part was originally manufactured. Make recommendations as to the type and extent of analysis that should be performed in the field. Understand the implications of the failure and the potential liability exposure of the company.
3. **Use your time efficiently.** Be prepared for the inspection. Plan your work carefully to obtain as much evidence as possible. Gather the right people and tools. Don't become distracted.
4. **Inspect failed components immediately.** Information relating to the failure should be compiled as soon and as thoroughly as possible. A rapid-response team should be in place. The longer the time between the failure and the inspection, the greater the risk those external influences will induce error into the analysis process.
5. **Preserve the evidence.** The ideal situation is to have information gathered on-site by a person or team familiar with failure-analysis methods. If you can't get to the failure site, have someone you can trust at the site report the situation to you or follow your explicit instructions. Photograph everything (especially in this age of digital photography), even details that seem like incidentals. Try to emphasize the importance of having quality tools at the job site, especially the right digital camera.

The one we use can photograph as close as 30 mm (1.2 inch) from an object with 8-megapixel resolution.

6. **Evaluate if field disassembly is a good idea.** In a perfect world, the assembly that failed would be undisturbed when the people assigned the task of field failure analysis arrive on the scene. Often, however, the assembly has already been taken apart, which may have been necessary to determine which component had failed. In some instances, the equipment has already been placed back into service. This usually makes the job of determining the root cause of the failure much more difficult.

   If asked by field personnel for permission to disassemble and return equipment to service before a thorough inspection by the failure-analysis team takes place, resist the urge to say yes. It may be necessary to ship the assembly back to the factory in one piece for a more controlled analysis where more specialized resources may be available. If field disassembly is required, the failure-analysis team should participate and not defer hands-on work to technicians or less qualified individuals.

7. **Document each step of the disassembly** whether in the field or in the laboratory. Do not hurry. Disassembly should stop while you inspect and document the condition of a component before proceeding to the next component.

8. **Do not form premature conclusions,** no matter how obvious the cause may seem. Concentrate on gathering facts (not speculation) along with all the evidence (and not just the evidence that may support a hastily drawn conclusion). Consider but don't be swayed by the opinions of others.

9. **Be patient.** Gather your thoughts and analyze the collected data. Don't let fatigue or conditions at the job site (wind, cold) become distractions or cause you to lose perspective. Be prepared for the environment in which you will find yourself.

10. **Document what you see and, perhaps of equal importance, what you don't see.** Remember that there is a reason for everything you are observing, and it may become important later on when you reconstruct the failure or when you consider the evidence as a whole. Taking photographs at each step of the disassembly is perhaps the best method of documentation, but it should be supplemented with notes of your observations and measurements of key dimensions.

11. **Control the investigation.** Make sure you come away with what you need. This includes items such as: service conditions (temperature, pressure, chemical or environmental exposure); time in service; circumstances surrounding the failure; cyclic conditions (if any); complete service his-

tory of both the failed component and assembly or machine that it is a part of (cleaning procedures, maintenance history); and related or similar failures in the same or similar applications.

12. **Secure the evidence for shipment back to the laboratory.** Identify which components need to be analyzed with more powerful techniques than might be available in the field. Take more evidence than you think you might need. Be aware of the special handling of failed components. Spare no expense in ensuring that the components will not suffer additional damage during transportation back to the laboratory for a more detailed analysis.

13. **Do not bow to unreasonable pressure.** Insist on privacy and don't become distracted. Do not report your findings until the investigation is complete and you've had a chance to think through the consequences of your conclusions. Failures can occur for a variety of reasons, including:
    - Overloading (tensile stresses, torsional forces and shear forces)
    - Impact (mechanical and thermal shock)
    - Wear (abrasive, adhesive, fretting, cavitation damage and erosion)
    - Fatigue (vibration)
    - Corrosion (uniform, pitting, galvanic, crevice and intergranular)
    - Stress corrosion cracking (hydrogen and sulfide embrittlement)
    - Heat treatment (improper hardening, tempering, quenching and cryogenic treatment)

THE BEST-LAID PLANS

Unfortunately, it's not unusual for a failed part or assembly to arrive back at the plant or lab accompanied by cryptic and/or incomplete information. If this occurs, or if adequate resources were not available to do a thorough field investigation, remember that this increases the uncertainty of your conclusion. Learn from the experience and educate others on the proper techniques for next time.

Preparing the Sample for Analysis

The number-one concern among metallurgists is that of improper sample preparation leading to incorrect characterization of the microstructure. This is one of the reasons you will find metallurgists hesitant to comment on the work of others.

The goal of sample preparation is to reveal the true microstructure of the material being analyzed. To achieve this, one must be aware of the following.

- The particular type of analysis or examination required determines the specific requirements for surface preparation. For example, a macroetch to approximate case depth does not require the same number of

steps or degree of preparation as trying to measure the depth of decarburization at the surface.
- ❖ Preparation can be stopped when the surface is acceptable for a specific examination. (Note: This is not an excuse to skip steps or use poor technique.)
- ❖ Avoid inducing thermal damage or other defects during sample preparation. For example, the method(s) required to section a sample can be critically important.
- ❖ The least number of steps and the shortest possible time is the goal. Keep in mind we must achieve the required results in the most cost-effective manner.
- ❖ Sample preparation must always be done anticipating that more sophisticated testing may be required.
- ❖ Be consistent in order to obtain reproducible results.

The basic requirements for proper sample preparation include:
- ❖ Freedom from deformation, which entails removing enough material to overcome the influence of sectioning, grinding and polishing (Fig. 12.4.2)
- ❖ Absence of thermal damage
- ❖ Flatness
- ❖ Freedom from scratches or other surface anomalies (e.g., pits, smearing, pullout, cracking). Cleaning the sample thoroughly before grinding and between each grinding and polishing step is absolutely necessary.

Careful sample preparation (to avoid the creation of so-called artifacts) is essential to obtain clear, accurate microstructural data. The importance of deformation-free sample preparation can be summarized as follows.
- ❖ Details in soft ductile phases (such as ferrite in steels) may be hidden by smeared or damaged surfaces.
- ❖ Hard constituents such as carbides or brittle inclusions may be fractured and removed (chip out) from the surface of interest.
- ❖ Soft, friable constituents such as graphite in cast irons may be washed out.
- ❖ Fine precipitates may be removed, leaving pits that could be misinterpreted as porosity.
- ❖ Critical edges representing the cross-sectioning of a surface or pore walls of powder-metal parts may be rounded, causing a loss of visual information.
- ❖ Hard constituents in highly dissimilar materials in relief may prevent accurate image analysis.

FIGURE 12.4.2 | Disturbed metal layers that sample preparation must remove[8]

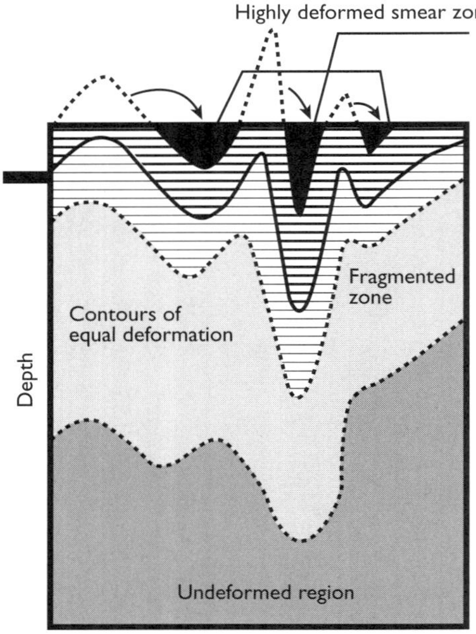

Documentation of incoming samples is absolutely critical. What follows is some of the data that should be collected.

- Origin, nominal composition, hardness and purpose of the analysis should be known and recorded.
- Any hazardous properties or sensitivity of the sample material should also be known so that steps may be taken to protect the operator or material from harm.
- Make a sketch or take a macrophoto showing the location and orientation of the cross sections to be made.
- Color code the sample mounts according to material or process (e.g., red and black for ferrous carburized, yellow and black for ferrous hardened, green and black for nonferrous, etc.).
- Use a code to label each mount. The code should be permanently engraved on the side of the sample mount, never on the top or bottom.
- Log all mounts in a permanent electronic file or bound logbook.
- Have all applicable standards (heat-treat specifications, reference prints, etc.) in a reference library and reference them in all reports.

The basic steps involved in the preparation of a sample for metallurgical examination (Figs. 12.4.3-12.4.6) are reviewed in detail elsewhere (Section 16.3) for those interested in the particulars.

12 | TESTING

FIGURE 12.4.3 | Typical cutoff saw (courtesy of Struers Inc.)

FIGURE 12.4.4 | Typical mounting press (courtesy of Struers Inc.)

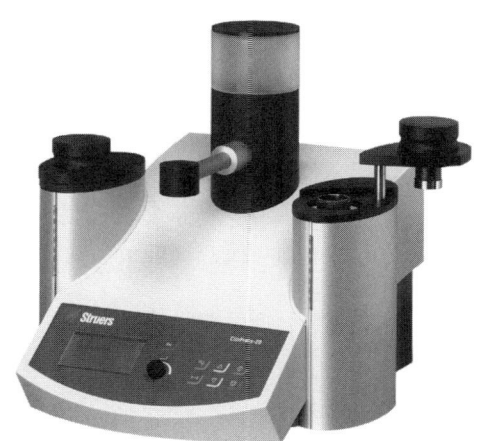

FIGURE 12.4.5 | Typical grinder and polisher (courtesy of Struers Inc.)

a. Grinder

b. Polisher

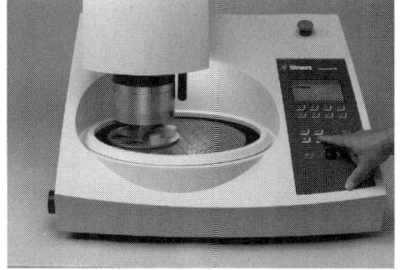

FIGURE 12.4.6 | Typical etchant station (courtesy of Struers Inc.)

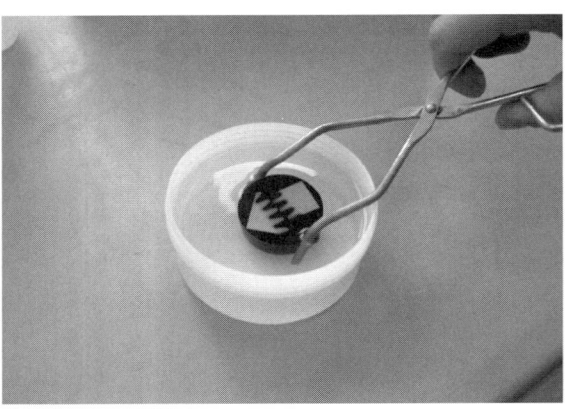

Summary

Metallurgical evaluation is one of the most important aspects of testing a component's suitability for its intended service application. To acquire the skills necessary to do the analysis right takes time and a great deal of practice. To interpret the results of the analysis takes an understanding of heat treatment as well as metallurgy. As such, individuals who have knowledge in both areas should always support these efforts.

Equally important is to ensure that the techniques used do not confuse or mislead the analysis effort. Metallurgical evaluation is painstaking work, with attention to detail and curiosity part of the necessary skill set.

REFERENCES

1. *Practical Failure Analysis*, ASM International
2. Wikipedia (www.wikipedia.org)
3. *ASM Handbook, Volume 12 Fractography*, ASM International, 1987
4. University of Virginia (www.sv.vt.edu)
5. Bhattacharyya, V.E., *Fracture Handbook*, 1979
6. Nishida, Shin-Ichi, *Failure Analysis in Engineering Applications*, Butterworth Heinemann Ltd., Jordan Hill, 1992
7. *Handbook of Case Histories in Failure Analysis*, Khiefa A. Esaklul (Ed.), ASM International, 1992
8. Petzow, Gunter, *Metallographic Etching*, ASM International, 1978, pg. 9
9. Vander Voort, George, *Metallography Principles and Practices*, ASM International, 1984
10. Wulpi, Donald J., *Understanding How Components Fail*, ASM International, 1985

11. Lawn, B.R. and T.R. Wilshaw, *Fracture of Brittle Solids*, Cambridge University Press, 1975
12. *ASM Handbook, Volume 11: Failure Analysis and Prevention*, R.J. Shipley and W.T. Becker (Eds.), ASM International, 2002
13. Fontana, M.G. and N.D. Greene, *Corrosion Engineering*, 3rd *Edition*, McGraw-Hill Book Company, 1985
14. Uhlig, H.H., *Corrosion and Corrosion Control*, John Wiley & Sons, 1963
15. Alan Stone, Aston Metallurgical Services Co. Inc. (www.astonmet.com), technical and editorial review, private correspondence
16. William Durako, Materials Engineering Inc. (www.materials-engr.com), private correspondence
17. Debbie Aliya, Aliya Analytical Inc. (www.itothen.com), technical and editorial review, private correspondence
18. George Vander Voort, Vander Voort Consulting (www.georgevandervoort.com), technical contributions and private correspondence
19. Craig Darragh, AgFox LLC, technical editing and private correspondence

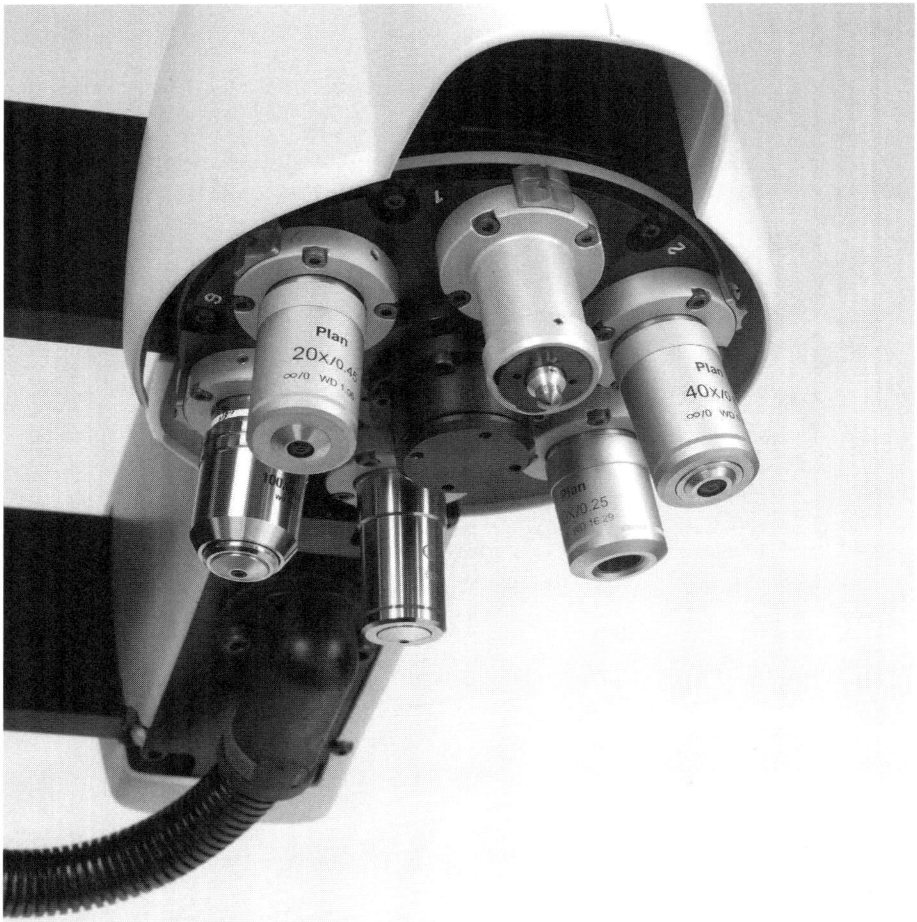

HARDNESS MATTERS

Take a closer look at DuraScan

Designed for fast and advanced automatic testing using high-quality optics and ecos Workflow™ software, DuraScan hardness testers give you shorter turnaround time, higher repeatability, versatility and ease of use.

Take a closer look on www.struers.com

12.5
FAILURE ANALYSIS

When a component part fails, it is important to ask why and then strive to determine the factors that might have contributed. Gathering all possible information about the damage event and performing a thorough failure analysis is a critical first step in the process. For the heat treater, this type of information helps us create a set of do's and don'ts, which are invaluable in avoiding a recurrence of the problem.

Failure Mechanisms

In simplest terms, a failure is the inability of a component part to perform its intended function (Fig. 12.5.1). Components in service experience different types of conditions/environments, damage mechanisms and applied loading, including tension, compression, bending, torsion and mixed modes (combinations). The failures that result may be categorized in a broad sense as those related to fracture, wear, corrosion and/or dimensional change (distortion). As heat treaters, we must also consider that residual stresses can often play an important role.

FIGURE 12.5.1 | Failed ASTM A490 bolts from structural beams in a power plant (courtesy of George Vander Voort Consulting)

TYPES OF FRACTURE: MACROSCOPIC SCALE[4]
Applied loads (Fig. 12.5.2) may be unidirectional or multidirectional in nature and occur singularly or in combination. The result is a macroscopic stress state comprised of normal stress (perpendicular to the

surface) and/or shear stress (parallel to the surface). These in combination with the other application conditions result in one of four primary modes of fracture: dimpled rupture, cleavage, decohesive rupture and fatigue.

TYPES OF FRACTURES: MICROSCOPIC SCALE

Virtually all engineering metals are polycrystalline. As a result, the two basic modes of deformation/fracture (under single loading) are shear and cleavage (Table 12.5.1). The shear mechanism, which occurs by sliding along specific crystallographic planes, is the basis for the macroscopic modes of elastic and plastic deformation. The cleavage mechanism occurs very suddenly via a splitting action of the planes with very little deformation involved. Both of these micro-mechanisms primarily result in transgranular (through-the-grains) fracture.

TABLE 12.5.1 | Differences between shear and cleavage fracture[2]

Metric	Shear	Cleavage
Movement	Sliding	Separation by snapping or tearing
Occurrence	Gradual	Sudden
Deformation	Yes	No
Behavior	Ductile	Brittle
Fracture appearance	Dull and fibrous	Bright and sparkling
Fractographic feature	Dimpled rupture	Cleavage

FIGURE 12.5.2 | Examples of fracture loading/opening modes (Type I = tearing, Type II = shear via edge sliding, Type III = shear via screw sliding)[2]

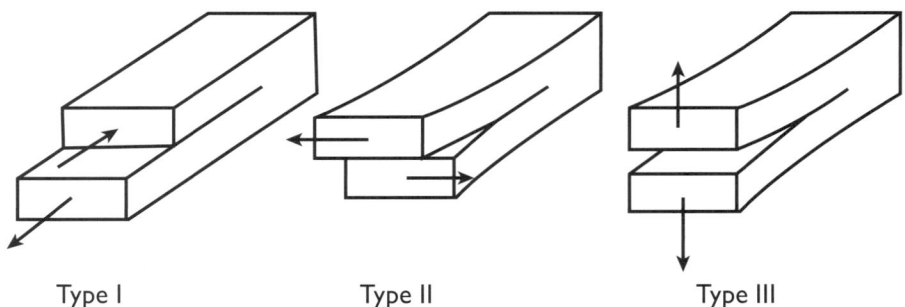

Type I   Type II   Type III

Ductile and Brittle Fractures

Numerous factors influence whether a fracture will behave in a ductile or brittle manner (Table 12.5.2). In ductile materials, plastic deformation occurs when the shear stress exceeds the shear strength before another mode of fracture can occur, with necking typically in evidence before final fracture. Brittle fractures occur suddenly and exhibit very little, if any, deformation before final fracture.

TABLE 12.5.2 | Typical characteristics of ductile and brittle fractures[2]

Parameter	Ductile characteristic	Brittle characteristic
Temperature	Higher	Lower
Rate of loading	Lower	Higher
Type of loading	Torsion	Tension or compression
Geometry	Absence of stress concentrators	Presence of stress concentrators
Size	Smaller, thinner	Larger, thicker
Pressure	Higher	Lower
Material strength	Lower	Higher

Ductile fractures typically have the following characteristics: considerable plastic or permanent deformation in the failure region; and dull and fibrous fracture appearance.[1]

Brittle fractures typically have the following characteristics: lack of plastic or permanent deformation in the region of the fracture; the principle tensile stress is perpendicular to the surface of the brittle fracture; and characteristic markings on the fracture surface pointing back to where the fracture originated.[1]

When examined under a scanning electron microscope, fracture surfaces can exhibit entirely dimpled rupture (i.e., ductile fracture), entirely cleavage (i.e., brittle fracture) or both, although one or the other may dominate. Other fracture modes include intergranular fractures, combination fractures and fatigue fractures.

Fatigue Fractures

Beachmarks (aka stop marks, arrest marks, clamshell marks, conchoidal marks) are a classic fracture feature (Fig. 12.5.3) found during failure analysis, and they usually identify fatigue failures by their presence. They manifest themselves as visible ridges and are indicative of interruptions in crack propagation. Beachmarks (as well as striations) identify the position of the tip of the fatigue crack at any given point in time. They are formed by plastic deformation at the crack tip and by differences in the time of corrosion (when present) in the propagating crack. Beachmarks expand outward from the fatigue origin and are often circular or semicircular in appearance. They will not be present if the part saw only brief interruptions in service.

FIGURE 12.5.3 | A crack surface illustrating multiple initiations, each with their own fatigue propagation (courtesy of Aliya Analytical Inc.)

Wear

Wear (Table 12.5.3) is a type of surface destruction that involves the removal of material from the surface of a component part under some form of contact produced by a form of mechanical action. Wear and corrosion are closely linked, and it is important not only to evaluate the failure but also take into consideration design and environment as well as have a good understanding of the service history of a component.

TABLE 12.5.3 | General categories of wear[2]

Type of wear	Key characteristic(s)	Solution(s)
1. Abrasive	Material removal from the surface (cutting).	Increase surface hardness; remove particles causing abrasion; replace worn parts.
a.) Erosion	Material removal, grooving or channeling of the surface.	Round corners; eliminate bends; change flow characteristics.
b.) Grinding	Cutting, plowing or grooving of the surface.	Eliminate high-stress locations such as points and edges; use coatings or case hardening.
c.) Gouging	Battering or impact damage by hard or abrasive products.	Replaceable parts; substitution of more resistant material.
2. Adhesive	Micro-welding of sliding interfaces due to frictional forces, scoring, scuffing, galling, seizing and tearing.	Lubrication; use of smooth surfaces, chemical films, insoluble contact materials.
3. Fretting corrosion	Wear/oxidation of mating surfaces under vibrational load (slight relative motion).	Eliminate or reduce vibration; reduce slip at the interface; lubrication; elastomeric materials in the joints.
4. Contact stress fatigue	Pitting/spalling under rolling and/or sliding motion (e.g., bearings, gears).	Improve metallurgical properties (e.g., cleanliness), change geometry and improve finish; lower loads or stresses.
a.) Subsurface-origin	Pitting/spalling initiating subsurface in regions of maximum shear stress.	Cleaner steel to eliminate hard, brittle inclusions and subsurface defects; improve load distribution.
b.) Surface-origin	Multiple causes including sliding- and rolling-induced (small surface) pitting; dents, poor finish, poor geometry.	Improve geometry, surface finish, lubrication and load distribution; minimize sliding.
c.) Subcase-origin	Initiates below the case in case-hardened or induction-hardened parts when case depth is inadequate or a damaging discontinuity is present close to the surface.	Increase case depth and/or core hardness.
d.) Cavitation	Pitting by collapse of vapor bubbles at dynamic liquid-metal interfaces; enhanced by vibration and corrosion.	Minimize vibration and vapor formation; increase stiffness; improve surface finish; increase hardness.

Corrosion

Corrosion is the destruction of a material or component by the actions of chemical or electrochemical reactions with the component environment. The major types of corrosion include: galvanic action, uniform corrosion, crevice corrosion, stress-corrosion cracking and corrosion fatigue. The mechanisms and effects created by each of these are well documented.[15,16] It is critical to understand that the effects of corrosion are present to some degree in every failure analysis. This is one of the reasons why protecting fracture surfaces is so critical to performing a proper failure analysis.

This brief introduction to the types and mechanisms involved with the failure of component parts in service is intended to invite the reader to learn more about this subject as it relates to his/her specific area of interest. One of the important takeaways is that no product failure should be treated lightly, and determination of the (single) root cause is essentially impossible.

ANALYSIS METHODS AND TOOLS

If our analysis is accurate, one of the fracture mechanisms discussed in this section will be consistent with the cause. If the mechanism is not properly understood or the analysis is incomplete or inadequate, all contributory factors (and thus the root cause) will not be properly identified, which will most likely make corrective action ineffective.

Role of Fractography

Fractography is the term used to describe the study of fracture surfaces. Fractographic methods are routinely used to help identify factors that facilitated a fracture. One of the goals of fractography is to establish and examine the origin of cracks in an attempt to reveal the cause(s) of crack initiation.

Initial fractographic examination is commonly carried out on a macro scale using the naked eye and then low-power stereo microscopy and oblique-lighting techniques (e.g., low-angle, often from various sides) to identify the extent of cracking, the loading geometry that actually created the crack and the likely origins. Optical or scanning electron microscopy is then used to pinpoint the nature of the failure and the causes of crack initiation and growth (if the loading pattern is known).

Common features that may or may not be present at crack initiation are inclusions; voids or holes in the material; contaminants; and stress concentrators such as dents, bruises, coarse surface finishes, sharp corners, insufficient radii or other abrupt changes in cross section. Lines on fracture surfaces often show crack direction. Some modes of crack growth leave characteristic marks on the surface that identify the mode of crack growth and origin on a macro scale (e.g., beachmarks or striations on fatigue cracks).

Also revealing are areas that exhibit subcritical cracks (i.e., cracks that have not grown to completion). They can indicate that the material was faulty when loaded or, alternatively, that the sample was overloaded at the time of failure. Other clues may be present such as cusps, which typically form where brittle cracks meet.

Fracture Modes at the Micro Scale

Fracture in most engineered materials occurs either by intergranular (along the grain boundaries) or transgranular (through the grains) fracture paths. As previously mentioned, the type of loading experienced produces different types of stresses in the material. This results in different modes of fracture. These can be classified as dimpled rupture, cleavage, decohesive rupture and fatigue.

DIMPLED RUPTURE

Dimpled rupture (aka microvoid coalescence or MVC) is essentially transgranular – ductile in nature – and this type of fracture exhibits cuplike depressions (Fig. 12.5.4) commonly referred to as dimples. Dimpled rupture is the microscale mode of fracture when a single load creates the crack. When a material is put under uniaxial tensile loading, equiaxed dimples that have complete rims appear.

Under tear loading, the dimples are elongated, the rims of the dimples are not complete and the dimples are in the same direction as the loading when viewing the mating fracture surfaces. Shear loading has the same features as tear loading except the dimples are in opposite directions on the mating surfaces. Microvoids grow during plastic flow of the matrix, and microvoids coalesce when adjacent microvoids link together or the material between microvoids experiences necking.

FIGURE 12.5.4 | Example of an equiaxed dimpled rupture-type failure (courtesy of Aliya Analytical Inc.)

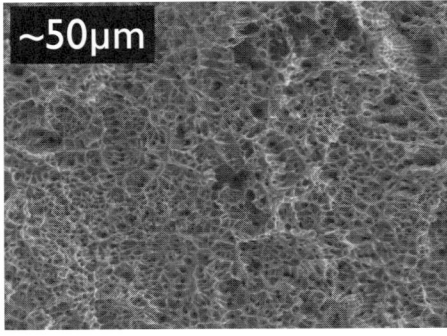

CLEAVAGE

Being a low-energy fracture propagating along well-defined crystallographic planes, cleavage (Fig. 12.5.5) is also essentially transgranular and brittle in nature. It is normally flat with shallow features. Cleavage occurs in brittle or ductile materials and in situations of compression with triaxial stress (i.e., where strain is not possible). Cleavage is more likely in situations of high stress along multiple axes with a high rate of deformation and at low temperature.

Characteristics of cleavage are cleavage steps (i.e., a cleavage facet joining two parallel cleavage fractures); feather markings (i.e., very fine, fan-like markings on a cleavage fracture); chevron patterns; herringbone structures; tongues and microtwins; Wallner lines (i.e., distinct V-shaped by pattern intersection of two groups of parallel cleavage steps, primarily found in brittle materials); and cleavage with ductile tear ridges. These features are only visible in commercial material through use of a scanning electron microscope.

FIGURE 12.5.5 | Example of a cleavage-type failure (courtesy of Aliya Analytical Inc.)

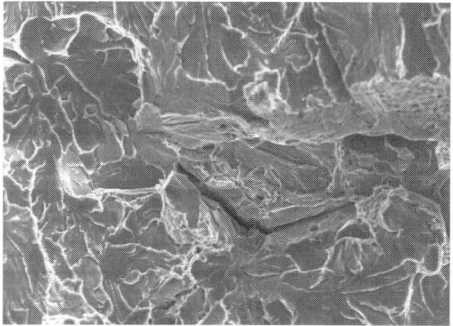

DECOHESIVE RUPTURE

By definition, decohesive rupture (Fig. 12.5.6) is intergranular in nature. Classical quench cracks and classical hydrogen embrittlement as well as creep at high temperatures over long time periods are commonly intergranular. At room and cold temperatures, intergranular cracks are indicative of brittle cracking. In the case of high-temperature creep, there may be considerable visible deformation associated with intergranular features. Note that there are areas of microvoids mixed in with the primarily intergranular ("rock candy") features in Fig. 12.5.6.

FIGURE 12.5.6 | Example of a decohesive rupture-type failure (courtesy of Aston Metallurgical Services Co. Inc.)

FATIGUE

Fatigue failures are essentially transgranular brittle fractures at the macro scale. They are the result of repetitive or (high or low) cyclic loading with nominal loading below the yield strength. The characteristic visible fracture features (clamshell marks, beachmarks, etc.) show the progress of the propagation of the crack-front over time.

Tools of the Trade
SEM

The scanning electron microscope (SEM) uses a focused beam of high-energy electrons to generate a variety of signals from the surface of a specimen. These signals reveal information about the specimen. The importance of the SEM to failure analysis can hardly be overemphasized, although the human eye remains even more crucial to any failure analysis.

When equipped with certain specialized detectors, features revealed may include external morphology (texture), chemical composition, crystalline structure and orientation. In most applications, data is collected over a selected area of the sample surface, and a two-dimensional image is generated that displays spatial variations in these properties. Areas ranging from approximately 1 cm to 5 microns in width can be imaged in a scanning mode using conventional SEM techniques (magnification ranging from 20X to approximately 30,000X; spatial resolution of 50-100Å).

The SEM is also capable of performing analyses of selected point locations on the sample. This approach is especially useful in qualitative or semi-quantitative analysis by helping to determine chemical compositions (using EDS), crystalline structure and crystal orientations (using EBSD).

TEM

The transmission electron microscope (TEM) operates on the same basic principles as the light microscope. However, it uses electrons as a "light source" because their lower wavelength makes it possible to obtain a resolution 1,000 times better than with an optical light microscope. Generally used to study ultra-thin sections (50-60 nanometers), the TEM allows the user to observe and analyze internal microstructures. Stained areas of the sample absorb or scatter the beam, producing dark spots. Unstained areas appear light.

STEM

The scanning transmission electron microscope (STEM) combines the principles of transmission electron microscopy and scanning electron microscopy and is increasing in popularity for failure analysis. Like the TEM, STEM requires very thin samples and looks primarily at a focused beam of electrons. One of its principal advantages is that it enables the use of other signals that cannot be spatially correlated in TEM, including secondary electrons, scattered-beam electrons, characteristic X-rays and electron energy loss.

STEM looks like a TEM but produces images one spot at a time (like in the SEM) rather than all at one time (like in the TEM). The STEM technique (like the SEM) scans a very finely focused beam of electrons across the sample, which is correlated with beam position, to build a virtual image in which the gray level at the corresponding location in the image represents the signal level at any location in the sample. Its primary advantage over conventional SEM imaging is the improvement in spatial resolution but not in surface topography.

OTHER TECHNIQUES

The most important aspect of failure analysis for the heat treater to understand is that tools and investigative practices exist to help us pinpoint where and how a component failed. If a problem with heat treatment is the major contributing factor, the information obtained by failure analysis allows informed decisions regarding changes to recipes, processes or equipment as part of the corrective action.

Failures can be traced to deficiencies in design, materials, processing, product characteristics and quality; known and unknown application factors; and human error. Examples include excessive distortion, buckling, ductile or brittle fracture, creep, rupture, cracking, fatigue, shock, wear, corrosion, misalignment, poor geometrical design and literally hundreds of other factors. Whatever the source, it is important to recognize that it is impossible to separate the product from the process. As such, the material, design, processing and applications are all interrelated.

When considering ways to prevent failures, one determines the factors involved and whether they acted alone or in combination with one another. We ask ourselves questions such as "Which of the various failure classifications were the most important contributors?" "Was the design robust enough and the safety factors properly chosen to meet the application rigors imposed in service?" Having a solid engineering design – coupled with understanding the application, loading and design requirements – is key to avoiding failures. If failures do happen, we must know what contributed to the damage.

FIGURE 12.5.7 | Failure triangle: external influences – application factors; internal influences – manufacturing factors; inter-relationships[17]

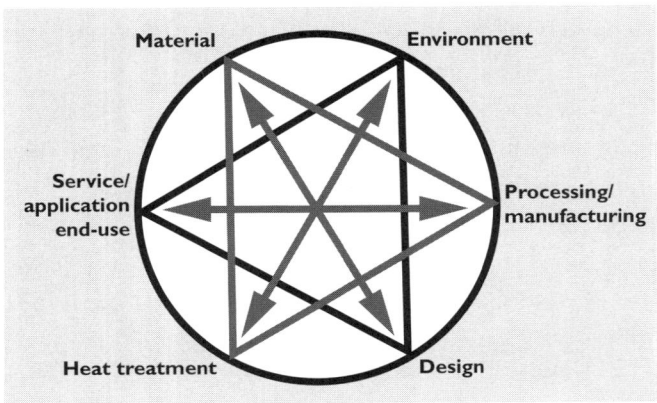

Notes:

[a] Application factors relate environment, service/application end use and design.

[b] Manufacturing factors relate material, processing/manufacturing and design.

[c] Inter-relationships exist between all the factors (indicated by arrows).

Failure Classifications

There are a multitude of types of failures and failure classifications, each requiring a detailed analysis of both the external and internal influences that contribute to either the success or failure of a product. Failure triangles (Fig. 12.5.7) help us visualize these basic classifications and the types of interactions that might take place when failures occur. In most cases, failures happen when one or more of these variables act alone or in combination with one another. These include the following.

SERVICE/APPLICATION END-USE FACTORS

The forces and conditions a part is subjected to in service (Table 12.5.4), coupled with an understanding of the factors that might contribute to a failure, are important considerations. Service conditions may change over time, making this a difficult area to predict.

MATERIAL FACTORS

The type and form of material selected for a given application (Table 12.5.5) are critical to its performance in a given application.

DESIGN FACTORS

Fitness for purpose is the principal focus of design (Table 12.5.6) and must be innately valid such that the expected service life, service conditions and loading can be safely accommodated.

PROCESSING FACTORS

Manufacturing practices must be robust. If not properly performed, they can also contribute to product failures (Table 12.5.7).

HEAT-TREATMENT FACTORS

Heat-treatment issues can contribute to failures in a number of ways (Table 12.5.8). Caution must be observed to avoid blaming heat treatment for revealing a problem the root cause of which may lie outside its contribution to the condition observed.

ENVIRONMENTAL FACTORS

Environmental-induced factors (Table 12.5.9) can play a significant role in field failures and are one of the most difficult aspects to anticipate or control.

TABLE 12.5.4 | Typical service/application influences[2]

Issue	Examples
Overloading	Improper application of load (offset loads), inadequate stress analysis and unexpected loads or conditions
Operating/exposure conditions	Temperature, vibration, impact and collision
Contamination	Chemical, mechanical and electrical
Maintenance or repair issues	Improper maintenance, servicing or substitution of improper parts
Type of forces	Bending, axial load, static load and fatigue
Human error	External influence to the form-fit-function

TABLE 12.5.5 | Typical material influences[2]

Issue	Examples
Alloy design/selection	Use of known engineering performance criteria
Form/condition	Forging, casting, bar, plate and wire
Microstructural characteristics	Grain size, matrix microconstituents and secondary phases
Chemistry	Primary compositional elements as well as residual element effects
Properties	Mechanical (strength/toughness, fatigue strength), physical or metallurgical (hardenability, formability, machinability, weldability)
Internal quality	Inclusion types, amounts, sizes, morphologies and distribution

TABLE 12.5.6 | Typical design influences[2]

Issue	Examples
Design factors	Stress state/type, loading (static or dynamic) and stress risers (geometrical and/or metallurgical)
Raw material	Form, hardness, mechanical/physical/metallurgical properties and cost
Performance characteristics under specific conditions	Duty cycle, impact loading, offset loading, misalignments, transient or unexpected loads and environmental factors
Safety factors	Anticipation of unforeseen service conditions

TABLE 12.5.7 | Typical processing (manufacturing) influences[2]

Issue	Examples
Equipment type and availability	Use of wrong equipment or processes
Manufacturing methods	Improper machining (feeds and speeds), improper selection of pre- and post-heat treatment or other manufacturing operations
Assembly methods	Use of improper tools or poor techniques introducing undesired stress into a component through forced fit-up, nicks, dings, etc.
Welding or joining methods	Introduction of residual stress, voids or cracks (surface or subsurface) and/or undesirable microstructures
Improper heat treatment	Case/core properties; introduction of distortion or undesirable residual stress
Surface finish	Proper or improper use of grinding, polishing, peening or coatings
Residual stress	Proper or improper use of heat treatment or surface treatments (shot blasting/shot peening)

TABLE 12.5.8 | Typical heat-treatment influences[2]

Issue	Examples
Heating and/or cooling	Shape changes and volumetric expansion
Changes in residual stress state	Differential expansion due to thermal gradients and transformational (phase change) induced stress
Quenching	Improper or inadequate quenching (wrong quenchant, poor agitation, improper equipment design) leading to distortion and/or cracking
Hardness	Improper hardness for a given application
Microstructure	Issues related to or arising from grain growth, microcracking, microsegregation, internal oxidation, alloy depletion, formation of high-temperature transformation products, decarburization, retained austenite, precipitation of unwanted phases and excessive carbide formation
Post-heat-treatment methods	Improper post-heat-treatment operations such as grinding, straightening, shot blast/shot peening or plating
Equipment/process selection	Use of the wrong equipment; process introduces unwanted variability

TABLE 12.5.9 | Typical environmental influences[2]

Factor	Examples
Humidity/moisture	Accelerates failure due to corrosion
Chemical attack	Corrosion (stress, fatigue, microbial) or embrittlement (hydrogen, liquid metal)
Lubrication	Viscosity, temperature, oxidation state and additive condition
Wear	Erosion, galling, seizing and pitting
High temperature	Warping, creep, oxidation and localized melting
Site changes	Introduction of unexpected conditions such as vibration, shock impact or exposure to external elements (radiation)

Summary

Accurate record keeping and careful documentation of failures if/when they occur is of critical importance to assist in determining the root cause of a particular product failure and to avoid its reoccurrence. Metallurgical laboratories and experts in failure analysis can then contribute their knowledge and expertise to the problem in order to establish the root cause and determine how it can be avoided in the future.

REFERENCES

1. Debbie Aliya, Aliya Analytical Inc., technical contributions, editing and private correspondence
2. Wulpi, Donald J., *Understanding How Components Fail*, ASM International, 1985
3. Herring, Daniel H., "Failure Analysis from the Heat Treater's Perspective," *Heat Treating Progress*, July/August 2002
4. Herring, Daniel H., "The Do's and Don'ts of Field Failure Analysis," *Industrial Heating*, January 2006
5. Herring, Daniel H., "The Role of Metallurgical Analysis in Solving Heat-Treat Problems," *Industrial Heating*, May 2011
6. Lawn, B.R. and T.R. Wilshaw, *Fracture of Brittle Solids*, Cambridge University Press, 1975
7. *ASM Handbook, Volume 11: Failure Analysis and Prevention*, R.J. Shipley and W.T. Becker (Eds.), ASM International, 2002
8. *ASM Handbook, Volume 12 Fractography*, ASM International, 1987
9. Bhattacharyya, V.E., *Fracture Handbook*, 1979
10. Nishida, Shin-Ichi, *Failure Analysis in Engineering Applications*, Butterworth Heinemann Ltd., Jordan Hill, 1992
11. *Handbook of Case Histories in Failure Analysis*, Khiefa A. Esaklul (Ed.), ASM International, 1992

12. Errichello, Robert, "The Ten Commandments of Gear Failure Analysis," *Gear Technology*, September/October 2003
13. *Practical Failure Analysis*, ASM International
14. "Analyzing Failures of Metal Components, Part One & Part Two," Key to Metals (www.keytometals.com)
15. Fontana, M.G. and N.D. Greene, *Corrosion Engineering*, 3rd Edition, McGraw-Hill Book Company, 1985
16. Uhlig, H.H., *Corrosion and Corrosion Control*, John Wiley & Sons, 1963
17. Herring, Daniel H., "Failure Triangles – A Diagnostic Tool," *Industrial Heating*, May 2012
18. Craig Darragh, AgFox LLC, private correspondence and technical editing
19. Alan Stone, Aston Metallurgical Services Co. Inc. (www.astonmet.com), technical and editorial review, private correspondence

12.6
NONDESTRUCTIVE TESTING METHODS

Nondestructive testing (NDT) is an important tool to help us ensure product safety, quality and reliability. Defects, whether visually evident or internal to the component part, can have devastating consequences to service life and potentially safety. Every effort should be made to find out if a potential problem has occurred during the manufacturing process (including heat treatment) well before a product is placed into service. In addition, product liability will be minimized, which will save time, resources and cost.

Types of NDT Methods
The most common types of NDT methods are:
- Visual inspection
- Dye (liquid) penetrant
- Magnetic-particle inspection
- Radiographic (X-ray) testing
- Eddy current
- Ultrasonic testing
- Leak testing
- Acoustic emission
- Infrared thermal imaging

Each method has advantages and limitations (Table 12.6.1), and more than one technique is often needed to identify the root cause of a failure and quantify the extent of the problem.

TABLE 12.6.1 | Comparison of NDT methods[3]

NDT method	Principal advantages	Main limitations
Visual inspection	Convenient Inexpensive Fast Requires minimal preparation	Subjective Inspector-dependent Surface inspection only Direct examination method
Dye penetrant	Inexpensive Can be used on irregular shapes Minimal equipment required	Messy Surface flaw detection only Requires nonporous surface Ventilation required
Radiographic testing	Wide material applicability Requires minimal preparation Full volumetric examination Permanent record	Radiation hazard Requires two-sided access to sample High skill level required
Eddy current	Speed Reliable with modern machines No human factor Permanent record Part contact not necessary	Difficult to interpret High mechanical precision required Flaw orientation critical Part must be electrically conductive Surface condition important Can lack specificity
Ultrasonic testing	Speed Sensitivity Portability Full volumetric information	Difficult to interpret Smooth surface required Flaw orientation important Complex geometry difficult
Leak testing	Cost Speed High sensitivity	System must be isolated Closed system required Root cause not identified
Acoustic emission	Remote and continuous Inaccessible flaws detected Permanent record	Contact required Multiple contact points Difficult to interpret
Infrared thermal imaging	Portability Remote sensing technique Permanent record Quantitative or qualitative	Expensive Influenced by ambient conditions Surface and material affect results Error a function of distance

Each of these methods will be reviewed with respect to how they work, where they are used and what can be found as a result of their use.

VISUAL INSPECTION

Visual inspection is the most basic and widely used method for the examination of parts. It is used to detect cracks and imperfections – so-called surface abnormalities. It can be performed by virtually anyone. It is subjective in nature, however, so it is critical to develop standards or use comparative aids (photographs, reference parts, etc.) A good visual inspection helps us focus on areas that require closer inspection by other, possibly more revealing methods and is the place to start before performing any other type of NDT test.

Always begin by preparing an area where the visual inspection will take place. It should be well lit and allow easy manipulation of the part so that all surfaces can be easily viewed. Some inspections (especially those involving surface appearance) are best conducted in natural light, so this is an important consideration as well.

Start by inspecting all surfaces in the "as-received" condition. If the part is broken, avoid the temptation of trying to fit components back together. (This can damage fracture surfaces and make subsequent failure-analysis work more difficult.) As the examination proceeds, the surface may need to be cleaned or treated in some manner to highlight a defect in more detail. All observations should be documented and adequate time allowed for a thorough inspection.

Visual inspection typically uses a wide variety of very simple tools, such as:
- Lights (natural, fluorescent, black, artificial)
- Magnifiers (loops, table lenses, 3X-50X stereomicroscopes)
- Mirrors (flat, convex, concave)
- Measurement devices (rulers, tapes, calipers, micrometers)
- Probes (fiberscopes, borescopes)
- Cameras (video, digital, film)

DYE PENETRANT

Liquid-penetrant testing uses a dye or other light-sensitive (visible or fluorescent) fluid to examine nonporous surfaces for defects (cracks, flaws) that may be invisible to the naked eye. Since dyes flow into these imperfections by capillary action, it is critical that the surfaces (including the crack area) be clean and dry and free of oils and residual cleaning fluids. These may interfere with the ability of the dye to wet the surface.

In a typical process, the part is sprayed with a liquid that penetrates any cracks or cavities on the surface. Then the liquid is allowed to soak into the component's surface. After soaking, the excess liquid penetrant is wiped from the surface and a developer applied. The developer is usually a dry white powder, which draws the penetrant out of any cracks by reverse capillary action to produce a visible mark on the surface. Since these colored indicators are broader than the actual flaw, they are more easily visible. Porosity, leak paths, seams and laps can also be inspected using this technique. Fluorescent penetrants (Fig. 12.6.1) are normally used with an ultraviolet lamp to enhance sensitivity.

FIGURE 12.6.1 | Penetrant inspection revealing a circumferentially orientated heat-treat indication (arrow) approximately 55.6 mm (2.19 inches) in length

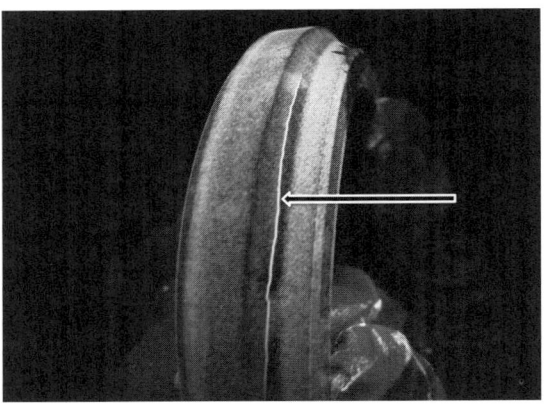

Common problems with this type of testing are the failure to allow adequate time for the dye to form a film over the surface and the failure to remove excess penetrant from the surface.

MAGNETIC AND MAGNETIC-PARTICLE INSPECTION

Inspecting a part by use of a magnet is a technique often forgotten given the modern tools available, but it can be an effective way to look at different materials or even evaluate retained austenite levels in certain stainless steels.

Magnetic-particle inspection is a combination of two nondestructive testing methods: magnetic-flux leak testing and visual testing. It uses a basic principle of magnetism – namely that if any defects are present on or near the surface, the defects will create a leakage field (Fig. 12.6.2). After the component has been magnetized, iron particles – in a dry or wet suspended form – are applied to the surface of the magnetized part. The particles will be attracted and cluster at the flux-leakage fields, thus forming a visible indicator that the inspector can detect.

FIGURE 12.6.2 | Leakage field in the area of a crack[12]

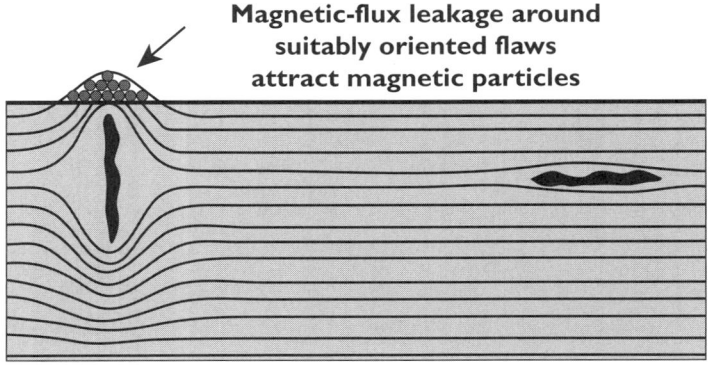

12 | TESTING

The basic components that make up magnetic-particle equipment are: yokes (either permanent magnets or electromagnets), coils (multi-loop windings) and prods (to pass current directly to the part).

RADIOGRAPHIC TESTING

Radiographic methods are often associated with components such as a casting, forging or weldment used in critical applications, including the nuclear industry. This method subjects a part to X-ray (gamma-particle) radiation. Defects are detected by differences in radiation absorption in the material as seen on a shadow graph displayed on photographic film or a fluorescent screen.

In radiography, the process to produce an image is quite different from that of photography. The camera is actually a radiation source, while the film is not placed inside the camera but instead on the opposite side of the object being imaged. The radiation is not reflected to the film but rather passes through the object and then strikes the film. The image on the film is dependent on how much of the radiation makes it through the object and to the film. Differences in the type of material and the amount of material that the X-rays must penetrate are responsible for the details in the image.[8]

Cracks can be detected in a radiograph only if the crack is propagating in a direction that produced a change in thickness that is parallel to the X-ray beam (Fig. 12.6.3). Cracks will appear as jagged and often very faint irregular lines. Cracks can sometimes appear as "tails" on inclusions or porosity.

FIGURE 12.6.3 | Radiographic crack detection[8]

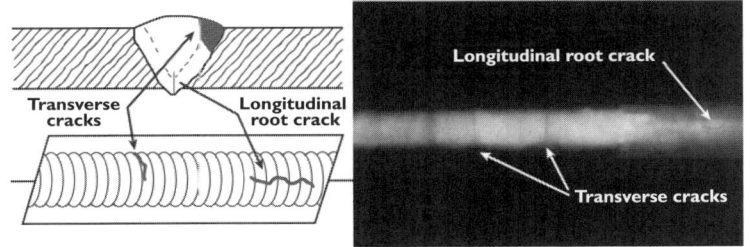

EDDY CURRENT

The two principal purposes for eddy-current testing are to investigate material properties and detect surface flaws. Eddy-current testing is an indirect method and involves the principle of electromagnetic induction. Electric current induced by an alternating magnetic field produces eddy currents. Parts to be inspected are placed within or adjacent to an electric coil. A high-frequency electric current is applied, and the primary field around the coil causes eddy currents to flow into the part.[3]

Material properties testing (Fig. 12.6.4) using the eddy-current method involves the relative permeability property of metals (and sometimes conductivity) to verify that a component is the correct alloy and/or that it has achieved the desired properties from heat treatment, including surface hardness, case hardness and case depth. Testing can verify that the correct microstructure has been achieved because it is the microstructure that determines the relative permeability and conductivity. For example, annealed structures (Fig. 12.6.5a) and hardened structures (Fig. 12.6.5b) can be detected and sorted.

FIGURE 12.6.4 | Typical eddy-current setup for materials property evaluation (courtesy of ALD Thermal Treatment)

FIGURE 12.6.5 | Microstructure testing[4]

a.

b.

a. Annealed microstructure (ferrite and pearlite)

b. Hardened microstructure (tempered martensite)

Surface-flaw testing using the eddy-current method utilizes the conductivity property of metals. Surface flaws disturb the flow of electrical currents that have been generated on the surface of the part being tested. Changes in electrical fields associated with the disruptions in the currents are detected by the eddy-current instrumentation to reveal the surface flaw. Cracks can be found on finished parts using this technique.

ULTRASONIC TESTING

Ultrasonic testing (UT) uses high-frequency sound energy to conduct examinations and make measurements. Sound waves are propagated across the part, and their behavior (velocity and attenuation) is a function of material properties. As these properties change during a test (due to a crack or pore), the recorded signal changes. A change in properties can be observed through analysis of these signals.[10] Since ultrasonic sound travels inefficiently through air, a liquid coolant (e.g., a drop of liquid such as glycerin, propylene glycol, water or oil) is placed between the transducer and the part at the point of measurement. UT is used for measuring part thickness, flaw detection and material analysis.

Ultrasonic-thickness gauges work by tracking the time it takes for an ultrasonic pulse to reflect off the far side of a sample. Such measurements require access to only one side of the part. Therefore, there is no need to cut parts in order to measure an interior surface.

Using the same sound-reflection principles, ultrasonic flaw detectors (Fig. 12.6.6) look for echoes that result from cracks, voids or other discontinuities. An ultrasonic pulse will travel through solid, homogeneous material (such as a gear) until it encounters a boundary (such as a crack) with a different material. Flaw detectors display information on the echo's amplitude and position that can be used to reliably identify and categorize flaws. A trained operator can easily locate hidden flaws by viewing the echo pattern from a reference part and then looking for differences in the patterns received from test pieces. Weld inspection is an important application for ultrasonic flaw detectors.

FIGURE 12.6.6 | Typical ultrasonic flaw detector (courtesy of Olympus Corp.)

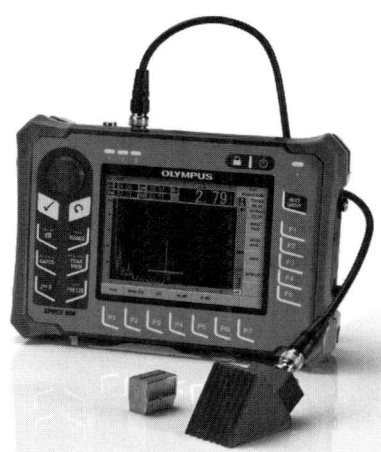

As the physical structure of a material changes so will the way in which sound waves pass through it. Ultrasonic material analysis generally involves looking at parameters such as sound speed, sound attenuation or scattering and frequency content of echoes. These parameters help to analyze or quantify material properties, including elastic modulus, density, grain structure, crystallographic orientation or polymerization patterns. Because sound-transmission properties of different materials vary, ultrasonic material analysis is a comparative process. Generally, a test is set up using reference standards representing the range of conditions to be quantified.

A typical UT inspection system (Fig. 12.6.7) consists of several functional units, such as the pulsar/receiver, transducer and display devices. A pulsar/receiver is an electronic device that can produce a high-voltage electrical pulse. Driven by the pulsar, the transducer generates high-frequency ultrasonic energy. The sound energy is introduced and propagates through the materials in the form of waves. When there is a discontinuity (such as a crack) in the wave path, part of the energy will be reflected back from the flaw surface. The reflected wave signal is transformed into an electrical signal by the transducer and displayed on a screen. Information about the reflector location, size, orientation and other features can be obtained from the signal.[10]

FIGURE 12.6.7 | Ultrasonic detection process[13]

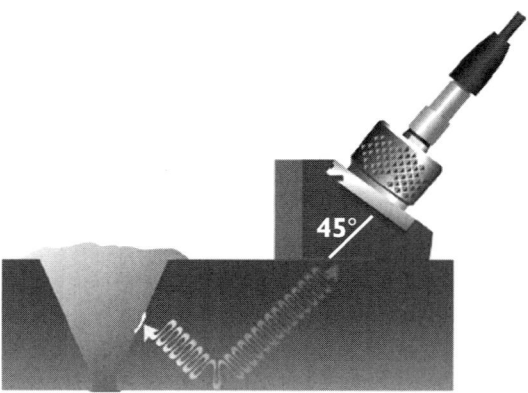

LEAK TESTING

Leak testing is the branch of NDT that is concerned with the escape of liquids, vacuum or gases from sealed components or systems. The three most common reasons for performing a leak test are content loss, contamination and reliability.

In order to effectively leak test an object, a number of questions should be asked. What is the purpose of the test? Should it locate every leak of a certain size? Should it measure the total leakage from the test object without regard to leak number or location?

After the purpose of the test has been defined, the selection criteria most often utilized is whether the test object is under pressure or vacuum. Many methods are reliable under only one of these conditions. Another important criterion is the size of the test object. As the size of the object expands, for example, electronic methods for locating leaks in pressurized units become increasingly impractical due to stratifying of the tracer and the slowness with which the detector probe must be moved. By contrast, mass-produced, hermetically sealed items that are accessible only on an external surface are an area where electronic methods excel.[5]

There are a number of leak-testing methods. Each has its own advantages and disadvantages as well as its own optimum sensitivity range. However, not all methods are useful for every application. Some of these methods involve:

- Bubble testing
- Electronic gas detection
- Mass flow
- Mass spectrometer (helium leak detector)
- Pressure differential/decay
- Ultrasonic leak detector

By applying a number of selection criteria, the choice can often be narrowed to two or three methods, with the final choice being determined by special circumstances or cost-effectiveness.

ACOUSTIC EMISSION

Acoustic emission (AE) is sound waves emitted by microcracks as they are created or move (Fig. 12.6.8). The sound waves propagate through the sample and are recorded by an acquisition system that continuously listens to the sample. Processing techniques similar to those used by seismologists are applied in order to determine information concerning the AE abundance, locations and mechanics. This leads to an analysis of the cracks that have created them.[11]

FIGURE 12.6.8 | Microcrack fracture detection by acoustic emission[14]

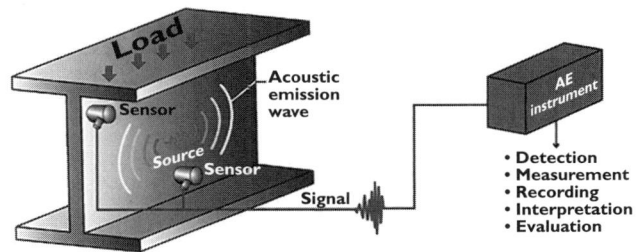

AE technology can determine the amount and intensity of fracturing as it is occurring as well as delineate regions of damage, namely microcrack distributions (mapped three-dimensionally over time to describe damage accumulation), crack coalescence and macrofracture propagation. In addition, AE mechanisms (fault-plane solutions and moment tensors) can determine microcrack orientations and failure properties (tensile crack growth, frictional sliding, grain crushing).

In combination with ultrasonic testing, damage to parts can be quantified and failure mechanisms understood.

INFRARED THERMAL IMAGING

The basis of infrared thermal imaging is quite simple. All objects emit heat or infrared (electromagnetic) energy, but only a very small portion of this energy is visible to our eyes. The infrared wavelength is longer than that of visible light but shorter than a radio wave. In order to see the heat being emitted, an infrared camera must be used. The camera detects this invisible thermal energy and converts it to a visible image. Special training is required to properly interpret these images.

Mechanical and electrical systems are two basic sources of infrared energy. All mechanical systems generate thermal energy during normal operation, which allows infrared thermography to evaluate their operating condition. One of the biggest problems in mechanical systems is excessive temperature. This excessive heat can be generated by cooling degradation, material loss, blockages or an excessive amount of friction (caused by wear, misalignment, over- or under-lubrication and misuse). In mechanical applications, thermography is more useful for locating a problem area than for indicating the root cause. Equipment such as vibration analysis, oil analysis and ultrasound can be employed to further determine where the problem actually lies.

The first step in evaluation is to determine a normal thermal signature. Afterward, any deviation from this normal signature will provide evidence of a developing problem. For example, thermography is ideally suited for locating hot spots prior to a component failing (Fig. 12.6.9).

FIGURE 12.6.9 | Thermographic analysis – furnace power leads showing potential hot spot (courtesy of Thermal Engineering Solutions)

Infrared thermography and ultrasonic leak detection can often complement each other. For example, a faulty electrical connection will produce detectable ultrasound before it generates enough heat to be detected by thermographic imaging. Likewise, thermography can highlight hot spots that ultrasound equipment may never detect.

Summary

Nondestructive testing is an invaluable tool in both preventing failures and analyzing why they occurred. Each method described in this section has its relative merits and limitations, so the selection of the tool (or tools) that best suits a particular application is the key to success.

REFERENCES

1. Alan Stone, Aston Metallurgical Services Co. Inc. (www.astonmet.com), technical and editorial review, private correspondence
2. National Physical Laboratory (www.npl.co.uk)
3. Trimm, Marvin, "An Overview of Nondestructive Evaluation Methods," *Practical Failure Analysis*, Volume 3 Issue 3, ASM International, June 2003
4. NDT Resource Center (www.ndt-ed.org)
5. Herring, Daniel H., "Failure Analysis, Nondestructive Testing Methods," *Industrial Heating*, April 2007
6. Herring, Daniel H., "The Do's and Don'ts of Field Failure Analysis," *Industrial Heating*, January 2006
7. The Open University (www.open.ac.uk)
8. NDT Resource Center (www.ndt-ed.org)

9. Thermal Engineering Solutions (www.tempsens.com)
10. Nelligan, T., "Measuring with Sound," *Quality Digest*
11. ESG Solutions (www.esg.ca)
12. Iowa State University, Institute for Physical Research & Technology (www.iprt.iastate.edu)
13. Legion Hollow & Associates (www.legionhollow.com)
14. MISTRAS Group, Inc. (www.mistrasgroup.com)

13 | MAINTENANCE REQUIREMENTS AND PRACTICES

13.1 PLANNED PREVENTIVE-MAINTENANCE PROGRAMS

The maintenance of heat-treatment equipment is a point of major emphasis, and this is especially true for atmosphere furnaces. Therefore, it is important to discuss what is required for furnace maintenance, as well as how to perform it, while providing useful tips and practical techniques to simplify the work and make sure that it is done correctly.

Why preventive maintenance?

We begin by underscoring the importance of the role of maintenance in the success of a heat-treatment department and, more specifically, how planned preventive maintenance is one of the key components in helping to manage the overall cost of an operation (Fig. 13.1.1).

FIGURE 13.1.1 | The importance of maintenance in total cost containment[4] (modified from the original)

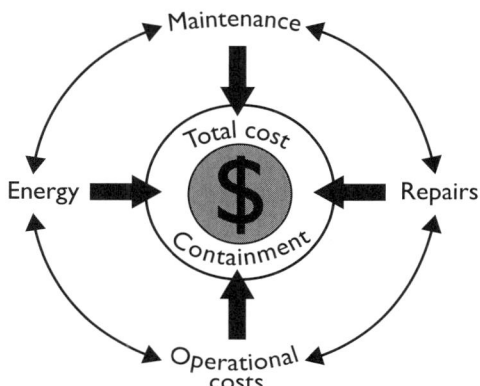

Accepting the Inevitable

Maintenance is a fact of life for heat-treat equipment. In general, the cost of maintenance increases dramatically as the operating temperature increases and/or the process environment becomes more severe (e.g., carburizing vs. hardening). This remains true in atmosphere furnaces, which are often operated near their maximum temperature ratings.

As with all equipment, some styles and designs require more attention than others. It is interesting to note, however, that construction of heat-treat equipment can often be classified as heavy duty or light duty, which impacts the degree of maintenance required. Of course, if any furnace is operated outside its design limitations, this almost always translates to a need for more extensive maintenance.

A great deal of money can be spent – and wasted – if careful thought and clear understanding of the equipment design and the extent of the repair is not taken into account. It is critical to spend the time up front to determine the root cause of why a component failed in order to avoid disastrous bottom-line consequences.

Proper maintenance maximizes uptime productivity, and the utilization of planned preventive-maintenance programs results not only in better equipment reliability but in improved process repeatability and control. This is essential to producing good parts with consistent metallurgical, mechanical and physical properties.

Once management understands, accepts and budgets for maintenance expenditures, the operation of all heat-treat equipment (especially atmosphere furnaces) becomes far more reliable.

Which type of maintenance program is right for me?

Over the years, two distinct styles of maintenance practices have emerged in the heat-treatment industry, namely planned preventive-maintenance programs and repair-as-needed strategies. While each approach has been found to require similar costs to perform, the trend today is toward planned preventive-maintenance programs to better match the production demands of modern manufacturing.

A planned preventive-maintenance program involves the following basic activities: planning the activity; executing the plan; evaluating the results of the maintenance effort; and revising the plan to make it better going forward. This is the so-called PEER system, which normally includes the following activities:

- Monitoring component usage times (via hour meters)
- Distinguishing normal from abnormal operating conditions
- Determining root cause when problems do occur
- Providing a complete explanation of repairs (what, where, why, when and how)
- Establishing mean time between failures of critical components
- Identifying essential spare parts (via critical spare inventories)

13 | MAINTENANCE REQUIREMENTS AND PRACTICES

Planning for Success

It is critical to follow these 10 steps for success in any planned preventive-maintenance program.

1. Set a realistic goal for uptime productivity of a particular heat-treatment furnace.
2. Understand the external constraints imposed on the operation with respect to issues such as:
    a. Equipment usage
    b. Budget
3. Understand why the equipment needs to be serviced. In other words:
    a. How is the unit working now?
    b. How should the unit operate under normal conditions?
4. Tailor the plan to meet realistic expectations.
    a. Identify critical spares and have them in stock.
    b. Understand which spares must come from an original equipment manufacturer (OEM) and which spares can be purchased from third-party suppliers.
5. Divide the work effort into manageable tasks supported by specific disciplines.
    a. Focus on those components or assemblies (internal or external) that are critical to the functionality of the operation.
    b. Do an exterior and interior review and observe how components interact.
6. Understand the interaction of functions and focus the maintenance efforts on the entire system being serviced. (What components and mechanisms comprise the load-transfer system, and how do they interact with one another?)
7. Put the repair information into usable (i.e., searchable) and retrievable form.
    a. Review needs with management.
    b. Get feedback through team meetings.
    c. Revise the plan as needed.
8. Establish a mean time between failures for key components.
    a. Conduct cause-and-effect analyses.
    b. Determine the root cause of a failure (don't just fix the obvious).
9. Be disciplined.
    a. Realize the benefits by having a carefully structured, rigorously adhered-to program. This is not punishment but prevention.
10. Do the job right (or not at all).
    a. Have the tools and supplies on hand to succeed.

Recordkeeping

Good electronic recordkeeping is an important element of planned preventive-maintenance programs. An electronic record, as well as a detailed maintenance logbook, should be kept at the equipment and every action noted. A key aspect of these logs is that they should be reviewed periodically to look for trends, identify problem areas and gain a better understanding of the types of maintenance (and personnel skills) that are required. An accurate log not only reflects the maintenance activity but also provides details and data critical to future success. "Replaced fan bearing" is not the type of logbook entry that will be helpful moving forward.

Having the correct spare parts, detailed furnace construction drawings (access to which must often be negotiated with the OEM at time of equipment purchase) and the right tools on hand is essential to minimizing downtime. Support by the plant engineering staff, OEM and/or a third party is often necessary. Remember, successful planned maintenance is a team effort.

What happens if you buy a used furnace or inherit a piece of equipment from another location? First, try to gather as many facts as possible about its history. For example, find out when it was built and for whom and for what application it was originally intended. Ask the OEM to look at his spare-parts records to identify both the type and frequency of parts sales. This is a good first indicator of places to look for problems. Obtain a copy of the instruction manual and as many detailed prints as possible (there is often a fee for such items).

Next, completely inspect the unit inside and out. Look for the obvious, and ask yourself the reason for everything you see (and don't see). Last, don't rush the equipment into service. Measure and record items such as the speed of moving components (e.g., door speed, load-transfer times and elevator speed). Once operational, record gas usage, current draws, incoming voltage and power. Install hour meters, energy monitors and other recording devices. In other words, learn as much about the normal operation of the equipment as possible before putting it in service so that you can monitor changes and quantify degradation that occurs over time.

Establishing a Preventive-Maintenance (PM) Plan

Divide and conquer. Begin by understanding the heat-treat process(es) you will be asking the unit to perform and compare these to the design ratings/limitations of the equipment, which include:

- ❖ Temperature rating
  - Normal and maximum operating temperature
  - Cyclic operating conditions
  - Idling conditions

- ❖ Load rating
  - Load size including volume or weight limitations
  - Load distribution and the necessity for load ballast
  - Maximum and minimum gross load weight as a function of temperature
- ❖ Atmosphere requirements
  - Type and function of gas(es) – process and heating
  - Gas flow rate, internal furnace pressure, etc.
- ❖ Quench requirements
  - Type of quenchant
  - Volume of quenchant (if a liquid) in relation to the gross load weight
  - Height of the quenchant above the workload
  - Quenchant temperature and rate of rise on quenching
  - Flow characteristics of the quenchant in the tank and around the part
- ❖ Special requirements
  - Baskets and fixturing
  - Quench restrictions
  - Access and site ports
  - Coolant (water) systems
- ❖ Design-specific features
  - Unique (custom) features
  - Support/ancillary items (heat exchangers, water-circulating systems, etc.)

Next, understand the external constraints being placed on the equipment (usage, budget, etc.). These factors are important in tailoring your plan to meet expectation. Then take the time to divide the equipment into logical sections so that the maintenance on each of these areas focuses on the components or assemblies that are critical to their functionality (and ultimately that of the entire machine).

Walk around the exterior and inspect the interior. (Note: Confined-entry training/permits and other safety regulations may be required.) Observe how all components interact. This takes a surprisingly short amount of time and yields a significant amount of information. Finally, put this information into a usable form (such as a spreadsheet), review it with management and implement your planned preventive-maintenance program.

Remember that feedback and refinements to the plan will occur constantly. Make sure that the reasons for the changes are captured in the documentation for later use, and make the system independent of changes that will inevitably occur in either the maintenance department, heat-treat department or management.

Establish a mean time between failures for critical components, and be sure to conduct a cause-and-effect analysis whenever a part fails prematurely. Constantly revise the frequency-of-repair column on your maintenance spreadsheet to reflect items such as number of cycles or less-than-optimal operating conditions. Measuring some aspect of how a particular component is operating and comparing it to the manufactured specifications often enhances this effort. While time consuming, it is surprising how often one finds a component running at or above its normal rating, necessitating more frequent maintenance or possible redesign.

In order to realize the benefits of a planned preventive-maintenance program, irrespective of the type of equipment being used, a carefully structured, well-disciplined and rigorous plan must be created, implemented and followed. It is amazing how (years later) plants struggle to perform maintenance on older equipment because of poor past recordkeeping.

Finally, understand that the frequency of maintenance (i.e., interval between routine repairs) is highly dependent on factors such as:

- ❖ The type and number of heat-treating processes performed
- ❖ Skill level of the operators and maintenance personnel
- ❖ Equipment design
- ❖ Quality of prior maintenance
- ❖ Type of spare parts used
- ❖ Type of support structure (e.g., training of maintenance personnel)

Heat-Treat Checklist

Here is a partial list of questions that should be asked in order to determine if your heat-treat department or outside heat treater has their process under control and is operating equipment that is properly maintained.

- ❖ Is the overall operation in control?
    - Are written instructions, operating procedures and all rules and regulations being utilized on a daily basis, or do they exist only for show?
    - Are all procedures understood by the workforce or only by management?
    - Are the heat-treat practices being effectively used?
    - Is maintenance planned, or does it occur only when machinery breaks?
    - Are the repairs patchwork fixes or permanent solutions?
    - What types of quality-control checks are being made on the furnace (daily, weekly, monthly, semiannually or annually)? What is needed versus what is planned?

13 | MAINTENANCE REQUIREMENTS AND PRACTICES

- ❖ How effective is the cleaning of parts?
  - Are the incoming parts clean?
  - How are they being cleaned?
  - How effective is the cleaning method?
  - How well is it controlled?
  - How often is it monitored?
  - Is a bath chemistry check performed? If so, how often?
  - How often are the washers monitored for proper concentration and pH?
  - How are the washers being cleaned?
  - Are oil skimmers in use, and are they properly maintained?
- ❖ How often is the furnace inspected?
  - What method(s) is used?
  - What are the criteria for acceptance?
  - How effective are the inspections?
  - Are they frequent enough?
  - Do the process parameters remain in a steady state, or do they fluctuate?
  - What method is being used to check the furnaces for leaks?
  - What types of thermocouples are being used, and are they adequate for the temperature range being run?
  - When maintenance is performed on thermocouples, are their insertion depths per OEM's recommendations?
  - Has the insertion depth of the thermocouples changed?
  - When was the last time a temperature uniformity survey (of the workload area) was performed?
- ❖ How effective is the quench system? (Fig. 13.1.2)
  - How is the quench gas or oil being monitored and controlled?
  - Is the degree of agitation sufficient for the quenching operation being performed? How is the quench gas being introduced into the chamber, and are there any signs of damage?
  - How often is the quench medium sent out for analysis?
  - How is the quench media checked for particulates?
  - Is the motion of the elevator (e.g., batch integral-quench furnaces) smooth and quick (2-3 seconds being typical)?
  - How often is the quench tank serviced?
  - How much drag-out (removal) of quench media occurs?
  - What is the transfer time of the workload to the quench?
  - What type of quench-tank maintenance is performed, and how often?

❖ How are loads being prepared? Are trays, baskets, screens and other fixtures/racks being inspected?
- Are loads volume-limited or weight-limited?
- What is the transfer time of the workload to the quench?
- Is proper inspection and maintenance of tray sensors being performed? (If they are cracked or sticking, trays will cause jam-ups or otherwise damage the interior of the furnace.)

FIGURE 13.1.2 | Examples of part surfaces after oil quenching

a. Unacceptable part surface appearance after heat treatment – part staining, sludge buildup, decarburized surfaces (courtesy of Houghton International Inc.)

b. Acceptable part surface appearance after heat treatment – parts are clean; typically gray/green in appearance in the as-quenched condition with no evidence of staining, soot, scale or other surface contamination (courtesy of Bodycote Inc.)

a.
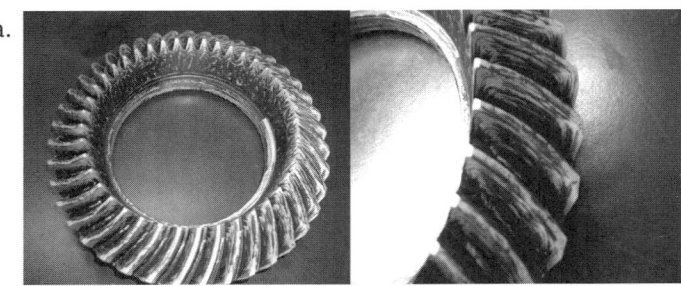

b.

Summary

Once a sound maintenance strategy has been established, the actual maintenance work can begin.

REFERENCES
1. Herring, Daniel H. and Gerald D. Lindell, "Periodic Atmosphere Furnace Maintenance: Part One – Preventative Maintenance Plans," *Heat Treating Progress,* July/August 2009
2. Herring, Daniel H., *Atmosphere Heat Treatment, Volume I*, BNP Media, 2014
3. Herring, Daniel H., "Equipment Maintenance Presentation," Furnaces North America, October 2014
4. DualTemp Mechanical & Refrigeration (www.dualtempmechanical.com)

13 | MAINTENANCE REQUIREMENTS AND PRACTICES

13.2
MAINTENANCE OF ATMOSPHERE FURNACE COMPONENTS

Let's focus on the specifics of what items require maintenance, when and how maintenance should be performed, and why it is necessary to do the work in an organized and careful manner.

First, it is important to recognize that atmosphere furnaces come in all shapes and sizes (Figs. 13.2.1, 13.2.2), and while they have some common features and maintenance needs, most furnaces will require unique maintenance solutions based on their design, age and the process(es) being performed in them. Thus, it is critically important to understand the particular needs of the atmosphere furnaces in your shop to have an effective maintenance program.

FIGURE 13.2.1 | Typical batch-style atmosphere furnaces

a. Box furnace (courtesy of Nutec Bickley)

b. Box furnace with retort (courtesy of L&L Special Furnace)

c. Pit furnace (courtesy of SECO/WARWICK Corp.)

d. Bell furnace (courtesy of Lindberg/MPH)

e. Integral-quench furnace (courtesy of AFC-Holcroft)

f. Car-bottom furnace (courtesy of J.L. Becker, a Gasbarre Furnace Group Company)

a.

b.

FIGURE 13.2.2 | Typical continuous-style atmosphere furnaces

a. Mesh-belt furnace (courtesy of Can-Eng Furnaces International)

b. Roller-hearth furnace (courtesy of Ipsen)

c. Pusher furnace (courtesy of Tenova/LOI)

d. Screw conveyor furnace (courtesy of Surface Combustion Inc.)

e. Rotary-drum furnace (courtesy of SECO/WARWICK Corp.)

f. Rotary-hearth furnace (courtesy of Aichelin Heat Treatment Systems)

13 | MAINTENANCE REQUIREMENTS AND PRACTICES

e.

f.

The most common components that require maintenance are:
- Alloy belts (mesh and cast link)
- Burners
- Combustion systems and radiant tubes
- Doors and cylinders
- Fans
- Flame curtains
- Flowmeters
- Furnace alloy
- Furnace-atmosphere generators
- Furnace-atmosphere controls
- Gauges (pressure, temperature, etc.)
- Heat exchangers
- Heating elements
- Instruments and controls
- Insulation (refractory, fiber, etc.)
- Motors, pumps and agitators
- Oxygen (carbon) probes
- Quench chutes and eductors
- Quench tanks and quenchants
- Oil/polymer/salt
- Water/brine/gas
- Sample and site ports
- Seals and gaskets
- Sensors and switches
- Thermocouples

Maintenance Overview

While each of the items in the previous list is worthy of a detailed discussion, here are a few experience-based maintenance tips designed to avoid the most common problems being reported by users (Fig. 13.2.3).

FIGURE 13.2.3 | Common atmosphere-furnace maintenance items

a. Alloy (mesh) belt (courtesy of Surface Combustion Inc.)

b. Burner (courtesy of Eclipse Inc.)

c. Gas train (courtesy of Eclipse Inc.)

d. Furnace door

e. Furnace fan

f. Flame curtain (courtesy of Aichelin Heat Treatment Systems)

a.  b.

c.  d.

e.  f.

g. Flowmeter and mass-flow control devices (courtesy of Atmosphere Engineering and MKS Instruments)

h. Furnace alloy (courtesy of Safe Cronite)

i. Furnace-atmosphere controls (courtesy of Super Systems Inc.)

j. Quench solenoid and related components (courtesy of Ajax TOCCO Magnethermic)

k. Water-to-oil heat exchanger

l. Electric heating elements

13 | MAINTENANCE REQUIREMENTS AND PRACTICES

g.

h.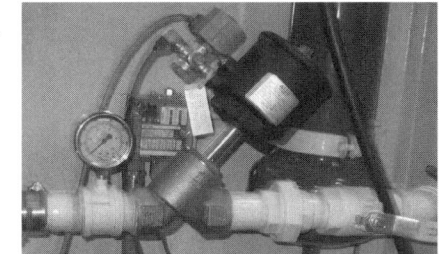

i.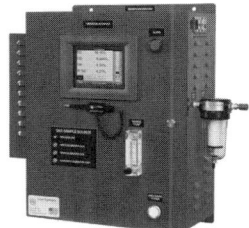

j.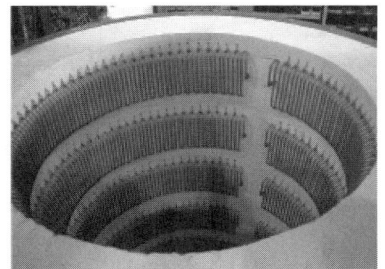

k.

l.

m. Instruments and controls (courtesy of Yokogawa)

n. Furnace insulation

o. Oil agitators

p. Oxygen probes (courtesy of Super Systems Inc.)

q. Seals and gaskets (courtesy of Packings & Insulations Corp.)

r. Limit switches, electric eyes and load movement sensors

m.

n.

o.
p.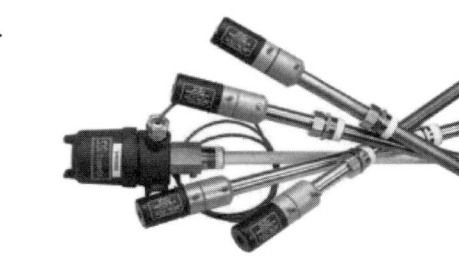
q.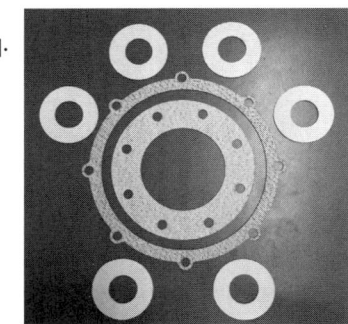
r.

What follows are observations, comments and maintenance tips for various devices associated with atmosphere furnaces.

COMBUSTION SYSTEMS

- ❖ Burners (Fig. 13.2.4), combustion blowers and gas trains
  - The air/fuel ratio control in most modern burners is analogous to a carburetor.
  - Ratio is controlled by mechanical or pressure devices that suffer from hysteresis and constantly changing parameters such as air temperature, barometric pressure, clogging air filters and application pressures.
- ❖ Tune the control loop frequently, and set the firing sequence of the burners to control uniformity.
- ❖ Devices associated with controlling burner ratios must be given extra attention. This includes valves, linkages, regulators and regulator impulse lines.
  - Inspect these devices monthly for proper settings and function.

13 | MAINTENANCE REQUIREMENTS AND PRACTICES

FIGURE 13.2.4 | Anatomy of a high-performance, high-velocity burner (courtesy of Eclipse Inc.)

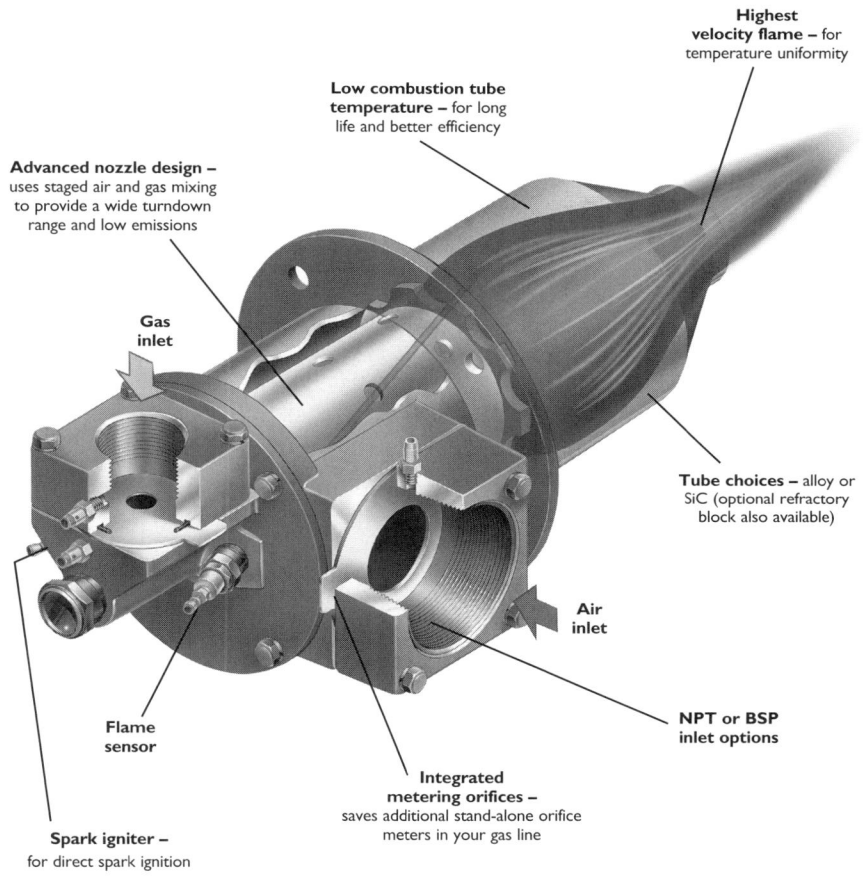

DOORS AND CYLINDERS

- ❖ Outer doors should be checked for smooth motion, damaged or bent cylinder rods and atmosphere leakage.
    - Cylinder rods should be checked for evidence of warpage or damage; chains should be taut but not tight to the point of binding.
    - Cylinder seals should be checked for evidence of leakage.
    - Pilots and flame supervision (UV, flame rods) should be checked daily for functionality.
    - Flame curtains should be properly adjusted to cover the entire width and height of the door opening.
- ❖ Inner doors should be checked for smooth motion, proper sealing to the mating refractory throat, gap spacing and proper positioning into the door jams.

- When the furnace is under atmosphere, check the area around the inner door with a gas wand for atmosphere leaks.
- Door lifting-cylinder rods or chains should be checked for evidence of rod warpage, damage and/or proper tension.
- Cylinder seals should be checked for evidence of leakage.

FANS

- Fans should be both statically and dynamically balanced before installation.
  - Fan speed is critical. Do not assume it is correct; check it.
  - Fan rotation is important and depends on the style (design) of the fan.
- Check that there is adequate cooling to the fan bearings.
  - Some fans have water-cooled bearings and others have air-cooled bearings while still others have rifle-bored water-cooled shafts.
- Most fans use special bearings and special grease, so be aware of this when replacing these items. In some cases, the OEM may have special instructions to the vendor that are not communicated by part number.
- Inspect for casting defects or fabrication issues, especially in welds.
- Once in service, check periodically for evidence of cracking (especially in the hub area), surface attack, drooping blades or melting.
- Check for vibration.
  - Have adequate structural supports.
- Check for cracks or other damage.
  - Welds and areas joining thin and thick sections are particularly vulnerable.

FLAME CURTAINS

- Flame curtains should be adjusted so that they cover the entire width and height of the door opening.
  - The goal is to introduce products of combustion into the furnace rather than air.
- The air-gas mixture should be adjusted so that the flame curtain runs slightly rich.
  - Avoid a yellow or sooty flame.
- A high-pressure pilot (as opposed to lazy flame) with flame supervision or an electric igniter should be positioned to ignite the flame curtain, especially on furnaces with doors.

FLOWMETERS

- Be sure to look at the nametag on the flowmeter to confirm that it is calibrated for the gas being used and at the right temperature and pressure.
  - Use correction factors provided by manufacturers to obtain accurate flow values.
  - As a general rule, flowmeters used in heat-treating applications are designed for a maximum temperature of 65°C (150°F) and an operating pressure no higher than 345 kPa (50 psig).
- Be sure that the flow reading is in the middle of the flow range (i.e., 25-75% of the scale value).
- Most mass-flow devices and mass-flow controllers have an accuracy of ±1% of full scale and a repeatability of ±0.25% of full scale. The accuracy of these devices is influenced by flow calibration, which ensures starting-point accuracy.

FURNACE BELTS (MESH AND CAST LINK)

- Precondition high-temperature belts prior to use.
  - For example, run mesh belts at 815°C (1500°F) for 2-4 rotations at a very slow speed (12.5 mm/minute; 0.5 inch/minute) prior to taking them to final temperature and starting production. Drive adjustments might be necessary.
- Measure belt speed by timing the actual belt movement. Look for jerky motion.
- Monitor belt tracking. Be sure the belt is not moving off center. Look for wear and/or distortion of the cross links.
- Flip and reverse belts every month when possible.
- Check that the belt is properly supported over its length.
- Front-end and rear-end drive systems must be adjusted differently, and different belt tension may be required. The zero point of loading is different for these two styles of drives.
- Check the tension of the belt frequently or as dictated by production usage.
  - Do not over-tension the belt.

GAS GENERATORS AND ATMOSPHERE SYSTEMS

- Have maintenance lists in place for items that need per shift, daily, weekly, monthly, bimonthly, semiannual and annual maintenance.

- Understand the particular requirements for each type of generator or atmosphere gas system in use.
    - For endothermic-gas generator systems, check the air/gas ratio and check for evidence of sooting (catalyst, heat exchanger) daily.
    - For exothermic-gas generator systems, monitor the air/gas ratio, water temperature, carbon monoxide and percent combustibles.
    - For ammonia dissociators, monitor inlet and outlet pressure to the retort and the ammonia supply level in the outside storage tank (refill at 20%).

GAUGES AND SENSORS
- Whenever possible, install manometers to sense changes in internal furnace pressure and monitor recovery time to positive pressure, especially during transfer to the quench in sealed-quench systems.
- All gauges should be calibrated or designated as "reference only."
- Pressure gauges should be checked for accuracy and replaced if the fluid level is not near the top of the gauge or if there is evidence that the gauge has been over-pressurized.

HEAT EXCHANGERS
- Oil-to-water heat exchangers:
    - Check water flow and pressure (inlet and outlet) daily.
    - Check tube bundles quarterly for evidence of attack or deterioration that could lead to water leaks.
- Oil-to-air heat exchangers:
    - Check the amperage draw on cooling-fan motors.
    - Keep cooling fans clean and free of dirt or other debris buildup.

HEATING ELEMENTS
- Check element connections and terminals. They must be tight and fit properly. Also, check resistance to ground.
- Check the element coils or loops for adequate spacing of elements on their supports.
- Check the supports and/or terminal holes through the insulation to prevent sagging or damage to the elements.
- Match resistivity of silicon-carbide elements in pairs when installing.
- Pack around elements with ceramic fiber if necessary to help maintain a good atmosphere seal.
- Use the lowest transformer tap voltage that will maintain the desired furnace operating temperature.

- ❖ Check amperage when in operation.
- ❖ Inspect for green rot, a type of corrosion phenomenon that takes place in heating elements made of nickel-chromium and occurs in the temperature range of 870-1040°C (1600-1900°F).
  - The elements appear green in color, which is a result of preferential oxidation of the chromium by the furnace atmosphere when it is oxidizing to chromium and reducing to nickel.
  - Inspect for mechanical damage or element collapse, especially if elements are exposed to a high carbon-potential atmosphere (>0.50% C).
  - Changes in electrical characteristics and/or mechanical collapse (softening) or melting can occur, especially above 1095°C (2000°F).

INSTRUMENTS AND CONTROLS

- ❖ Control, over-temperature and recording instrumentation should be calibrated with a frequency that ensures the proper accuracy. Many companies are now required to meet stringent pyrometry standards.
  - The latest revision of AMS 2750 (Pyrometry) should be followed.
- ❖ Programmable logic controllers typically do not require frequent maintenance. However, be sure that a backup copy of the ladder logic and programming sequence is available and in a format accessible by modern PCs.
  - Relays and relay logic should be updated if practical.

MOTORS, PUMPS AND AGITATORS

- ❖ Motors should be checked daily for amperage draw.
  - If appropriate, check belt tension weekly.
  - Agitators and pumps should be checked for proper rotation, current draw, speed and vibration.
- ❖ Pumps should be checked once a month for priming speed, capacity, noise in pump casing, gaskets and O-rings, shaft-seal fluid leakage, hoses, hose washers and suction strainer, and the engine (e.g., crankcase oil level, spark-plug condition, air cleaner, noise, RPM, carburetor adjustment).

OXYGEN (CARBON) PROBES

- ❖ Be sure that the oxygen (carbon) probe sits in a location where it will be exposed to the true furnace atmosphere.
  - Normal furnace insertion depth is 100-150 mm (4-6 inches) into the chamber.

- ❖ Be sure the reference air system (vibratory or diaphragm pump) is working properly.
  - Failure to do so will result in false readings.
- ❖ Be sure a separate supply of burn-off air is provided to the probe.
  - Failure to do so will often result in premature failure (due to carbon buildup on the tip of the probe).

QUENCH CHUTES AND EDUCTORS
- ❖ Quench curtains should be properly set so that the flow from both sides meets in the middle of the chute.
  - Avoid cooling the quench-chute area in salt quench tanks because salt will solidify, interfering with the quench.
- ❖ Quench-chute eductor settings are one of the most important yet most overlooked aspects of a continuous furnace. Improper settings result in:
  - Furnace atmosphere and maintenance problems
  - Part quench issues
  - Safety concerns
- ❖ Atmosphere flow rate and internal pressure (in the quench chute and in the furnace) should be monitored on a continuous basis.

QUENCHING (MOLTEN SALT)
- ❖ The bath should be rectified when the oxide level exceeds 0.005%.
- ❖ Desludging and filtering must be done on a frequent basis. Remove and/or dislodge buildup from around the tank and heating elements. Typically, high-temperature tanks should be desludged twice per shift and low-temperature tanks should be desludged every hour.
- ❖ When salt lines are going to be idle for an extended period of time (over the weekend), reduce the salt temperature for better separation and remove contaminants or high-temperature salt from low-temperature salt baths.

QUENCHING (OIL)
- ❖ Quench-oil level should be checked daily.
  - Not having the proper oil level can have disastrous safety consequences.
- ❖ A water-detection system should be installed and checked daily in every oil quench tank.
  - Remember, oil floats on top of water, so sampling should take place from the bottom of the tank.

- ❖ Quench oil should be analyzed by an outside lab monthly to quarterly (usage-dependent) to determine its condition and suitability for use.
    - Be sure you understand how to interpret the quench-oil analysis report.

QUENCHING (POLYMER)
- ❖ Concentration should be checked daily by a refractometer or similar device.
    - Be sure that you properly interpret the scale reading.
    - If the concentration is out of control, rusting can quickly occur.
- ❖ Bath agitation must be carefully controlled to avoid aeration and loss of heat transfer.
- ❖ Bath temperature should be maintained in order to avoid reaching the polymer/water separation temperature (typically around 70°C/160°F) on quenching.
- ❖ Check that the bath is not contaminated (oils, metal fines, etc.).
- ❖ Polymer bath solution should be analyzed by an outside lab monthly or quarterly (usage dependent) to determine its condition and suitability for use.
    - Be sure you understand how to interpret the polymer-quench analysis report.

QUENCHING (WATER, BRINE, CAUSTIC)
- ❖ Tank systems should be checked daily.
    - Remove scale and buildup from the tank.
    - Clean filters, strainers and separators.
    - Monitor fluid levels.
    - Check that proper agitation is maintained throughout the tank.
- ❖ Spray systems should be checked daily.
    - Check for clogged nozzles.
    - Check to be sure that the flow from the nozzles results in a uniform pattern over the work.

RADIANT TUBES
- ❖ Alloy radiant tubes
    - Check for pinhole leaks by either pressurizing the tubes or by going to high fire and observing if smoke exits the exhaust leg.
    - Select the proper alloy-tube material for the service application (not all alloys are created equal).
    - Inspect for evidence of mechanical damage due to creep, thermal expansion, embrittlement, thermal fatigue, thermal shock, mol-

ten-metal embrittlement or corrosion/attack (oxidation, sulfidation, carburization, nitriding, chlorides/salts or metal dusting).
- Alloy tubes tend to bend inward over time and should be rotated 180 degrees a minimum of once a year.

❖ Ceramic (SiC) radiant tubes
- Provide adequate room for thermal expansion in and around the tubes. Sectional inner tubes may deform over time.
- Avoid situations where physical damage or thermal shock may occur, especially during installation.

REFRACTORY

❖ Insulating firebrick
- Check for carbon penetration into the brickwork by periodically removing a brick and looking at its cross section.
- Carbon deposits will continue to diffuse into the refractory over time. They will appear either black (significant attack) or gray (slight attack). If found, plan more frequent furnace burnouts. If severe, rebricking will be necessary.
- Use low-iron (<0.5% Fe) brick.
- The rating of the insulating firebrick should be at least 150-260°C (300-500°F) higher than the maximum furnace operating temperature. The higher the rating of the refractory, the poorer its insulating value.

❖ Ceramic-fiber insulation
- Inspect the insulation for damage or areas where the insulation is no longer present. If a coating (rigidizer, reflective thermal barrier) is used, check for evidence of erosion or delaminating away from the surface.
- Inspect the fiber for evidence of sagging, separation (gaps) or crumbling. Brittle fibers are evidence that the temperature rating of the fiber may have been exceeded.
- Fiber insulation comes in a variety of densities. Select the proper density for the application end use (temperature rating, atmosphere, etc.)

SAMPLE AND SIGHT PORTS

❖ Sample ports must be kept clean and free of deposits such as soot buildup.
- Sample ports should be checked daily and cleaned (rod out) as necessary.

❖ Sample ports should be located for easy connection of furnace atmosphere-control devices.

- ❖ Sample ports should be located in areas that represent the true condition of the parts being processed.
- ❖ Keep the sight-port glass clean and the manual blast gate valve in good repair. Caution: Do this when the furnace is cold. Do not perform maintenance on the sight port with the furnace hot or full of combustible atmosphere.
  - Peep sights are used to view into the heating chambers of furnaces and/or to view critical areas inside the furnace (such as areas where load transfers take place).
  - Peep sights have been converted into shim pullers in some furnaces.
- ❖ When the furnace is cold, check the viewing area of the sight port to be sure it is at the proper angle.
- ❖ The area around the sight port and all access points (ladders, etc.) should be kept clear.

SEALS AND GASKETS
- ❖ Select the proper gasket material that is suitable for both the exposure temperature and the atmosphere that might be present.
- ❖ Use high-temperature silicon-rubber sealant (e.g., GE RTV 106) where applicable.
  - Maximum temperature rating is 315°C (600°F)
  - Exposure to high temperature causes the silicon rubber to dry out and become hard/brittle. If this happens, you know that the area in question is being exposed to higher-than-normal temperatures, indicating that a more serious problem may exist.

SENSORS AND SWITCHES
- ❖ Solenoid valves should be checked weekly for functionality.
- ❖ Limit switches, proximity switches and electric eyes should be checked daily for functionality.
  - Bent arms should be replaced.
  - Dirty lenses should be cleaned or shielded.

SOOTING
- ❖ Pressure gauges should be installed (and values monitored) before and after heat exchangers. Endothermic-gas generators and tubes should be designed so they can be opened and cleaned out easily since they can clog with soot in a matter of minutes.
- ❖ Flowmeters should be disassembled and the inner workings cleaned of soot accumulation.

- Transmission lines from the gas generators to the furnace(s) should contain drop legs that can be opened and cleaned quarterly and the lines blown free of soot accumulation once every five years.

THERMOCOUPLES AND PROTECTION TUBES
- Thermocouples that are located too close to heat sources, the furnace walls or the workload will not sense or control true furnace temperature.
- The frequency of thermocouple replacement (unless mandated by specifications such as AMS 2750) should be as follows:
  - High-temperature furnace operating above 760°C (1400°F) or furnaces running combustible atmosphere every 6 months to 1.5 years.
  - Low-temperature furnace operating below 760°C (1400°F) every 2-3 years.
- Thermocouples and protection tubes must be properly positioned inside the furnace – an exposure of at least 20 times the thermocouple's diameter – for the most accurate readings.

Maintenance Planning

When performing maintenance, it is important to have a written plan defining the specific task to be performed and outlining the reason why a particular task is necessary. A work order should be issued and the work signed off upon completion (includes testing to ensure that the repair was successful). The following conditions should be met before any repairs are undertaken:

- Power should be switched off for any repairs not directly involved with the electrical systems, controls or instrumentation.
  - Lockout/tagout procedures should be in place and checked by all personnel performing the maintenance.
- Disconnect all utilities, including gases, water and air. Lockout/tagout procedures should be followed.
- The furnace should be cool, less than 50°C (120°F).
- The furnace door(s) should be in the open position and secured so that they cannot be closed.
- Check that the furnace environment is safe and that adequate ventilation is in place and functioning properly if working inside the unit.
  - The oxygen level must be checked and confirmed to be safe for human exposure before entry into the unit by both the safety team and the personnel performing the maintenance functions.
- Wear protective clothing, including safety glasses and safety shoes.

- Be sure that all confined-entry procedures are thoroughly understood and followed without exception.
- Use the buddy system.

Maintenance Safety

Safety is **mandatory** and cannot be compromised during any maintenance activity. Safety interlocks must never be bypassed, and verification that all potentially hazardous energy sources have been isolated and disabled is a necessary first step in the maintenance process (NFPA 86 and NFPA 70).

Lockout/tagout procedures are required to disable machines or equipment during maintenance to prevent injury as part of OSHA code (Regulation 1910.147). An atmosphere furnace may have different places in which electrical power must be disconnected: a single main electrical disconnect (i.e., circuit breaker) for the entire furnace or power supplied from several electrical sources each with a separate disconnect device. The electrical drawings for the specific unit in question should be reviewed and physical inspections should be conducted in the event that undocumented changes have taken place.

In addition, atmosphere furnaces may also have pneumatic or hydraulic systems, which include sources of compressed air, inert gases and process (reactive) gases. These must be isolated from the system.

All safety interlocks present in normal equipment operation should be tested on a regular basis to ensure proper operation. These include:

- Over-temperature instrumentation (test monthly)
- Process interlocks (test semiannually)
- Water interlocks (test semiannually)
- Air interlocks (test semiannually)

Summary

The goal of any maintenance performed on an atmosphere furnace is to return the unit to full operational service. Making sure that all problems are uncovered, the root cause of the problem has been correctly determined and repairs don't involve Band-Aid fixes are critical to achieving this goal. Be sure to document the problems and their solutions.

REFERENCES

1. Herring, Daniel H., "Equipment Maintenance Presentation," Furnaces North America 2014, October 2014
2. Herring, Daniel H., *Atmosphere Heat Treatment, Volume I*, BNP Media, 2014

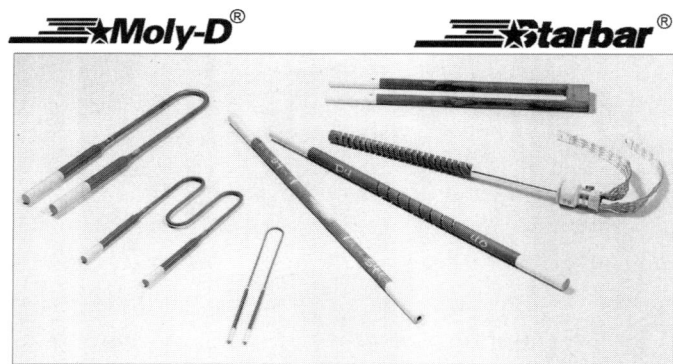

13.3
MAINTENANCE TECHNIQUES FOR ATMOSPHERE FURNACE SYSTEMS

This section will focus in greater detail on servicing major areas common to many atmosphere heat-treatment furnaces, such as furnace fans, flowmeters, mesh belts and drive systems, pilots and flame curtains, and thermocouples.

Furnace Fans
Fans in all types of furnaces and ovens are critical to the successful performance of the heat-treatment operation. There are many styles and types, but all of them must be properly designed for their intended use and, of equal importance, properly maintained. Fans are often considered a necessary evil in most heat-treating furnaces. If they are ignored or overlooked, disastrous consequences have been known to happen. So, take the time to understand where and how they are being used, and you will be rewarded with hours and hours of time to do other things. All fans face a variety of common problems that must be overcome.

TYPES
There are a number of different fan styles that are used throughout the heat-treating industry, each selected for use with a particular furnace design and application requirement. These include blowers, squirrel-cage fans, radial-blade fans (Fig. 13.3.1), and axial- and semi-

axial-blade fans. These are typically divided into two general classes: centrifugal fans and propeller (axial-flow) fans. This classification is functional. The fundamental difference between them involves the way in which air or atmosphere is moved.

FIGURE 13.3.1 | Circulation pattern of a radial-blade fan in a pit-style furnace (courtesy of SECO/WARWICK Corp.)

a. Circulation pattern

b. Fan detail

a.
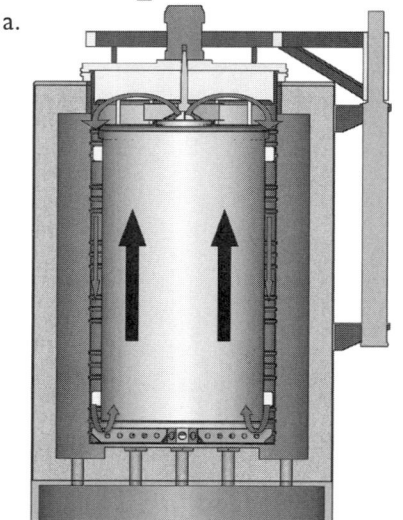

b.
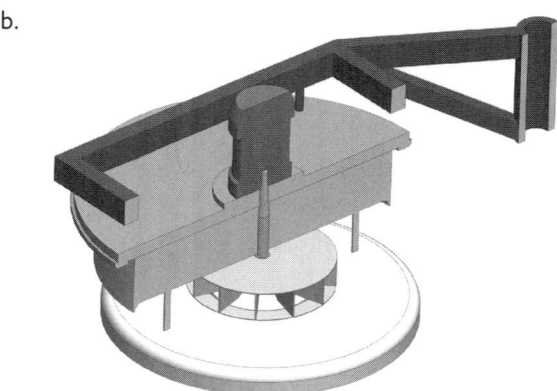

MECHANICAL PROBLEMS

Mechanical problems typically involve items such as bearings, belts, motors and wiring. Should these components fail, it is important to evaluate why they failed or why they failed prematurely. It is often a good idea to determine the mean time between failures (MTBF) of these components. Their failure may be a symptom of a much deeper problem – a design, material or performance issue.

One of the central mechanical issues often involves bearings. Many designs use special bearings or special lubricants, so substitution of a standard bearing may cause premature failure and unexpected downtime. In one instance, a well-intentioned manufacturer modified standard bearings by substituting under-sized balls and packed them with special high-temperature grease but failed to inform his customers, who were baffled as to why the bearings failed prematurely when they purchased the same part number from a catalog.

It is important to also remember that fan performance must match the OEM's specifications after a repair. If not, either process or quality issues may arise after repairs unrelated to the "mechanical" performance of the fan are complete.

Finally, it is not uncommon to find designs with top and bottom bearings spaced too far apart. This causes the fans to wobble or vibrate in service, which shortens their life. Vibration meters can be used to provide an accurate vibration number. Years ago, one placed a penny on the fan housing, and the vibration was considered excessive if it vibrated off.

POWER FAILURES

A hidden problem in many fan applications, especially for fans operating at high temperature, is that on restart after a power failure the torque (stress) exerted on the fan blades can be enormous given their weakened strength at temperature. A good rule is to never stop a fan at temperature, although there are exceptions (e.g., when opening a pit-furnace cover to transfer a load). The restart force can be great enough to literally tear the blades from the hub or cause cracks to form (this happens far too often). Backup power supplies or soft-start features avoid many of these problems.

CONSTRUCTION

Fan design is a function of the working environment, and how a fan is constructed often makes a huge difference. Some fans are one-piece castings, others consist of two-piece assemblies, and cast blades being pressed or welded onto a fabricated shaft are common. Other fans are made only of fabricated components. The method of welding and the service environment should be carefully matched. It is common to find weld cracks in the area of the hub. Cracks grow rapidly. If discovered, irrespective of their size, the fan should be replaced immediately. Remember, fans should be inspected whenever the furnace is cooled down. Carburizing and high-temperature applications are the most problematic.

SPEED

Many atmosphere-control, uniformity and stability problems are directly attributed to fan performance. One interesting example involves a continuous carbonitriding furnace with an atmosphere stability problem so severe that carbon was observed dropping out of the atmosphere in the area of the furnace belt-support rolls, which would lock up. The unit had to be shut down every 2-4 weeks for maintenance. In addition, the burn-off at the front end was erratic – alternately sucking air in and violently blowing out endothermic gas. Variable case depths and case uniformity were also observed. The problems disappeared when it was discovered that the fans were not operating at the proper speeds or rotating in the proper direction, and the rotation and speed were adjusted back to manufacturer's recommended settings.

BALANCING

It is important to know if the fan you use has been statically balanced, dynamically balanced (and at what speed) or both. A number of problems with short fan life in integral-quench and pusher furnaces have been directly traced to the failure of the manufacturer to balance the fan at full rated speed.

In one noteworthy case, fan problems persisted for over seven years despite design and material changes until the retirement of the person in the balancing department, who, as it turned out, didn't see the necessity to balance the fans at speed even though the instructions clearly indicated that he should.

DIRECTION

Amazingly, some fans can be run backward for most of their life while metallurgists and heat treaters question poor quality and equipment performance. Other fans literally unscrew if run in the wrong direction and have been found atop baskets of parts after their removal from the furnace. Few fans are designed to run in either direction, and a simple rotational check can avoid major headaches.

TESTING

One of the most valuable tests that can be conducted to judge fan performance is a simple streamer test. With the furnace cold, place a test fixture (often simply the TUS fixture or empty baskets simulating the load arrangement) in the furnace with streamers attached in the same location as you would place temperature uniformity survey thermocouples. Pieces of brightly colored yarn between 100-150 mm (4-6 inches) in length properly secured to the fixture work well. Seal the opening to the furnace with a piece of Plexiglas® (the furnace door must be safely secured), turn the fan on and observe the streamer pattern using a work light. Areas of poor circulation are easy to spot (the streamers hang limp), and the overall circulation pattern is easily observable. This practical test yields a surprisingly large amount of information.

Alternately, or in addition to the streamer test, velocity measurement devices can be used to quantify the velocity in various locations of the furnace. This type of test is extremely useful for measuring flow in and out of air ducts in tempering furnaces and ovens.

MATERIALS OF CONSTRUCTION

Cost and availability are important issues with respect to deciding which alloy and design to use. Not all fan alloys are created equal, and many material choices depend on both fan size and operating conditions. A common tendency based on price, availability or familiarity is to universally use a standard alloy such as 330 (35% Ni, 15% Cr) and not consider other compositions.

For example, sulfur present in some atmospheres will eat away nickel and leave a metallic sponge of porous material behind. Using a material with at least 25% Cr will increase the thickness of the oxide layer on the alloy fan surface. The increased chromium will also react during furnace (air) burnouts or to oxygen leaks by re-forming these oxides. Having the right chemistry improves resistance in areas most susceptible to hot gaseous corrosion. The increased hot strength of a more suitable alloy grade will extend the performance life.

TYPE OF ATMOSPHERE

Hot gaseous corrosion and metal dusting (Fig. 13.3.2) are two of the phenomena that can often occur with fans running in chemically active furnace-atmosphere environments (e.g., carburizing). These problems are often more severe after repair or reinstallation of a fan due to damage to (or missing) refractory in the area immediately adjacent to the fan shaft or due to the introduction of a small air leak. Furnace roofs have been known to overheat and steel shells warp if insulation is missing or a furnace plug has been damaged, causing bearing problems, excess vibration and abnormally short fan life.

FIGURE 13.3.2 | Catastrophic carburization (aka metal dusting) of a furnace alloy fan shaft

COOLING ISSUES

Water-cooled fan bearing housings are notorious for leaks due to porous castings and poor furnace insulation in the area of the fan (allowing excess heat to reach the jacket). Water leaking into a furnace can radically change the dew point inside of it. Air-cooled designs avoid many of these problems and are becoming more popular for this reason.

Flowmeters and Maintenance

Once installed on our atmosphere furnace(s), flowmeters (Fig. 13.3.3) are often just taken for granted. It is assumed that they can read accurately forever. In addition, very little maintenance is routinely performed on flowmeters. This can often lead to serious flow errors and potential process or safety issues that compound themselves over time.

Flowmeters should be inspected, cleaned and have the float inside the housing inspected and oil replaced on a routine basis. The internals should always be carefully inspected for bent rods and cleaned (e.g., soot buildup or foreign matter collecting on the weights can throw off the accuracy of the readings), and they should be calibrated against a known flow source.

We will now address some of the more common questions about flowmeters.

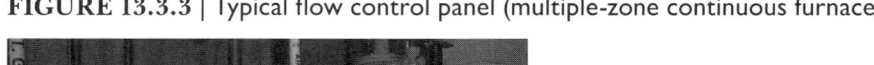

FIGURE 13.3.3 | Typical flow control panel (multiple-zone continuous furnace)

WHAT IS A FLOWMETER?

A flowmeter (aka flow scope) is a device for measuring the flow of gases or liquids (as anyone running a nitrogen/methanol system knows). There are actually two different methods of measuring flow: by volumetric means (Fig. 13.3.4a) and by mass-flow techniques (Fig. 13.3.4b). While aware of both, heat treaters are probably most familiar with the volumetric-flow devices. The principle behind them involves the displacement of the gas volume over time. Atmosphere

13 | MAINTENANCE REQUIREMENTS AND PRACTICES

furnaces, gas generators and combustion systems typically use these types of devices. Mass flow involves measuring the weight of a gas and is more commonly used when specific volumes of gas are required, such as on vacuum furnaces.

FIGURE 13.3.4 | Electronic flowmeter

a. Electronic flowmeter (courtesy of Super Systems Inc.)

b. Mass-flow controller (courtesy of MKS Instruments)

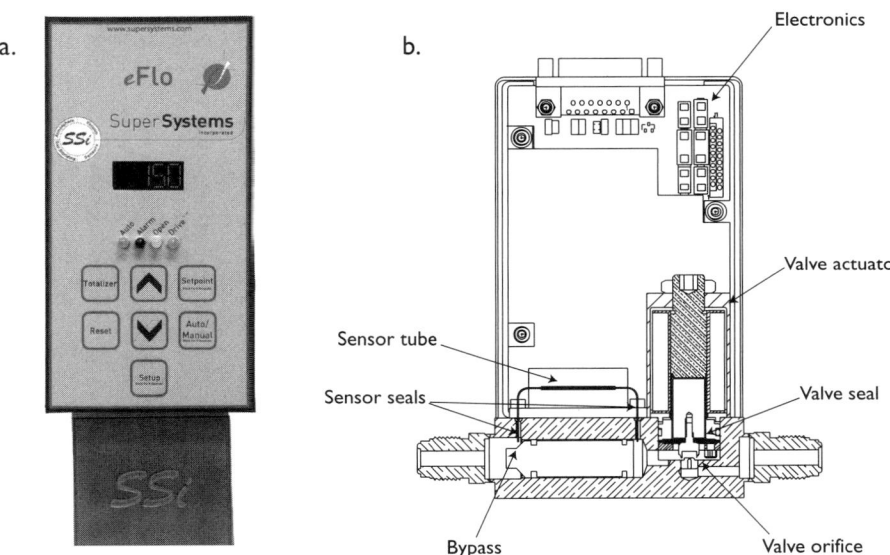

HOW MANY DIFFERENT TYPES OF FLOWMETERS ARE THERE?

In addition to flow scopes and mass-flow meters, other common flow-measurement types include orifices, rotameters, positive-displacement meters, electromagnetic meters, ultrasonic (Doppler-effect) devices, turbine meters, wedge flow devices, impact meters and turbine meters.

WHAT ARE THE FEATURES AND ADVANTAGES OF THE MOST COMMON FLOWMETER TYPES?

Variable-area flowmeters allow:
- ❖ Mechanical flow measurement with only a single moving part, ensuring measurement reliability
- ❖ Application versatility and availability of a variety of construction materials, inlet and outlet sizes and types
- ❖ Easy installation with generally no straight pipe requirements
- ❖ Low pressure drops
- ❖ Linear scales, allowing easy flow-measurement interpretation

❖ Electronic output availability, preserving the benefits of mechanical flow measurement

Tapered-tube rotameters allow: low instrument cost (when glass or plastic metering tube is used) and accuracy at very low flow rates.

Slotted-cylinder flowmeters allow: a flow range of 25:1 since flow-measurement accuracy is determined by the precision of the slot manufacturing operation; instrument specifications can be changed by field replacement of the slotted tube and float without having to re-pipe the flowmeter body; ability to handle high flows and pressures; and improved immunity to the effects of pulsating flows, with no minimum back-pressure.

Limitations common to both tapered-tube and slotted-cylinder variable-area flowmeters are that vertical mounting is required and they contain moving parts.

The user should also be aware that the accuracy of mass-flow meters and mass-flow controllers is determined by two factors: flow calibration and repeatability. Proper instrument calibration ensures starting-point accuracy. Repeatability is the measure of continuous performance-to-specification over the lifetime of the device. Most mass-flow meters and mass-flow controllers have an accuracy of ±1% of full scale and a repeatability of ±0.25% of full scale.

IS IT EASIER TO CONTROL A GAS OR A LIQUID?

Interestingly, liquids are easier to measure and control because of their small compressibility. For most volumetric-flow applications, the incoming pressure in liquid systems does not need to be closely controlled. By their very nature, liquids can be captured easily and measured to a high degree of accuracy. On the other hand, gases – because of their compressibility – require more complex sensing and control methods.

WHAT IS THE ACCURACY RANGE OF A GAS FLOWMETER?

When measuring gas flows in heat-treatment applications there is an important distinction between the operating range of a flowmeter and the design range when purchasing a new meter. Plan to operate a flowmeter in a range not below 25% and not above 90% of the flow scope's scale capacity. In other words, if your flowmeter is rated for 0-60 m^3/hour (~2,100 cfh), accuracy is ensured when the flow is 14-45 m^3/hour (500-1,600 cfh). Flows outside these limits are not considered accurate.

A good rule of thumb for sizing a flowmeter is to purchase your meter "in the middle third;" that is, size the flowmeter so that the actual flow will be no less than one-third and no higher than two-thirds of the scale you select. This gives you the ability in actual operation to compensate for unexpected changes in flow requirements that may occur. Process requirements and operating conditions often change over the life of a heat-treating furnace, sometimes dramatically, and you want your gas measurement to remain accurate.

SHOULD I HAVE MY FLOWMETERS RECALIBRATED?

If a change of operating conditions is permanent, such as the desire to constantly operate at a different pressure, then recalibration of the flow measurement device is strongly recommended. As a rule, flowmeters used in heat-treating applications are designed for a maximum temperature of 65°C (150°F) and an operating pressure up to 345 kPa (50 psig). However, application-specific flowmeters have maximum operating pressures outside these ranges.

DO I NEED TO MAINTAIN MY FLOW DEVICES?

All flowmeters eventually require maintenance. It is a sad truth that some units require more maintenance than others, so this factor should be considered when a unit is selected. In most heat-treatment operations, however, the equipment manufacturer has already made that choice for you, so understanding what maintenance is required and when it should be performed is of paramount importance.

Flowmeters have moving parts and require internal inspection, especially if the fluid is dirty or viscous. In furnaces using endothermic gas, for example, the flowmeters often become contaminated with soot (carbon) and must be cleaned by CAREFULLY disassembling the flowmeter and cleaning all internal moving parts as well as replacing the dirty fluid in the flowmeter tube. This involves isolating the flowmeter or performing maintenance when the unit is shut down and must be done in a safe manner because many of the gases involved are asphyxiants as well as being flammable, toxic and possibly life-threatening.

Remember also that electromagnetic flowmeters and all flow-measurement devices that use secondary instruments (such as pressure sensors to actuate a control valve or send a signal to a remote source) must be periodically inspected, calibrated, repaired and/or replaced. Improper location of the flowmeter itself, the secondary sensor or readout devices can result in measurement errors and hidden costs.

DO I REALLY NEED TO LEARN ABOUT MY FLOWMETERS TO BE IN CONTROL, STAY IN CONTROL, OPERATE SAFELY AND KEEP THE COST OF MY OPERATION AS LOW AS POSSIBLE?

Yes. Hopefully, this discussion has helped reinforce this idea. Now go out today and check your flow devices.

TABLE 13.3.1 | Specific-gravity conversion factors (courtesy of Waukee Engineering, a member of United Process Controls)

Gas formula	Gas that meter is calibrated for	Specific gravity	Air	Acetylene	Ammonia	Ammonia dissociated	Argon	Butane	Carbon dioxide	City gas	Endothermic cracked	Endothermic cracked (lean)	Endothermic cracked (rich)	Forming gas	Helium	Hydrogen	Natural gas	Nitrogen	Oxygen	Propane
	Air	1.00	1.00	1.060	1.302	1.838	.851	.704	.809	1.302	1.302	1.00	1.085	1.072	2.692	3.343	1.240	1.021	.951	.811
C_2H_2	Acetylene	.907	.953	1.00	1.240	1.754	.811	.760	.770	1.240	1.240	.953	1.033	1.021	2.564	3.613	1.181	.972	.906	.772
NH_3	Ammonia	.59	.768	.806	1.00	1.414	.654	.540	.621	1.00	1.00	.768	.833	.824	2.068	2.914	.953	.784	.731	.623
	Ammonia dissociated	.296	.543	.570	.707	1.00	.462	.382	.439	.707	.707	.543	.589	.582	1.462	2.060	.674	.554	.517	.440
Ar	Argon	1.38	1.175	1.233	1.529	2.163	1.00	.827	.950	1.629	1.629	1.175	1.274	1.259	3.162	4.456	1.457	1.199	1.116	.952
C_4H_{10}	Butane	2.02	1.421	1.492	1.850	2.617	1.210	1.00	1.149	1.850	1.850	1.421	1.542	1.524	3.826	5.391	1.763	1.451	1.352	1.152
CO_2	Carbon dioxide	1.529	1.236	1.298	1.610	2.277	1.053	.870	1.00	1.610	1.610	1.236	1.341	1.326	3.329	4.690	1.534	1.262	1.176	1.002
	City gas	.59	.768	.806	1.00	1.414	.654	.540	.621	1.00	1.00	.768	.833	.824	2.068	2.914	.953	.784	.731	.623
	Endothermic cracked	.59	.768	.806	1.00	1.414	.654	.540	.621	1.00	1.00	.768	.833	.824	2.068	2.914	.953	.784	.731	.623
	Endothermic cracked (lean)	1.00	1.00	1.050	1.302	1.841	.851	.740	.809	1.302	1.302	1.00	1.085	1.072	2.692	3.793	1.240	1.021	.951	.811
	Endothermic cracked (rich)	.85	.922	.968	1.200	1.698	.785	.649	.746	1.200	1.200	.922	1.00	.988	2.482	3.497	1.144	.941	.877	.747
	Forming gas	.927	.963	1.011	1.253	1.773	.820	.677	.779	1.253	1.253	.963	1.044	1.00	2.592	3.652	1.194	.983	.916	.780
He	Helium	.138	.371	.390	.484	.717	.316	.261	.300	.484	.484	.371	.403	.398	1.00	1.409	.461	.379	.353	.301
H_2	Hydrogen	.0895	.264	.277	.343	.485	.224	.185	.213	.343	.343	.264	.296	.274	.710	1.00	.387	.269	.251	.214
CH_4	Natural gas	.65	.806	.846	1.050	1.484	.686	.567	.652	1.050	1.050	.806	.874	.864	2.170	3.058	1.00	.823	.767	.654
N_2	Nitrogen	.98	.980	1.029	1.276	1.804	.834	.689	.792	1.276	1.276	.980	1.063	1.060	2.638	3.717	1.215	1.00	.932	.794
O_2	Oxygen	1.105	1.105	1.104	1.369	1.935	.895	.740	.850	1.369	1.369	1.105	1.140	1.127	2.830	3.987	1.304	1.073	1.00	.852
C_3H_8	Propane	1.522	1.234	1.295	1.606	2.271	1.050	.868	.906	1.606	1.606	1.234	1.338	1.323	3.321	4.680	1.530	1.259	1.174	1.00

13 | MAINTENANCE REQUIREMENTS AND PRACTICES

WHAT AFFECTS MY GAS MEASUREMENTS?

If knowing the proper flow rate is important to you, it is imperative to be aware that a change in temperature, pressure or specific gravity of the gas for which the meter was calibrated will cause a serious error in the indicated scale reading. It is quite common in a heat-treat shop to find flowmeters operating at different pressures and temperatures than they have been calibrated for.

One of the most common problems seen in heat-treat shops is that operating personnel and supervisors are unaware of the consequences of a flowmeter that has been calibrated for use with one gas while having another gas flowing through it. The change in specific gravity of the two gases is the principle factor that must be taken into account when switching gases. Specific gravity is the ratio of the density of the gas under consideration to the density of dry air at standard temperature and pressure – 25°C (77°F) and 0.1 MPa (14.7 psi).

To calculate the actual flow rate of gas being metered, multiply the indicated scale reading of the flowmeter by the factor shown (Table 13.3.1).

Equation 13.3.1 allows us to calculate the actual flow when a change in specific gravity occurs.

13.3.1) $\quad Fa = Fi \times \sqrt{\dfrac{SG_1}{SG_2}}$

where:
Fa = actual flow
Fi = flow indicated scale reading
SG_1 = nameplate (calibrated) specific gravity
SG_2 = specific gravity of gas to be used in the flowmeter

HOW DO I COMPENSATE FOR CHANGES IN TEMPERATURE AND PRESSURE?

A change in temperature or pressure of the gas for which the meter was calibrated will cause a serious error in the indicated scale reading. However, there is an easy way to calculate the effect of these changes.

Temperature Compensation

Equation 13.3.2 allows us to calculate the actual flow when a change in temperature occurs. Common values (Table 13.3.2) are easily calculated.

13.3.2) $\quad Fa = Fi \times \sqrt{\dfrac{T_1}{T_2}}$

where:

Fa = actual flow

Fi = flow indicated scale reading

T_1 = nameplate (calibrated) temperature + 460 (English system) or nameplate (calibrated) temperature + 273 (Metric system)

T_2 = new temperature + 460 (English system) or new temperature + 273 (Metric system)

TABLE 13.3.2 | Temperature conversion factors

Temperature, °C (°F)	Scale multiplier
10 (50)	1.02
15 (60)	1.01
21 (70)[a]	1.00
26 (80)	0.99
32 (90)	0.98
38 (100)	0.97
42 (110)	0.965
50 (120)	0.96
55 (130)	0.95
60 (140)	0.94
65 (150)	0.93

Note: [a] Flowmeter calibrated for 21°C (70°F)

Pressure Compensation

Equation 13.3.3 allows us to calculate the actual flow when a change in pressure occurs. Common values (Table 13.3.3) are easily calculated.

13.3.3) $\quad Fa = Fi \times \sqrt{\dfrac{P_2}{P_1}}$

where:

Fa = actual flow

Fi = flow indicated scale reading

P_2 = new inlet pressure + 14.7 (English system) or new inlet pressure + 101.3 kPa (metric system)

P_1 = nameplate (calibrated) inlet pressure + 14.7 (English system) or nameplate (calibrated) inlet pressure + 101.3 kPa (metric system)

TABLE 13.3.3 | Pressure conversion factors

Pressure, kPa (psi)	Scale multiplier
3.4 (0.5)[a]	1.00
6.9 (1)	1.03
13.8 (2)	1.06
20.7 (3)	1.10
27.6 (4)	1.13
34.5 (5)	1.16
69.0 (10)	1.30
103.4 (15)	1.42
137.9 (20)	1.53
206.9 (30)	1.75
275.8 (40)	1.93
344.8 (50)	2.10

Note: [a] Flowmeters are typically calibrated for 3.4 kPa (0.5 psi).

If it is necessary to compensate for both temperature and pressure, Equation 13.3.4 should be used.

13.3.4) $\quad Fa = Fi \times \sqrt{\dfrac{P_2}{P_1} \times \dfrac{T_1}{T_2}}$

Internal Furnace Pressure Measurement

Most heat treaters do not know what the internal pressure is, or should be, inside their atmosphere furnace (Section 13.6). Also, many fail to understand why this is an important value to monitor, especially when it comes to safety or process control. As far as this writer is concerned, there is no excuse not to have internal-pressure monitoring capability on every heat-treating furnace running protective atmosphere. The amount of information that can be gained far outweighs the cost of adding these types of sensors to our furnaces.

Most atmosphere furnaces, whether they be batch or continuous, normally operate at internal pressures in the range of 0.0025-0.025 kPa (0.01-0.10 inches of water column), with the upper value the targeted operating pressure. There are exceptions, such as when performing deep-case carburizing in a pit-style furnace and in other very tight furnace systems where running at 0.050-0.075 kPa (0.20-0.30 inches of water column) is desirable. To put this pressure range in perspective, an internal pressure of 0.025 kPa (0.10 inches of water column) is about equivalent to the pressure exerted on a flat table by 10 $1 bills stacked one atop another.

Common misconceptions are that increasing the flow alone will automatically increase the pressure or that simply adding more weights to a furnace-atmosphere burn-off exhaust flapper will, in and of itself, force the furnace pressure higher. Pressure will always be lower in poorly sealed batch furnaces or in a continuous furnace open on both ends.

Furnace Mesh Belts and Belt Drives

Mesh belts in continuous furnaces come in all shapes, sizes, materials and weaves, and they are used for diverse applications such as case hardening, brazing, sintering, austempering and glass-to-metal sealing. Belts run at temperatures from near ambient to several thousand degrees. They are also used for part conveyance in water, oil, brine, polymer and salt quench tanks.

Mesh belts are exposed to a multitude of furnace atmospheres ranging from air to 100% hydrogen. They are also exposed to oxidation, sulfidation, carburizing and nitriding. They operate in environments having dew points below -73°C (-100°F) to above 38°C (+100°F). Mesh belts are expected to perform well beyond normal life expectations. From a maintenance perspective, there are a few tips to help them survive and perhaps allow them to last even longer.

When operating a mesh-belt conveyor furnace (Fig. 13.3.5), the goal should be to maintain a consistent temperature profile and uniform belt loading for any given part number over time. Premature or abnormally short belt life is often a sign of overloading, misapplication, abuse, neglect or ignorance. Our goal should be to extend belt life by 33% or better.

FIGURE 13.3.5 | Typical mesh-belt sintering furnace (courtesy of C.I. Hayes, a Gasbarre Furnace Group Company)

a. Front end

b. Back end

COLOR IMAGE SECTION

FIGURE 9.5.27 | 200-mm-diamter (8-inch-diameter) bar of SAE 4150 entering spray water quench for hardening

FIGURE 9.6.14 | Endothermic-gas generator chemistry

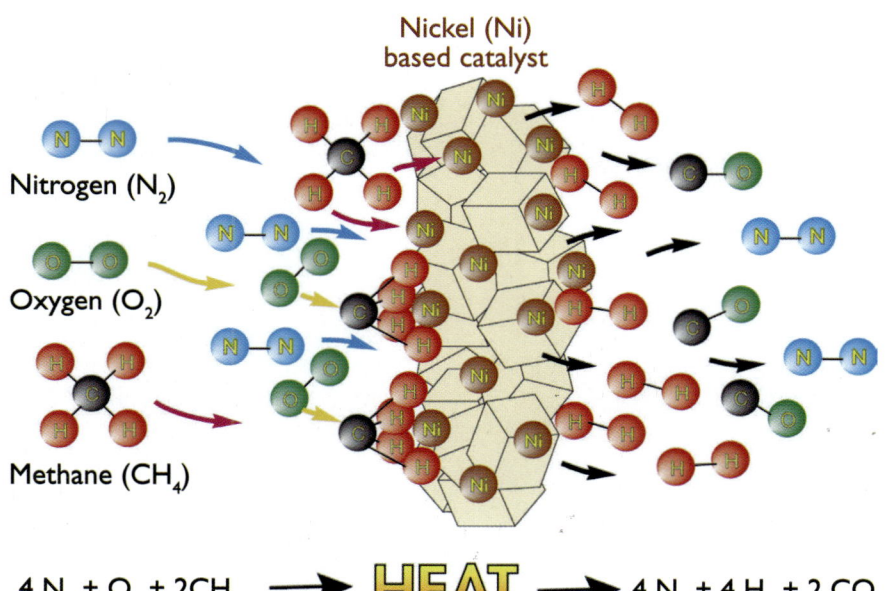

$4 N_2 + O_2 + 2CH_4 \longrightarrow$ **HEAT** $\longrightarrow 4 N_2 + 4 H_2 + 2 CO$

FIGURE 10.2.6 | Example of free quenching

FIGURE 10.2.16 | Large mill pinion being lowered into a polymer quench

FIGURE A | Quench sequence. Outer door and inner door still closed, operator poised to pull load onto brine quench tank elevator after furnace pushes load to the outer door. (courtesy of Phoenix Heat Treating Inc.)

FIGURE B | Quench sequence. Outer and inner door open. Furnace is pushing load forward. Load visible through the flame screen. Endothermic gas burning off. (courtesy of Phoenix Heat Treating Inc.)

FIGURE C | Quench sequence. Operator is quickly pulling load at 860°C (1575°F) forward over the quench elevator of the brine quench tank. (courtesy of Phoenix Heat Treating Inc.)

FIGURE D | Quench sequence. Load in position over brine and ready to be lowered into the quenchant. Operator is moving the handling bar out of position. (courtesy of Phoenix Heat Treating Inc.)

FIGURE E | Load just prior to quench. (courtesy of Phoenix Heat Treating Inc.)

FIGURE F | Load being rapidly lowered into the brine quench. (courtesy of Phoenix Heat Treating Inc.)

FIGURE G | Quench sequence. Load of SAE 4130 material nearly fully submersed into the brine quench. (courtesy of Phoenix Heat Treating Inc.)

FIGURE H | Load of Ti-6Al-4V alloy parts after solution treatment in air at 955°C (1750°F). In position to be lowered into a water quench tank. (courtesy of Phoenix Heat Treating Inc.)

FIGURE I | Endothermic gas exiting a pit-style furnace upon opening the lid. (courtesy of Phoenix Heat Treating Inc.)

FIGURE J | An 815-kg (1,800-pound) load of large SAE 4140 alloy bolts emerging from a pit-style furnace after austenitizing at 845°C (1550°F) for crane transfer to an oil quench. (courtesy of Phoenix Heat Treating Inc.)

FIGURE K | Load of bolts being lowered into the oil quench. Quench oil ignites as parts make contact with the oil and will subside as parts are fully submerged. (courtesy of Phoenix Heat Treating Inc.)

FIGURE L | Baskets filled with SAE 4340 alloy bars ready for loading into an integral-quench furnace. Outer door is just beginning to open. Flame curtain has ignited. (courtesy of Phoenix Heat Treating Inc.)

FIGURE M | Workload entering an integral-quench furnace through the flame curtain. (courtesy of Phoenix Heat Treating Inc.)

FIGURE N | Load in vestibule of an integral-quench furnace. Outer door ready to close before inner door opens for transfer to heating chamber. (courtesy of Phoenix Heat Treating Inc.)

FIGURE 12.2.1 | Typical scratch (file) hardness testing arrangement

FIGURE 12.6.9 | Thermographic analysis – furnace power leads showing potential hot spot

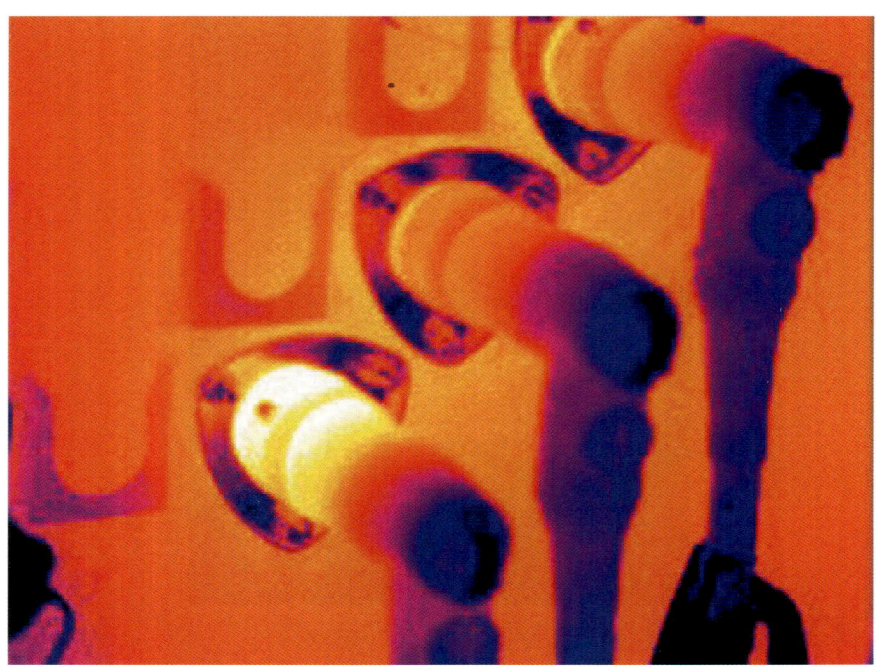

FIGURE 13.6.5 | Copper-steel-stainless steel coupons

FIGURE 13.6.6 | Copper-steel-stainless steel coupons

FIGURE 13.6.7 | Copper-steel-stainless steel coupons

FIGURE 13.6.8 | Copper-steel-stainless steel coupons

FIGURE 14.1.1 | Endothermic gas igniting during load discharge from an atmosphere box-carburizing furnace

FIGURE 14.3.8 | Furnace atmosphere produced by products of combustion from high-velocity self-recuperative burners firing inside a box-type furnace

TIPS TO IMPROVE BELT LIFE

Here are 10 tips you can follow to improve belt life.

1. Know your belt speed by actually timing the belt movement. Do not assume it is what is stated in the instruction manual or what is displayed on a digital indicator. Calibrate the speed control (typically a magnetic or inductive pickup device) as often as your temperature instruments.
2. Determine your belt loading over time. Improperly adjusted vibratory-feed systems and weigh scales that do not read correctly are two of the most common causes for improper loading of furnace belts. Load as uniformly as possible to evenly distribute wear across the bottom of the belt and help to prevent camber and other belt distortion problems.
3. Avoid using skid plates or belt guide rollers to assist belt tracking. These items tend to do more harm than good. Where skid plates must be used, consider coating them with laminated plastic strips. Observe the motion of the belt. It should be smooth – not jerky – and maintain a consistent speed and tracking over time. Remember, a metal mesh belt has flexibility along its length, semi-rigidity across its width and rigidity in its thickness.
4. Work with your equipment or belt suppliers to select a belt weave and belt alloy best suited for your process and parts. Be aware that some belts using heavier wire or larger cross rods actually decrease furnace throughput. Avoid upturned-edge belts whenever possible.
5. Flip and/or reverse the belt at frequent intervals when signs of wear or camber become evident. Camber in a high-temperature belt may be either convex or concave or, in some cases, it may develop with one edge of the belt leading the other edge. It is frequently an indicator of other problems, such as uneven loading, uneven cooling, improper drive tension or a combination of these factors. Only flip and reverse the belt when it is sufficiently flexible to ensure that reversing it does not create a fatigue problem.
6. Be sure that the belt is properly supported over its length. Watch for signs of premature belt failure, such as abnormal distortion of the cross rods, flat spots, deterioration of belt edges, buildup of foreign residue and tracking problems.
7. Check the tension on the belt frequently or as dictated by production usage. Some systems rely on springs to maintain tension, so check their length. Other systems rely on cylinders, so check that plant air pressure does not fluctuate significantly.
8. Have your belt drive system analyzed by a furnace manufacturer or belt company at least once a year.

9. Keep loading consistent. Many parts are loaded on the belt in such a way as to leave space along the edges to prevent parts from moving off the sides. This concentrates loading over a smaller square area and may affect belt selection or life.
10. Understand that maintenance is different between front-end drive systems having pinch and tangential rolls (for tracking) and rear-end drive systems having only pinch rolls.

OTHER IMPORTANT CONSIDERATIONS

Belts at ambient temperatures or in quench tanks should be run for a minimum of three revolutions with necessary belt tracking adjustments made and cycling repeated prior to loading or filling the tank. All considerations for use, such as the environment (e.g., dirt, chips) or expansion (e.g., molten salt), should be anticipated before putting the belt in service.

For high-temperature belts, two frequently overlooked or ignored procedures are break-in and stress relief. Take the time to do both. Though opinions differ on break-in procedures and how best to perform them, most agree that the furnace should be brought up slowly (<150°C/300°F per hour) to an operating temperature of 815-870°C (1500-1600°F), and the furnace belt should be allowed to make 2-4 complete revolutions through the furnace at the lowest practical belt speed setting, if possible.

The goal is for every portion of the belt to see this temperature range for at least 4-6 hours with no load to ensure proper seating of the spirals and to allow for initial movement of the rods. Belt tracking adjustments should be made during and after this procedure.

Stress relief should be performed immediately after the break-in procedure is complete. For belts in furnaces operating below 900°C (1650°F), a stress relief at 10°C (50°F) above the normal operating temperature is required for a sufficiently long period of time so that every portion of the belt reaches this temperature for at least one hour. For belts in furnaces operating above 900°C (1650°F), the furnace should be brought into the 925-955°C (1700-1750°F) range for one complete revolution through the furnace at the lowest practical belt speed setting. The goal is for every portion of the belt to see this temperature for at least 1-2 hours with no load to prevent excessive grain growth and embrittlement.

After the stress-relief treatment is performed, the furnace should be brought to normal operating temperature and the belt held at this temperature for a minimum of two complete revolutions through the furnace at the lowest practical belt speed setting. The goal is to hold the belt at operating temperature under atmosphere with no load for a minimum 4-6 hours before use. The belt should be allowed to operate unloaded for as long as practical (up to 100 hours) to increase the creep strength of the belt.

Belt tracking is another critical issue. This begins and ends with a properly adjusted drive system. The vast majority of belt failures are either due to overloading or poor tracking belts.

Proper belt tension (Fig. 13.3.6) is critical, as illustrated by a study from a major belt manufacturer on copper brazing furnaces operating at 1120°C (2050°F). The results are reportedly applicable to all types of high-temperature belts and make the point that every effort should be made to maintain operating tension at the lowest practical level. Either the OEM or the belt supplier usually determines the allowable operating ranges.

Various devices, from springs to air cylinders, are used for tensioning, but they are seldom checked for conformance to original specifications. If the furnace was not purchased new, it is even more important to know how the drive was sized, where the zero point of loading is, and what the proper drum adjustments and belt tension settings are.

FIGURE 13.3.6 | Effect of belt tension on service life[5]

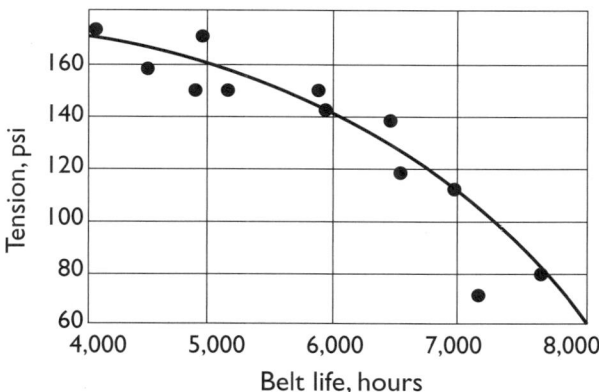

When possible, keep the belt clean (free of oxidation and scale) by running it under protective atmosphere above 760°C (1400°F). In certain situations, such as copper brazing, it has been reported that pre-oxidizing the belt can extend its service life where copper or other braze alloys attempt to penetrate the surface of the belt.

Finally, the condition of your mesh belt is a clear indication of the health of your mesh-belt furnace. Insist on extra time to properly take care of this vital component in your heat-treating operation. It will save time and money.

Pilots and Flame Curtains

Far too many heat treaters do not understand the function, design, adjustment or safe operating requirements of their pilots and flame curtains.

FUNCTION

Flame curtains placed across furnace openings (Fig. 13.3.7) have several functions:

1. To produce a vertical stream of products of combustion across the full width and height of the door opening, thereby minimizing both the infiltration of room air into the furnace chamber and disruption of the furnace atmosphere inside.
2. To serve as a positive source of ignition of the combustible atmosphere that escapes through the door when it is open.

FIGURE 13.3.7 | Properly adjusted integral-quench furnace flame curtain (courtesy of Phoenix Heat Treating Inc.)

Pilots that ignite these flame curtains must be robust and oriented such that the flame curtain will ignite immediately upon activation. All gas pilots are susceptible to atmospheric conditions in the heat-treat area. "Lazy flame" pilots are particularly prone to being extinguished, which creates the potential of raw gas entering the area.

DESIGN

Most flame-curtain systems are engineered to utilize your plant's natural gas, propane or aerated-propane supply regulated down in pressure to a stable 0.99-1.50 kPa (4-6 inches of water column or 2.30-3.45 psig). Air is supplied from either the plant's high-pressure air supply adjusted to 550-690 kPa (80-100 psig), a small combustion-air blower or by tapping into the existing combustion-air blower if your system is gas-fired.

For high-pressure systems, the air line should include a Venturi-type inspirator, which converts a relatively small flow of high-pressure air to a much larger flow of air at low pressure. A globe-type valve in the air line, ahead of the solenoid valve, is the shutoff valve used when shutting down the furnace and flame curtain. A needle-type valve in the air line, downstream of the solenoid valve, is the flow-adjusting valve. Having been properly set at the time of initial start-up of the furnace, the setting of this valve normally will not be changed.

The flame-curtain gas and air-solenoid valves are connected in parallel to the door position-sensing limit switch. Thus, as the door just begins to open, both solenoid valves open simultaneously and the curtain lights off and continues to fire until the door has been closed. Because the curtain is engineered with a very high firing rate, it generates considerable heat. To prevent possible damage from heat distortion to the door frame and/or door plate, the door must never be held open any longer than required to turn on and initially adjust the flame curtain or to subsequently load or unload the furnace.

Whether you have a high- or low-pressure system, operators must routinely check that the flame curtain fully covers the door opening and alert supervisors and/or maintenance personnel if/when adjustments are needed. Variation in plant air pressure and drafts generated by changing plant conditions are the most likely reasons flame curtains go out of adjustment.

OPERATING SAFETY REQUIREMENTS

Here are four key operating safety requirements for most furnaces with respect to flame curtains.

1. Whenever routinely starting up a furnace and before opening the door to turn on and verify adjustment of the flame curtain, you must first turn on the flame-curtain pilot, ignite it and verify it to be burning with a good, strong blue flame. Be sure that drafts or other sources of air do not move the pilot flame into a position where it cannot ignite the flame-curtain air-gas mixture when the door opens.
2. Whenever the furnace is in operation, you must never open the door without first verifying the flame-curtain pilot is burning with a good, strong, stable blue flame.
3. Whenever you wish to leave the furnace idling at temperature overnight or over a weekend with atmosphere flowing, you must verify that the flame-curtain pilot is burning and properly positioned before you leave.
4. Whenever you shut down and the furnace atmosphere has been removed, make absolutely certain that the flame-curtain gas and air-line shutoff valves are fully closed. You must also do this if you leave the furnace idling at temperature with atmosphere not flowing.

TYPICAL INITIAL STARTING AND ADJUSTING PROCEDURES

The following procedure is typically required to initially turn on and adjust a high-pressure flame curtain.

1. Remove the protective cap over the air/gas ratio adjusting screw in the air-gas mixer on the end of the air-line inspirator. Then turn the screw in all the way clockwise until you feel it "bottom."
2. With the furnace door fully closed, open the shutoff cock in the curtain gas line and the small cock in the pilot gas line and, without delay, light the pilot. Adjust the pilot so that it burns with a good, strong, stable blue flame. A blue flame without orange or yellow tips indicates the pilot is burning at close to the required 10:1 air/gas ratio (perfect combustion) on natural gas, 25:1 on propane or 14:1 on aerated propane. If the pilot is equipped with a supervising flame rod and flame-detection relay, the relay contacts are now closed.
3. Fully open the air-line shutoff valve ahead of the solenoid valve. Then stand as far away from the furnace door and flame curtain as possible and open the door. As the door begins to open, the gas and air solenoid valves open but gas does not yet flow.
4. Set the air-line flow-adjusting valve downstream of the solenoid valve so that the air-pressure gauge reads 35-75 kPa (5-10 psig).
5. Verify that the pilot is still burning with a strong flame and start turning the air/gas ratio adjusting screw in the air-gas mixer to start gas flow to the curtain burner. Continue until the burner "lights-off" and the essentially raw-gas flame extends in height to approximately the top edge of the door opening (or the crown of the door vestibule arch). Note: If the flame curtain cannot cover the entire door opening, the system is under-sized and must be upgraded *before* operating the furnace.
6. Without delay, and only if required to make the flame transparent enough so that you can see through it, further open the curtain needle-type air-flow adjusting valve downstream of the solenoid valve to add more primary air to the mixture. As you add air, the height of the flame will lessen somewhat and the flame will become less orange or yellow in color.

    Proper air/gas ratio adjustment will leave the flame curtain burning very rich (at approximately a 1.2:1 ratio). If not enough air is added, the flame curtain will produce soot (carbon deposits) around the door, which is undesirable and indicates an adjustment to the air/gas ratio is needed.
7. Without delay, close the furnace door to turn off the flame curtain.
8. Verify that the flame-curtain pilot is still burning with a strong blue flame of adequate size (Table 13.3.4). Make any adjustments required to obtain a proper flame.
9. Lock the air/gas ratio adjusting screw in the mixer in position with the jam nut and replace the protective cap.

ROUTINE STARTING AND ADJUSTING PROCEDURE

Here are the required procedures to routinely turn the flame curtain on.

1. Verify that you have already brought the furnace to above 760°C (1400°F); that you have started atmosphere flow into the furnace; that you have observed atmosphere to be burning into the furnace chamber off the end of the inlet pipe with a good, strong flame; and that you have fully closed the furnace door.
2. Without delay, fully open the flame-curtain gas shutoff cock ahead of the solenoid valve and light both the flame-curtain pilot and the pilot to the furnace-atmosphere burn-off can, if supplied. Verify both pilots are burning with good, strong, stable blue flames.
3. Fully open the curtain air-line shutoff valve ahead of the solenoid valve.
4. Stand as far away from the furnace door and flame curtain as possible and open the door. As the door begins to open, verify that the curtain lights-off with a flame of adequate height to cover the full width and height of the door opening and has proper transparency.
5. Without delay, close the furnace door to turn off the flame curtain.

TABLE 13.3.4 | Maximum flame-curtain gas and air demands

Flame-curtain length, inches	Natural gas, m³/h (scfh)	Propane, m³/h (scfh)	Aerated propane, m³/h (scfh)	Air, m³/h (scfh)	Air, m³/min @ 550 kPa (cfm @ 80 psig)
24	6.6 (234)	2.6 (94)	4.6 (164)	0.04 (1.39)	0.0025 (0.09)
30	8.3 (293)	3.3 (117)	5.8 (205)	0.05 (1.71)	0.0031 (0.11)
36	9.9 (351)	4.0 (140)	7.0 (246)	0.06 (2.17)	0.0039 (0.14)
48	13.3 (468)	5.3 (187)	9.3 (328)	0.08 (2.80)	0.0050 (0.18)
60	16.6 (585)	6.6 (234)	11.6 (410)	0.10 (3.57)	0.0065 (0.23)

Thermocouples and Thermocouple Wire

Accurate temperature control of heat-treatment furnaces and ovens depends in large part on the proper choice of thermocouples as well as ancillary items such as extension (lead) wire, protection tubes and connectors. A list of the common types of thermocouples and their application uses can be found elsewhere in the book (Section 16.4). Specifications such as AMS 2750 (Pyrometry) have helped by focusing attention on the critical role of these important temperature sensors.

In addition, correct installation techniques and good maintenance procedures are a must. The most common operational problem with furnaces today is inaccurate temperature control due in large measure to the fact that the thermocouples have been in the furnace longer than their normal life expectancy.

WHAT IS A THERMOCOUPLE?

Thermocouples (Fig. 13.3.8) used to sense temperature in heat-treating furnaces are a type of electrical sensor that consists of two dissimilar metals joined together that produce an output when subjected to a difference in temperature. The joined end that is placed inside the furnace is called the hot (or measuring) junction. The end attached to the connector outside the furnace is called the termination end, and the end attached to the instrumentation is commonly called the cold end. An electromotive force (EMF) is generated and measured in millivolts, which are proportional to the difference in temperature between the hot and cold ends.

The different types of materials used to construct thermocouples produce different output signals. Extension wire, which must match the thermoelectric characteristics of the thermocouple, is used to connect the termination end to the temperature instrumentation. Extension wires can either be specified as thermocouple-grade, which is calibrated over the entire thermocouple temperature range, or extension grade, which is calibrated to a specific temperature, typically 94°C (200°F).

It is important to avoid creating additional junctions by splicing wire lengths together (this can cause erroneous readings). It is necessary to provide a separate metal conduit housing for the extension lead wire since it isolates or shields the lead wire from stray voltage signals, such as those produced by other electrical or power wiring or nearby machinery.

FIGURE 13.3.8 | Typical thermocouples and thermocouple assemblies (courtesy of Pyromation Inc.)

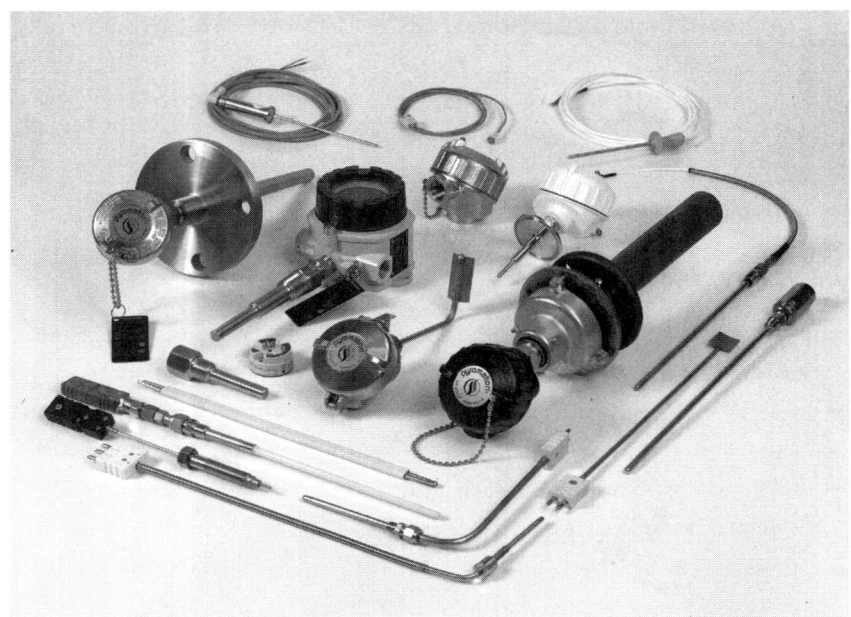

13 | MAINTENANCE REQUIREMENTS AND PRACTICES

WHAT IS THERMOCOUPLE WIRE?

Thermocouples, from the point of sensing temperature (the so-called hot junction) to the point where the signal is measured (the so-called cold junction), are made of two dissimilar metals joined together at the sensing end. At the cold junction, the millivolt value provided by the thermocouple represents the difference in temperature of the sensing end as compared to the cold-junction end (also called the reference end).

WHAT IS EXTENSION WIRE?

Extension wire is used to extend a thermocouple's signal back to the instrument reading the signal. The extension-grade wire typically has a lower operational temperature limit (normally ambient temperature) in which the wire may be used. As a rule, thermocouple wire may be used as extension wire, but extension wire may not be used in the sensing point (or probe part) of the thermocouple.

WHAT IS THE MAXIMUM LENGTH OF A THERMOCOUPLE?

The two main factors in determining usable thermocouple length are total loop resistance and prevention of electrical signal noise. Since thermocouples are made from dissimilar wire, the resistance will vary based on the type and the wire diameter and length. The allowable loop resistance (typically under 100 ohms) is affected by the input resistance of the amplifier circuit to which it is attached. As a general guideline, 20 AWG or thicker wire is adequate for runs up to 30 meters (100 feet). Thermocouple wire creates a low-voltage signal and should not be run near power wires, motors, etc. To help minimize noise pickup, thermocouples used in industrial furnace applications are almost always run in separate metal conduit.

HOW TO INSTALL THERMOCOUPLES

The one common feature of all furnaces is the fact that within their insulated chambers there are horizontal planes of even heat flow – areas of equal temperature – called isotherms. Factors that cause these temperature planes or gradients to vary include uneven heating, inadequate circulation, uneven distribution of the workload within the furnace, improper location of the heating source and the like.

In order for a thermocouple to properly sense furnace temperature, it must be oriented parallel to these isotherms and installed into the furnace a minimum of 20 thermocouple diameters. For example, a 3-mm-diameter (1/8-inch-diameter) thermocouple will need to be inserted 63.5 mm (2.5 inches) into the furnace chamber.

There are three basic types of construction for thermocouples (Table 13.3.5). Each has advantages and disadvantages. For example, ceramic-beaded thermocouples lack flexibility compared to base-metal thermocouples, which can have various types of insulation placed directly on the wire. This makes them extremely flexible.

By contrast, ceramic-beaded thermocouples can often be used at higher operating temperatures than many of their base-metal counterparts. Mineral-insulated, metal-sheathed thermocouples provide excellent performance and can be used as replacement elements to ceramic-beaded, base-metal elements in many cases. Metal-sheathed cable is offered in many sizes and materials, and it can be optimized to help provide long-lasting stable temperature measurements.

TABLE 13.3.5 | Common thermocouple construction techniques[12]

Construction method	Type
Ceramic beaded	High-purity (99.9%) alumina
	Mullite (70% Al_2O_3 + 30% SiO_2)
	Steatite
Insulated	Plastic (Polyvinyl, Teflon, etc.)
	Glass braid
	Ceramic fiber
Metal sheathed	Stainless steel
Mineral insulated: MgO, Al_2O_3	Inconel Specialty high-performance alloys

WHY PROTECTION TUBES?

A thermocouple is often inserted into a protection tube. Protection tubes (Table 13.3.6) are used to shield thermocouples from contaminants and/or mechanical damage. The heat treater must be aware that all types of thermocouple protection tubes can crack or distort, potentially damaging or exposing the thermocouple to the environment they were intended to protect it from. In addition, contamination from handling/touching the wires, oils and dirt inside the protection tube, and contamination buildup on the outside of the protection tube are some of the many factors that can cause thermocouple error.

TABLE 13.3.6 | Types of protection tubes used in the heat-treating industry[9]

Material	Maximum usage temperature	Typical applications	General comments
Steel	538°C (1000°F)	Annealing, drying, stress relief, tempering	Useful in a variety of low-temperature applications in non-corrosive environments.
Stainless steel (446)	1095°C (2000°F)	Hardening, nitriding, salt baths	Good resistance to corrosion at high temperatures; highly resistant to sulfur attack.
Inconel 600/601	1150°C (2100°F)	Carburizing, hardening, nitriding, gas generators, salt baths	Most common materials in use today. Greater mechanical strength than 446 SS; excellent resistance to corrosion and oxidation at high temperatures. Inconel 601 is more resistant to the influence of sulfur than 600. Both materials are subject to hydrogen embrittlement.
Metal/ceramic	1370°C (2500°F)	Sintering and high-temperature heat treatment	Superior oxidation resistance; thermal conductivity equal to that of stainless steel. Typical composition: 77% Cr / 23% Al_2O_3
Mullite	1540°C (2800°F)	Annealing, forge furnaces, metallizing ceramics	Impervious to gases at high temperatures; good thermal shock but poor mechanical shock; often necessary to provide secondary tube protection; vertical orientation recommended. Typical composition: 65% Al_2O_3 / 35% SiO_2
Silicon carbide	1650°C (3000°F)	Soaking pits, kilns	Excellent thermal conductivity for quick response to temperature changes; eliminates possibility of iron pickup; can be used as a secondary protection tube for resistance to thermal shock. Typical composition: 90.0% SiC / 9% SiO_2
Alumina	1870°C (3400°F)	Kilns	Fair resistance to thermal and mechanical shock; impervious to gases up to 1760°C (3200°F). Typical composition: 99.7% Al_2O_3

THERMOCOUPLES IN THE HEAT-TREAT INDUSTRY

The heat-treat supervisor, heat treater or metallurgist, along with the OEM or maintenance department, should be responsible for the selection of the type of thermocouple. They should also make sure that it is in the correct position within the furnace (both location and insertion depth) to accurately sense and control furnace temperature. Thermocouples located too close to heating or insulation sources or too close to the workload itself will not represent true furnace temperature. Temperature uniformity checks of the workload areas must factor in deviation from the control thermocouple. Although temperature offsets are allowed in some instances, this practice is highly discouraged.

The majority of heat-processing applications in the metals industry, including virtually all heat-treating processes, occurs in the range of -185°C (-300°F) to 1650°C (3000°F). No one type of thermocouple can span this entire range, and quite often we attempt to use a particular thermocouple well beyond its normal temperature range just because it's available. This practice should be avoided since it affects both the accuracy for the special application and, once exposed to abnormal conditions, may alter its accuracy when used for normal applications. As strange as it seems, it is not uncommon to find the wrong type of thermocouple being used to control a critical process.

Thermocouples must be checked regularly for accuracy against a known certified standard (probe) thermocouple and replaced on a periodic basis, usually every six to nine months depending on the severity of the end-use application. This calibration must take place while the thermocouple is installed in its normal operating location for reliable measurements. The operating life of a thermocouple depends on its operating temperature, time at operating temperature, ambient temperature, cyclic range (high to low temperature variation) and the influence of contaminants either on an exposed thermocouple or on the protection tube itself.

INDUSTRY STANDARDS

Thermocouple wire is supplied according to various industry standards (Table 13.3.7).

TABLE 13.3.7 | Typical industry standards for thermocouples

Organization	Standard number	Description
ASTM	E230	American Society for Testing Materials
ANSI	MC 96.1	American National Standard Institute
IEC	584 – 1/2/3	European Standard (International Electrotechnical Commission)
DI	EN 60584 – 1/2	Deutschee Industrie Normen
BS	4937.1041/EN 60584 – 1/2	British Standard
NF		Norme Francaise
JIS	C 1602, C 1610	Japanese Industrial Standard
GOST	3044	Russian Commonwealth Standard

Finally, thermocouples are not a set-it-and-forget-it technology. They require constant monitoring and confirmation of accuracy to ensure that the temperature being sensed and controlled is both precise and accurate. Remember, the quality of your products is highly dependent on these simple and relatively inexpensive devices, which makes their selection, care and replacement critical to your success.

13 | MAINTENANCE REQUIREMENTS AND PRACTICES

Summary

The empirical methods and techniques presented in this section are intended not only to provide a deeper understanding of how to set up and adjust the components discussed but also to reduce the maintenance work involved. If you take the time to make the necessary adjustments and go through the various steps to simplify the job, the time between servicing will be lengthened.

REFERENCES

1. Herring, Daniel H., "Equipment Maintenance Presentation," Furnaces North America 2014, October 2014
2. Herring, Daniel H., *Atmosphere Heat Treatment, Volume I*, BNP Media, 2014
3. *North American Combustion Handbook, 2nd Edition*, North American Manufacturing Company, Cleveland, Ohio
4. Braziunas, Vytas and Daniel H. Herring, "A Flowmeter Primer," *Heat Treating Progress*, March/April 2004
5. Thermal Systems Engineering (www.tempsens.com)
6. Herring, Daniel H., "The Flame Curtain – Function, Adjustment and More," *Industrial Heating*, March 2008
7. Baukal Jr., Charles E., *Industrial Burners Handbook*, CRC Press, 2004, p. 483
8. ASTM E230 (Standard Specification and Temperature-Electromotive Force (EMF) Tables for Standardized Thermocouples), ASTM International.
9. "The Right Thermocouple Makes A World of Difference," The Cleveland Electric Laboratories, white paper
10. *Volume 4: Heat Treating*, Metals Handbook, 10th Edition, ASM International
11. Wang, T.P., "Thermocouples for Special Applications," Proceedings of the International Conference: Equipment and Processes, 1994, ASM International
12. Nanigian, J., "Improving Accuracy and Response of Thermocouples in Ovens and Furnaces," Proceedings of the International Conference: Equipment and Processes, 1994, ASM International
13. Herring, Daniel H., "What is a Thermocouple?," *Heat Treating Progress*, March 2003
14. Omega Engineering (www.omega.com)
15. Kanthal Corp. (www.kathal.com)
16. Kary Peterson and Patrick Weymer, private correspondence
17. Eric and Jason Jossart, Atmosphere Engineering Co. (www.atmoseng.com), private correspondence
18. Jack Titus, AFC-Holcroft, technical contributions and private correspondence

19. Ralph Poor, Surface Combustion, technical contributions and private correspondence
20. Robert Brodeur, C.I. Hayes, a Gasbarre Furnace Group Company, technical contributions and private correspondence
21. James T. LaFollette, GeoCorp Inc., technical contributions and private correspondence
22. Mark Greenwood, Pyromation Inc., technical contributions and private correspondence
23. Daniel Reardon, Abbott Furnace Company, technical contributions and private correspondence

13.4

MAINTENANCE OF QUENCHANTS AND QUENCH BATHS

The goal of quenching is to produce the desired hardness, strength and toughness in a component part while minimizing distortion and residual stress. Uniformity of heat extraction by the quench medium is critical, and the mechanical, physical and metallurgical properties developed are directly related to the material and the cooling rates within the part.

For this reason, quenchants and quench tanks deserve special attention when it comes to maintenance due to the fact that their performance is of such vital importance during the heat-treatment process.

Quenchant Types and Their Maintenance

Quenchant selection is an important consideration in the performance of a quench system, and it defines to a large degree the type, frequency and difficulty of maintenance required. There are many choices of quenchants and a variety of (performance-driven) product variations, which are most often determined by the material and end-use requirements of the component parts being processed. These choices include:

- ❖ Brine/caustic
- ❖ Water (cold, warm, boiling)
- ❖ Polymer (water-like or oil-like)
- ❖ Oil (fast, medium, low or slow)
- ❖ Gas (sub-atmospheric to ultrahigh pressure)
- ❖ Molten salt
- ❖ Air (still, moving, mist)
- ❖ Furnace cooling

The use of oil quenching is an example of the challenges quenchant selection presents to maintenance personnel. First, there are different manufacturers of these products, some using different base oils and blending methods. All are designed to create versatility by providing different quenching speeds: high, medium, low and slow (i.e., marquenching) oils. The result is different physical properties for each oil (e.g., viscosity, recommended temperature range for usage) from each manufacturer.

Second, each choice often requires different agitation requirements, circulation settings, filtration and accelerators (i.e., speed enchancers). Degradation characteristics and drag-out volumes also vary as does the servicing needs of in-process monitoring devices.

Third, maintenance on the tank is often messy, as with any liquid. Special precautions and procedures may need to be used. Temporary fluid storage may also be required.

In addition, ancillary devices will require differing degrees of maintenance. Examples include heat exchangers (air-to-oil, water-to-oil), filters, centrifuges, smoke-abatement devices (e.g., fume hoods, ventilation systems), burn-off stacks (including eductors), tank measuring devices, oil storage containers, oil transport containers, fire-prevention systems (e.g., Cardox® or CO_2 systems), nitrogen-blanketing piping over the oil, spill management and environmental/hazardous-waste handling.

Also, the mixing of quenchants, either from various suppliers or different products from the same supplier, within a quench tank is discouraged because these products often have different formulations. While adding make-up oil is acceptable, at some point the tank should be drained, cleaned and recharged with new chemistry.

Finally, routine testing of these oils (monthly, quarterly or on a more frequent basis) is mandatory from a safety (flash point, water, contamination), performance (quench speed, sludge content) and deterioration or oxidation (total acid number, precipitation number) standpoint. These tests help determine some of the maintenance steps that will be required.

QUENCH-TANK MAINTENANCE

The most important considerations in all forms of quenching are achieving the required quench rates for the material (to achieve required properties), even heat extraction from all part surfaces (to manage distortion and the residual stress state in the material), stability over the operating range (to avoid separation or stratification), minimizing bath contamination and maintaining a low pollutant potential.

As such, the design of the quench tank (Section 10.6) and maintenance of the quenching system play a major role in the success of any heat-treatment operation. It is also important, therefore, to understand how the quench tank was/is intended to be used in service (i.e., what type of loads and/or component parts will be quenched into the bath and what properties must be achieved). The quench-system design is normally fixed by the manufacturer, so measuring the key process parameters and keeping the system operating at peak performance becomes the main focus of our maintenance efforts.

One of the issues many heat-treatment shops face, however, is that intentional or unintentional changes are made to the design of the quench tank over time, often in an effort to improve performance, without adequate benchmarking of the capabilities (or limitations) of the original design. If all aspects of the quenching process are not taken into consideration, the result is often an underperforming quench system.

In all quench systems, changes to the rate of temperature rise on quenching, unusual noises emanating from the tank area, excessive vibration, or abnormal evolution of flames or liquid is a clear indication that the system is in immediate need of repair and perhaps in imminent danger of destroying itself. If it looks, sounds, smells or feels different than it did the day before, the system should be taken out of service and maintenance repairs implemented.

Quench tanks are divided into two main classifications: open and integral (or sealed). The routine maintenance of these tanks often includes confirmation of proper liquid levels in the tank; checking items such as current draw, rotation and speed of pumps and agitators; cleaning of filters and strainers; checking on the proper operation of tank heaters; checking for smooth and rapid movement of elevators powered by pneumatic or hydraulic systems or overhead cranes; lubrication of bearings; and cleaning/descaling operations.

Finally, one of the most underutilized tools available to maintenance personnel is the use of computational fluid dynamics (CFD) for modeling quench-tank problems or to model the flow pattern within the tank as a function of load/part configuration. The information gained from such models will better focus the maintenance effort on specific trouble spots and should definitely be considered before any quench-tank modifications are implemented. The model developed can be improved by feedback from empirical tests conducted.

BRINE/CAUSTIC AND WATER QUENCH SYSTEMS (SECTION 10.4)

The vast majority of brine/caustic and water quench tanks are typically open-style tanks. As such, corrosion of the tank at the fluid/air interface and in areas where splashing can occur is an additional maintenance issue. During operation, the system should be checked for unusual foaming/frothing of the quenchant, noise (signs of cavitation from the agitators or gas entrapment in pump lines) and adequate movement of fluid across the surface of the tank (or a change of flow direction, which may be an indication that baffles have shifted or agitators/pumps are not circulating properly).

Sodium chloride (NaCl) at concentrations around 5-10%, sodium hydroxide (NaOH) at concentrations around 3-5% and certain proprietary blends are used for brine and caustic solutions. These tanks are normally operated in the 17-38°C (65-100°F) range. Solid deposits of these chemicals in various areas of the tank are to be expected, and these areas must be routinely cleaned. Appropriate safety precautions must be observed when performing maintenance on these tanks.

Water tanks run from ambient temperature to 100°C (212°F) in certain applications, such as quenching of aluminum castings. While these temperatures are not excessive, steam rising from the tank and higher-than-normal shell temperatures means that special precautions may be in order.

Another indicator of the need for maintenance of these systems is the presence of an unusual odor or smell. This often indicates the presence of algae, mold or other biological (bacterial/fungal) agents. Countermeasures include the use of biocides (often discouraged), but the keys are cleanliness, increased oxygen content (e.g., solutions should be kept agitated and rust/solids removed from the bath) and good filtration. Sand filters, for example, are often more effective than bag filters, which become a breeding ground for biologicals.

Finally, if these tanks are not used routinely or are not drained and emptied for extended maintenance periods, there is a distinct possibility that accelerated corrosion (rusting) can take place. External and internal surfaces should be thoroughly dried, and appropriate steps must be taken to avoid flash rusting if high humidity is present.

Water mist systems are used in a number of applications for quenching/cooling of workloads. These systems should be checked to ensure that the fog produced is in the form of a fine mist as opposed to large droplets of liquid. Often placing a piece of Plexiglas or one's hand at a distance equivalent to where the load would be is a good method of observing that the system is performing properly.

Water-spray quench systems – like those used for the heat treatment of bar, pipe and tubular products – are a good example of an operation that can benefit from good maintenance practices. The ability to properly maintain and control key process variables in these systems (e.g., pressure, flow, temperature) will ensure proper metallurgical transformation and achieve required mechanical or physical properties. Having video of the system's performance when operating properly can help spot problems later on.

Temperature rise of the product after exiting the system is usually a sign of trouble. Some of the more common problems reported include:

❖ Clogged or improper spray nozzles producing inadequate spray coverage (as a function of position of the product in the quench). A common misunderstanding is that all spray nozzles are the same. In point of fact, there are hundreds of different types of nozzles that each produce different spray patterns, coverage areas and intensities.

❖ Misaligned spray pattern (as indicated by the inability to prevent the formation of a secondary vapor phase, thus exposing both the length and the end of the product to reduced quench severity). This can often be detected by the presence of laminar flow across the surface of the product or ineffective application of the volume of water supplied.

❖ Systems with marginal (or undersized) water flow, pressure and/or temperature; systems lacking independent regulation and control.

Here are a few tips on what to look for.
1. Adequate water and heat-exchange capability should be provided to maintain 30°C (85°F) water temperature on a 35°C (95°F) day.
2. Check that the quench-ring assembly is large enough in circumferential diameter to allow a uniform spray pattern from the nozzles.
3. Individual water supply hoses (with quick disconnects), larger in capacity than the individual quench ring, should be provided to each ring in each quench-ring assembly to add flexibility in adjustment of the quench severity.
4. A common water-supply manifold can be used, but each independent water-supply line coming from the manifold should have a pressure and temperature indicator and be equipped with a ball-type adjustment valve.
5. Nozzles positioned at the tips of smaller-diameter pipes should extend outward from the quench ring toward the product. Depending on nozzle design, a 15- to 55-degree angle oriented in the direction of bar travel is recommended. A manual changeover of the extension pipes may be required when running certain product diameters.
6. High-density spray quench nozzles should be employed. Depending on the nozzle selected, consider staggering the nozzles (ring to ring) and/or placing them at variable distances from the product.
7. The areas between sets of quench rings *must* have some type of quench flow, either from spray rings with nozzles or a flood-quench tunnel placed in these locations.
8. Individual quench pumps for each quench-ring assembly or flood-quench tunnel are recommended. The horsepower and capacity should be designed to supply adequate quench pressures to each quench-ring

assembly. Common industry values for pressure are in the range of 620-820 kPa (90-120 psig) and in the range of 3,400-4,900 lpm (900-1,300 gpm) per quench-ring assembly for flow.

POLYMER QUENCH SYSTEMS (SECTION 10.2)

Contamination of polymer quench tanks can be a major concern because it will change the cooling characteristics of the bath. Since the vast majority of applications involve open tanks, airborne dirt and dust, oils left on parts, carbon (in the form of soot) and other solids can enter the bath. They can be detected by testing. On small tanks, a simple cover over the tank when not in use can dramatically extend the life of the polymer.

Quench tanks use various polymer solutions (e.g., polyalkylene glycol, or PAG). The quench rate is affected by temperature, agitation, concentration, contamination and overall degradation of the bath. Therefore, the areas of the system that affect these parameters should be the focus of our maintenance efforts. A few simple tests can often yield a wealth of information about the bath. These tests include:

- Visual observation
- Refractive index
- Viscosity
- Conductance
- Separation temperature
- Corrosion inhibitor concentration

Visual observation of the polymer involves extracting a sample from the tank and simply observing it. Oil contamination (which leads to non-uniform quenching, distortion and cracking) can be seen simply by looking for oil floating on the surface or as a milky-white emulsification.

Polymer quenchants can be readily controlled by concentration, which is measured by use of a refractometer (Fig. 13.4.1) or kinematic-viscosity instrument (Fig. 13.4.2). They are used to determine the refractive index.

FIGURE 13.4.1 | Refractometers (courtesy of Houghton International)

FIGURE 13.4.2 | Kinematic-viscosity instrument (courtesy of Houghton International)

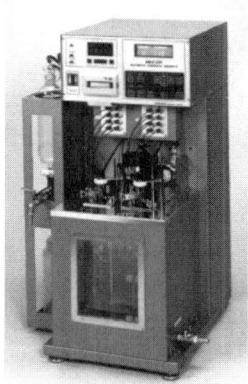

There are a number of different types of refractometers available (% Brix, refractive index, Abbe). Each has different scales, applicability ranges, accuracy, repeatability and suitability for use in the heat-treat shop. Most refractometers used in the heat-treatment industry are designed to be read in % Brix, which allows direct conversion to the refractive index (Fig. 13.4.3). The % Brix is the percentage of sucrose in water at -7°C (20°F).

The main advantage of a refractometer is that it is a small hand-held instrument that is simple to use on the shop floor. The test itself is quick and easy to perform – a drop of tank solution is placed on the instrument, which is then held up to a light source before a reading is taken on the scale provided. Results, however, are prone to sample contamination. Refractometers require a minimum of quarterly calibration (by kinematic viscosity), a fact almost always overlooked by those who use the devices.

FIGURE 13.4.3 | Refractive index vs. % Brix (courtesy of Houghton International)

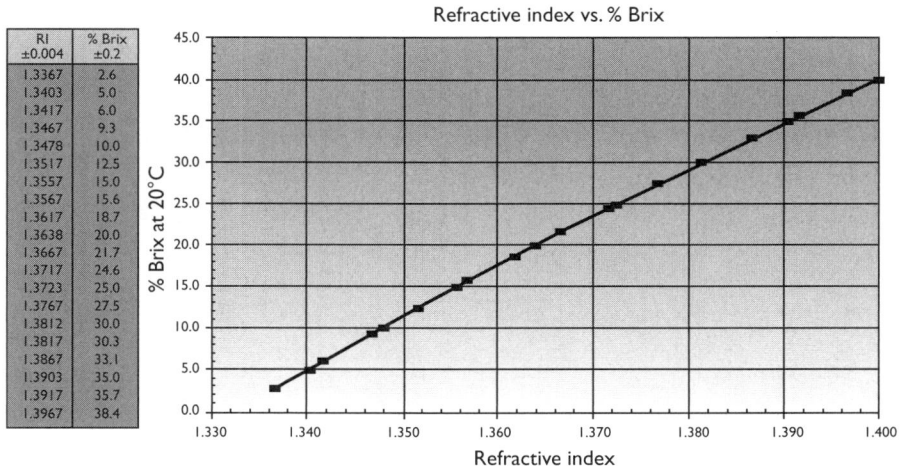

Kinematic viscosity is the preferred method, but it is a test often performed in a laboratory environment. It is used for reference measurements of concentration and is relatively immune to contamination.

Viscosity can be tested using a Cannon-Fenske tube, a stopwatch and a constant-temperature bath. This type of test can tell you how quickly your polymer is degrading.

Conductivity meters can be used to determine the presence of salts, which raise the overall conductivity. Simply dip the instrument's probe into the solution.

Lastly, separation temperature can be determined by heating the solution on a laboratory hot plate. When the fluid becomes so cloudy that the thermometer placed in the solution to measure its temperature is no longer visible, the separation temperature has been reached. It is important to note that degradation of the polymer causes the separation temperature to rise. A 2-4°C (4-7°F) rise over the life of the bath is considered normal. A larger increase or sudden change is cause for concern.

If the concentration of polymer systems is allowed to vary below prescribed limits, accelerated tank corrosion often results, which is why concentration measurements are so important. Finally, many polymers use sodium nitrite in the solution as a corrosion inhibitor. Corrosion kits can test for this chemical, typically by a color comparison technique.

Other more-involved testing (e.g., biological contamination tests) is often done by the polymer supplier, who can also detect changes in the cooling-curve behavior. This is why it is so important to have quarterly checks (at a minimum) of your polymer bath (Fig. 13.4.4).

FIGURE 13.4.4 | Typical polymer quenchant report (courtesy of Houghton International)

Date sampled	2/4/02	1/23/02	1/14/02	12/26/01
Appearance	opaque tan	opaque grey	hazy tan	hazy pink-yellow
Viscosity @ 38°C (100°F), cSt	1.03	1.1	0.89	0.97
Concentration by viscosity, %	4.8	5.7	3.1	4
Refractometer read	1	1.2	0.7	0.8
Current factor	4.8	4.75	4.4	5
pH	8.6	7.6	8.1	8.2
Oil, %	none	none	none	none
Sediment, %	0.01	0.04	0.02	0.01
Corrosion inhibitor, %	0.35	0.32	0.35	0.26
Problems reported	none	none	none	none
	Concentration (%)	Current factor	pH	
Microbiological Data:				
Bacteria / ml: - 10 K				
Fungi / ml: - <10				

OIL QUENCH SYSTEMS (SECTION 10.2)

Oil quenching is one of the most popular forms of quenching in the heat-treatment industry due to its unique combination of flexibility and overall cost. Oil quenching is performed in both open tanks and integral-quench tanks. The fact that vaporization of the oil occurs on quenching and that ignition of these oil vapors during the quench operation is a possibility heightens the concern over conducting proper maintenance on these tanks.

Some common maintenance requirements for oil quench tanks include:

- ❖ Check the oil level at least once a day (or more frequently if the loads are such that a large amount of oil drag-out occurs with the load).
  - Many quench tanks are equipped with dip-stick-style indicators with high/low level indicators. Others have sight tubes with similar indications.
- ❖ Ensure the water detection system on the quench tank is functioning properly.
- ❖ Pull an oil sample from the tank for testing. Remember, oil floats on top of water, so sampling should take place from the bottom of the tank – not just where it is easy to take a sample.
  - Quench oil can be poured into a graduated cylinder and placed in a vibration-free area to allow separation to occur. This type of test should be done daily and is effective at detecting both the presence of water in the oil as well as the amount of particulate matter (e.g., soot, scale, sludge) present. It is one of the most effective tests for judging overall cleanliness of the oil. Some organizations keep samples for several months (between quarterly analyses) for comparative purposes and to observe contamination of the oil.
  - Quench oil should be analyzed by an independent outside laboratory monthly or quarterly (or more frequently) to determine its condition and suitability for use. (Be sure that you understand how to interpret the quench-oil analysis report.)
- ❖ Check the oil circulation pattern and level of oil rise when the pumps or agitators are turned on. Flow measurement or height indicators should support these visual observations. A video of the flow pattern is a simple yet highly effective way of comparing flow characteristics day-to-day, week-to-week and month-to-month.
- ❖ Check the agitation system for excessive vibration and correct if necessary.
  - Vibration meters are available to quantify changes to the quench system. An older method was to place a penny on the agitator or pump housing. If it vibrated off, the system was out of balance.

- Vibration may or may not be due to the agitator or pump. It may indicate a problem with the draft tube, directional flow baffles or other structural members.
❖ Measure the rate of rise of oil temperature in the quench tank after quenching a full load.
- In a properly designed integral-quench tank, the quench temperature increase should not exceed 22°C (40°F). For open tanks, this value is dependent on the application and both the gross load weight and the mass of the component part(s) being quenched.
❖ Measure the current draw on the oil heaters.
❖ Measure the current draw on the oil agitators.
- This measurement alone does not guarantee adequate flow. There have been instances where agitator propellers have fallen off the shafts, which are simply spinning.
❖ Check for part staining. This may indicate that the oil is breaking down or fractionating in service.
❖ Change filters frequently. Inspect the contents of the filters to judge the degree of sludge or contamination present.
❖ Observe the characteristics of the flames that rise from the (open) quench tank during a quench with a known load weight and crane/transfer system speed.
- If the flames are abnormally high, this is one indication that the quench circulation, flow or flow pattern may have changed. Problems have been traced to malfunctioning agitators, propellers, warped and distorted internal-flow devices, water in the oil and cavitation or improper crane (down) speed.
❖ For integral-quench tanks, observe the characteristics (including the height) of the atmosphere burn-off flame (including the duration of negative pressure).
- If the flames are abnormally high, this is one indication that the quench circulation, flow or flow pattern may have changed. Problems have been traced to malfunctioning agitators, propellers, warped and distorted internals, water in the oil and cavitation.

Cooling-Curve Interpretation

Cooling curves, cooling-rate curves and oil analyses are provided by quenchant suppliers to assist heat treaters in better controlling their quenching process. Interpreting the information provided on these forms is important to making informed decisions on changes that may be required to the quench bath.

Cooling curves and cooling-rate curves (Fig. 13.4.5) not only tell us how the oil behaves through the three stages of quenching but also characterize the quenching behavior. In other words, these types of cooling curves measure our "quenching power" and tell us how the oil was performing on the day of the test. When looked at over time and quantified by part test data, they become a powerful predictive tool.

FIGURE 13.4.5 | Typical cooling curves and cooling-rate curves for new vs. used quench oil (courtesy of Houghton International)

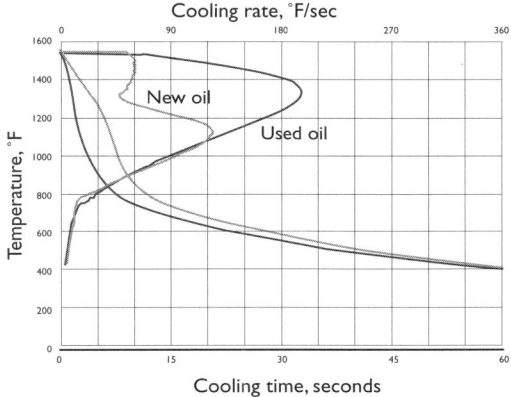

The basic shape of a cooling curve is set by the quenchant formulation. Over time, we see shifts in the cooling curves. As we build up a history of how the oil is changing, we can tell how different conditions (e.g., contamination, oxidation, drag-out, agitation) affect quench performance (Fig. 13.4.6).

FIGURE 13.4.6 | Typical cooling curves and cooling-rate curves for new oils (courtesy of Houghton International)

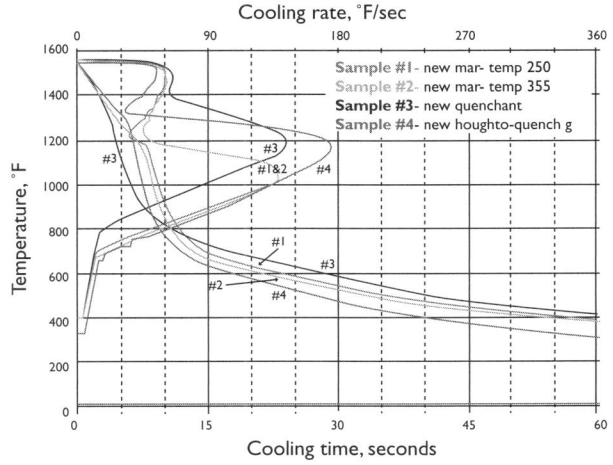

Immersing a heated thermal probe into an unagitated, heated quenchant and generating time/temperature data as the probe cools typically generates cooling curves. Standard test procedures are based on ASTM D-6200-01 (Standard Test Method for Determination of Cooling Characteristics of Quench Oils by Cooling Curve Analysis) and ISO-9950 (Industrial Quenching Oils – Determination of Cooling Characteristics – Nickel Alloy Probe Test Method). These methods allow comparison of new oil results from supplier to supplier and from plant to plant all over the world.

How to Benefit from Cooling-Curve Data

Cooling curves are normally determined under controlled conditions in the laboratory as opposed to monitoring quench tanks in real time. Also, the choice of cooling probe further limits our ability to interpret the results. When quenching parts in the heat-treat shop, the quenching power of the oil and the heat-transfer characteristics vary not only over the surface of an individual part but within the workload.

For example, the transition between the vapor phase and the nucleate-boiling phase can take place at different times at different points on various part surfaces throughout the load. Even oxidation of the part surface is known to change the cooling performance. Therefore, it is critical for the heat treater to monitor part hardness, distortion and other properties to understand how his/her particular quench tank performs as his/her quench oil ages and changes.

Quenching Speed and Other Useful Data

Quarterly checks of cooling-curve data from both new and used quench-oil samples (Tables 13.4.1-13.4.5) revealed that while the quench oil in an integral-quench furnace had cooling characteristics similar to the previous quarter's tests, it was significantly slower than a new oil or oil installed in another (open) quench tank. While the new oil and open tank samples fit the category of a "fast" quenchant, this particular IQ furnace had fallen in the speed category to a "slow" oil. Quenching performance will be different for similar materials, parts and loads quenched in these two tanks. Adding an accelerator in this case moved the cooling curve of the oil in the IQ furnace closer to that of the new oil.

13 | MAINTENANCE REQUIREMENTS AND PRACTICES

TABLE 13.4.1 | Classification of quench oils

Quench oil type	Cooling rate, °C/s (°F/sec)[a]	GM Quench-O-Meter nickel ball test, seconds[b]	Typical operating temperature range, °C (°F)
Fast	> 91.4 (165)	8-10	32-95 (90-200)
Medium	69.2-91.3 (125-165)	11-14	95-150 (200-300)
Slow	49.8-69.2 (90-125)	15-20	150-230 (300-450)
Marquench	49.8-69.2 (90-125)	18-25	150-230 (300-450)

Notes:

[a] Per ASTM D 6200

[b] Per ASTM D 3520

TABLE 13.4.2 | Maximum cooling-rate history, °C/s (°F/sec)

Description	1st quarter	2nd quarter	3rd quarter	4th quarter	1st quarter
New oil	49.8-69.2 (90-125)	97.4 (175.8)	93.5 (168.7)	93.5 (168.7)	95.2 (171.9)
Used oil (IQ furnace)	72.3 (130.4)	73.8 (133.2)	63.3 (114.2)	66.3 (119.7)	67.7 (122.2)
Used oil (open tank)	93.0 (167.9)	93.8 (169.3)	90.5 (163.2)	91.3 (164.7)	93.0 (167.8)

TABLE 13.4.3 | Characteristic temperature history, °C (°F)

Description	1st quarter	2nd quarter	3rd quarter	4th quarter	1st quarter
New oil	670 (1240)	672 (1242)	680 (1256)	682 (1260)	675 (1245)
Used oil (IQ furnace)	632 (1171)	632 (1171)	605 (1121)	618 (1144)	625 (1157)
Used oil (open tank)	687 (1268)	688 (1270)	690 (1274)	694 (1281)	695 (1283)

TABLE 13.4.4 | Cooling-rate history at 315°C (600°F), °C/s (°F/sec)

Description	1st quarter	2nd quarter	3rd quarter	4th quarter	1st quarter
New oil	20.9 (37.3)	19.5 (34.9)	20.0 (35.7)	18.2 (32.5)	19.4 (26.7)
Used oil (IQ furnace)	7.7 (14.2)	7.6 (13.9)	7.6 (14.0)	7.6 (14.1)	8.0 (15.0)
Used oil (open tank)	7.4 (13.4)	7.4 (13.6)	8.4 (15.8)	7.5 (13.8)	7.2 (13.1)

TABLE 13.4.5 | H-value history

Description	1st quarter	2nd quarter	3rd quarter	4th quarter	1st quarter
New oil	0.37	0.37	0.38	0.37	0.36
Used oil (IQ furnace)	0.27	0.27	0.24	0.25	0.26
Used oil (open tank)	0.34	0.34	0.35	0.34	0.35

Relationship of Physical Properties of Quenching Oils to their Performance

Quench oil should be routinely analyzed (quarterly, or monthly if heavily used) to determine its performance characteristics, and the testing laboratory or oil supplier's report (Fig. 13.4.7) should be carefully scrutinized since it contains information about the physical-property characteristics of the oil. Oil analysis uses standard test methods (Table 13.4.6), but in order to gain deeper insights into the meaning of the test results, as opposed to just comparing them with previous results, we need to understand what each category is telling us.

FIGURE 13.4.7 | Typical quench-oil report

	02-190	01-1299	01-983	01-615
Date sampled	2/5/02	11/13/01	8/21/01	5/29/01
Water (%) < 0.1	0.001	0.006	0.018	0.015
Visc. @ 100°F (SUS) 75 - 100	92	91.6	91.3	90.6
Flash point (°F) > 335	350	350	350	350
Sludge (%) < 0.20	0.01	0.01	0.02	0.01
Precipitation No. < 0.15	0.01	0.01	0.01	0.01
GMQS @ 80°F (sec) 7-10	8.9	8.9	8.6	8.8
Problems reported	none	none	none	none

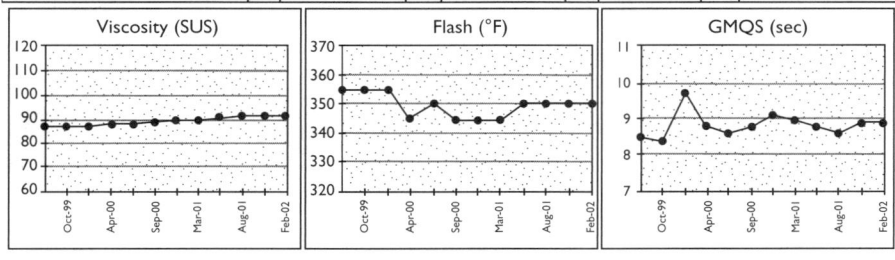

TABLE 13.4.6 | Quench-oil test standards

Test	ASTM method
Viscosity	D445
Flash point	D92
Fire point	D92
Water	D95/D6304
Naturalization number	D974
Ash	D484
Conradson carbon residue	D189
Precipitation number	D91
Sludge	D91
Specific gravity	D287
Quenching speed (GMQS)	D3520
Quenching speed (cooling curve)	D6200

HIGH GAS-PRESSURE QUENCH SYSTEMS

A number of manufacturers are moving toward adapting gas-pressure quenching technology to atmosphere furnaces (e.g., pusher or integral-quench-style furnaces). Given that these systems have the potential to operate at high pressure (up to 20-25 bar in heat-treat shops and 40 bar in laboratory environments), maintenance again takes center stage. Static and dynamic balancing of fans, avoiding excess vibration, using sealed motors in particular environments (e.g., carburizing), recognizing that welding on or changing pressure vessels is not allowed by ASME code, and modifying heat exchangers or piping are a few of the areas that become critical maintenance concerns.

AIR OR GAS QUENCH SYSTEMS

Cooling of workloads in still or moving air, in inert gas or in a protective atmosphere (such as over the elevator or in the top cool of an integral-quench furnace) is routine in heat treatment. While the focus of maintenance is often on the cooling fans (e.g., lubrication, belts, bearings, fan speed) and water (or oil) jackets (e.g., flow, hot spots, temperature rise, avoidance of steam buildup), a few other precautions are noteworthy.

First, alignment and warpage may be a concern for structural components due, for example, to cooling of a hot load on the elevator mechanism of an integral-quench furnace.

Second, certain designs have load-transfer chains or spring-loaded brake mechanisms on the elevators. Springs can anneal, making the brakes ineffective and/or the chains bind. This creates problems with proper positioning of workloads being transferred to the quench – the worst case resulting in a jammed load or quench-oil fire.

Finally, gas movement and the creation of negative pressure in a furnace can create the danger of an imminent explosion. For example, changing from endothermic to (rich) exothermic gas, but not compensating for the difference in the characteristics of these gases, can produce a fire-breathing, banshee-howling condition in an integral-quench furnace resolved only in one instance by shutting off a top-cool atmosphere cooling fan and switching back to endothermic gas.

MOLTEN SALT-BATH SYSTEMS

Maintenance of molten salt baths is necessary to control the quality of the products being hardened and tempered. Some of the more common activities performed on salt-bath equipment are highlighted.

Low-Temperature Preheat Baths

Low-temperature salt baths used for preheating prior to hardening in the high-temperature bath often use a salt of similar composition to that of the high-heat bath in order to avoid contamination of the high-heat bath due to carryover of the salt. Maintenance of these baths is similar to that of the high-heat bath and is typically done on the same frequency (for convenience if no other reason).

High-Temperature Baths

High-temperature salt-bath tanks typically use a chloride-based salt to heat the work to the austenitizing temperature. The salt is continuously subjected to oxidation by air, moisture from the environment and contaminants on parts placed into the bath. Contamination of the high-temperature bath with low-temperature salt from another salt operation (other than preheating) must also be avoided. Such contamination frequently occurs from unwashed fixtures and tank cleaning tools. The quality of the bath must be properly maintained in order to minimize decarburization of the work.

The quality of the bath can be measured by two methods: shim stock (to determine carbon content) and total sodium oxide. If the shim carbon is low or the sodium oxide level is high, the bath should be rectified to remove these impurities. Rectifier pellets (Fig. 13.4.8) are typically used to remove the sodium oxide from the salt.

FIGURE 13.4.8 | Basket loaded with rectifying tablets for loading into high-temperature tank

Low-Temperature Quench Baths

Many low-temperature salt baths use a nitrate/nitrite salt to quickly cool the work to the desired austempering, martempering or tempering temperature. By the very nature of the operation, each load of work that is heated in the high-temperature bath also carries over some amount of high-temperature salt. The high-temperature salt contaminates the quench bath, which affects the heat-transfer rate and the resulting hardness.

Salt drag-out and salt carryover from tank to tank is a fact of life with all molten-salt operations. The amount of salt drag-out depends on the mass and configuration of the load and parts being quenched and the fixtures or conveying system used, but it is usually in the range of 50-100 g/m^2 (0.1-0.2 lbs/ft^2).

Contaminants such as metallic debris from the parts and carbonates that form during service may build up in the bath. Agitation keeps fine metallic particles in suspension. Particle concentration should not exceed 0.5% because the quench severity can be affected. Similarly, if the level of carbonate exceeds its solubility limit, quenching speed will be adversely affected.

A process known as desludging the bath can remove both metallic particles and carbonate. In high-volume operations, desludging of the high-temperature bath is done as often as twice per shift, while desludging and filtering of the low-temperature quench bath may need to be done every hour. This normally consists of removing and/or dislodging all buildup from around the outside of the tank and the heating elements (Fig. 13.4.9). The best method is to lower the bath temperature, turn off agitation and heat, and then allow the contaminants to settle to the bottom of the quench tank, where they can then be scooped out. (Caution: Desludging tools must be *dry* prior to use.)

FIGURE 13.4.9 | Area around heating elements after cleaning

When a salt line is going to be idle for an extended period of time (such as over a weekend), a common practice is to set the low-temperature bath to 230°C (450°F) to allow for better separation and removal of the high-temperature salt. A properly maintained tank should have a dark brownish/black appearance (Fig. 13.4.10) around 260°C (500°F). If the salt has an orange/yellow color, it still has a large amount of contamination from the high-temperature salt.

FIGURE 13.4.10 | Proper low-temperature bath appearance

If addition of water is a common practice, its content should be checked periodically and adjusted to ensure the required quench severity. In addition, the pump and metering system must be carefully checked and maintained since water reacts violently with molten salt.

The melting point of the bath also needs to be occasionally checked since it normally changes very little over time. A significant increase may indicate thermal breakdown of the salt, which is usually due to accidental overheating. The cause of overheating should be investigated and properly rectified.

Electric heating elements or radiant tubes should be periodically checked for deterioration due to the corrosive nature of the salt bath. Heating elements should be properly submerged and operating at the correct power settings, voltage and amperage (current), all of which are commonly recorded during operation. RA446 is a common heating-element material.

Salt pots (if used) should be checked for excessive distortion or signs of corrosive attack. Cast steel or alloy pots are used in some small operations, and the choice of one material over another is often a life-versus-cost decision. Scaling of steel pots should be carefully checked.

Finally, careful consideration should be given to those situations in which the molten salt bath is allowed to cool down and solidify in the tank. Damage can be done to internal components during the solidification process if one does not take expansion and contraction into consideration. Removal of molten salt into heated holding tanks must be done *very* carefully, keeping worker safety foremost in mind both during the transfer and for the duration of time in temporary storage.

Summary

No matter what quenchant is used or how confident we are in its performance, routine maintenance and testing is important, as is our ability to interpret test results so that we can make informed judgments as to how we can best control our quenchants and quenching practices. It is also important to remember that where the sample is extracted from the quench tank is of equal importance to the test results themselves.

REFERENCES

1. Totten, George E., C.E. Bates and N.A. Clinton, *Handbook of Quenchants and Quenching Technology*, ASM International, 1993
2. Bodin, J. and S. Segerberg, "Measurement and Evaluation of the Quenching Power of Quenching Media for Hardening," Proceedings of the First International Conference on Quenching & Control of Distortion, ASM International, September 1992
3. Gajen Dubal (retired), Park Metallurgical Corp. (www.heatbath.com), private correspondence
4. Herring, Daniel H., "A Review of Factors Affecting Distortion in Quenching," *Heat Treating Progress*, December 2002
5. Totten, George E. and Glenn M. Webster, "Maintaining Polymer Quenchants," *Advanced Materials & Processes*, June 1996

6. Wachter, D.A., G.E. Totten and G.M. Webster, "Quenching Fundamentals: Maintaining Quench Oils," *Advanced Materials & Processes,* 1997, No. 2, p. 48AA-48CC
7. Mackenzie, D. Scott, "Quenchant Selection, Maintenance and Care," Houghton International, Heat Treating Atmosphere & Quench Systems Best Practices Seminar, Air Products & Chemicals, 2007
8. Mackenzie, D. Scott, Houghton International (www.houghtonintl.com), private correspondence
9. *Sugar Analysis-ICUMSA,* F. Schneider (Ed.), International Commission for Uniform Methods of Sugar Analysis (ICUMSA), 1979
10. Mackenzie, D. Scott, "Care and Proper Maintenance of Polymer Quenchants," Houghton International, 2009
11. Herring, Daniel H., "Oil Quenching Part One: How to Interpret Cooling Curves," *Industrial Heating,* August 2007
12. Herring, Daniel H., "Oil Quenching Part Two: What is Your Quench-Oil Analysis Telling You?," *Industrial Heating,* September 2007

13.5

MAINTENANCE OF SENSORS AND CONTROLS

The choice of sensors or measurement devices for the myriad of furnace atmospheres used for control and troubleshooting in the heat-treatment industry depends to a large degree on ease of use, longevity, the type and amount of information required and/or cost. Regardless of the selection criteria, proper maintenance and calibration of these devices ensure their success.

Typical applications for furnace-atmosphere sampling instruments include the measurement of endothermic or exothermic gas, nitrogen/methanol or nitrogen/hydrogen atmospheres, dissociated ammonia and hydrogen atmospheres, oxygen levels, and plant air systems.

Maintenance for the most commonly used sensors will be discussed in this section.

Oxygen (Carbon) Probes

Oxygen (carbon) probes (Fig. 13.5.1) are in-situ control devices that can be located directly inside the furnace and exposed to the atmosphere environment, inside the retort of a gas generator or inside a thermal chamber (aka thermal well) into which furnace atmosphere is introduced. These are often referred to as self-heated probes. Oxygen probes are currently the de-facto standard for carbon control in the heat-treatment industry.

FIGURE 13.5.1 | Typical oxygen (carbon) probe (courtesy of Super Systems Inc.)

INSERTION

As a general rule, an oxygen probe should be inserted into a furnace so that 100-150 mm (4-6 inches) of the probe end extends into the heated chamber and is exposed to a representative sample of the furnace atmosphere. Probes should be located so as not to interfere with load movement and should not extend into the workload area or be closer than 50-100 mm (2-4 inches) to the workload. The probe location should be such that it is not directly influenced by incoming atmosphere from an atmosphere inlet port(s) and should also be in a location where the chamber temperature is a minimum of 760°C (1400°F) and a maximum of 1370°C (2500°F).

Probes can be mounted in either a horizontal or vertical orientation. Ideally, the probe should be located near a furnace thermocouple. Some manufacturers offer probes with internal thermocouples. Outside the furnace, the head of the probe should not be exposed to temperatures higher than 94°C (200°F).

Probes can be inserted into the furnace when it is cold (i.e., under 150°C/300°F) or hot. If probes are to be inserted hot, care must be taken to avoid thermal shock and cracking of the ceramic. The following procedure[2] is commonly used having predetermined the insertion depth by either measuring the probe that has been removed from the furnace, reviewing a cross-sectional drawing of the furnace in the location of interest or having established the insertion depth by installing the probe when the furnace was cold. Do not attempt this procedure if the insertion depth is not known.

Specifically:

1. Measure 150 mm (6 inches) from the end of the probe sheath and mark with a felt-tip pen. Mark the remainder of the probe in 25-mm (1-inch) graduations.
2. Carefully insert the probe from the first mark to the 150-mm (6-inch) mark. If the probe is inserted into a compression fitting, make sure the seal rings or O-ring is between the compression-fitting body and the nut. Wait five minutes while the probe warms up.
3. Begin inserting the probe at a rate of about 25 mm (1 inch) every 1-2 minutes until the probe has been inserted into the desired length. Slower rates should be used on furnaces operating over 925°C (1700°F).
4. Continue until the probe is at the desired insertion depth.
5. Tighten (by hand) the compression-fitting body and nut on the probe sheath.
6. Make the connections for the probe millivolt signal and, if provided, with the integral probe thermocouple.
7. Connect the reference air and burn-off air.

When operating an oxygen probe, it is highly desirable to determine the carbon monoxide (CO) value of the atmosphere so that the default value (20% CO) in the controller can be changed to the actual value. This value will change from day to day and must be adjusted so that daily monitoring of the furnace atmosphere with a three-gas analyzer improves the accuracy of carbon control. It is helpful, although not mandatory, to install a shim puller or use a percent-carbon analyzer to compare probe values with actual percent-carbon readings.

TROUBLESHOOTING

There are several key components in all atmosphere-control systems. When a problem arises, it is important to identify the cause with minimum effort and expended time. The following procedure is designed to aid in that process.

The starting point for any troubleshooting procedure is to properly identify the symptom that necessitates it. The cause of the symptom can often be determined by answering some basic questions. Is this a start-up problem, or has the system been operating under control? If it is a start-up problem, it is necessary to establish that all system components have been properly connected and configured for the application.

If the system has been operating properly and there has been either a gradual or sudden change in the control performance, it may conceivably be a problem with the probe. In order to establish the correct performance of the carbon sensor, resist the temptation to remove the sensor from the furnace. Begin by contacting the manufacturer and discussing the problem.

All of the tests outlined here must be done while the sensor is located in the furnace, at temperature and exposed to a reducing atmosphere. If the values are not similar, the following steps can be taken.

First, if possible, perform a shim-stock analysis and/or check the furnace atmosphere with a three-gas analyzer or dew-point instrument and compare carbon-potential values of the atmosphere. If the values are close (recognizing that each measurement device may yield a slightly different number), the problem is likely not the probe.

Next, use the following procedure, which can be performed on most probes.[1]
1. Verify that all cables between the sensor and the controller are firmly connected and making good contact with the probe and controller terminals. Verify polarity.
2. Verify that the reference-air supply is connected to the reference-air fitting.
    a. This is normally the fitting closest to you when you face the probe. It has been found on occasion that the reference air has been connected in error to the burn-off air fitting, causing low readings.

3. Check that the reference air is flowing. Disconnect the air supply at the probe and hold it up to your hand or submerge it in a cup of water. Bubbles verify the flow.
4. Verify that no air is flowing into the burn-off fitting by submerging the burn-off tubing in a cup of water.
    a. Flow can occur, for example, if the burn-off air pump is subject to external vibration.
5. Perform a leak test on the probe. This test can detect a cracked or broken (substrate) tube.
    a. Verify that reference air is flowing at 0.015-0.05 m^3/hour (0.5-2.0 scfh).
    b. Turn off the reference air for one minute and read the probe output in millivolts.
    c. Turn the reference air back on and note the change in millivolts. It should not display more than a 5 millivolt increase.
6. Check that the controller's control factor (COF) is set to the proper value.
    a. This factor is also referred to as process factor, furnace factor, CO factor, circulation factor and calibration factor.
    b. The factor may require adjustment to eliminate any offset or discrepancy between the indicated carbon potential and the actual achieved result in the workpieces or shim stock.
7. Check the sensor temperature and millivolt output via independent (digital) devices.
    a. Confirm that the calibrator agrees with the indicated values on the controller with one sensor and one T/C lead disconnected. If not, there is most likely a controller calibration problem or a cable problem.
8. Does the probe millivolt signal return to within 1 millivolt of its original value in one minute as measured by a digital VOM after it has been shorted for 5 seconds? If it does not, go to step 10.
9. Perform a probe impedance (resistance) test. This is one of several electrical tests that determine the electrical integrity and reliability of the probe. Some contemporary controllers can perform it. If yours does not, conduct this simple test.
    a. At process temperature, disconnect the controller cable at the probe millivolt output and measure the millivolt value with a volt-ohm meter. Shunt the signal with a 100-kilohm resistor.
    b. After 10 seconds, read the new millivolt value, divide the original value by the new value, subtract 1 from the result and multiply by the value of the shunt resistor (=100K).

c. The calculated value is the sensor resistance in kilohms, which should be less than 25 kilohms.
10. If the problem is not corrected by probe and/or furnace burnout (normally described in the probe manual), the problem may indeed be a faulty probe, in which case the supplier should be contacted for repair or replacement of the device.
   a. Caution: Even though you suspect a faulty sensor, **do not** remove the probe from a hot furnace at a rate faster than 50 mm/minute (2 inches/minute). Cool the sensor on an insulating medium (e.g., firebrick) to avoid thermal shock. This will help prevent damage that is expensive to repair.

Sampling Systems

Furnace-atmosphere measuring devices that rely on the extraction of a gas sample from the furnace or generator (e.g., infrared analyzers, dew-point instruments) to provide gas analysis are highly dependent on the quality of the sampling system for the accuracy of their measurements. A typical gas sampling system (Fig. 13.5.2) consists of the following parts:

- Sample port
- Sample lines – typically copper or stainless steel tubing
- Sample cooler
- Sample pump
- Flow meters/regulators/valves
- Filters – in-line
- Connection to measurement device – typically Tygon® tubing
- Sample measurement device – fixed or portable

Some sampling systems are set up for discrete (i.e., single-point) source sampling, while other systems are set up for continuous (i.e., multiple-location) sampling. A continuous-sampling arrangement accesses a number of ports one at a time in a repetitive timed-duration manner. These systems often utilize a flow of dry nitrogen or inert gas running through the sample lines, which are not actively being sampled to prevent condensation and false readings.

Let's review the function of each and their maintenance requirements.

FIGURE 13.5.2 | Typical gas-atmosphere sampling system (courtesy of Super Systems Inc.)

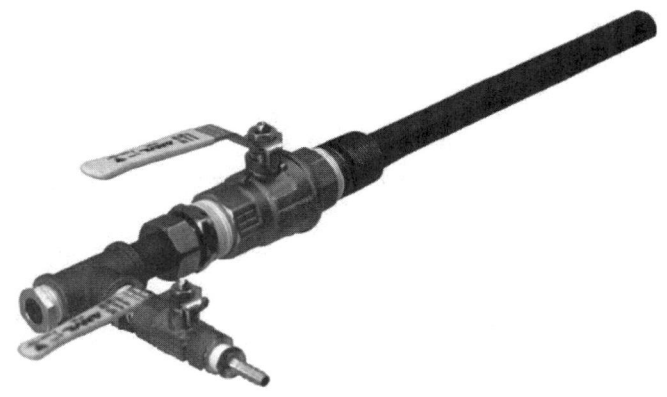

SAMPLE COOLERS

Accurate measurement requires a stable dew point. If the incoming gas sample is not cooled to room temperature, not only will we risk damage to the analyzer but possibly introduce unwanted measurement errors. Sample coolers are available from a variety of manufacturers, many with condensate-removal devices (automatic drains for pressurized systems or stand-alone pumps). It is common in a heat-treat shop to see copper or stainless tubing wrapped into multi-turned coiled loops to increase the residence time of the gas stream, which allows natural cooling to drop the gas to ambient temperature. Small-finned air-to-gas heat exchangers in the piping line close to the sample port have also been used. These simple solutions avoid an unstable dew point, which will not only bias the measurement but increase both analyzer and system maintenance in most cases.

SAMPLE PUMPS

Sample gas must be transported through the entire sampling and conditioning system at a steady pressure and flow. The sample's chemistry can be altered if the pressure is too high, and variable pressure may induce measurement errors. Key features include avoiding interference from condensate and using corrosion-resistant materials of construction.

FLOWMETERS/REGULATORS/VALVES

Flowmeters, regulating valves, shutoff and selection valves are part of any sampling system. These devices control the flow rate of the sample gas to the analyzer.

FILTERS

In-line filters are common in most sampling systems and often consist of more than one filter type. A bowl filter with a 50- to 100-micron cotton element is commonly used to capture large particulate matter (soot, dirt, scale) and water, while a fine silk-fiber 10-micron filter element placed close to the analyzer catches fine particulate. One of the common mistakes with filters is trying to clean them to save dollars while putting analyzers costing thousands of dollars at risk. Throw the filter elements away and replace them with new ones – unless they are wet from moisture, in which case they can be dried by passing clean dry nitrogen through the filter but not the analyzer.

NDIR Infrared Analyzers

The most common type of infrared analyzer system (Section 9.7) is the three-gas analyzer, which measures carbon monoxide (CO), carbon dioxide (CO_2) and natural gas (CH_4) in the furnace atmosphere. Multi-gas infrared analyzers can be supplied to measure excess oxygen in the combustion system and hydrogen in furnace atmospheres. Here are the typical operating ranges for each of these gases.

- CO: 0-30%
- CO_2: 0-2%
- CH_4: 0-15%
- O_2: 0.01-25%
- H_2: 0-100%

Proper sample handling is crucial to maintaining the integrity of the gas sample delivered to the analyzer. These units are suitable for installation on the shop floor or as portable devices. When not in use they should be stored in a location at ambient temperature, typically 17-26°C (65-80°F).

The ideal flow rate for most sampling systems is typically 0.02-0.04 m³/hour (1.0-1.5 scfh). In-line filters help prevent soot and other contaminants from entering the instrument and require periodic replacement. Similarly, water can collect in the sample tubing due to condensation from the rapid cooling of the hot furnace atmosphere and perhaps even in the sample filter. This requires immediate maintenance of these items.

ZERO AND SPAN

Changes in the temperature, pressure or flow rate of the gas sample can necessitate recalibration of the gas analyzer. The calibration procedure involves setting the electronic zero and span adjustments before performing a gas calibration with a certified calibration gas of known composition.

The instrument should be turned on and allowed to warm up and stabilize (typically 15 minutes). A zero gas – nitrogen or argon – is introduced into the analyzer at a flow rate of 0.05-0.18 m^3/hour (2-6 scfh). After a stabilization period of a few minutes (i.e., once the sample cell has purged completely), switch the instrument to its lowest range and rotate the zero control on each channel so that a zero reading on the appropriate meter or digital display is shown.

After zeroing the instrument, begin flowing a gas containing gases known to be measurable/detectable by the analyzer. The flow rate and pressure used for the gas calibration should be approximately the same as the expected sample gas flow rate and pressure. Each channel of the analyzer should be individually zeroed and spanned.

CALIBRATION

Some infrared analyzers allow for field calibration, and others require factor calibration. Calibration is warranted anytime there is a significant change in ambient or sample gas conditions. A calibration applies a known signal to the instrument under test with errors being easily detected (the error being the algebraic difference between the indication and the actual value of the measured variable). Span errors, zero errors, combinations of the two and linearization errors are typical examples.

Routine maintenance of infrared analyzers typically consists of the following:

- ❖ Daily – adjust zero and span controls once a day to ensure optimum instrument performance.
- ❖ Quarterly – with the power disconnected, carefully remove the sample cell and inspect for contamination of both the sample cell and sample window. It may be necessary to disassemble and clean the cell with isopropyl alcohol. A cotton swab can be used to clean the sample window, which can be dried with a soft, lint-free cloth or tissue.
- ❖ As required – the exterior of the analyzer can be wiped with a dry rag or cleaned with a mild detergent. Solvents should not be used.

When they do occur, problems (Table 13.5.1) are well known and easily corrected.

TABLE 13.5.1 | Common problems and troubleshooting steps

Problem	Probable cause
Motor rotation detection signal fault.	• Motor is faulty. • Motor rotation detector circuit is faulty.
Zero calibration fails or is out of range. Amount of zero calibration (indication value) is over 50% of full scale.	• Zero gas was not used or set at the wrong flow rate. • Leaks exist in the lines, valves or fittings. • Dirty cell. • Detector is faulty. • Optical balance is misadjusted.
Span calibration is not within the allowable range. Amount of span calibration (difference between indication value and calibrated concentration) is over 50% of full scale.	• Span gas was not used. • Zero calibration was not properly performed. • Calibrated concentration setting does not match actual cylinder concentration. • Leaks exist in the lines, valves or fittings. • Dirty cell. • Detector sensitivity has deteriorated.
Measured values fluctuate too much during zero and span calibration.	• Calibration gas is not supplied. • Time for flowing calibration gas is too short.
Calibration is abnormal during auto calibration.	• Circuitry is faulty. • Dirty cell.
Output cable connection is improper.	• Wiring is detached between analyzer and interface module. • Wiring is disconnected between analyzer and interface module.
Drift	• Sample gas flow rate is improper. • Leaks exist in the lines, valves or fittings. • Dirty cell. • Optical system requires cleaning.
Readings are abnormally high or low.	• Moisture in the sample lines. • Contaminants (soot, dirt) present in the sample. • Leaks exist in the lines, valves or fittings.
Readings do not increase.	• Zero and span calibration needed. • Sample gas flow rate is improper. • Leaks exist in the lines, valves or fittings.

Dew-point Instruments

Dew-point devices (Section 9.7) have the ability to detect moisture from -100°C (-150°F) to +27°C (+80°F), although most (but not all) types used in the heat-treatment industry operate in reduced ranges. Instruments that measure the dew point in furnace atmospheres require calibration to ensure that they continue to read accurately over time.

As previously discussed, proper sample handling is crucial to maintaining the tight moisture control demanded by our industry. Mounting the dew-point instrument as close as practical to the measurement point improves the accuracy of readings.

Before operating a dew-point instrument and following all safety protocols, carefully open each sample port on the furnace or generator to be monitored and remove (typically by passing a small metal rod through the port) all soot accumulation from the sample port and all moisture from the sample lines prior to attaching the sample tubing to the measurement device. This will involve passing dry nitrogen through the lines in some instances. For carburizing, this process may have to be repeated after each shift or even during a processing run itself depending on the carbon potential of the atmosphere. Other tips include:

- ❖ For panel-mounted units, install air conditioners in the instrument panel if the ambient temperature of the cabinet exceeds 40°C (105°F).
- ❖ For portable units, store the analyzer in a location at ambient temperature of 17-26°C (65-80°F) for at least two hours prior to operation.
- ❖ Stop the sample process if the ambient temperature of the analyzer exceeds 40°C (105°F).
- ❖ Maintaining proper sensor temperature will prevent the premature failure of the sensor. The operating temperature of most sensors should remain below 55°C (130°F) at all times.
- ❖ Do not operate dew-point instruments in environments where corrosive gases are present, including but not limited to ammonia, sulfur dioxide, sulfur trioxide, chlorine or fluorine gas or acids (e.g., hydrochloric).
- ❖ In dissociated-ammonia atmospheres where residual ammonia is present (typically in the range of 50-250 ppm), recognize that the life of the analyzer may be shorter than normal and that the readings obtained may be different given the fact that both water vapor and ammonia are dipole molecules and cause a shift in the readings of the analyzers.
- ❖ Erroneous readings will occur in the short term. Long-term sensor and internal component damage will result if contaminants such as soot or particulates exist in the gas sample or if excessive water is present.
  - Soot/particulate contaminants: When taking a sample from a furnace or generator, care should be taken to reduce the amount of soot that enters the instrument. The filters will trap these particles, but cleaning the sample line before attaching the instrument will increase the life of the filter. Furnace ports can be rodded out or burned off by carefully pumping air through them while hot or, if possible, by removing the sample tube from the heat and mechanically cleaning them. Generator ports should be opened before the instrument is attached to allow any particulate buildup to be blown out. It is also helpful to tap on the port while it is being blown out to eject any loose particles before the instrument is attached.

    If soot is allowed to collect on the dew-point sensor in the instrument, it could result in higher readings. This soot will also retain moisture that can corrode the sensor over time. It may be possible in some units to remove the sensor tip and clean it, often by rinsing it in isopropyl alcohol. The power should be off while this is done, and the power should remain off for at least 30 minutes after

13 | MAINTENANCE REQUIREMENTS AND PRACTICES

this procedure to allow all of the alcohol to completely evaporate.
- Water/moisture contamination: When a furnace or generator is being started up or cooled down, the resulting gases will contain unusually high amounts of carbon dioxide. When these gases cool, moisture will precipitate out and become condensation inside the sample tubing assembly. Even if the furnace or generator is operating normally, residual moisture may still be present in the sample tube or plumbing system. In the same way that the ports are checked for soot (see previous), they should be checked for moisture before attaching the instrument. This is especially important when taking a sample from a generator since the sample port is usually preceded by a significant amount of plumbing. All traces of moisture should be eliminated before attaching the instrument. Failure to do so will result in erroneous measurements and could result in damage to the analyzer.

The first signs of moisture in the instrument are often visible condensation in the tubing or moisture collecting in the filter. Unusually high dew-point readings result. The upper range of the sensor is +26°C (+80°F), so if that value is displayed on the instrument it is probably due to the presence of moisture. If this moisture is not removed, it will cause the sensor tip to corrode and eventually require the sensor to be replaced.

To remove moisture from the instrument, the filter should be removed (since it will probably be wet), and an inert gas such as nitrogen or argon should be introduced for as much time as it takes to dry it out. To test if the unit is operating properly after dry-out, verify the ambient dew point against a weather station that will report the ambient dew point for your area. If the displayed reading is within three degrees of the reported dew point when the instrument is taken outside, then all of the moisture has probably been successfully removed. The wet filter and sample tubing can be reattached after they have been completely dried out, or they can be replaced.

Be sure that the measured dew point is below ambient levels before it is stored to prevent the possibility of moisture damaging the instrument. If necessary, dry nitrogen or argon can be used to purge the instrument after use.

Finally, care should be taken not to drop or handle the analyzer in a rough manner (if it is a portable unit). Also, avoid vibration. One should always operate these instruments in a manner consistent with their intended use.

Other Sample Measurement Devices

The analyzers discussed here normally require very little maintenance, but certain steps should be taken to ensure they operate reliably.

OXYGEN ANALYZERS

Another common sample device installed on industrial furnaces is an oxygen analyzer. Other than calibration, most oxygen analyzers only require checking of the batteries and sensor to ensure continued operation. Avoid exposure to excessive heat or vibration and store them, covered, in a room-temperature environment – 17-30°C (65-85°F) – when not in use.

Test batteries regularly and replace them immediately when all battery-strength indicator bars are missing. Most of these devices only use alkaline batteries (check the product manual). Recalibrate after replacing the batteries. If the analyzer is not used for a period of 30 days or more, the batteries should be removed from the device.

Check the sensor for leaks or condensation; cable for split, cracked or frayed insulation; and connections for tightness and dryness. Troubleshooting (Table 13.5.2) is usually straightforward.

HYDROGEN ANALYZERS

In many cases, the furnace atmosphere consists of a gas mixture, which is some variant of oxygen, carbon monoxide, carbon dioxide, methane and hydrogen. As such, measuring hydrogen by use of thermal conductivity is complicated by the presence of these other gases (because each gas contributes its own thermal-conductivity value to the overall thermal-conductivity value of the mixture). Hydrogen analyzers (Fig. 13.5.3) are typically factory-calibrated devices that require no additional maintenance other than batteries.

TABLE 13.5.2 | Troubleshooting guide for oxygen analyzers[3]

Symptom	Why	What to do
New sensor responds slowly or drifts.	If the sensor is new and was just removed from its sealed bag, it may need to run for several hours.	Wait 1-2 hours and recalibrate.
Sensor will not read below 22% after calibration in 100% O_2.	Calibration in 100% was invalid or the room air was enriched with excess oxygen.	• Recalibrate using dry gas. • Make sure that at least 150 mm (6 inches) of tubing is attached to the exhaust side of the tee adapter to prevent back filling. O_2 flow rate should not exceed 5 l/min. • Oxygen concentration at the sensor is significantly higher than 21%. Take the analyzer to a well-ventilated area and check the reading again. • Try calibrating with a known good sensor; if this fails, contact customer support.
The sensor does not react to changes in oxygen concentration, or the readings are unstable and drifting.	• Water may have condensed on the sensing surface. • Electrical interference is disrupting the electronics.	• Remove the sensor from tee adapter and unscrew the plastic flow diverter. Using absorbent tissue or a cotton swab, gently wipe off the sensing surface inside the threaded portion of sensor assembly. Do not damage the sensing surface, and do not leave any tissue/swab residue on the sensing membrane. • Relocate unit away from sources of electrical noise such as two-way radios.
The display is flashing "√ SENSOR."	• The unit has detected a fault in the signal from the sensor. • Sensor has expired. • The sensor has been exposed to a gas containing less than 18% oxygen.	• Check sensor cable connections to ensure they are completely inserted into the mating connector and the capture nut is firmly in place. • Expose the sensor in a 100% oxygen atmosphere.
The oxygen reading fluctuates or appears to be incorrect.	The sample pressure may be changing.	• Make sure there is no restriction on the exhaust side of the sensor during calibration. If the reading changes with flow, the sensor is pressurized or there may be a leak in the system. • If a high degree of accuracy is desired or the concentration of O_2 being analyzed is in excess of 40%, calibration with 100% is recommended.
The unit has stopped working and the LCD is displaying alphanumeric characters.	Some units are equipped with an electronic "watch dog," which analyzes the circuitry within the unit for potential faults and renders the unit inoperable until the condition is corrected.	• Disconnect the batteries and inspect the contacts for corrosion. Reconnect the batteries. If the unit functions properly, calibrate the unit and reset the alarm values. • Try a new set of batteries. • Increase the distance between the unit and any source of radio frequency interference. The sensor cable is a prime source of pickup because it can act like an antenna. Relocate the sensor cable and, if possible, change its coiled length to "de-tune" its antenna effect. Placing the cable in a different position may also help.
No display	• Batteries expired. • Bad battery connection.	• Check/replace batteries. • Check battery connections and calibrate.

FIGURE 13.5.3 | Hydrogen analyzer mounted on a circuit board (courtesy of Super Systems Inc.)

To provide an accurate hydrogen analysis, a number of analyzers subtract the values of the interfering gases from the thermal-conductivity results. For example, when the non-hydrogen gas comprises 100% of the mixture, the hydrogen reading is 0%. Conversely, when the hydrogen gas comprises 100% of the mixture, the non-hydrogen reading is 0%. The region in the middle represents various mixtures of the gases present. This region may not be inherently linear. However, algorithms in hydrogen analyzers have the capability to linearize this condition and allow accurate measurement of hydrogen in multiple-gas mixtures.

COMBUSTION ANALYZERS

Combustion analyzers fall into a family of products intended for analyzing oxygen, combustibles and methane in flue gases for combustion control, safety and thermal NOx reduction.

Some systems measure oxygen and total combustibles or carbon monoxide by using a system designed to draw a sample from a stack or duct over a relatively short distance to a pair of sensing cells. The first cell is a zirconia sensor for measuring the oxygen. The second is either a platinum catalytic bead sensor for total combustibles or an electrochemical sensor for carbon monoxide. The sensor chosen depends on the accuracy and the measurement levels required and whether total combustibles or carbon monoxide is the gas to be measured. Other systems use probes (Fig. 13.5.4a) or are in-situ-type devices (Fig. 13.5.4b).

FIGURE 13.5.4 | Combustion analyzers in use

a. Portable combustion analyzer (courtesy of MacLean-Fogg)

b. Panel-mounted combustion analyzer (courtesy of Super Systems Inc.)

a.

b.

Why measure flue gas?

Ideally, every burner should mix a precise ratio of air and fuel, and the fuel should burn at a stoichiometric ratio (i.e., combustion yielding only heat, water vapor and carbon dioxide). In reality, this rarely happens. Burners age, mixing is imperfect, calorific value of fuel varies, firing rates vary and the weather changes from day to day. Any of these factors can alter the amount of air required for safe and efficient combustion of fuel. In addition, too little air sends unburned fuel to the stack – a costly and potentially explosive condition. Too much air wastes the heat of combustion by allowing excess hot air to go up the stack and drives up cost.

Accurate flue-gas analysis minimizes fuel costs and reduces pollution in all combustion processes (Fig. 13.5.5). Incomplete combustion can lead to increased air pollution. The amount of excess air influences SO_3 and NOx formation. The correct air/fuel ratio minimizes the emission of pollutants.

FIGURE 13.5.5 | Typical combustion-control scenario[4]

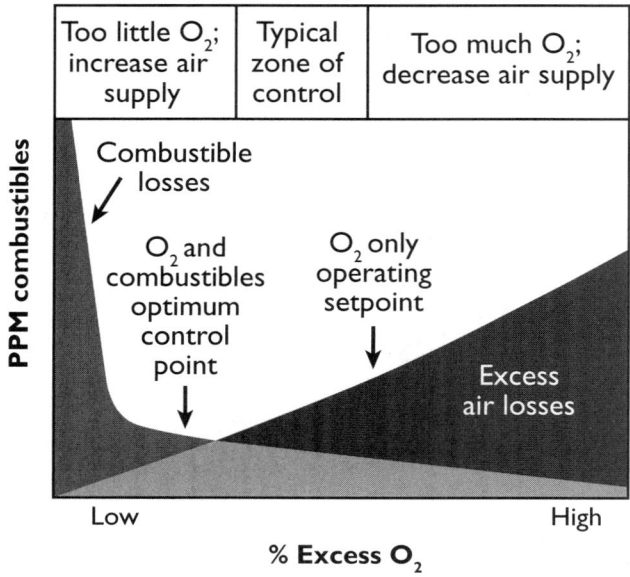

Summary

The maintenance of gas-analysis instruments and sampling systems is critical to obtaining reliable and repeatable readings. The vast majority of furnace-atmosphere control issues can be traced back to faulty sensors and improper sampling methods. Working on these systems takes a great deal of time and patience, but the reward is a process that will stay in control.

REFERENCES

1. Thompson, Stephen, "Carbon Sensor Troubleshooting," Super Systems Technical Data Sheet T4413
2. Oxygen Probe Installation and Operation Manual, Version 004, Furnace Control Corp., United Process Controls Group
3. Operating and Service Instructions for AX300 Portable Oxygen Analyzer, Teledyne Analytical Instruments, 2009
4. GE Sensing, FGA 200 Panametrics Flue Gas Oxygen Analyzer Bulletin 920-0358
5. James Oakes, Super Systems Inc. (www.supersystems.com), technical and editorial review, private correspondence

13 | MAINTENANCE REQUIREMENTS AND PRACTICES

SuperSystems
incorporated

How We Do Business.

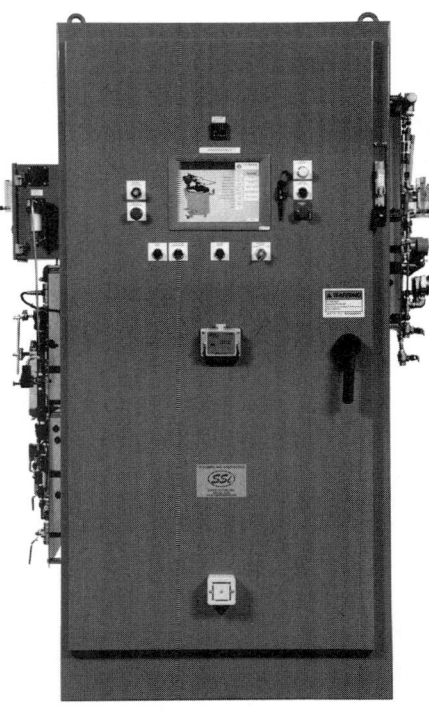

Heat treaters depend on process control for the highest quality and predictability.

Robust control depends on precision and expertise.

At SSi we have experience and the know-how to deliver the highest levels of precision in the industry, with innovative, cost-effective solutions customized to each client's specifications and unmatched technical support for the life of the product.

www.supersystems.com

13.6

FURNACE TESTING AND TROUBLESHOOTING

Maintenance is a complex subject and one that is often best learned by doing. For some specialized tasks and tests, however, the experience of others is often helpful in shortening the learning curve.

Tests for Leaky Radiant Tubes

Pressure changes that occur within a gas-fired furnace are often due to leaky radiant tubes. If a tube cracks, the combustion products will leak out and mix with the furnace atmosphere, creating a fluctuation in the internal atmosphere pressure in the furnace. If you suspect that you have a leaky tube, here are a few suggestions on how to find it from those who work on these problems every day. Remember that the burner, recuperator (if present) and all associated piping are very hot, and proper personal-protective equipment must be used to avoid injury.

It is important to recognize that leaks in radiant tubes generally begin as a pinhole or small crack and gradually increase in size. Symptoms include an erratic (oscillating) carbon potential (i.e., oxygen-probe millivolt values) or a furnace dew point that slowly creeps up higher and higher. In either case, the atmosphere-control system usually calls for more enriching-gas additions to hold the setpoint value. As the leak worsens, still more enriching gas is needed until the point when you may not be able to hold the desired carbon-potential setpoint. Parts may show evidence of unusual soot deposits or even decarburization despite large amounts of enriching gas in use.

Pinhole leaks are difficult to find with conventional techniques because any carbon dioxide (CO_2) produced from combustion will react with endothermic gas present. Large leaks are generally easier to spot but have been known to open and close with changing temperature, making the task far more difficult. A recording system will assist in more quickly detecting upward or downward trends.

Here is one method for leak testing.

1. Lower the furnace temperature to around 815°C (1500°F).
2. Stop the flow of all enrichment gases (i.e., hydrocarbon gas, air, ammonia).
3. Record the millivolt output from the oxygen probe as the combustion system cycles on and off at setpoint. It is always a good idea to record the combustion-system output percentages as well. Note: The millivolt signal from the oxygen probe should be essentially a steady state (i.e., not oscillating wildly).
4. Raise the furnace temperature to 925°C (1700°F) and record the millivolt signal as the furnace heats to its new setpoint with the combustion system on high fire. Note: Under normal circumstances, as you raise temperature, the millivolt signal on the oxygen probe will steadily increase. If you have a leaky tube, you either won't see an increase or there will only be a minimal increase in millivolt values.
5. When you reach the new higher setpoint, and as the combustion system cycles on and off, record the carbon-potential millivolt values and the combustion-system output percentages.
6. Turn the temperature setpoint back to 815°C (1500°F) so that the combustion-system output is zero and record the probe millivolt values as the temperature drops. Note: The millivolt signal should steadily decrease. If it does not, or if there is only a minimal decrease, a leak will be present.
7. Repeat steps 1-6 again to confirm results. Note: If the furnace is equipped with a plunge-cool option (where air can be blown through the tubes), heat back to a setpoint of 925°C (1700°F) and activate the plunge-cool feature. If the millivolt values fall rapidly, there is a leaky tube.
8. If you suspect a leaky tube, you may be able to use a digital CO_2 analyzer to help confirm the leak. Watching the analyzer during firing cycles, there should be a trend of increasing CO_2 levels with increased burner firing.

Once a leak has been confirmed, the difficult part of isolating the bad tube begins. Systematically test each burner tube. This should be done while the furnace is running and a stable atmosphere is present. Use the following test procedure at each burner, one at a time.

1. Shut off the natural gas and air supply by closing the burner gas and air valves.
2. CAREFULLY loosen the exhaust elbow and insert a blanking plate to seal off the exhaust and retighten. If a recuperator is present, loosen the exhaust piping at its connection to the recuperator and remove it. Place a blank plate over the tube discharge leg and clamp it in place to seal the tube.
3. With the burner radiant tube shut off and sealed, watch the CO_2 analyzer and look for a cycling CO_2 value with burner firing cycles. If the cycling

disappears, that is the leaking tube. If, however, there is more than one leaky tube, the CO_2 content of the furnace will still be erratic. When this is the case, with one bad tube sealed and one or more still leaking, the CO_2 content will not rise as fast during a burner firing cycle. Analyzing the data will help locate the leaky tube.
4. Note: Testing tubes for leaks with pressurized air outside of the furnace is not recommended by a number of OEM manufacturers since the tubes are not designed as pressure vessels and will not normally withstand pressures greater than 35 kPa (5 psig).
5. A similar test with an oxygen meter is to shut off the gas to a burner and allow the tube to purge. When the oxygen level reaches approximately 18-20%, shut off the air. Watch the oxygen meter. If the level drops, this tube is suspect.

Recuperators can also be leak tested following the manufacturer's recommendations. This often involves pressure testing. If the pressure is lost immediately, the recuperator is leaking. A small leak is often permissible, but that recuperator should be marked for future testing or replacement.

Tests for Measuring Furnace Pressure

Use of a simple manometer (Fig. 13.6.1) is a fast, easy and inexpensive way to monitor furnace pressure. Digital displays, recorders and remote signal transmission simplify the task of accurate measurement and control. It is always valuable to monitor changes in pressure during normal operation (Fig. 13.6.2) and to determine the normal length of time a furnace is in a negative-pressure situation (e.g., after a load is discharged to the quench tank in an integral-quench furnace). Unusually long periods of negative pressure or changes in pressure are a good way to determine if furnace troubleshooting or maintenance is necessary.

FIGURE 13.6.1 | Pressure-monitoring system for a two-chamber integral-quench furnace with oil quench

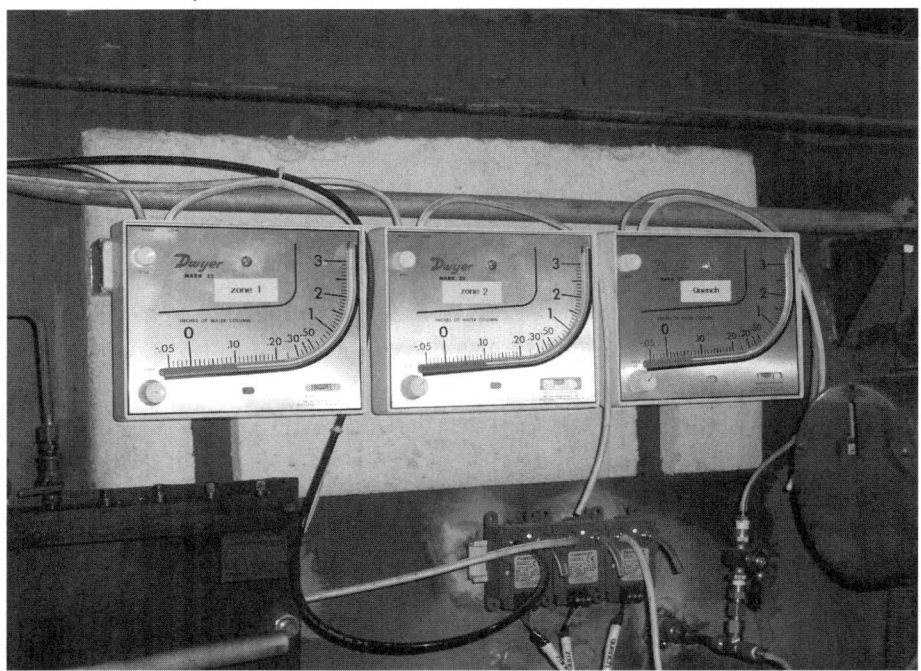

FIGURE 13.6.2 | Two-chamber integral-quench furnace pressure chart

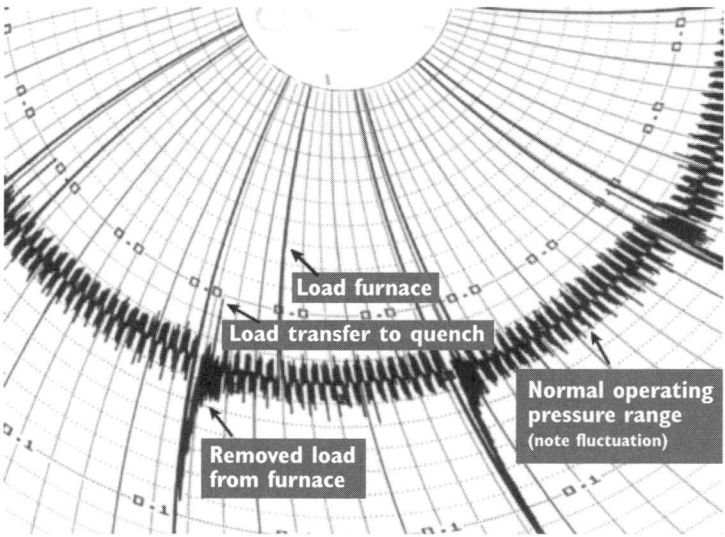

Tests for Furnace (Air) Leaks

We all know (or should know) that a furnace atmosphere must be present in sufficient volume *and* pressure to exclude air from entering the furnace atmosphere. Both parameters are important for safe operation, especially during

13 | MAINTENANCE REQUIREMENTS AND PRACTICES

door openings, load transfers and when the furnace is operating below 760°C (1400°F). In addition, if one is trying to precisely control the active furnace atmosphere to reach a specific carbon potential or dew point inside the furnace, changes in pressure can upset the delicate balance of the atmosphere.

Typically, leaks present below the furnace hearth will draw air in and those above the hearth will leak atmosphere out. Finding leaks above the hearth can often simply be done using a small torch to light off the combustible gas (observing all appropriate safety precautions!), but leaks below the hearth line are much more difficult to find. Techniques such as over-pressurizing the furnace (to force atmosphere out of these lower areas) or smoke-bomb testing are commonly used to locate them. Keep in mind a furnace leak that affects the atmosphere is typically considered a large leak, with a total area in the neighborhood of 480 mm^2 (0.75 inch^2) or greater.

TESTS FOR FURNACE TIGHTNESS (SMOKE-BOMB TEST)

A common atmosphere furnace (air) leak-checking procedure involves the smoke-bomb test. The following procedure will help determine if your atmosphere furnace has an air leak (radiant tubes, door seals, quench chutes, shaft seals, etc.). A symptom of a leaky furnace is if your process requires excessive amounts of enrichment (i.e., natural) gas added to the endothermic-gas (or nitrogen/methanol) atmosphere to maintain furnace dew point or carbon setpoint. Excessive internal sooting is another symptom of a leaky furnace.

- ❖ Step 1: Purchase smoke bombs (Fig. 13.6.3). This can be done through a number of industrial sources. Smoke bombs are typically rated by burn time, which is a function of the volume of gas they can produce (Table 13.6.1).

FIGURE 13.6.3 | Family of smoke bombs (courtesy of Superior Signal)

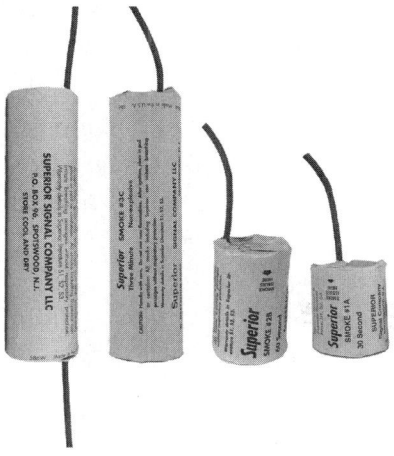

TABLE 13.6.1 | Smoke-bomb burning times and volumes (courtesy of McMaster-Carr)

Time duration	Gas volume, m³ (ft³)
30 seconds	110 (4,000)
1 minute	225 (8,000)
3 minutes	1,130 (40,000)

- ❖ Step 2: Prepare the furnace for testing. The furnace must be cold. When the test is under way, all openings (e.g., eductors, front end, access doors) must be positively sealed or closed off. Kaowool® can be used as a temporary seal material. If your furnace has a quench chute, do not seal it off. Turn off all fans and pumps. Hook up a compressed (dry) air line through a fitting on the furnace (be sure it is positively sealed) so that it has the ability to create a positive pressure inside the unit. Try to regulate the compressed pressure to around 200-415 kPa (30-60 psig), if possible. In extremely large furnaces, two air lines should be hooked up (one near the front, one near the back). Note: Be sure to figure out a way to locate the smoke bomb(s) within the furnace before igniting them.
- ❖ Step 3: Ignite the smoke bombs. You will see an instant billowing of gray-white smoke. The smoke media is zinc-chloride mist. Again, be sure the furnace is positively sealed. Turn the compressed air on after about two minutes. Within a matter of a minute or two, smoke will begin to exit the furnace through any furnace opening (leak).

For checking radiant tubes, shine a flashlight on the exhaust leg. If the tube is leaking, you will see small traces of smoke in the light beam.

Tests for Furnace (Air and Water) Leaks

In addition to monitoring the pressure inside the furnace (Section 13.2), we often need to know if the process problems we are experiencing are related to a water or air leak.

In the heating chamber of a continuous mesh-belt conveyor furnace (Fig. 13.6.4), for example, we typically monitor the dew point in the heating chamber(s) and monitor the parts per million of oxygen present in the cooling chambers. We often need an oxygen level in the cooling chambers of 20-50 ppm (maximum) for work to exit the furnace "bright."

FIGURE 13.6.4 | Mesh-belt conveyor sintering furnace with hypercooler for sinter hardening (courtesy of C.I. Hayes, a Gasbarre Furnace Group Company)

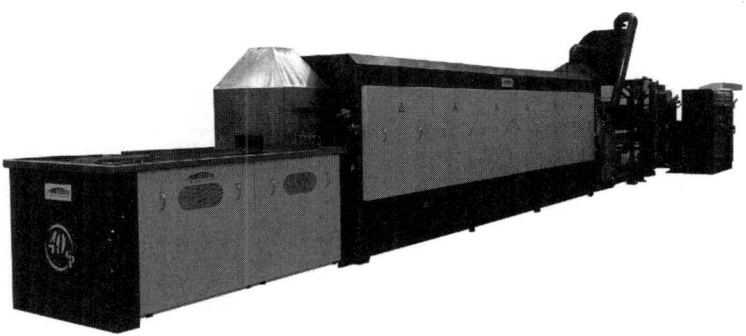

COPPER-STEEL-STAINLESS STEEL TEST

The question is often asked, "Do I have an air leak or a water leak in my furnace, and how can I tell them apart?" Fortunately, there is a really simple test procedure that yields very accurate results. Here's how it works.

Test Procedure

This test can be run in any style of furnace. For this example, we illustrate the procedure for a continuous mesh-belt furnace running a dissociated ammonia (75% hydrogen, 25% nitrogen) atmosphere. The test involves gathering clean, relatively thin cross-section samples made of steel, copper and (if available) stainless steel. Many people simply use steel shim stock and cut a small length of copper tubing. If necessary, sand the copper surface so that it is bright and shiny.

The furnace temperature is lowered to 980-1010°C (1800-1850°F), and the belt speed is reduced to ensure a dwell time in the high-heat and cooling chambers of at least 20 minutes. Note: Pure copper has a melting point of 1085°C (1984°F), so care must be taken not to approach this temperature.

Interpreting the Test Results

Results of the test (Table 13.6.2) are interpreted as follows. Steel parts will discolor (oxidize) in an atmosphere contaminated with either air or water. Copper will emerge shiny if water is present but will be discolored if air is present – often with black spots or streaks and sometimes pinkish in color if electrolytic tough-pitch copper is used for the test piece. Stainless steel will be gray if the air or water leak is a minor one and various shades of green (becoming darker and more vibrant) if the leak is more severe.

TABLE 13.6.2 | Results of copper-steel-stainless steel test[6]

Sample surface appearance			Conclusions
Copper	Carbon steel	Stainless steel	
Oxidized	Oxidized (grayish/black)	Oxidized (gray or green)	Air leak
Shiny/bright	Oxidized (grayish/black)	Oxidized (gray or green)	Water leak
Shiny/dull	Shiny/dull	Oxidized (light gray or light green)	Air or water leak (minor)

Upon exiting the furnace, the sample coupons should be visually examined for surface appearance (Figs. 13.6.5-13.6.8). It is best to do this in natural light.

FIGURE 13.6.5 | Copper-steel-stainless steel coupons

SEE FULL COLOR IMAGE ON PAGE K

Process parameters	Recorded data
Furnace temperature	1010°C (1850°F)
Atmosphere	Dissociated ammonia (saturated)
Dew point (high heat)	+2°C (+35°F)
Oxygen level (cooling chamber)	30 ppm
Belt speed	50 mm/min (2 inches/min)
Dwell time, approx. (high heat)	25 minutes
Dwell time, approx. (cooling chamber)	35 minutes
Results	
Copper	Bright, shiny
Steel	Grayish/black, dark
Stainless steel	Gray, dark
Conclusion	Water leak

FIGURE 13.6.6 | Copper-steel-stainless steel coupons

Process parameters	Recorded data
Furnace temperature	1010°C (1850°F)
Atmosphere	Dissociated ammonia
Dew point (high heat)	-33°C (-29°F)
Oxygen level (cooling chamber)	6 ppm
Belt speed	50 mm/min (2 inches/min)
Dwell time, approx. (high heat)	25 minutes
Dwell time, approx. (cooling chamber)	35 minutes
Results	
Copper	80% discoloration
Steel	Grayish/black, dark
Stainless steel	Gray, light
Conclusion	Air leak (minor)

FIGURE 13.6.7 | Copper-steel-stainless steel coupons

Process parameters	Recorded data
Furnace temperature	1010°C (1850°F)
Atmosphere	Dissociated ammonia
Dew point (high heat)	-43°C (-45°F)
Oxygen level (cooling chamber)	9 ppm
Belt speed	50 mm/min (2 inches/min)
Dwell time, approx. (high heat)	25 minutes
Dwell time, approx. (cooling chamber)	35 minutes
Results	
Copper	80% discoloration
Steel	Tan/golden brown, light
Stainless steel	Silvery, light
Conclusion	Air leak (minor)

FIGURE 13.6.8 | Copper-steel-stainless steel coupons

SEE FULL COLOR IMAGE ON PAGE K

Process parameters	Recorded data
Furnace temperature	1010°C (1850°F)
Atmosphere	Nitrogen (air-introduced)
Dew point (high heat)	+16°C (+62°F)
Oxygen level (cooling chamber)	70 ppm
Belt speed	100 mm/min (4 inches/min)
Dwell time, approx. (high heat)	12 minutes
Dwell time, approx. (cooling chamber)	18 minutes
Results	
Copper	100% discoloration (temper color) with black spots/patches in evidence
Steel	Gray, dark
Stainless steel	Gray, dark
Conclusion	Air leak (major)

Tests for Temperature and Temperature Uniformity

Of all the process variables in heat treatment, temperature is arguably the most influential. It is the application of temperature (heat or cold) that differentiates heat treatment from other manufacturing operations. Temperature causes our furnaces to degrade over time, and changes in temperature often inhibit their ability to accurately control the processes being performed in them.

One of the reasons for this is that a slight variation in temperature can result in significant (and unexpected) changes to the part microstructure and, hence, the resultant properties of the parts being treated. In other words, a small change in temperature produces a big change in the process. Another reason is that variations in temperature can interfere with our ability to repeat the process reliably and predict the outcome of the heat-treatment process selected.

Several types of tests can be conducted to determine the temperature inside a particular heat-treat furnace. These include the use of part/load thermocouples where practical (Fig. 13.6.9), ceramic shrinkage devices (Fig. 13.6.10), measurement tools such as in-furnace data-logging devices (Fig. 13.6.11) or performing/relying on temperature uniformity surveys (Section 11.4).

Each of these devices and techniques has advantages and limitations. Part/load thermocouples may require special considerations to extend the thermocouples out of the furnace chamber; ceramics require careful handling; in-situ measurement devices have time duration limits (Table 13.6.3); and performing temperature uniformity surveys requires considerable time. However, the benefit of knowing that the process temperature is uniform far outweighs any of these limitations.

FIGURE 13.6.9 | Part/load thermocouples in position on a load of brass wire for bright annealing (courtesy of Solar Atmospheres of Western PA)

FIGURE 13.6.10 | TempTab® ceramic shrinkage devices (courtesy of Orton Ceramic Foundation)

a. Fixtured on a load

b. Measurement tool

a.

b.
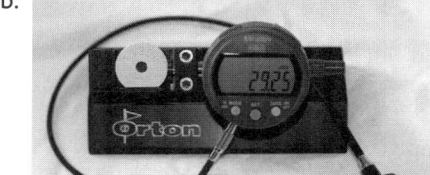

FIGURE 13.6.11 | Temperature-measurement package in position to survey a batch furnace (courtesy of Datapaq Ltd.)

TABLE 13.6.3 | Temperature duration for in-furnace data-logging devices

Temperature, °C (°F)	Time at temperature (Style I), hours[a]	Time at temperature (Style II), hours[b]
250 (480)	7.5	13.0
300 (570)	5.75	10.5
350 (660)	4.75	8.58
400 (750)	4.0	7.42
450 (840)	3.5	6.5
500 (930)	3.0	5.58
550 (1020)	2.83	5.0
600 (1110)	2.5	4.67
650 (1200)	2.25	4.24
705 (1300)	2.17	4.0
750 (1380)	2.0	3.75
800 (1475)	1.83	3.5
845 (1560)	1.75	3.25
900 (1650)	1.75	3.0
950 (1740)	1.5	3.0
1000 (1832)	1.5	2.58

Notes:

[a] Super Systems Model No. HB1012
[b] Super Systems Model No. HB1015

Finally, it is important to remember that each type of furnace is unique in regard to different methods of construction, heating/cooling capabilities and state of repair, so we need to monitor temperature to level the playing field. Over the years, most if not all heat treaters have heard of, become familiar with or performed system accuracy tests and temperature uniformity surveys on their atmosphere furnaces (Section 11.4). Meeting the procedural demands of the latest revisions of specifications, such as AMS 2750, and qualifying heat-treat shops to Nadcap and CQI-9 compliance is a task (although daunting at times) of critical importance.

Tests for Carbon Potential (Daily Carbon Checks)[7]

The carbon potential of a furnace atmosphere can be reliably and accurately measured by using either a carbon-potential analyzer (Fig. 13.6.12a) or a shim-stock test. These tests not only verify the accuracy of carbon (oxygen) probes, infrared (three-gas) analyzers and dew-point instruments, they also help determine if the carbon potential of the furnace atmosphere is varying over time. Both types of tests are direct-measurement techniques and do not rely on equilibrium equations or calculations in one form or another to determine the percent carbon in the furnace atmosphere.

FIGURE 13.6.12 | Carbon-potential analyzer setup (courtesy of Super Systems Inc.)

a. Test instrument

b. Test coil in end of insertion rod

c. Test coil being inserted in furnace

a.
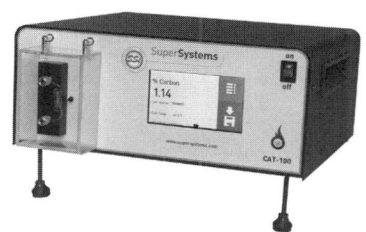

b.
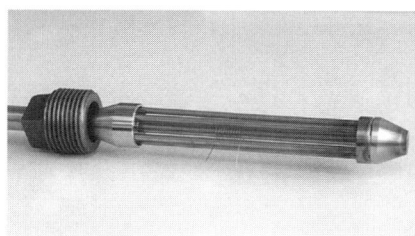

c.

CARBON ANALYZER METHOD

A test coil (Fig. 13.6.12b) is placed inside the furnace atmosphere using an insertion rod (Fig. 13.6.12c), where it is exposed to both the furnace temperature and atmosphere for approximately 30-40 minutes. Both furnace temperature and atmosphere should be held constant during the period the test coil is exposed. Metallurgical changes caused by carbon diffusion in the coil affect its electrical resistance, which can be measured after the coil is removed from the furnace and slow cooled (not quenched) to ambient temperature.

Using the baseline electrical resistance and carbon content of the untreated wire, the addition (or depletion) of carbon in the heat-treated wire can be measured and digitally displayed. The accuracy of the carbon potential is reportedly 0.03% or better in the 0.1-1.3% carbon range. Since nitrogen absorption influences the outcome, the instrument should not be used for carbonitriding or when ammonia additions are made to the furnace atmosphere.

SHIM-STOCK METHOD

The shim-stock test also provides a snap shot of the furnace-atmosphere characteristics at the time the shim is exposed. It can also be performed during the actual part-processing cycle. The success of this test is due to the rapid absorption of carbon (or carbon and nitrogen) into a thin shim-stock material, which in effect eliminates carbon diffusion as a variable in the process.

The shim-stock test is applicable to processes such as hardening, carburizing and carbonitriding.

Shim stock should be purchased from very-low-carbon annealed sheet, typically SAE 1005, 1008 or 1010 material in a thickness of 0.10 mm (0.004 inch), 0.25 mm (0.010 inch) or 0.38 mm (0.015 inch). Test coupons should be cut to size (Table 13.6.4), depending on the analysis method employed.

TABLE 13.6.4 | Shim-stock size recommendations

Analysis method	Test coupon size		Remarks
	Surface area, cm² (in²)	Strip size, mm (inches)	
Weight gain	25-35 (3.75-5.5)	30 x 75 (1.25 x 3) or 35 x 100 (1.375 x 4)	3 mm (⅛ inch) diameter punched holes in each end are sometimes used to wire shim to rod
Combustion	2.25 (15)	38 x 38 (1.5 x 1.5)	6 mm (¼ inch) diameter punched hole, 6 mm (¼ inch) from edge for handling

Before handling the shims, the individual responsible for the test should be sure that their hands are clean and dry. One of the most common sources of error is due to oils, grease or dirt from dirty hands or a dirty work area. The use of clean white gloves is recommended.

Note: If the combustion method is to be used, do not clean shims with solvent. If necessary, use grit blasting or other abrasive techniques to get a clean surface, removing all oxides and other residues. If the weight-gain method is to be used, the coupons can be thoroughly cleaned with solvent (e.g., alcohol or acetone), rinsed with water and wiped clean. Allow the shim to dry thoroughly.

Analysis Methods

Methods for analysis of shims in common use include the combustion method, weight-gain method and spectrographic analysis. Each method requires a slightly different procedure to ensure accuracy of results.

In the combustion method, three shims are typically hung from a rod and inserted into the furnace through the furnace wall or roof. The insertion depth (Table 13.6.5) must be adequate to allow the furnace atmosphere to circulate freely around them and to position the shim in a location that is representative of the load in the furnace. A typical position is 50-75 mm (2-3 inches) away from the side of the workload (batch equipment) or 50-75 mm (2-3 inches) above the work (continuous equipment).

TABLE 13.6.5 | Shim insertion depth

Furnace style	Minimum insertion depth, mm (inches)
Batch furnace	127-178 (4-7)
Continuous furnace	178-255 (7-10)

After allowing adequate dwell time (Table 13.6.6), the shims should be pulled from the furnace and quickly water quenched. They can also be slow cooled in a similar manner to the weight-gain method described later in this section. Appropriate cautions should be observed so that personnel are not exposed to burning (i.e., flammable) furnace atmosphere, the hot test piece or the hot insertion rod.

TABLE 13.6.6 | Shim dwell time[4]

Temperature, °C (°F)	Austenitizing time, minutes
845 (1550)	45-65
870 (1600)	30-50
900 (1650)	20-40
925 (1700)	15-30
955 (1750)	10-25

In the weight-gain method, shims are weighed and then loosely wrapped and secured around a rod before being inserted into the furnace through a specially constructed shim-puller arrangement (Fig. 13.6.13). The insertion depth must be adequate to allow the furnace atmosphere to circulate freely around them and to position the shim in a location that is representative of the load in the furnace.

FIGURE 13.6.13 | Typical shim-puller arrangement (courtesy of Super Systems Inc.)

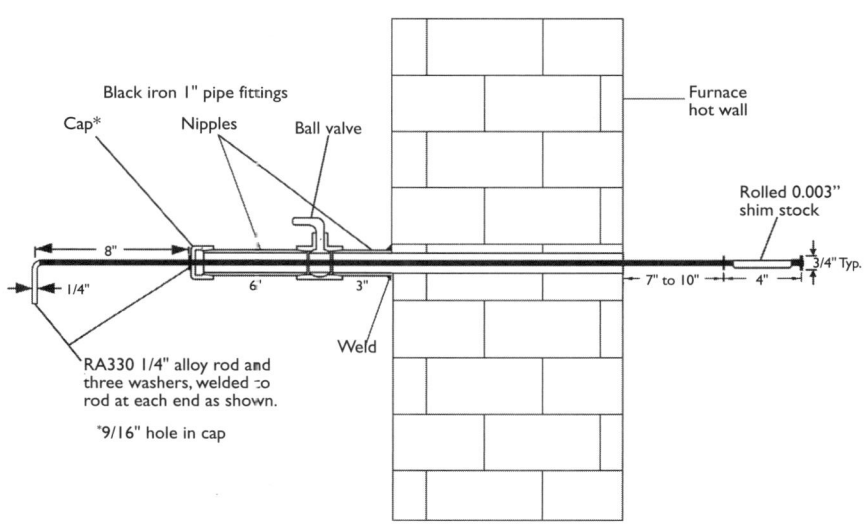

After allowing adequate dwell time, the shims should be pulled out of the furnace chamber and allowed to cool for an appropriate length of time (Table 13.6.7) in a black-body environment surrounded by furnace gas or nitrogen. This is typically done by pulling the shim to a position between the valve and end cap shown in Figure 13.6.13. Again, appropriate cautions should be observed so that personnel are not exposed to burning (i.e., flammable) furnace atmosphere, the hot test piece or the hot insertion rod.

TABLE 13.6.7 | Minimum shim cooling time (weight-gain method)[5]

Temperature, °C (°F)	Dwell time for 0.10 mm (0.004 inch) shim thickness, minutes	Dwell time for 0.25 mm (0.010 inch) shim thickness, minutes	Dwell time for 0.38 mm (0.015 inch) shim thickness, minutes
845 (1550)	89	353	450
900 (1650)	56	156	274
955 (1750)	37	101	175

The long-established combustion method for carbon (or carbon and nitrogen) determination is often the referee if different analysis methods yield different results. This method typically requires a sample weight of approximately 1 gram to be placed in the carbon determinator (Fig. 13.6.14). If the approximate carbon potential is known, calibrate the equipment against standards in this range.

FIGURE 13.6.14 | Carbon determinator for shim-stock analysis by the combustion method (courtesy of LECO Corp.)

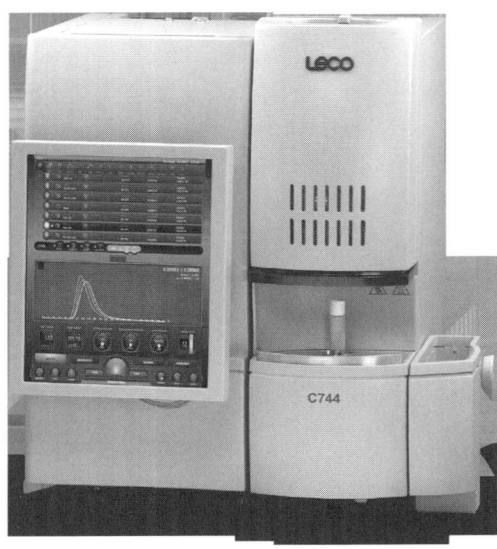

For the combustion method, analyze all three shims run and use the average of all readings in the interpretation of results. If one reading varies significantly from the other two (greater than 4-5 points), throw it out and average the other two. If two readings are significantly different, throw all readings out and require a new set of samples to be run.

For the weight-gain method, the shim is weighed before and after exposure to the furnace atmosphere. The gain in weight is determined by use of a highly accurate analytical balance and involves measuring the difference in the shim weight before and after processing. The carbon potential of the atmosphere is then calculated (Equation 13.6.1).

13.6.1) $$C_p = \frac{(W_g \times 100)}{W_f} + \%C_o$$

where:
   C_p is the carbon potential of the atmosphere
   W_g is the weight gain (the difference in before and after)
   W_f is the final weight of the test coupon
   C_o is the original carbon percentage of the test coupon

Interpretation of Results

Shim-stock analysis determines the maximum carbon potential of the furnace atmosphere up to the limit of saturation of carbon in austenite (Table 13.6.8). Generally, the carbon content of the shim as determined by these methods is higher than the surface carbon of the parts due to factors including mass, alloying elements, diffusion characteristics and the like. Exceptions are steels that contain high percentages of carbide-forming elements. These steels have higher surface carbon than the shim stock.

TABLE 13.6.8 | Limit of solubility of carbon in austenite

Temperature, °C (°F)	Theoretical saturation limit, %C
845 (1550)	1.05
870 (1600)	1.11
900 (1650)	1.20
925 (1700)	1.33
955 (1750)	1.40

The shim-stock test is applicable to carbonitriding since the weight gain produced by the adsorption of both carbon and nitrogen from the furnace atmosphere may be treated as though it were carbon alone; that is, the same calculations are used as those for a carburizing atmosphere.

Precautions

The following precautions should be noted.

1. The shim should always be exposed to a representative sample of the furnace atmosphere (away from areas such as atmosphere inlets or heating sources) and be placed in an area of unobstructed flow. If the shim is being used to correlate to another atmosphere-analysis device (such as an oxygen probe or infrared analyzer), the shim should ideally be placed in close proximity – within 150 mm (6 inches) – of the device or sample port.
2. The shim must be exposed to the furnace atmosphere for the proper length of time – not too long or too short but adequate to allow for saturation to the limit of carbon in austenite for the temperature selected.
    a. If the shim is exposed for too short a period of time, a carbon gradient will occur within the part, and the average value will be lower.
    b. If the shim is exposed for an extended period, massive carbides may develop, resulting in excessively high readings.
3. The shim should be placed in an area where the temperature of the shim and the temperature of the surrounding furnace are within acceptable limits, preferably ±5.5°C (±10°F).
4. The furnace being tested should be at temperature and the furnace atmosphere given enough time to stabilize.
5. If the shim is cooled in the shim-puller assembly, care must be taken to avoid decarburizing the shim and obtaining a false (low) reading. The shim should be cooled under nitrogen or by the furnace atmosphere. (Sooting must be avoided.)
6. For continuous furnaces, the carbon potential indicated by the shim represents only the zone or area tested and may be influenced by adjacent zones or areas.

A properly exposed and cooled shim should appear as follows.

1. Shims containing 1.30% carbon or less should be bright and clean (scale-free). A slightly oxidized or discolored surface indicates improper technique. The weight-gain method will produce a reading that is slightly high because of the increase in weight of the oxide.
2. Shims containing approximately 1.40% carbon or greater will have a dull, matte-gray finish. Depending on the actual carbon content, the shim surface may be coated with soot (loose, dry carbon). This is a normal appearance caused by an atmosphere of high carbon potential. These deposits should be blown or wiped off prior to weighing the shim.

Tests for Hardness and/or Case-Depth Uniformity

One of the least utilized but most highly effective tests that can be performed to determine the uniformity of the furnace and furnace atmosphere is using either component parts or test coupons strategically located in all areas of the furnace. Test coupons can be manufactured from the same material as the parts or be a specific material chosen as a standard (the latter being done in cases where a variety of materials are being processed, such as in a commercial heat-treat shop).

The samples are then located in the corners and center of the load (or in every basket in the load) and set up like the 5 pattern on a die. After processing, the parts are checked for hardness, case depth (if applicable) and microstructure. This simple test can tell you the maximum variation in hardness and/or case-depth uniformity throughout the load. This is an early indication that equipment or process problems may exist.

Tests for Excessive Vibration

Excessive fan or agitator vibration will destroy the functionality of the furnace or quench tank and damage key components, leading to major maintenance repairs. Consequences of excessive vibration (Fig. 13.6.15) can be profound.

Fans have been known to literally disintegrate in furnaces, separating blades from their shafts; destroying radiant tubes and electric heating elements; and embedding themselves in refractory walls. In quench tanks, they can lead to poor oil-flow velocity through the workload; change the oil-flow pattern; destroy the effectiveness of the quench; and (in open tanks) change the degree, intensity and duration of the resultant flame evolution on quenching.

Vibration meters (e.g., Fluke's Model 810, SKF's Machine Condition Advisor or IRD Mechanalysis Model 808 Troubleshooter) can be used to quantify the amount of vibration present at any point in the cycle. The old method was to place a coin on the fan or agitator housing. If it vibrated off, the vibration was too excessive.

FIGURE 13.6.15 | Consequence of excessive vibration (courtesy of SECO/WARWICK Corp.)

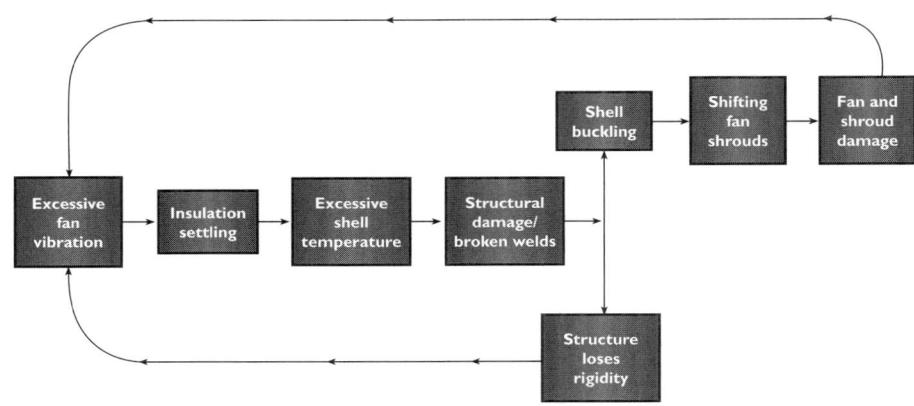

Summary

There is no excuse for poor heat treatment, particularly poor heat treatment due to failure to understand what maintenance is required on a particular furnace or oven. Testing to determine the root cause and subsequent corrective action ensures success. Resist the temptation to ignore a problem or use Band-Aid fixes, which all too often become permanent.

REFERENCES
1. Herring, Daniel H. and Gerald D. Lindell, "Periodic Atmosphere Furnace Maintenance Part Two: Lessons Learned," *Heat Treat Progress*, November/December 2009
2. Herring, Daniel H., *Atmosphere Heat Treatment, Volume I*, BNP Media, 2014
3. *ASM Metals Handbook, Volume 4: Heat Treating*, ASM International, 1991, pp. 589-590
4. Lotze, Thomas H., "Shim Stock Analysis," Technical Data Sheet T4408, Super Systems Inc., February 2001
5. "Ensure Atmosphere Carbon Potential," *Heat Treating*, April 1993
6. Copper/Steel/Stainless Steel Test Chart, Abbott Furnace Company, 2007
7. "Tool for Atmospheric Carbon Potential Analysis," *Advanced Materials & Processes*, June 2014
8. Herring, Daniel H., "Furnace Atmosphere Analysis by the Shim Stock Method," *Industrial Heating*, September 2004

13 | MAINTENANCE REQUIREMENTS AND PRACTICES

13.7
PROCESSING ISSUES AND THEIR CONTRIBUTION TO MAINTENANCE

The frequency and extent of required atmosphere furnace maintenance is significantly influenced by how we operate our equipment. Production demands, the choice of process, loading arrangement and selection of process variables (e.g., temperature, time, carbon potential) are examples of factors that affect maintenance operations. How well parts are cleaned or how braze alloy or stop-off paints are applied can also make a big difference.

In addition, equipment availability for service is a huge factor. It is important, therefore, that we understand what is required of the process versus what is desired of the processes we run. (This is equivalent to differentiating between needs versus wants when we go to purchase something.)

An uptime productivity in the heat-treat shop of 85% is considered a realistic target for most well-run operations (captive or commercial). This number has been found to vary from the upper 70s to the low 90s depending on the type of equipment in use (e.g., light-duty vs. heavy-duty), how hard it is pushed in service, the robustness of the maintenance program and operational demands.

Some organizations have an annual or biannual shutdown period for maintenance, while other plants schedule one day a week for maintenance activities. Whatever frequency is chosen, one must resist the temptation to rush or cut short the maintenance activities. Experience has shown that failure to allow adequate time to maintain equipment results in unplanned downtime, often at the most inopportune time.

A few examples focused around the presence of carbon in the furnace environment will suffice to illustrate these points.

Sooting

Soot formation inside atmosphere heat-treatment furnaces should be taken very seriously. Methods must be devised to either prevent its formation or limit its deleterious effects.

CARBON PENETRATION INTO REFRACTORY

The presence of soot raises many red flags, one of them being that carbon from an unstable furnace atmosphere is rapidly absorbed into refractory linings, which are incredibly porous. Infrequent or improper burnouts (so-called flash burnouts) give a false sense of security because they address only the refractory's near-surface layer. Once absorbed, carbon will continue to diffuse deeper and deeper into the refractory (Fig. 13.7.1). The penetration halts in the temperature range of approximately 480-705°C (900-1300°F).

As more and more carbon is absorbed, the refractory loses its thermal properties and becomes conductive. The result can be damage or melting of alloy components that extend into or through these conductive layers. Heating elements, terminal ends and support hangers in electric furnaces are particularly vulnerable as are atmosphere inlets, sample tubes and even oxygen probes (Fig. 13.7.2). Gas-fired furnaces are not immune because radiant tubes are also susceptible to carbon attack and metal dusting (Section 4.6).

Many heat treaters believe box and integral-quench furnaces are only vulnerable to the effects of soot if they are performing deep-case carburizing cycles. This is not true. Furnaces running shallow case depths and neutral atmospheres for medium- or high-carbon steels (e.g., 0.60-0.80% C) are also at risk.

FIGURE 13.7.1 | Carbon-rich subsurface layer inside a refractory wall

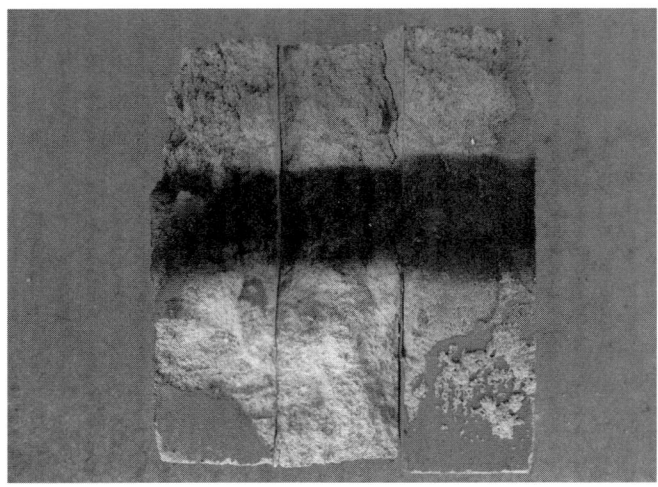

FIGURE 13.7.2 | Oxygen probe and its alloy protection tube destroyed by melting at the exact location of the carbon-rich subsurface layer

The use of low-iron brick (<0.5% Fe) helps to retard carbon absorption. In all cases, a robust and frequent burnout procedure, in which air reacts with the soot (carbon) to form carbon dioxide and carbon monoxide gas, must be established to keep or return the furnace to good working order. Samples taken from the refractory walls are the best way to evaluate the effectiveness of any burnout operation.

There are several ways in which this furnace burnout can be accomplished. Each has advantages and limitations, and it is important to recognize that air burnouts may cause extensive damage to equipment if done improperly. Extreme care must be exercised to avoid this situation. For integral-quench or pusher-style carburizing furnaces running endothermic gas, the following methods have been found to be effective.

- Method 1: Raise the dew point of the furnace atmosphere to around +21°C (+70°F) with air additions and maintain this condition over a long period of time (typically 24-72 hours).
- Method 2: Remove the furnace atmosphere and add a fixed volume of air, entering the furnace through a flowmeter, for a prolonged period of time (typically 12-36 hours). This is often combined with opening and closing of the furnace door(s).
- Method 3: Remove the furnace atmosphere and use an air lance or wand (under highly reduced pressure) directed at locations in the furnace interior with heavy soot deposits. This is normally followed by the introduction of air through a flowmeter for a period of time (typically 4-12 hours). Caution: This is the riskiest of the air burnout methods because of the potential to significantly damage the furnace interior, including an alloy fan if present. Only highly experienced personnel should be allowed to attempt this procedure.

Again, care must be taken to prevent overheating of the furnace during an air burnout. For this reason the furnace temperature is usually lowered to around 845-870°C (1550-1600°F), and the process is stopped if the furnace temperature rises by 38°C (100°F) or more. Most people are not aware that carbon (soot) burns at over 2480°C (4500°F) – high enough to melt through any of the materials in the furnace!

Furnace burnouts must be done before carbon has had a chance to build up in the refractory. They must also be frequent enough to ensure carbon diffusion is not occurring. One way to measure the effectiveness of the current burnout procedure is to extract a small section of insulation and inspect it on a frequent basis. If at any point it is determined that the insulation is absorbing an unacceptable amount of carbon, either increase the time of the burnout or the number and/or duration of burnouts or use some combination of the methods explained previously.

Soot not only affects alloy life, it deposits on the work and is then carried into the quench tank, where it negatively influences the performance and life of the quenchant.

Finally, it is a good idea to monitor the carbon potential of the furnace atmosphere during operation (via shim stock, turn bars or other methods). Atmosphere systems must be properly calibrated, and control devices such as oxygen probes must be complemented by three-gas analyzers and/or dew-point meters. Also, an automatic burnout of the oxygen probe during processing is mandatory to make sure the probe does not soot up and give false readings. In addition, the reference air system should be checked to ensure it is operating properly. Oxygen probes should be changed based on manufacturer's recommendations.

Sooting in Various Furnace Types

As has been previously stated, maintenance practices are often dependent on the type of furnace in the heat-treat shop. A few examples will serve to illustrate the different maintenance activities required. What is important for the reader to understand is that whenever a maintenance task becomes repetitive, one should ask themselves if there is a better solution to the problem than perpetual maintenance. Failure to do so often results in considerable time and money spent for repairs.

BATCH FURNACES

Integral-quench (IQ) furnaces are some of the most common and complex furnaces in the heat-treatment shop. They can suffer from myriad problems, from carbon buildup in the heat chamber to load-transfer problems and quench-oil-related issues (e.g., oil drag-out, oil contamination or deterioration). Units that use electric heating elements, for example, are particularly vulnerable should carbon (soot) impregnate the brickwork and create a current path leading to shorting or melting of the elements. Similarly, gas-fired radiant-tube furnaces are susceptible in the area where the tube is embedded in the insulation (Fig. 13.7.3).

FIGURE 13.7.3 | Radiant tube damaged by soot buildup in the refractory

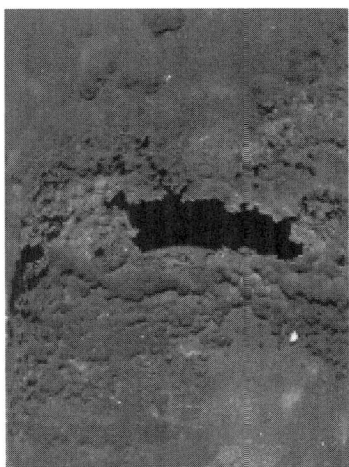

To keep furnaces as soot-free as possible, first make sure that the operators are running the right gas flows (i.e., not too rich) and that the enriching-gas additions being used are limited over the course of the total cycle to an acceptable percentage (usually 10-15%) of the total gas flow. It is noteworthy that most automatic systems add enriching gas in short increments of high flow. These should be timed to determine exactly how much enrichment is actually entering the furnace. In many cases, the peak flow should be limited to prevent too much enriching gas from entering the chamber in too short a time.

Next, performing a routine furnace burnout is dependent on use. As frequently as once a week to no longer than once every other month is strongly recommended for any integral-quench furnace operating above 0.50% carbon and running with an endothermic or nitrogen/methanol atmosphere enriched with either natural gas or propane.

Soot not only will shorten alloy life, it will also deposit on the work and wind up in the quench tank, where it negatively influences the performance of the oil and contributes to staining and overall life issues.

CONTINUOUS FURNACES

Several examples of problems in continuous furnaces highlight areas of maintenance.

Mesh-Belt Conveyor Furnaces

Mesh-belt furnaces running endothermic-gas atmosphere are not immune to issues with sooting. Using any generated atmosphere requires a careful understanding of the conditions that may result in a loss of atmosphere control followed by heavy and uncontrolled sooting (Fig. 13.7.4). Fluctuating dew points, either at the generator or the furnace, are often a clear indication of an unstable atmosphere condition.

For example, a roll-supported mesh-belt carbonitriding furnace having four control zones – where zone 1 was operating at 920°C (1690°F), zone 2 at 915°C (1675°F), zone 3 at 900°C (1650°F) and zone 4 at 870°C (1600°F) – was forced down for maintenance every 2-4 weeks due to binding of the rolls from soot buildup.

The carbon deposits built up in the area where the rolls pass through the furnace insulation. This was due in part to an unstable atmosphere and in part to conditions in which a carbon reversal reaction took place (Section 9.3).

The solution involved the following steps:

- ❖ The true carbon potential inside the furnace was determined by running low-carbon "hockey pucks" through the furnace together with the parts. These 32-mm-diameter (1.25-inch-diameter) by 6.4-mm-thick (0.25-inch-thick) discs were subsequently analyzed by spectroscopy for surface carbon content. The results were compared with values obtained by readings from oxygen probes, infrared (three-gas analysis) and dew point. It was determined that the atmosphere-sensing equipment was not providing accurate data to the control system. Subsequent investigation revealed the sampling system was not at the proper insertion depth into the furnace. Further, the oxygen-probe control factors (tuning constants) were based on incorrect information.
- ❖ The incoming part condition was reviewed. Parts were extremely wet, oily and had a phosphate coating on their surface in some cases. Although this did not affect the overall metallurgy of the final result, these contaminants made control of the furnace atmosphere extremely erratic.
- ❖ The quench-chute eductor flow was measured and adjusted to ensure that a positive pressure was maintained in the furnace.
- ❖ The atmosphere flow pattern in the furnace was changed by modifying the endothermic-gas flow rates, limiting the amount of enriching gas (natural gas and ammonia) allowed to enter individual zones and making sure that valves were set and functioning properly.
- ❖ The fan settings (speed and rotation) were returned to manufacturer's settings.
- ❖ Maintenance procedures with respect to the rollers were thoroughly reviewed to ensure proper techniques were being used and that the manufacturer's required expansion clearances were being observed.

These actions allowed the furnace to operate for a minimum of six months without roll maintenance.

FIGURE 13.7.4 | Soot accumulation to a depth of 25 mm (1 inch) depositing on an external furnace load table in less than one hour due to a furnace atmosphere that is out of control.

Pusher Furnaces

Pusher furnaces are by no means immune to maintenance. By their very nature, pushers are designed for high-volume production in a 24/7 operating mode. As such, anything that can cause damage inside the heating chamber threatens production and must be designed so that maintenance can be performed on routine shutdowns, typically every six months to one year. This puts added pressure on the maintenance department to have adequate spares on hand (despite the fact they don't know for sure exactly how extensive the repair process might be) and to accomplish the work in a expeditious manner. Typical pusher problems that require maintenance include:

- ❖ Alloy deterioration, especially if high carbon-potential atmospheres or high operating temperatures are used.
  - • Metal dusting in areas around fan shafts, sample ports and thermocouple wells, as well as wear of chain guides and alloy hearth components (especially in older units with alloy hearth designs)
- ❖ Internal tray transfer mechanisms that normally sit at right angles in the furnace as well as tray flippers and limit sensing rods.
- ❖ Load jam-ups (Fig. 13.7.5) due to pusher/puller chain problems, misaligned or misadjusted side transfer or linear pushers, or index rods resulting in placement of trays in the wrong positions.

- ❖ Operator error (e.g., trying to load too many trays in the furnace).
- ❖ Hearth (silicon carbide or other refractory materials) wear over time.
- ❖ If trays/baskets become deformed or crack over time, they have a tendency to get stuck in the furnace, often in the worst possible locations, causing peripheral damage as they are cleared.

FIGURE 13.7.5 | Jack arch repair (courtesy of Surface Combustion Inc.)

All other concerns for pusher furnaces are the same as for batch furnaces (atmosphere seals around fans, radiant tubes, tube failures, etc.). For example, one of the most serious problems is making sure that radiant-tube leaks are detected early and either replaced or repaired. If this is not done, compositional variations in the furnace atmosphere will occur, which affect case uniformity. If enough tubes are shut down, temperature uniformity may be compromised.

Rotary-Hearth Furnaces

The frequent door openings on rotary-hearth furnaces create both thermal instability and atmosphere problems. For example, the furnace is prone to heavy sooting in a rotary running a nitrogen/methanol atmosphere enriched with natural gas or propane to maintain 0.80% surface carbon. Atmosphere control is improved by weekly air burnout cycles. A typical burnout procedure entails:

- ❖ The furnace is purged with nitrogen and the temperature lowered to 845°C (1550°F).
- ❖ The burnout starts with a low airflow rate (typically 0.28-0.70 m^3/hour, 10-25 scfh) and then gradually increased in flow, first to 1.4 m^3/hour (50 scfh) and then to 2.8 m^3/hour (100 scfh) if no temperature spikes or other adverse effects from soot burn-off are observed.
- ❖ The length of time required can vary from four hours to all weekend, and the volume of air used depends on the amount of soot buildup.

Poor Quench-Tank Circulation

In the heat-treat industry, we often talk about having 1 gallon of oil per pound of steel being treated. This oil capacity, however, is not always an assurance of success. For example, parts run in continuous furnaces that discharge parts into quench chutes may see problems with low hardness or staining due to breakdown (fractionation) of the oil in a small localized area, lack of proper heat extraction or poor oil circulation in the quench chute (Fig. 13.7.6). Cooling systems should be sized to handle the heat extraction and should be free of copper and other materials known to be catalysts for oxidation of oil products.

FIGURE 13.7.6 | Oil breakdown on quenching – fractionation (cracking) due to poor bath agitation, creating solid pieces of carbon

Many endothermic generators are operated in a dew-point range of +4.5°C to +7°C (+40°F to +45°F). Lowering the endothermic-gas generator dew point aids in reducing the amount of hydrocarbon enrichment-gas additions required at the furnace but increases the frequency of maintenance on the generator itself.

Effects of Metal Dusting (aka Catastrophic Carburization or Carbon Rot)

War is being waged between the life of internal components and the destructive forces of temperature and atmosphere within a heat-treatment furnace. The weapons of the enemy are numerous and include hot gaseous corrosion in the form of carburization, oxidation, sulfidation and nitridation. At stake for the heat treater is the cost of lost production and downtime for maintenance.

To withstand these attacks, heat-resistant iron-nickel-chromium alloys are chosen for components such as radiant tubes, fans, heating elements, roller rails and rollers, chain guides and atmosphere inlet tubes (Section 4.6).

Although we tend to focus on resisting the effects of high temperature, furnace parts in the intermediate temperature range of 450-800°C (840-1475°F) are also suspect. At these temperatures, materials exposed to carbonaceous atmospheres are particularly susceptible to a phenomenon called metal dusting (aka catastrophic carburization), which causes premature failures by rapidly wearing away the surface of the alloy until mechanical failure occurs.

We seldom think about this intermediate temperature range because our alloys are very strong at low temperatures and most heat-treating processes take place at elevated temperatures. However, exposure to intermediate temperatures can occur in transition areas, such as where alloy parts penetrate insulating refractories or where there are temperature differentials (e.g., the underside of inner doors or where chain guides exit the heating chamber).

Since these heat-resistant alloy parts are often the most expensive furnace components, heat treaters should have an understanding of how they can be attacked by metal dusting and what can be done to extend their life by minimizing or preventing it.

WHAT IS METAL DUSTING?

Metal dusting is an erosion process in which a metallic component disintegrates into a dust of fine metal and metal-oxide particles mixed with carbon.

Generally, metal dusting occurs in a localized area. How rapidly the disintegration progresses is a function of temperature, the composition of the atmosphere and its carbon potential, and the material. Other significant factors include the geometry of the system, reaction kinetics, diffusivities of alloy components, specific-volume ratio of new and old phases, and ultimate plastic strain.

Metal dusting usually manifests itself as pits or grooves on the surface or as an overall surface attack in which metal thickness can be literally reduced to that of foil. For example, a cast HT-alloy plate was mounted underneath a refractory-lined inner door on an integral-quench furnace. The atmosphere passed underneath the door and into the quench vestibule. Plate thickness was reduced from 12.5 mm (0.5 inch) to less than 0.75 mm (0.03 inch) in a little over two months.

HOW DOES METAL DUSTING OCCUR?

In general, metal dusting of ferrous alloys proceeds via the formation and subsequent disintegration of metastable carbide. The first step in the process is adsorption of the gaseous phase on the surface of the metal; the more reactive this phase, the easier it decomposes or is catalytically decomposed (in the case of iron) on the surface. This step is followed by diffusion of carbon atoms from the surface into the bulk metal.

As a result, there is a continuous buildup of carbon within the surface layer.

As this layer becomes saturated with carbon, a stable-carbide, metastable-carbide or an activated-carbide complex forms. This then grows until it reaches a state of thermodynamic instability, at which point it rapidly breaks down into the metal plus free carbon.

It is at this stage that the metal disintegrates to a powder as the result of plastic deformation and subsequent fracture in the near-surface layer. The process is controlled by internal stresses due to phase transformation. In other words, competition between stress generation and relaxation exceeds the ultimate strength in this near-surface layer and fracture occurs.

THE MECHANISM EXPLAINED

The first stage of metal dusting occurs when carbon monoxide decomposes to carbon dioxide and carbon on the surface of the alloy (Equation 13.7.1).

13.7.1) $2CO \rightarrow CO_2 + C$

This reaction tends to proceed at temperatures between 315 and 730°C (600 and 1350°F). In the second stage of metal dusting of ferrous metals, iron-carbide particles form and grow (Equation 13.7.2).

13.7.2) $3Fe + C \rightarrow Fe_3C$

These precipitates then deteriorate, followed by the rapid disintegration of the bulk metal. For example, the inside diameter of a cast HT-alloy radiant tube shows (eutectic) carbide particles having agglomerated and formed continuous carbide networks (Fig. 13.7.7).

Coarsening of secondary carbide has also occurred. It has been theorized that cementite in the iron-carbon system is metastable up to a particular concentration and precipitate size, after which it reverts to iron plus carbon.

FIGURE 13.7.7 | Surface and subsurface attack by metal dusting (courtesy of Aston Metallurgical Services Co. Inc.)

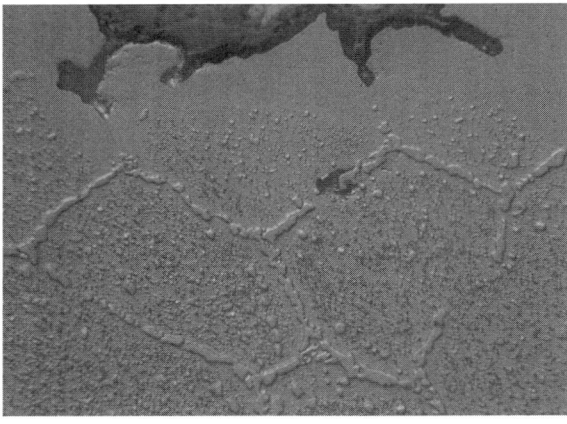

1000X   Kallings Reagent

HOW TO AVOID METAL DUSTING

What can be done to minimize metal dusting of alloy furnace components? First, determine which areas of the furnace experience intermediate temperatures and then establish an alloy inspection schedule. Next, ensure that areas where alloy components transition or pass through refractories are well-insulated. This will make it more difficult for the furnace atmosphere to penetrate into areas that see intermediate temperatures. (Failure to do this is a major mistake that is often made during repair or rebricking of an atmosphere furnace.)

Operating practices for heat-treatment furnaces should also be reviewed. Idling furnaces at reduced temperatures with continuous endothermic atmosphere flow over long weekends or other periods of inactivity should be reconsidered if metal dusting is the primary mode of alloy failure. Cycling furnace temperature is never a good idea, and it will accelerate the metal-dusting process.

Finally, take a close look at the alloys selected for specific applications. It is of little benefit to use less-expensive materials if they need to be replaced more often. Suppliers have a great deal of information about metal dusting and other failure mechanisms, and they can offer guidelines for alloy selection. For example, the addition of silicon can have a beneficial effect on resistance to metal dusting. A rule of thumb states that metal dusting will be reduced if the alloy can satisfy this criterion: $\%Cr + 2 \times \%Si > 24\%$.

Note that sulfur, either as an additive in the material or present in hydrogen-sulfide gas, can inhibit or even prevent metal dusting by suppressing the nucleation of graphite. Unfortunately, sulfur is also extremely harmful to nickel-bearing alloys due to the formation of a low-melting-point eutectic.

Another approach to avoid metal dusting is to use a coating, especially in transition areas. Note that surface oxides, which generally take the form of a chromium-rich duplex layer in the intermediate temperature range, should be effective, but they are reduced by the furnace atmosphere. Ceramic, or "frit," coatings or aluminum diffusion coatings are effective. Metal dusting can still occur at defects in the coating, however, resulting in pitting.

Remember, a proactive approach is absolutely necessary to help the alloy components inside heat-treatment furnaces resist attack by metal dusting or any other destructive force.

Summary

Carbon-related problems in atmosphere heat-treatment furnaces can be avoided. The secret to success is to catch the problems early. Furnaces seldom fail catastrophically; they tend instead to deteriorate over time. As such, we have ample warning of an impending problem if we know the warning signs. The key is to observe your furnace during normal operation and pay immediate attention when there is any change or shift from what you expect. Remember, problems do not get better or go away on their own. They only get worse if we ignore them or postpone taking corrective action. Maintenance is truly a necessity in every heat-treatment shop.

REFERENCES

1. Herring, Daniel H. and Gerald D. Lindell, "Periodic Atmosphere Furnace Maintenance Part Two: Lessons Learned," *Heat Treat Progress*, November/December 2009
2. Herring, Daniel H., "Soot," *Industrial Heating*, January 2012
3. Herring, Daniel H., *Atmosphere Heat Treatment, Volume I*, BNP Media, 2014
4. Herring, Daniel H., "What To Do About Metal Dusting," *Heat Treat Progress*, August 2003

14 | HEALTH, SAFETY, ENERGY AND THE ENVIRONMENT

14.1 HEALTH

Heat treatment is a dangerous occupation (Fig. 14.1.1).[1] As such, worker health and safety (Section 14.2) must be of paramount consideration. Heat treaters and visitors to heat-treat shops can be exposed to a plethora of dangers, from cuts by sharp objects to strains from heavy lifting as well as bumps, bruises and broken bones from falls or falling objects.

The workplace is often hot and humid or cold and breezy. Physical labor may produce dehydration or aggravate other medical conditions. Furnace atmospheres bring with them risk of explosion, burns, toxicological effects and asphyxiation. Exposure to various chemicals in solid, liquid and/or gaseous forms presents even more risks. All of these can have short- or long-term effects on our health. While we don't claim to be in a position to offer medical advice, some of the more insidious conditions are discussed here in an effort to heighten awareness.

FIGURE 14.1.1 | Endothermic gas igniting during load discharge from an atmosphere box-carburizing furnace (courtesy of MET-TEK Inc.)

While personal protective equipment (PPE) requirements may vary from plant to plant, heat treaters must have eye protection and safety shoes and wear long-sleeve shirts and pants when in the heat-treat shop. For specialized work, face shields, heat-resistant gloves and thermal body protection may also be required. Everyone should be comfortable knowing the appropriate way to use PPE and be sure that each device fits properly. A recent industrial study found that almost 60% of workers suffering eye injuries were not wearing eye protection at the time of the accident, and many of these injuries were caused by small flying objects or sparks.

In the U.S., manufacturing companies are required to follow the standards published by the Occupational Safety and Health Administration (OSHA). Workers should be familiar with these as well as all company health and safety rules and regulations.

Common-sense shop rules contribute to the health and safety of workers. These include, but are not limited to, the following:

1. Wear safety glasses with side shields, safety shoes, long pants/shirts and face masks or thermal clothing if appropriate. Insist your visitors do the same.
2. Don't eat, drink out of open containers or handle food in the heat-treat shop without washing your hands first.
3. Be sure the ventilation system is adequate and avoid breathing in any type of fumes.
4. Follow lockout/tagout procedures when working on equipment.
5. Understand and obey confined-space regulations.
6. Obey all company safety rules.
7. Understand who and how to call in case of an emergency.
8. Understand your role in the company emergency plan.

Dangers of Furnace Atmospheres

Operators, supervisors, maintenance personnel and visitors to a heat-treat department or shop must be trained in the proper use of (and risks associated with) furnace atmospheres and gases to which they may be exposed. As part of this training, emergency procedures, medical contacts and information about the risks must be clearly communicated. In addition, warning signs should be posted in English, Spanish and other languages as appropriate.

The effects of exposure to the more common heat-treatment atmospheres (Table 14.1.1) should be known to all as well as any specialty gases or conditions with which someone would come in contact. Appropriate sensors and warning signals – audible and visual – should be provided, such as industrial CO (Fig. 14.1.2) and hydrogen detectors in the shop and in shop or plant offices. Proper handling and securing of pressurized gas cylinders is another focus area.

14 | HEALTH, SAFETY, ENERGY AND THE ENVIRONMENT

FIGURE 14.1.2 | CO detector in a manufacturing shop located outside the heat-treat department

a. CO detector and emergency instructions

b. Close-up of CO emergency instructions

a.
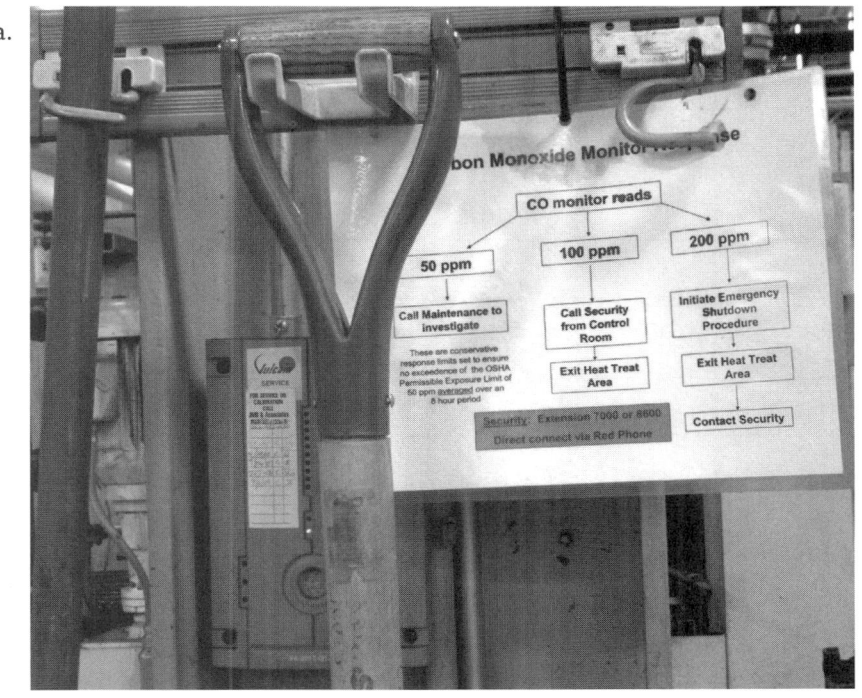

b.
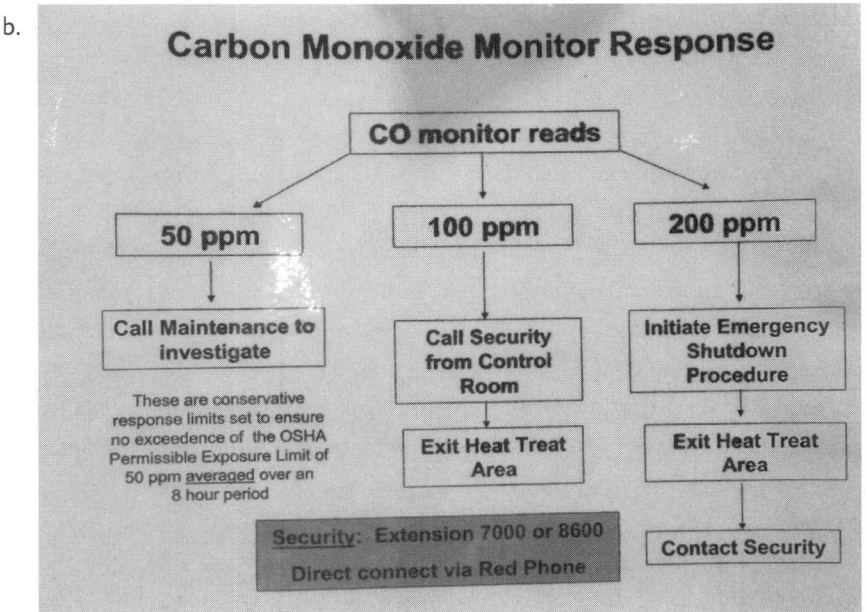

TABLE 14.1.1 | Common heat-treatment atmosphere gas constituents[1]

Gas species	Symbol	Flammable?	Toxic?	Asphyxiant?	Atmosphere function[a]
Ammonia	NH_3	Yes	Yes	Yes	Source of nitrogen (nitriding, carbonitriding)
Argon	Ar			Yes	Inert gas, purging or quenching
Carbon dioxide	CO_2		Yes[b]	Yes	Oxidizing or decarburizing
Carbon monoxide	CO	Yes	Yes	Yes	Carburizing, mildly reducing
Helium	He			Yes	Inert gas, quenching
Hydrogen	H_2	Yes		Yes	Strongly reducing
Methanol (methyl alcohol)	CH_3OH	Yes	Yes	Yes[c]	Source of CO or H_2
Natural gas[d]	CH_4	Yes		Yes	Source of carbon; strongly carburizing and deoxidizing
Nitrogen	N_2			Yes	Purging or blanketing gas[e]

Notes:
[a] Most common application(s)
[b] At high concentrations
[c] Liquid at room temperature, vapor is an asphyxiant.
[d] Purity levels of commercial methane are not guaranteed.
[e] Not inert in all instances.

Physiological Effects of Carbon Monoxide

When most of us think about health and safety in the heat-treat shop, we tend to focus on things that are hot, heavy or dangling overhead. The threats we can see, smell, hear or perhaps even taste are most often dealt with quickly. However, the old adage, "what you can't see won't hurt you" does not apply in our industry. Carbon monoxide is one example.

Carbon monoxide (CO) is an odorless, colorless, toxic gas present in many heat-treat shops from furnace atmospheres and products of (incomplete) combustion. We have a fairly sophisticated understanding of the mechanism by which CO poisoning occurs.[2]

A simple way to think about how CO interacts with the human body is to remember "fast in, slow out." Normally, oxygen is transported from the lungs to red blood cells. This process occurs when oxygen atoms bond to an iron atom at the center of a complex protein molecule known as oxyhemoglobin, which is a fairly unstable molecule that decomposes in the intercellular spaces to release free oxygen and hemoglobin. The oxygen is then available to carry out metabolic reactions in cells, reactions from which the body obtains energy.

If CO is present in the lungs, this sequence is disrupted. CO bonds with iron in hemoglobin to form carbonmonoxyhemoglobin, a complex protein molecule somewhat similar to oxyhemoglobin. Carbonmonoxyhemoglobin, however, is a more stable compound than oxyhemoglobin. It has much less tendency to break down when it reaches cells, and it continues to circulate in the bloodstream in its bound form. As a result, cells are unable to obtain the oxygen they need for metabolism, and energy production dramatically decreases. The clinical symptoms of CO poisoning described as follows are manifestations of these changes.

The symptoms of CO gas poisoning, defined as levels of 10% or higher absorbed gas in the bloodstream, include (in descending order of danger):

- Headaches
- Dizziness
- Irritability
- Confusion/memory loss
- Disorientation
- Muscle aches/poor reaction time
- Balance/coordination problems
- Nausea and/or vomiting
- Difficulty breathing
- Chest pain
- Swelling of the brain (cerebral edema)
- Convulsions/seizures
- Coma
- Death

CO (Table 14.1.2) has powerful effects even at very low-dosage exposures and often begins with flu-like symptoms, such as being tired, having achy muscles, a headache that just won't go away, eye strain and even a runny nose.

TABLE 14.1.2 | Physiological effects of carbon monoxide[4]

Concentration, ppm	Physiological effect
100	Allowable limit for an exposure of several hours.
400	One hour of exposure at this level will be without appreciable effect.
600	Low-dosage symptoms begin after one hour of exposure.
1,000	Severe side effects, but often no permanent damage after exposure of one hour.
1,500	Dangerous health consequences for exposures of one hour or longer.
4,000	Fatal for exposure of less than one hour.

When CO enters the body, it is absorbed into the bloodstream. This prevents oxygen absorption and the transfer of oxygen to vital organs such as the heart, central nervous system and brain. The heart responds by beating more rapidly and irregularly and by decreasing blood pressure. In extreme exposure conditions, a life-threatening neurological condition results due to the destruction of brain cells.

Monitoring the heat-treat environment coupled with proper fresh-air ventilation is the key to combating this problem. One of the most important considerations in the treatment for CO exposure is the immediate recognition of the problem. That's why CO detectors should be installed in every heat-treat shop. Does yours have one?

Fresh air is very important to recover from CO exposure. Once detected or suspected, the following actions can be taken.

1. Move the victim(s) to fresh air.
2. If the victim(s) are experiencing any gas-poisoning symptoms, call for trained paramedics and activate your company's emergency plan.
3. Warn others and ventilate the affected area.
4. Monitor the victim(s) for respiratory problems.
5. Try to ascertain the source and shut down suspect equipment until trained professionals can assess what is wrong.

Physiological Effects of Ammonia

Even in small concentrations, ammonia causes severe side effects. It can cause death in larger amounts (Table 14.1.3). Ammonia in contact with water will form a strong base (i.e., alkaline chemical) called ammonium hydroxide. This is particularly noteworthy because areas of the human body such as the nose, throat, ears, eyes and lungs are moist, and gaseous ammonia can react instantly. This will cause burning or permanent damage. Care must be taken when moving, hooking up and storing ammonia cylinders either outside or inside (Fig. 14.1.3) the building.

14 | HEALTH, SAFETY, ENERGY AND THE ENVIRONMENT

TABLE 14.1.3 | Physiological effects of ammonia[4]

Concentration, ppm	Physiological effect
20	First perceptible odor (some individuals)
40	Slight eye irritation (some individuals)
53	Perceptible odor (almost all individuals)
100	Noticeable irritation (eyes and nasal passages) after a few minutes of exposure.
400	Severe irritation of the throat, nasal passages and upper respiratory tract.
700	Severe eye irritation; no permanent effect if exposure is limited (less than 30 minutes).
1,700	Serious coughing and bronchial spasms; less than 30 minutes of exposure may be fatal.
5,000	Serious edema, strangulation, asphyxia; almost immediately fatal.

FIGURE 14.1.3 | Typical in-plant ammonia tank hookup (courtesy of AmeriKen)

Dangers of Oxygen Deprivation

When dealing with any gas (including nitrogen) one must be aware of its inherent dangers and understand that the risk of asphyxiation is real. Know that oxygen may be displaced to a level of around 16% or less before symptoms begin to appear (Table 14.1.4).

TABLE 14.1.4 | Physiological effects of oxygen deficiency[5]

Concentration, %	Physiological effect
19.5	Minimum concentration for exposure by humans (OSHA).
19	Some adverse physiological effects occur, but they may not be noticeable.
15-19	Impaired thinking and attention. Increased pulse and breathing rate. Reduced coordination. Decreased ability to work strenuously. Reduced physical and intellectual performance without awareness. Can induce early symptoms in persons with coronary, pulmonary or circulation problems.
12-15	Poor judgment. Faulty coordination. Abnormal fatigue upon exertion. Emotionally upset. Increased pulse rate and respiration.
10-12	Very poor judgment and coordination. Impaired respiration that may cause permanent heart damage. Possibility of fainting within a few minutes without warning. Nausea and vomiting. Further increase in pulse and respiration; giddiness; blue lips.
< 10	Inability to move. Fainting almost immediately. Loss of consciousness. Convulsions. Death.

Summary

Health and safety are everyone's responsibility, and being aware of the dangers and taking steps to avoid incidents in the heat-treat shop (or anywhere in the plant) will make it a better place for all of us to work. Remember, nothing replaces common sense when it comes to avoiding injury. Learn and follow the company rules, and protect yourself at all times.

REFERENCES

1. Stratton, P.F. and M.S. Stanescu, "An Introduction to Atmosphere Furnace Safety," BOC Gases, Proceedings of the 22nd Heat Treat Conference and 2nd International Surface Engineering Congress, September 2003
2. Carbon Monoxide – Physiological Effects, Net Industries (http:/science.jrank.org/pages/1212/Carbon-Monoxide-Physiological-effects.html)
3. Herring, Daniel H., "Health and Safety in the Heat Treat Shop," *Industrial Heating*, April 2009
4. *ASM Handbook, Vol. 4B, Steel Heat Treating Technologies*, ASM International, 2014, p. 115
5. Air Products & Chemicals Inc., Safteygram 17, "Dangers of Oxygen-Deficient Atmospheres," 2014

14.2
SAFETY

Thermal processing is a dangerous occupation (Fig. 14.2.1).[1] As such, worker health (Section 14.1) and safety must be of paramount consideration.

FIGURE 14.2.1 | Worker pouring molten metal in a small foundry[1]

Safety Hazards in the Heat-Treat Shop

Worker safety and the safe operation of heat-treat equipment is both MANDATORY and NON-NEGOTIABLE, especially for equipment using any type of furnace atmosphere (combustible or noncombustible). There is no substitute for understanding the dangers inherent in the heat-treat shop and for taking the necessary safeguards to prevent accidents from happening. Safety and safety issues are a serious matter and should be treated as such by all individuals within a company.

RULES AND REGULATIONS

One real problem with most rules and regulations is that few people take the time to read them and/or are properly trained to fully understand them. Even though equipment suppliers must comply with industry standards and best practices, it is everyone's responsibility within a company to ensure that safe procedures are being followed before a piece of equipment goes into operation and throughout its lifetime. It's a must for the plant safety committee to have copies of the codes and standards and allow access to anyone who is interested in safe workplace practices.

Sources for commonly cited safety documents that apply to heat-treatment equipment come from the following organizations:

- ❖ OSHA (Occupational Safety and Health Administration) regulations
- ❖ NFPA international standards (e.g., NFPA 86, Standard for Ovens and Furnaces; NFPA 70: National Electrical Code)
- ❖ Insurance regulations (e.g., Factory Mutual/Industrial Risk Insurers)

PROACTIVE SAFETY PROGRAM

If your company does not have an official one, establish a sound, usable safety program within the heat-treat department and be sure it is integrated into the plant safety program. Some features to consider include:

- ❖ Identify one person as the safety coordinator and give that person the power and authority to inspect, identify and correct safety violations, as well as provide training and education to all employees. This individual should be empowered to stop a heat-treatment operation if deemed unsafe.
- ❖ Train all operators and maintenance personnel in the proper use of the atmosphere furnace systems and heat-treatment furnaces.
- ❖ Review and inspect all heat-treatment equipment to ensure compliance with all applicable safety standards.
- ❖ Provide names and contact information in case of emergency. This should include information about how to contact hospitals and physicians at all times when the plant is open.
- ❖ Provide adequate fire- and safety-protection equipment, and be sure it is operational and up to date.
- ❖ Insist on and inspect for use of personal protective equipment (PPE) when operating or maintaining the heat-treatment equipment. This includes safety glasses and shoes, gloves, hard hats, breathing apparatus and lockout/tagout devices. Also, discipline against violations regarding PPE.
- ❖ Maintain good records of any incident or accident, and investigate to determine a root cause. Take necessary corrective action as needed, including periodic follow-up checks.

WARNING SIGNS

One of the best safety measures to warn employees of potential hazards is to post adequate signs in key locations. These constant reminders reinforce the safety message on a daily basis. Although it is the employer's responsibility to interpret and explain all warning signs to employees who do not read or understand English, it is in everyone's best interest to train all of our coworkers in safety. Often misunderstood, sign classifications are reviewed here in order of importance.

- **Danger.** Indicates an immediate hazard that can produce serious injury or death. White letters on red background.
- **Warning.** Indicates a potentially hazardous situation that will cause severe injury or death if not avoided. Black letters on an orange background.
- **Caution.** Indicates a potentially hazardous situation that could cause severe injury but not necessarily death. Black letters on yellow background.
- **Emergency information.** Rules, health and general safety information. White letters on a green background.
- **General notice.** General information. White letters on a blue background.

RESPECTING ATMOSPHERES

C-A-U-T-I-O-N describes many furnace atmospheres in use today in heat-treatment shops or departments. They can be colorless, odorless and tasteless, and they can cause asphyxiation and/or death if inhaled. Adequate ventilation systems are mandatory in all heat-treat shops not only to exhaust these gases but also flue gases, vapors and other contaminants evolving from the furnace load or construction materials of the equipment itself. Permanent injury and/or death are a real concern even after very short exposure to small quantities of these materials. Long-term health problems may also occur.

Gases and liquefied compressed gases fall into the following categories:

- **Corrosive.** Products that react chemically and deteriorate materials (including skin) on contact.
- **Cryogenic.** Contact with these liquids causes "cryo burns" due to freezing of the skin. Rapid vaporization in confined spaces can cause asphyxiation.
- **Flammable.** When mixed with air, oxygen or another oxidant, these fluids burn or explode upon ignition.
- **High pressure.** A sudden release of pressure may cause serious damage to personnel or equipment.
- **Inert.** These displace oxygen in confined spaces and threaten life support.
- **Oxidant.** Gases that initiate and support combustion.
- **Toxic.** Substances that may chemically produce injurious or lethal effects.

With an asphyxiant, oxygen may be displaced up to a level of under 19% quickly and before symptoms appear (Section 14.1). Symptoms such as rapid respiration, diminished mental alertness and impaired muscular coordination indicate there may be a problem. Death occurs almost instantaneously when the oxygen content is below 10%. Examples of asphyxiants include:

- **Argon (Ar).** Heavier than air. Difficult to purge from pits and recessed areas.
- **Nitrogen (N_2).** Slightly lighter than air.
- **Helium (He).** Much lighter than air.
- **Hydrogen (H_2).** Much lighter than air. Accumulates in ceiling areas and can reach dangerous concentration levels. Roof sensors tied to automatic ventilation systems are recommended.
- **Methane (CH_4) or natural gas.** Very slightly lighter than air. Often confused with natural gas (typically 90 to 95% methane). Natural gas has an odorant added, but methane does not.
- **Propane (C_3H_8).** Heavier than air. Slight odor.

Explosiveness of the air-gas mixture is the primary hazard with some of these gases. Hydrogen, natural gas and propane should not be used near an open flame or heat source because they can self-ignite. They are very easily oxidized and react explosively in the presence of oxidizing substances.

Toxic gases that are especially dangerous include:

- Carbon monoxide (CO) is particularly noxious because it does not provide adequate warning of its presence, and its effects are cumulative. In addition, the bloodstream absorbs it, and it is life threatening. CO monitors should be provided in all heat-treat shops.
- Ammonia (NH_3) is a colorless gas with a strong, pungent odor and is lighter than air. It is usually transported liquefied by compression in cylinders. Ammonia is an extreme irritant to the skin, eyes, nose, throat and lungs. A moderately toxic inhalant, it can cause severe damage to the lungs when inhaled excessively. Not unexpectedly, ammonia is highly toxic when ingested orally. It is not in and of itself a fire hazard since it is inflammable under 650°C (1200°F). If ammonia leaks into an area exposed to heat, however, it can certainly be explosive. And remember that dissociated (cracked) ammonia has the properties of hydrogen and nitrogen listed previously.

GAS POISONING SYMPTOMS

Finally, it's important to be aware of the changes (either rapid or slow) in your physical condition. These include dizziness, headache, stiff neck, weakness and nausea. If any of these symptoms are detected, immediately institute company safety procedures. In many instances, if a person collapses, the safety policy may tell you not to attempt to rescue the person in question until after ascertaining that the area is completely safe. Most people find this disturbing, but the problem that overcame a fellow worker could also overcome you.

Company policy should be crystal clear on this issue. If you are not sure, ask questions and bring up numerous what-if scenarios in safety meetings. Always remember to think before acting. When possible, all personnel should be removed from the problem area, which should be sealed until proper ventilation is provided. In the event a person is overcome, call for medical attention immediately and implement plant safety procedures.

MAINTENANCE SAFETY

Prior to entering a piece of heat-treatment equipment to perform maintenance, some steps necessary to ensure safety include:

- Turn the equipment off.
- Shut off or disconnect all process-gas/purge lines.
- Put adequate physical restraints in place – block doors, chain cylinders, etc.
- Allow equipment to cool to room temperature.
- Shut off electrical power (lockout/tagout procedures should be used).
- Secure the area yourself; never assume that a coworker has done something.
- Purge the area with a sufficient quantity of air and continue to direct airflow into or at the area.
- Use an oxygen-sensing device to ensure a safe environment. WARNING: The refractory insulation used in many heat-treatment furnaces retains process gases that can be slowly released over time. An area that may be deemed safe initially may become hazardous or even life-threatening. Constant monitoring of the environment and providing a fresh air supply is critical. Awareness of the inherent danger is mandatory.
- Use the "buddy system," and never work alone. Be sure everyone in the department is aware of what is to be done, and check periodically that everything is OK.

Common Shop Hazards

Some of the most common hazards found in the heat-treat shop are highlighted. It is critical that the working environment minimizes risk from these hazards and that appropriate PPE be used at all times. All injuries should be reported immediately to your supervisor and the circumstances surrounding the injury fully investigated to avoid its reoccurrence.

- Hot/cold surfaces: Heat treatment involves the application of temperature (hot or cold), and virtually all metals can burn unprotected skin. Metal temperatures can be deceiving. Parts coming out of a tempering furnace, for example, may look like they are at room temperature when in fact they are at several hundred degrees. Cryogenic or deep-frozen parts may cause cryo burns. Similarly, the basket or fixture that held the parts may retain heat for many hours after processing. Furnace doors or openings allow heat and atmosphere to exit, possibly even burning flames. Open quench tanks with combustible quenchants (e.g., oils) can produce large evolutions of flame (i.e., fireballs) during the initial stages of quenching. Finally, many external surfaces of a furnace or oven may be high enough in temperature to cause serious injury. Through-members extending to the shell can cause a localized area to greatly exceed the overall OSHA shell temperature requirement of 60°C (140°F).

One suggestion that many heat treaters follow is to never handle parts without heat-resistant gloves and only after they have positioned a hand near enough (without allowing it to come in contact with) the part to feel if it is giving off heat or cold. It has been reported that the average person can touch and remain in contact for a few seconds with a part whose temperature is around 60°C (140°F) but can only very briefly touch a part whose temperature is around 82°C (180°F). Do not attempt to prove this!

- Pinch points: Slow- or fast-moving devices such as actuators or rotating devices (e.g., rolls, drives, work-transport cars) may have pinch points that can injure hands or cause loss of fingers if they are in the wrong location at the wrong time. Quench elevators or doors being serviced also present risks.
- Exposure to chemicals: Metal processing generally requires the use of specialized chemicals (e.g., large acid tanks used to pickle scale from steel surfaces, rust-preventive or black-oxide dip tanks, machining fluids, cleaning compounds, laboratory chemicals and chemicals used in the production process). Many chemicals will be labeled, but some are not. Remember, chemicals can be present in solid, liquid or gaseous form. Contact with them, even incidental in nature, should be dealt with in a manner defined by the company and the chemical supplier. Washing hands after exposure is always a good idea. Touching of body parts (e.g., eyes, or internal areas of the nose or ears) or handling of food before hands are thoroughly cleaned is never a good practice. Finally, do not eat, drink out of an open container or bring food into the heat-treat shop. (Note: This practice is so common it must be regulated by management.)
- Contact injuries: These types of injuries often occur due to unexpected contact with objects and/or equipment being moved or handled. Par-

ticular attention should be paid when objects are moving overhead or by others (tow motors being a prime example). The sights and sounds of the shop can be good indicators of potential danger.

- ❖ Slippery surfaces: Oils, lubricants, cutting fluids or oil-dry spread on the shop floor, and liquids used to clean floors can create slippery surfaces. Keeping floors clean should be a major priority. Visitors should be encouraged to stay within marked travel paths to minimize the possibility of injury and be appropriately cautioned before leaving the path and entering the work area.
- ❖ Confined-entry spaces: Some locations in a plant have limited means of egress (i.e., entry and exit). They should be clearly indicated by signage as confined-entry spaces. Nobody may enter a confined space (including certain pits and basements) without special training and only then by following special procedures. Putting one's head inside a furnace that is down for maintenance "just to take a quick look" or entering a furnace without permission to do so must be discouraged. An entry permit, special training and oxygen monitors are required as is the presence of individuals trained in confined-entry spaces.

Cylinder Storage and Use

Gas cylinders (Fig. 14.2.2) are a common sight in the heat-treat shop, and it is important to know how to handle and store them. Specifically:

- ❖ Gas cylinders should be stored in a well-illuminated and well-ventilated space away from combustible materials. OSHA requires that cylinders containing flammable gases be stored at least 6.1 meters (20 feet) from cylinders containing oxygen and other oxidants (or a fire-resistant wall be used with a rating of at least 30 minutes).
- ❖ A chain or another approved fastening method (e.g., bench or wall clamps) should be used to secure cylinders. While in storage, cylinder valve protection caps must be firmly in place.
- ❖ The steel cylinder temperature should not be permitted to exceed 51°C (125°F).
- ❖ Cylinder storage should be such that old cylinders are used first. Empty cylinders should be stored separately and clearly identified.
- ❖ All cylinders should be clearly labeled.
- ❖ Cylinders should be properly secured before usage and be clean and dry. If oil or grease is present on the valve of a cylinder containing oxygen or another oxidant, do not attempt to use it.

FIGURE 14.2.2 | Nitrogen purge cylinder setup in a typical heat-treat shop

Summary

Health and safety are everyone's responsibility, and being aware of the dangers and taking steps to avoid incidents in the heat-treat shop (or anywhere in the plant) will make it a better place for all of us to work. Remember, nothing replaces common sense when it comes to avoiding injury. Learn and follow the company rules, and protect yourself at all times.

REFERENCES
1. U.S. Department of Energy (www.energy.gov)
2. Stratton, P.F. and M.S. Stanescu, "An Introduction to Atmosphere Furnace Safety," BOC Gases, Proceedings of the 22nd Heat Treat Conference and 2nd International Surface Engineering Congress, September 2003
3. Carbon Monoxide – Physiological Effects, Net Industries (http://science.jrank.org/pages/1212/Carbon-Monoxide-Physiological-effects.html)
4. Herring, Daniel H., "Health and Safety in the Heat Treat Shop," *Industrial Heating*, April 2009
5. "Cylinder Storage and Use," Air Liquide America

14.3
ENERGY AND ENVIRONMENTAL ISSUES IN THE SHOP

In a past life, heat treatment had a reputation as an energy user and environmental abuser. Today, however, heat treaters conserve energy and resources whenever possible. This shift has not only been driven by regulation and a desire to maximize profits, but a profound sense that we have a responsibility to be good corporate neighbors.

Energy Usage and Conservation

Historically, heat-treatment furnaces have been designed to meet production and quality demands with little regard for energy conservation. This oversight in the equipment design and selection process has resulted in highly inefficient furnaces (particularly older ones). Given the 30- to 50-plus-year life span of most industrial furnaces, this is a significant oversight. Today, companies are looking at retrofitting older units and adding energy-efficient devices on new ones to reduce their energy footprint and save money.

RETROFIT EXAMPLES

In one study, the number of burners in the soak zone of a vintage continuous roller-hearth furnace was reduced without impact to product quality or throughput but with significant fuel savings (in the order of 25%).[2] In another example at that same plant, the burners in an older car-bottom furnace were upgraded to a high-velocity style, improving both throughput and energy. Savings were as high as 33% for some heat-treat cycles. In yet another example, reducing the number of burners and replacing them with regenerative ones in a rotary-hearth furnace combined with increased oxygen enrichment resulted in a 10% savings in natural gas.

Energy efficiency in a modern pusher furnace line (Fig. 14.3.1) can be achieved by:

- ❖ Use of high-quality and thicker furnace insulation to lower the furnace shell temperature, ensuring reduction of energy losses. Manufacturers report up to a 20% savings in energy from better insulation alone.
- ❖ Use of recuperative burners to lower exhaust-gas temperature and save up to 30% in natural gas costs (as compared to cold-air burners) while lowering emission rates.
- ❖ Use of the waste heat from the flue gas and quench-oil cooling system to heat the shop and/or low-temperature units such as washer tanks. Savings up to 70% during continuous production have been reported.
- ❖ Minimizing flow of and disturbances to the process-gas atmosphere by directional flow to improve metallurgical quality with reduced process-gas consumption.

Another present-day challenge with respect to energy, at least as far as North America is concerned, is the low cost of natural gas. This makes conversion to recuperative burners, for example, cost prohibitive. Despite this setback, energy-conservation measures are being adopted with long-term returns in mind.

FIGURE 14.3.1 | Energy-saving locations in a pusher furnace system (courtesy of Aichelin Heat Treatment Systems)

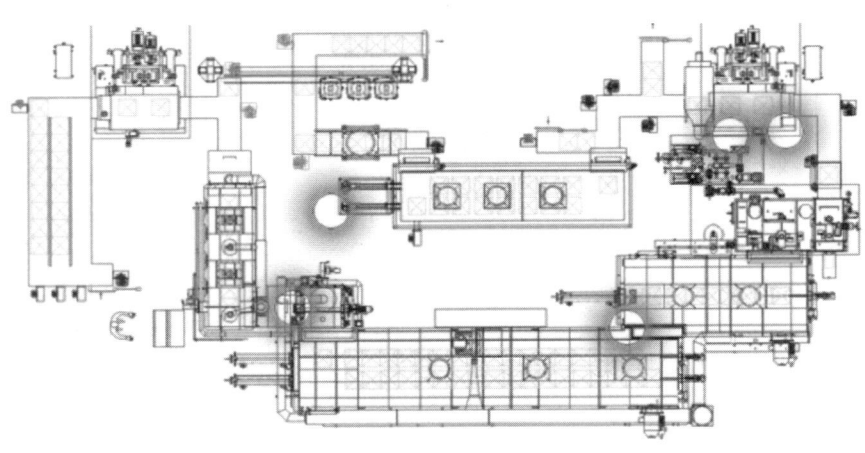

Another example is direct-fired box furnaces (Fig. 14.3.2) or car-bottom furnaces, which are often supplied with conventional high-velocity burners but can be upgraded to self-recuperative direct-fire burners (Fig. 14.3.3) with higher efficiency. The self-recuperative burner is used in combination with a pulse-firing system (i.e., the burner is on/off controlled) to achieve AMS 2750 temperature uniformity throughout the operating range of 425-1200°C (800-2200°F). At the same time the burner is on, an eductor built into the burner exhaust sends products of combustion through a built-in heat exchanger, allowing higher air preheat temperatures and higher efficiency.

Globally, energy availability and costs fluctuate significantly, but heat treaters have become more efficient both in their equipment and in their processes (i.e., cycles).

FIGURE 14.3.2 | Energy-efficient gas-fired box furnaces (courtesy of Nutec Bickley)

FIGURE 14.3.3 | Cross-sectional view of a self-recuperative burner (courtesy of Eclipse Inc.)

a. Single-ended radiant tube burner

b. Cross-sectional view

a.

b.

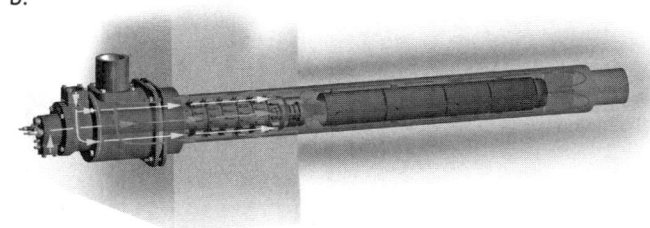

GATHERING ENERGY INFORMATION

An essential component of any successful energy-management program is a comprehensive detailing of energy use and its cost. Company energy bills should be treated as though the money was coming out of your own pocket. This requires keeping up-to-date records of monthly energy consumption and all related costs. Recording the information on a spreadsheet or similar record allows easy comparison and analysis for spotting anomalies. Data for each type of energy (gas, electricity, water, oil, etc.) shown as a separate record should be distributed to all functional departments. Also, a single energy unit (e.g., kwh or MMBtu) should be used to express the heating value of the various fuel sources to allow a meaningful comparison of fuel types and conservation.

CONDUCTING AN ENERGY AUDIT

An energy audit requires performing both efficiency tests and energy balances. Traditional focus areas with large potential for saving energy are heat-treatment equipment design and construction, burner combustion efficiency, power distribution of electrical systems and electric motors (power factor). Non-traditional areas that offer good potential for saving energy include cooling towers and pumping systems, oil skimmers, fans and ventilation systems, compressed gas storage and transmission, and steam systems.

Here are some obvious places to look for energy savings.

Gas-Fired Furnaces
- Recover hot exhaust gases through recuperation (Fig. 14.3.4).
- Properly adjust the air-to-fuel ratio and control specialty gases to the furnace (more easily accomplished than heat recovery).
- Install variable-frequency drives and controls on motors (especially combustion blowers).
- Control important parameters including air and gas flow rates; furnace temperature; operating hours and mode (continuous vs. intermittent); exhaust-gas composition; temperature blower sizes; and measured current (amperage) draw.

Electrically Heated Furnaces
- Control of furnace operating hours can dramatically lower demand charges, which can often be reduced by changing the energy provider.
- Install energy monitoring/control equipment (Fig. 14.3.5) based on time-of-day metering. If an automated control system is unaffordable, stagger equipment start-ups to minimize energy use.
    - Variable-speed motor control panels can automatically control the speed of electric motors (e.g., agitators, pumps on water systems).
    - Peak kW energy totals for the entire shop can be displayed on LCD screens located throughout a facility.
- Be aware of energy equivalents (Table 14.3.1) and alternative energy sources.
- Control important parameters including furnace operating schedule and mode (continuous vs. intermittent).

All Furnaces
- Upgrade furnace thermal insulation and cover heated tanks where possible.
- Heat treat parts only to required specifications or standards.
- Reduce fixturing and other nonessential material in the heat-treatment process.
- Calibrate furnace instrumentation.
- Control important parameters including wall temperatures and furnace operating schedule and mode (continuous vs. intermittent).

FIGURE 14.3.4 | Self-recuperative burner installed on a heat-treatment furnace (courtesy of Eclipse Inc.)

FIGURE 14.3.5 | Plant-wide energy monitoring device (courtesy of Solar Atmospheres Inc.)

a. Real-time power meter

b. Power-meter status light

Green means no problems with power either internal or external; yellow means you have exceeded the peak demand threshold; and red means a phase loss, power outage or low-voltage condition. When the red light goes on, e-mails are sent to management and the maintenance team for immediate responses.

a.

b.

TABLE 14.3.1 | Energy equivalents

Energy unit	Energy equivalent, kW (BTU)
Kilowatt-hour	1 (3,412)
Natural gas, 0.028 m³/h (1 cfh)	0.29-0.31 (1,000-1,050)[a]
Therm	29.3 (100,000)
Propane, 3.8 liters (1 gallon)	27.0 (92,000)[a]
Fuel oil (No. 2, diesel)	41.0 (140,000)[a]
Fuel oil (No. 6, bunker C)	44.0 (150,000)[a]
Gasoline, 3.8 liters (1 gallon)	38.1 (130,000)[a]
1 horsepower, electric	0.74 (2,515)
1 horsepower, boiler	9.7 (33,500)

Note:
[a] Varies with source

The cost of operating energy-consuming equipment is under intense scrutiny. Although individual devices can reduce energy, a systemic approach is best if cost savings are to be sustained. A number of obvious and not-so-obvious areas to find savings in heat treatment allow us to quickly ascertain where we should expend our limited resources. These include:

- Scheduling workload and production runs so equipment is full and operated when energy costs are most attractive
- Monitoring energy use and idling practices
- Adjusting combustion systems and electrical power devices (silicon-controlled rectifiers, saturable core reactors, variable reactant transformers)
- Optimizing cycles and process parameters
- Surveying furnace operation (heating and cooling rates, temperature and atmosphere uniformity)
- Inspecting temperature systems (controls, sensing devices, etc.)
- Measuring furnace efficiency (e.g., shell temperature measurements to determine insulation efficiency)
- Negotiating long-term contracts to reduce rates
- Analyzing electrical-system efficiency

Environmental Issues

Not too long ago, the sentiment in the U.S. heat-treatment industry was that the federal, state and local government-regulated environmental standards were all that were necessary and no additional measures were needed. This is no longer the case.

GOOD-NEIGHBOR POLICY

Companies performing heat-treatment operations are focused on being responsible neighbors. In one example, a fastener manufacturer meeting all mandated air-quality standards for capturing oily fumes from tempering ovens via bag-house filters (Fig. 14.3.6) was not satisfied they had done enough because of the close proximity of residential housing to their property.

In investigating the root cause of the problem, and recognizing that production demands and product type (small, nested loads dragging out large quantities of quench oil) could not change, they focused on: adding time to the drain cycle over the oil before discharge from the integral-quench furnace; improving their existing parts washer by changing time-temperature-chemistry; adding post-drying of parts; and investigating advanced cleaning technology for a new wash system (Section 7.2).

Fumes were reduced to the point that the bag-house filters were not necessary (but were left in place as a secondary containment measure). One of the hidden benefits was that the temper ovens, which were saturated with oil (Fig. 14.3.7) and presented a potential fire hazard, were cleaned up.

FIGURE 14.3.6 | Original bag-house smoke fume-abatement arrangement

FIGURE 14.3.7 | Oil-saturated oven walls

DANGERS OF NOx

NOx (pronounced "knocks") emissions is the generic term for a group of highly reactive gases that contain combinations of nitrogen and oxygen in varying amounts. Nitrogen monoxide (NO) and nitrogen dioxide (NO_2) are prime examples.

NOx is responsible for a wide variety of long-term health and environmental problems, so it is important for the heat treater to understand that most combustion processes, like those that take place in our gas-fired furnaces, can be a source of NOx emissions. These pollutants spread out over great distances, which means that problems associated with NOx are not confined to just the heat-treat shop. These effects include:

- ❖ Smog (ozone) – formed when NOx and volatile organic compounds (VOCs) react in the presence of sunlight.
- ❖ Dust (particulate material) – NOx reacts with ammonia, moisture and other compounds to form solid particles, which permeate the air we breathe.
- ❖ Acid rain – NOx and sulfur dioxide react with other substances in the air to form sulfuric, nitric and other acids, which fall back to earth as rain, fog, snow or dry particles.
- ❖ Oxygen-depleted bodies of water – NOx emissions increase the absorbed nitrogen level in water, upsetting the chemical balance of nutrients used to support life.

- ❖ Toxic compounds – NOx reacts in the air with common organic chemicals and the ozone to form a wide variety of toxic chemicals and byproducts.
- ❖ Greenhouse gases – Nitrous oxide (N_2O), a member of the NOx family, is a greenhouse gas believed to contribute to climate change.
- ❖ Global dimming – Nitrate particles and nitrogen dioxide block sunlight.

The latest burner technology – either direct-fired (Fig. 14.3.8) or in combination with silicon-carbide or alloy radiant tubes – optimizes ignition and burner behavior with reduced NOx emissions.

FIGURE 14.3.8 | Furnace atmosphere produced by products of combustion from high-velocity self-recuperative burners firing inside a box-type furnace (courtesy of Eclipse Inc.)

REDUCING NOx EMISSIONS

The amount of NOx emissions is a function of the type (composition) of fuel being burned, combustion conditions, burner design and flame temperature. "Thermal NOx" is the result of the conversion of atmospheric nitrogen and oxygen and is dependent on the reaction temperature, residence time and gas chemistry (stoichiometry). Long residence times at high temperature contribute to thermal NOx formation as does rapid mixing of combustion air (oxygen) with the fuel. "Fuel NOx," by contrast, is the conversion of chemically bound nitrogen in the fuel with oxygen. It also increases with rapid mixing.

Incomplete combustion products can usually be held to industry minimums by the proper operation of modern burner equipment (Fig. 14.3.9). There are also several strategies for reducing NOx that rely on either post-combustion treatment of the gas or upgrading to different types of burners. Do you know your shop's total NOx emissions? You should. Here are some methods to better control how much NOx is being emitted.

❖ Flue-gas recirculation (FGR) is a post-combustion technique that extracts a portion of the flue gas from the stack and mixes it with combustion air, thus lowering the peak flame temperature and thermal NOx formation (reportedly on the order of 50%). The addition of flue gas also reduces the oxygen content of the combustion air, thereby limiting the amount of oxygen available to react with the nitrogen. Flame instability and a decrease in the net thermal output limit the recirculation rate to about 15-25%. This technique is useful for natural gas and other low-nitrogen fuels. Recirculating flue-gas temperatures are generally in the range of 315°C (600°F) or less.

❖ Selective catalytic reduction (SCR) involves injecting ammonia into the flue gas upstream of a catalyst bed. The NOx and ammonia combine and subsequently decompose to produce elemental nitrogen and water. SCR systems typically operate in the range of 315-370°C (600-700°F).

❖ Selective noncatalytic reduction (SNCR) is another post-combustion control method that reduces NOx to elemental nitrogen and water vapor by injecting ammonia into the upper part of a combustion chamber or into a thermally favorable location downstream.

Upgrading burner designs is yet another way to maintain control over NOx emissions. Options include:

❖ Staged-air burners, which are a simple and inexpensive way to reduce NOx values by as much as 20-35%. Staged-air-burner systems divide the incoming combustion air into primary and secondary paths. All of the fuel is injected into the throat of the burner and combined with the primary air, which flows through the venturi and burns. In this fuel-rich zone, the fuel partially burns and the nitrogen is converted into reducing agents. These nitrogen compounds are subsequently oxidized to elemental nitrogen. Peak flame temperature is also lowered. In the secondary combustion zone, additional air is injected through refractory ports to complete combustion and optimize the flame profiles.

❖ Staged-fuel burners, which inject a portion of the fuel gas into the combustion air to reduce peak flame temperatures. The remainder of the fuel gas is injected into a secondary combustion zone through secondary nozzles. The combustion products and inert gases from the primary zone reduce the peak temperatures and oxygen concentration in the secondary zone, further inhibiting NOx formation. Staged-fuel burners can reduce NOx emissions by as much as 50-60%. This type of burner can operate with a small flame length and at lower excess-air levels than staged-air burners. The flame in

staged-fuel burners is about 1.5 times longer than those in standard burners.
- ❖ Ultra-low-NOx burners, which use a combination of staged-fuel burning and internal flue-gas recirculation. In these designs, the fuel gas pressure in combination with compressed air (or steam) is used to induce flue-gas recirculation within the burner.
- ❖ Low-excess-air burners, which work on the principle that low levels of excess air suppress NOx formation. These burners are typically a forced-draft design and employ a self-recirculating technique to produce a multi-staged combustion effect. It has been found that reducing excess air from 30% to 10% cuts NOx emissions by 30% in these systems.

FIGURE 14.3.9 | Cross-sectional view of an ultra-low-emission burner capable of limiting NOx emissions to less than 30 ppm at 3% oxygen (courtesy of Eclipse Inc.)

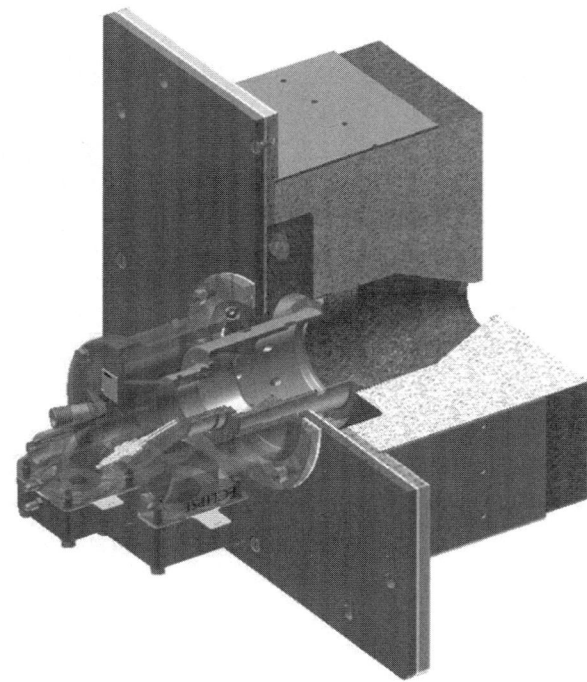

Water Quality

Water is used in most of our heat-treat shops for a variety of purposes. Examples include water-cooled bearings on fans and rolls; seals on pit-furnace covers; water-cooled jackets on continuous furnaces; water-cooled jackets for quench tanks, top or side cooling chambers, inner doors and plate coils; endother-

mic generator "top hats;" seals (e.g., oil seals on rotary-hearth furnaces); and makeup water for water systems.

Water quality requirements are often defined differently for open (Table 14.3.2) and closed (Table 14.3.3) systems. Open systems are typically more problematic because the issue of water quality varies. Water is often classified as soft or hard, depending on its mineral content. Soft water has an ideal hardness of approximately 120 ppm (7 grains/gallon). Hard water results in the formation of mineral deposits, which can lead to blockages in water systems (Fig. 14.3.3).

Furthermore, we must ensure that the water being discharged from our heat-treatment operations is clean and meets EPA standards. We must be especially careful to avoid cross-contamination from other sources in the shop (e.g., polymers, quench oils, chemicals, etc.).

TABLE 14.3.2 | Typical water requirements for open systems[5]

Description	Value
Hardness (calcium carbonate)	120-170 ppm[a] (7-10 grains/gallon)
Total suspended solids	10 ppm (0.58 grains/gallon)
Total dissolved solids	200 ppm (11.7 grains/gallon)
Iron	0.3 mg/L (3.7×10^{-5} ounces/gallon)
Aluminum	0.05-0.2 mg/L (6.1×10^{-6} – 2.4×10^{-5} ounces/gallon)
Copper	1.0 mg/L (1.2×10^{-4} ounces/gallon)
pH	7.0-8.0
Odor	3 threshold odor number
Conductance[b]	≤ 300 μS/cm
Maximum water temperature (inlet)	31°C (88°F)
System drain pressure	≤ 24.1 kPa (≤ 3.5 psig)

Notes:

[a] Grains per gallon is defined as 64.8 mg (1 grain) of calcium carbonate per 3.79 liters (1 U.S. gallon), or 17.1 ppm.

[b] Conductance is measured in Siemens (S). The more archaic term is mho.

[c] For best cooling efficiency and component longevity, the water supply should be treated to prevent corrosion and scale (controlled by phosphonate test, range 15-20 ppm), scum formation, algae and other biological agent buildup.

TABLE 14.3.3 | Water requirements for closed hydronic systems[5]

Parameter	Optimal conditions	Comments
Glycol freeze protection	30-50%	Below 20% can promote the growth of bacteria; above 50% will dramatically reduce heat-transfer ability. Glycol seepage can occur at O-rings and seals.
Corrosion inhibitor	[a]	Without the addition of nitrite or molybdate inhibitors, corrosion of the metallic compounds will begin and eventually lead to leaks.
pH	9-10.5	A pH below 9.0 will promote corrosion of steel; above 10.5 will promote the corrosion of brass and copper.
Conductivity	700-3,200 µS/cm	Conductivity above 3,500 µS/cm will cause the water to become physically abrasive and damage O-rings. Addition of chemicals to the water will raise the conductivity.
Hardness	100-300 ppm (5.8-17.5 grains/gallon)	Artificially soft water can be aggressive to the system. The use of unsoftened water is recommended. Do not use distilled or purified water.
Bacteria/mold	None	The growth of bacteria will cause erosion of seals, and the deposit of a bacterial slime will clog the system. Bacteria can attack the O-rings and cause premature failure. Glycol above 20% will kill any bacteria.

Note:

[a] Molybdate (100-150 ppm) or nitrite inhibitor (800-1,200 ppm)

FIGURE 14.3.10 | Flow blockage in the top cool of an integral-quench furnace, leading to sludge buildup

SALT DISPOSAL

Although modern salt quench baths do not give off toxic or hazardous fumes, good exhaust around the bath is highly recommended. This is particularly helpful during charging with fresh salt and during the quenching operation.

After quenching, parts and fixtures are typically immersed in an agitated hot-water bath, where most of the salt is dissolved and then rinsed in hot-water spray. Salt from wash water can be recovered by evaporation of its water content. What results is molten salt that is transferred to box-type metal containers, where it is allowed to freeze into blocks.

Following the recovery-and-reuse route eliminates disposal of wash water. The drawback is that it causes buildup of undesirable contaminants. Periodic adjustment of salt chemistry is required to maintain uniform quenching performance. This helps explain why many heat treaters still prefer to dispose of their wash water.

Although neither highly toxic nor flammable, quenching salt is classified as a hazardous material because of its oxidizing nature. The hazard is reduced considerably when salt is contained in wash water, and many local waste-treatment authorities permit discharge of wash water into their drainage systems. If permission cannot be obtained, the handling of wash water can be delegated to a waste disposal company. Sludge can be used as chemical landfill where permitted. Otherwise, it can be dissolved in water and treated the same as wash water.

Summary

Energy and environmental management is vital for everyone. Begin by identifying areas of high energy consumption and waste. Understanding energy costs and performing an energy audit are valuable first steps. The heat-treatment industry needs to take the lead in energy and environmental conservation to remain the most cost-effective and green technology choice to the marketplace.

REFERENCES
1. Herring, Daniel H., "Energy Conservation in Heat Treating," *Industrial Heating*, March 2004
2. Siefert, Mike and Nathan Abbound, "Energy Saving Initiatives," TimkenSteel, presentation at FIA|TECH Energy Workshop, 2010
3. Kratowicz, R., "Energy Conservation and Thermal Processing," *Plant Services Magazine*, January 2001
4. Herring, Daniel H., "Energy Saving Tips," *HOT TOPICS in Heat Treatment and Metallurgy*, Vol. 1, No. 2, December 2003
5. "Water Quality in Hydronic Systems," INFO 29, pp. 1-2, Heat Link Group (www.heatlinkgroup.com)
6. Gary S. Berwick, Dry Coolers Inc. (www.drycoolers.com), technical contributions and private correspondence
7. Arturo Arechavaleta, Nuctec Bickley (www.nutecbickley.com), techni-

cal contributions and private correspondence
8. Chet Allen, Eclipse Inc. (www.eclipsenet.com), technical contributions and private correspondence
9. Charles Miller, Solar Atmospheres of California (www.solaratm.com), technical contributions and private correspondence
10. Robert Hill, Solar Atmospheres of Western PA (www.solaratm.com), technical contributions and private correspondence
11. Doug Welling, Dominion East Ohio (www.dom.com), private correspondence

14 | HEALTH, SAFETY, ENERGY AND THE ENVIRONMENT

Vacuum Heat Treating

The clean, safe alternative to atmosphere heat treating.

Top 5 Reasons to Vacuum Heat Treat

1. Bright and clean parts, every time, no sandblasting or post-heat treat cleaning
2. No quench tank fires or oil disposal costs
3. No dangerous, explosive atmospheres
4. No smoke, soot, or risk of devastating fires
5. Clean and safe environment for workers

Vacuum Heat Treating Services
- *Vacuum Heat Treating*
- *Vacuum Brazing*
- *Vacuum Carburizing*
- *Vacuum Nitriding*

Vacuum Heat Treating Furnaces
- *Vacuum Furnaces*
- *Replacement Hot Zones*
- *Aftermarket Spare Parts*

SOLAR ATMOSPHERES
Souderton, PA USA
Hatfield, PA USA
Hermitage, PA USA
Fontana, CA USA
Greenville, SC USA
solaratm.com
1-855-WE-HEAT-IT (934-3284)

SOLAR MANUFACTURING
Souderton, PA USA
Telford, PA USA
solarmfg.com
267-384-5040

14.4 IDENTIFYING SOURCES OF POTENTIAL QUENCH-OIL FIRE HAZARDS

Oil quenching (Section 10.2) is one of the most common practices performed in the heat-treat shop. However, it also presents some of the greatest safety hazards and can result in personal injury, equipment damage, business interruption, environmental cleanup and possible litigation expenses. The most common sources of potential problems, including explosion or fire, are covered.

Water in Quench Oil

One of the most dangerous types of industrial fires can occur when the quench oil contains water. Even minute amounts of water anywhere above 1,000 ppm (0.10%) are cause for concern. Quench oil should be continuously monitored and routinely checked for the presence of water, and contaminated tanks should be pumped and separation techniques should be used. Because water will accumulate in the bottom of the quench tank, the sampling method must take this into account.

Integral-quench and pusher furnaces are especially vulnerable to this problem, but it is also seen in open quench tanks. The source of water contamination may come from a leaking water-to-oil heat exchanger, a leaking water jacket in the atmosphere-cooling section or any other area that uses water for cooling. Furnaces shut down for maintenance can easily be contaminated if the quench-tank area is open.

Oil Drag-out

Most furnace loads are allowed to drain over the oil prior to being removed from the quench tank. However, insufficient drain time or part/workload geometry may be such that a significant quantity of oil remains on the parts. This is called drag-out. The load could ignite when it passes through the door flame curtain, and the resultant fire is extremely dangerous to personnel trying to extinguish it and close the front door. In rare instances, insufficient quench time results in a load that is still above the self-ignition point, and a fire will erupt upon removal of load from the furnace.

Load Hang-up

One of the most dangerous types of furnace-related fires occurs when a load being transferred to the quench "hangs up" or becomes stuck partially in and partially out of the oil. The oil at the interface between the hot load and the surface generates huge amounts of oil vapor, and (in many furnaces) these fumes are carried out with the furnace atmosphere and ignited by a pilot. This results in an over-pressure situation often including a strong flame, or torch-like condition, that exists while the workload slowly cools.

This type of phenomenon can also occur when the quench elevator jams. In other words, the elevator's travel is restricted due to a critical failure of plant air pressure, hydraulics or when mechanical malfunction occurs. In some designs, low air pressure can cause the elevator to slowly lower into the quench oil, resulting in an unsafe condition.

Even the presence of a nitrogen safety purge system does not guarantee that this situation can be controlled. In one instance, an integral-quench furnace was responsible for burning down an entire plant when the initial fire spread to the adjacent plating department, igniting its fiberglass ventilation system. The operator was told to turn the furnace and nitrogen purge system off because the fire would burn itself out, an action that resulted in a catastrophe.

Exhaust Stacks

Fires in exhaust stacks are common occurrences. This may be due to improper design of the stack itself, improper airflow or improper maintenance. Oil fumes (Fig. 14.4.1), either carried out of the furnace or from tempering an improperly cleaned load, can condense in the exhaust stacks. Several ignition sources, such as the rapid evolution of furnace gases, can trigger a fire that is not easily controlled because of its ability to spread quickly in areas that are difficult to access with fire-suppressant materials. Awareness of the potential for a fire and appropriate maintenance steps are important in avoiding this situation.

FIGURE 14.4.1 | Oily fumes present in the heat-treat shop

Inner Door Open During Quench

To speed up transfer time from the heating chamber to the quench, especially for thin parts that are prone to heat loss, some furnaces allow quenching to take place while the inner door between the heating chamber and the quench tank is still open (or is being closed simultaneously with the lowering of the furnace elevator). Some of the most devastating plant fires have occurred in such a manner. The evolution of gas during the transfer in combination with the oil volatilization on quenching can combine to shoot flames up and out of the furnace.

High Surface-Area Loads

It is impossible to avoid running a workload having high surface area in certain circumstances. Fasteners and other small components that tend to "nest" very densely present the potential to volatilize a large amount of oil when quenched. These oil fumes can be carried out of the furnace and ignited. Even when precautions such as positioning and manning fire extinguishers have been taken, the unexpected can happen. In one particular industrial fire, flames shot more than 30 feet in the air, igniting the supposedly fire-retardant roof insulation and overwhelming the fire-suppression equipment on hand.

Oil Selection and Process Parameters

Depending on the heat-treating requirements of the parts, the type of quench oil selected is an important consideration. In addition, its use over a wide range of process variables – including temperature, agitation, viscosity, quench speed and contamination – must be considered. All of these variables affect oil vaporization, oil drag-out and oil-quench characteristics. In extreme cases where the variables are not properly controlled or are allowed to change over time, what was a normal and safe quenching operation may turn into a potentially hazardous one.

Inadvertent Oil Discharge

When "burning in," or introducing furnace atmosphere, into the quench areas of an integral-quench or pusher furnace, pockets of air may remain and, as they suddenly ignite, result in a pressure buildup on the surface of the oil. This presents the real danger of an oil "burp," or discharge, out of the quench tank. Even overflow pipes that discharge to containment drums may suddenly be over capacity and spill oil out into the pit area. In rare cases, the pressure buildup is so severe that oil is dumped out the front door, often igniting if a pilot has been lit or another source of ignition is present.

One particularly damaging plant fire was due to a relatively small floor fire that spread into a pit area that contained an estimated 75-100 mm (3-4 inches) of oil from repeated (and ignored) oil discharges from the quench tank. Extensive equipment and building damage resulted.

Irregularly Shaped Parts

When quenching irregularly shaped parts or loads (e.g., long tubes having hollow internal diameters) from pit furnaces into open quench tanks, oil "spouts" and flames can literally shoot up and out from the inside of the tubes. This forms an oil fountain in a spectacular, albeit dangerous, display. Factory ceilings have been known to catch fire due to the height and internal pressure developed. Extreme precautions are needed. Pay particular attention to areas immediately adjacent to the quench tank.

Oily Parts in Washers

Successful cleaning involves the correct choice of cleaning-solution chemistry together with the correct use of time, temperature and energy. Equipment design, age and use – as well as ancillary devices such as oil skimmers – are important to ensure that a minimal amount of oil will remain on the parts when they are transferred from the washer into the tempering furnace. Due to their inherent porosity, powder-metallurgy parts can retain up to 3% (by weight) oil despite the best washing practices available. Also, washer fires due to the presence of excessive amounts of oil floating on the surface have been reported.

Oily Parts in Temper Furnaces

If the tempering furnace or oven is operating above the flash point of the oil, introduction of an oily load may cause an immediate and uncontrolled explosion. Oil can accumulate over time even if the unit is operated below the flash point. Raising the operating temperature for another process or attempting to perform an elevated-temperature bake-out can result in a fire or explosion. This is one of the most common hidden dangers of oil quenching.

Summary

The use of oil and oil quenching is to be respected, not feared. Knowledge, training and common sense – in addition to an awareness of the dangers presented here and others specific to the shop performing the quenching operation – are critical to avoiding quench-oil-related fires. In addition, clear emergency instructions and safety training for heat-treat and in-house fire personnel will avoid (or at the very least minimize) quench-oil fire hazards.

REFERENCES
1. Herring, Daniel H., "Identifying Sources of Quench Oil Fires," *Industrial Heating*, November 2004
2. Dr. D. Scott Mackenzie, Houghton International, private correspondence

14.5
FURNACE WATER-COOLING SYSTEMS

A large number of heat-treatment furnaces utilize water-cooled components. Examples include the lower fan bearing housing on an integral-quench furnace and water-cooled rolls on a high-temperature roller hearth. Water is the most common cooling medium, but ethylene glycol or propylene glycol diluted in water is used in many closed-loop systems to protect from freezing. These glycol additives have a negative effect on heat transfer.

Types of Water-Cooling Systems
Three types of cooling systems dominate the heat-treat landscape, namely:
- City- or well-water (so-called one-pass) systems
- Cooling-tower (recirculated-water) systems
- Closed-loop systems (mechanical, ambient air or water-cooled)

City- or well-water systems (Fig. 14.5.1) were once the most common types of cooling systems. Today, cost and environmental concerns (including bans on discharge of water in drought-stricken areas) have made other choices more attractive. In many locations, water needs to be treated before discharge into sewer systems to meet various codes (even if it has only been heated by the system it is cooling). Discharging water into streams, retention ponds and ditches is restricted and sometimes illegal due to chlorides in the water. In addition, city or well water typically has high hardness, resulting in mineral deposit and scale buildup that reduces heat-transfer efficiency.

FIGURE 14.5.1 | Typical schematic of one-pass water-cooling system (courtesy of Dry Coolers Inc.)

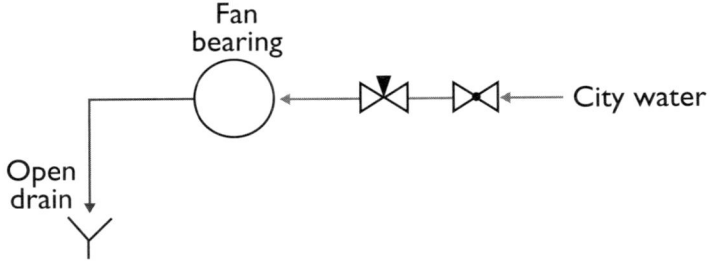

Cooling-tower, or recirculated-water, systems (Fig. 14.5.2) are also very common. Many users run tower water directly into the component being cooled. A cooling tower reduces water usage up to 95% or more compared to a city-water system. Cooling towers typically use 29°C (85°F) temperature water. A limitation of cooling-tower water can be the high level of dissolved and undissolved solids and oxygen, which leads to corrosion.

FIGURE 14.5.2 | Typical schematic of recirculating open cooling-tower system; filter system and chemical treatment are not shown (courtesy of Dry Coolers Inc.)

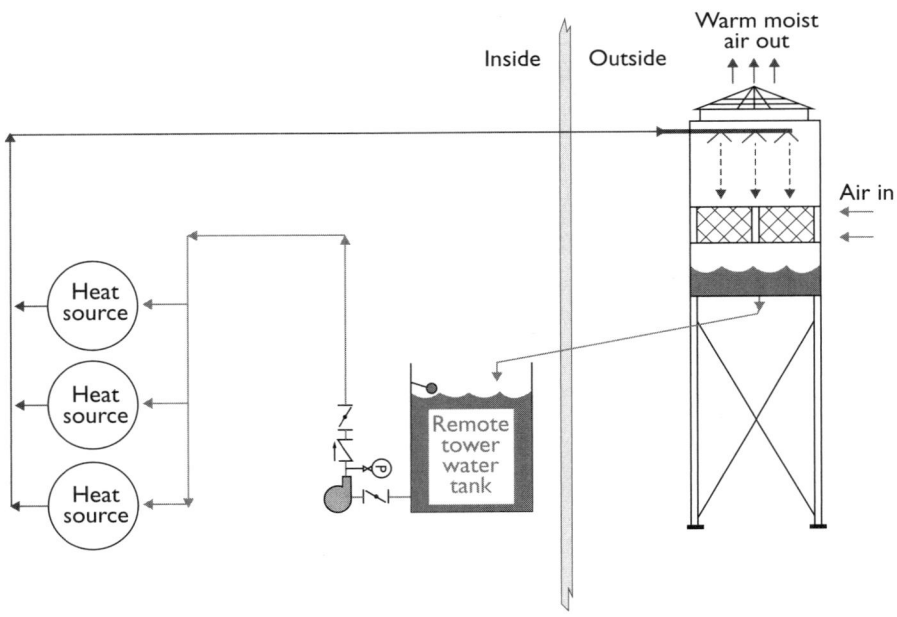

Cooling-tower water systems (Fig. 14.5.3) provide maximum protection from corrosion, which improves the overall life of components. While these systems are easy to maintain, they do require aggressive filtration and chemical treatment. Mechanical filtration uses strainers, centrifugal separators or side-stream sand filters. For chemical treatment, a variety of antifungal and bacteriological agents are added. Monitoring of pH and conductivity is essential. To minimize dissolved water content, a small percentage (about 1.5%) of the water flow is usually sent to drain and replaced with fresh water. This continuous bleed and refill helps keep the cooling-tower water chemistry in balance. Some communities now require treatment of the cooling-tower bleed water before dumping into the sewer.

Cooling-tower water systems often use intermediate heat exchangers on closed-loop systems. In this way, tower water is only exposed to the heat-exchanger circuit. Plate-and-frame heat exchangers and shell-and-tube heat exchangers are the most common types. It is important that only water and not glycol solutions run through the evaporative open cooling tower.

FIGURE 14.5.3 | Typical schematic of recirculating open cooling-tower system with intermediate heat exchanger (courtesy of Dry Coolers Inc.)

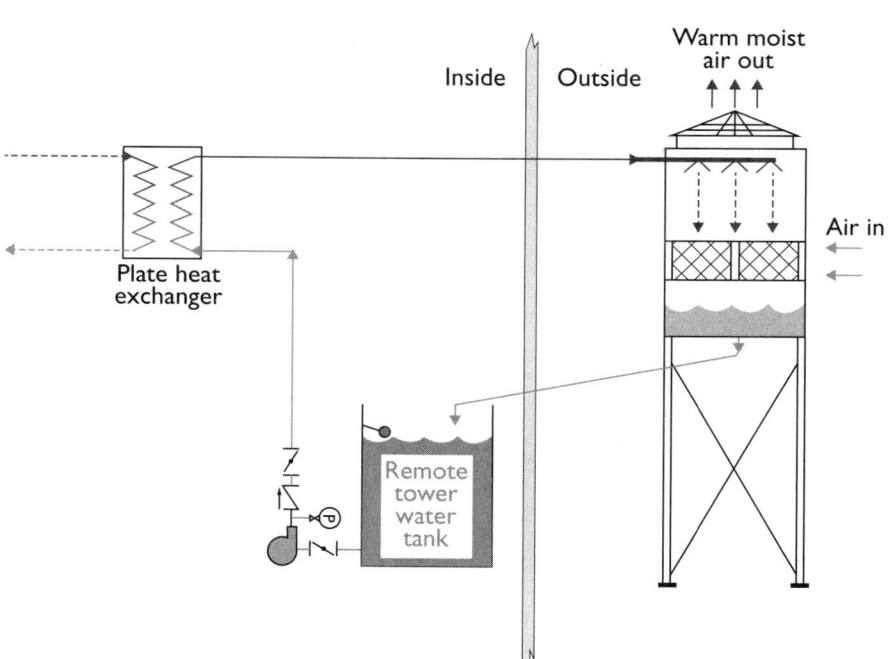

Closed-circuit evaporative cooling towers (Fig. 14.5.4) are often provided in warmer climates. This design uses galvanized or bare steel coils in the cooling tower that the process water/glycol runs through. Cooling water is then sprayed over the coil while a fan runs air over the water. This style allows glycols to be cooled within the cooling tower.

The principal advantages of this design are simplicity of the piping circuit and the fact that corrosion remains within the tower, where it is easily dealt with. In addition, no indoor sump is required. The disadvantages of this design are high initial capital cost, danger of freezing and inability to clean the internal coil. This type of cooling tower is very popular in Asia.

FIGURE 14.5.4 | Schematic of closed-loop evaporative cooling-tower system (courtesy of Dry Coolers Inc.)

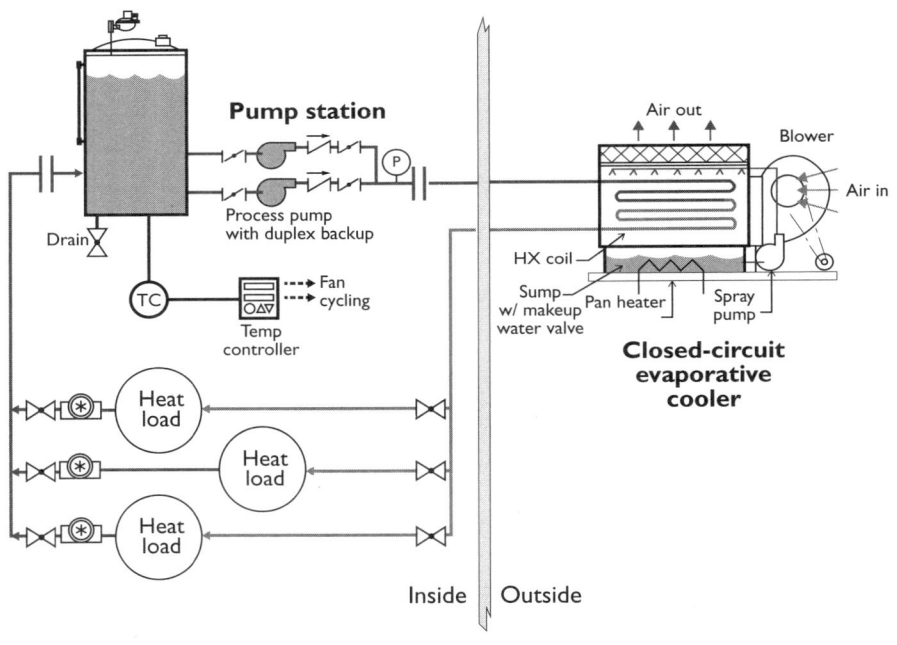

In colder climates, air-cooled closed-loop systems (Fig. 14.5.5) are preferred. Air-cooled systems require less maintenance and are less complex, but they are the most expensive from an initial-cost standpoint. These systems can generally cool the process water to within 5.5°C (10°F) of ambient air temperature; that is, cool the water to 38°C (100°F) on a 32°C (90°F) day.

FIGURE 14.5.5 | Schematic of closed-loop air-cooled system (courtesy of Dry Coolers Inc.)

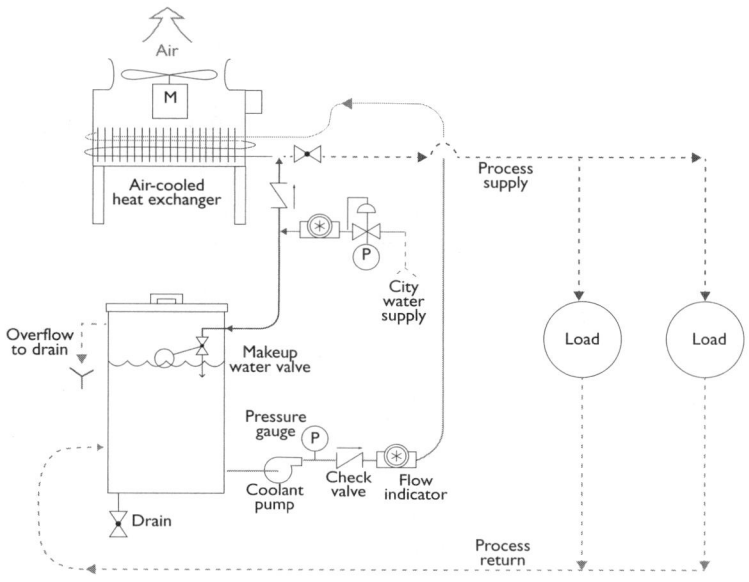

Facilities that already have cooling-tower water can choose a simple closed-loop system with an isolating heat exchanger (Fig. 14.5.6). This separates cooling-tower water from the process water. The advantage of these designs is that there is no fan, and they require very low maintenance because they can be readily cleaned or replaced.

FIGURE 14.5.6 | Schematic of closed-loop water-cooled system (courtesy of Dry Coolers Inc.)

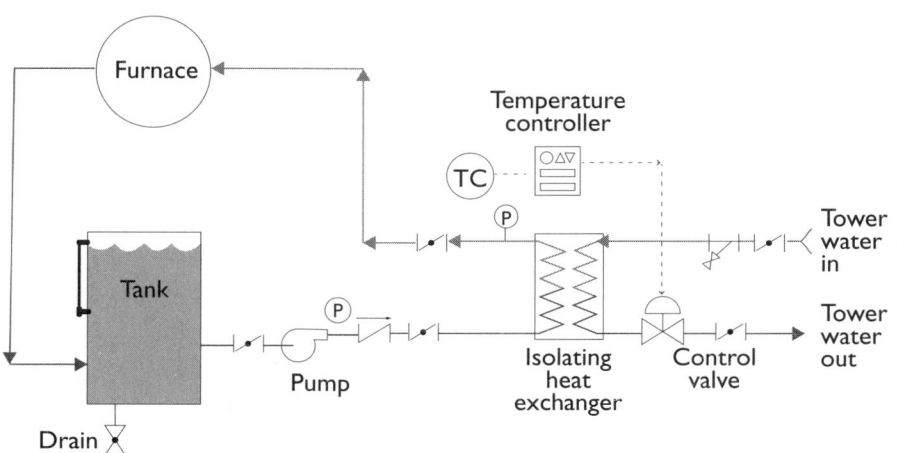

Closed-loop air-cooled systems only require an initial charge of a sodium-nitrite-based water-treatment chemical to minimize corrosion. The nitrite concentration is then checked quarterly. If properly maintained, minimum scale and corrosion will be present. Filtration of the cooling water is not needed in these systems, but it can be done with a simple strainer.

Water is not discharged from the system, but it is a good idea to install both a pressure-reducing valve and solenoid valve to allow city water to run through the system in the event of a power failure and drain out the top of the expansion tank. Air-cooled systems work best when the ambient air temperature is 32°C (90°F) or less.

Maintenance of Cooling Systems

Cooling-water maintenance requirements often depend on whether the system is an open-loop or closed-loop type. For an open cooling tower, where evaporation is utilized, good water quality is essential (Table 14.5.1). Water maintenance is often performed by outside companies or specialists due to the complex nature of treatment.

TABLE 14.5.1 | Water-quality requirements for open cooling towers[2]

Property	Recommended level
pH	7.0-8.0
Total dissolved solids	1,500 ppm maximum
Conductivity	2,400 micromhos
Chlorides	250 ppm maximum as Cl (410 ppm as NaCl)
Hardness as $CaCO_3$	30-750 ppm
Alkalinity as $CaCO_3$	500 ppm maximum

Closed-loop water systems typically have far fewer maintenance requirements. Many water-quality companies sell nitrite and non-nitrite-based water additives that are added at the time of initial fill. Water quality checks are then done either monthly or quarterly. Simple water-drop tests with color changes are used (similar to a home swimming pool test). If the nitrite level gets too low, it's simple to add more. A competent and qualified water-treatment specialist is often relied upon to recommend the best product.

An interesting point about internal scaling is the solubility of one of the most common minerals in city or well water, namely calcium carbonate ($CaCO_3$), which produces a whitish buildup in pipes and furnace components. The solubility of calcium carbonate in water decreases with increasing temperature (Fig. 14.5.7); that is, it exhibits reverse solubility. Use of closed-loop water systems avoids this concern.

FIGURE 14.5.7 | Calcium-carbonate level as a function of temperature[4]

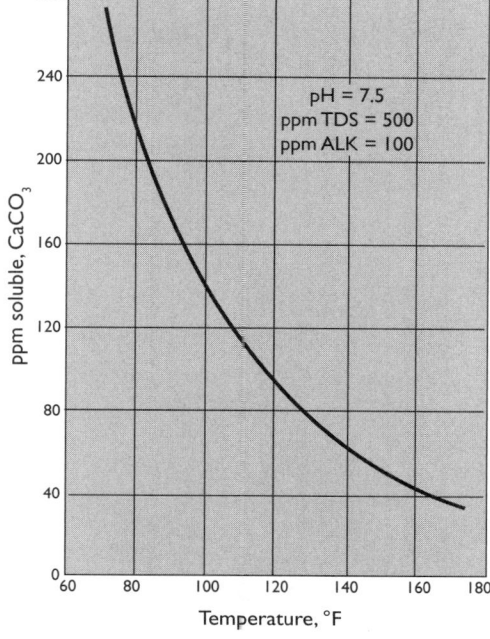

Flow and Pressure Indication

Typically, valves and pipes leading to and from various components that are being cooled need indicators to ensure there is both flow and pressure. Sight flow indicators (Fig. 14.5.8) are available that have a rotating impeller viewed through a clear window to ensure water is flowing. These are installed either before or after the component and should be periodically cleaned so that operators and maintenance personnel can readily observe them.

FIGURE 14.5.8 | Sight flow device (courtesy of Dry Coolers Inc.)

For low-flow sensing and alarming, paddle flow switches (Fig. 14.5.9) and thermal-dispersion flow switches are used. Paddle switches are inexpensive and readily available, but a disadvantage is that the paddle may stick in the same position after years of use. Thermal-dispersion flow switches have no moving parts and work by sensing the cooling effect of the water flow. These switches should be installed on the outlet, not the inlet, to know if water is leaking into your component.

FIGURE 14.5.9 | Paddle switch (courtesy of Dry Coolers Inc.)

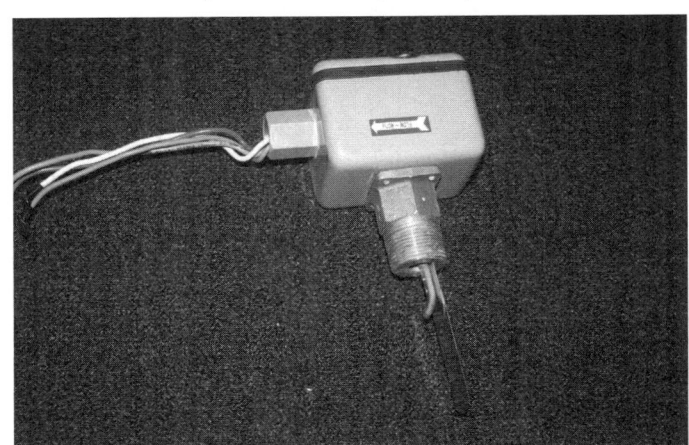

A pressure gauge can be helpful on the inlet side of the component. Note the normal pressure of the system so that any increase in pressure over time suggests the component in question may be plugging up. High pressure means low flow, and low pressure means high flow.

Design of Cooling Systems

In order to design a water-cooling system, one must begin by consulting the original equipment manufacturer (OEM) or device manufacturer to obtain minimum, maximum and normal flow requirements along with the temperature and pressure required for the component(s) to be water cooled. Once these requirements are known, a qualified water-system supplier can be consulted to help design the system.

Be aware that older furnaces (those built before 1980) were often designed for use with city water running colder in the winter and hotter in the summer. Some furnaces used recirculated direct cooling towers typically designed for 30°C (85°F) cooling water. If it is not well maintained, scale buildup will occur in the piping to the furnace over time (Fig. 14.5.10). A modern system should be designed for 35-40°C (95-104°F) cooling water, allowing for the option of air-cooled systems.

FIGURE 14.5.10 | Pipe scale (courtesy of Dry Coolers Inc.)

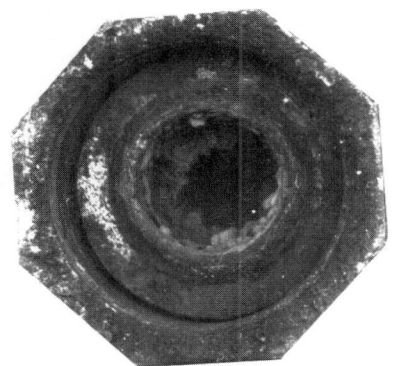

One can either measure water flow (preferred) or estimate it by looking at the size of the piping (Table 14.5.2). Pressure can be similarly measured.

TABLE 14.5.2 | Example of flow estimation from pipe size[a]

Pipe size (DN), mm	Pipe size (IPS), inches	Flow, m³/h (gpm)
15	0.5	1.36 (6)
20	0.75	2.72 (12)
25	1	4.31 (19)
32	1.25	7.50 (33)
40	1.5	10.0 (44)
50	2	16.6 (73)
65	2.5	22.7 (100)
80	3	36.3 (160)
100	4	62.5 (275)
125	5	100 (440)
150	6	143.1 (630)
200	8	252.1 (1,110)

Note:
[a] Flow calculations are based on a speed in pipe of 1.8 m/second (6 feet/second).

Summary

Water cooling of heat-treatment furnaces has been around for a very long time. Many furnaces still employ drain boxes that rely on visual observation of water flow. The industry is moving rapidly toward cooling-water systems as a positive method for ensuring adequate cooling and safety.

REFERENCES
1. Herring, Daniel H., *Vacuum Heat Treatment*, BNP Media, 2012
2. Gary S. Berwick, Dry Coolers Inc. (www.drycoolers.com), technical contributions and private correspondence
3. Baltimore Aircoil Co. (www.baltimoreaircoil.com), technical reference library
4. General Electric (www.ge.com)

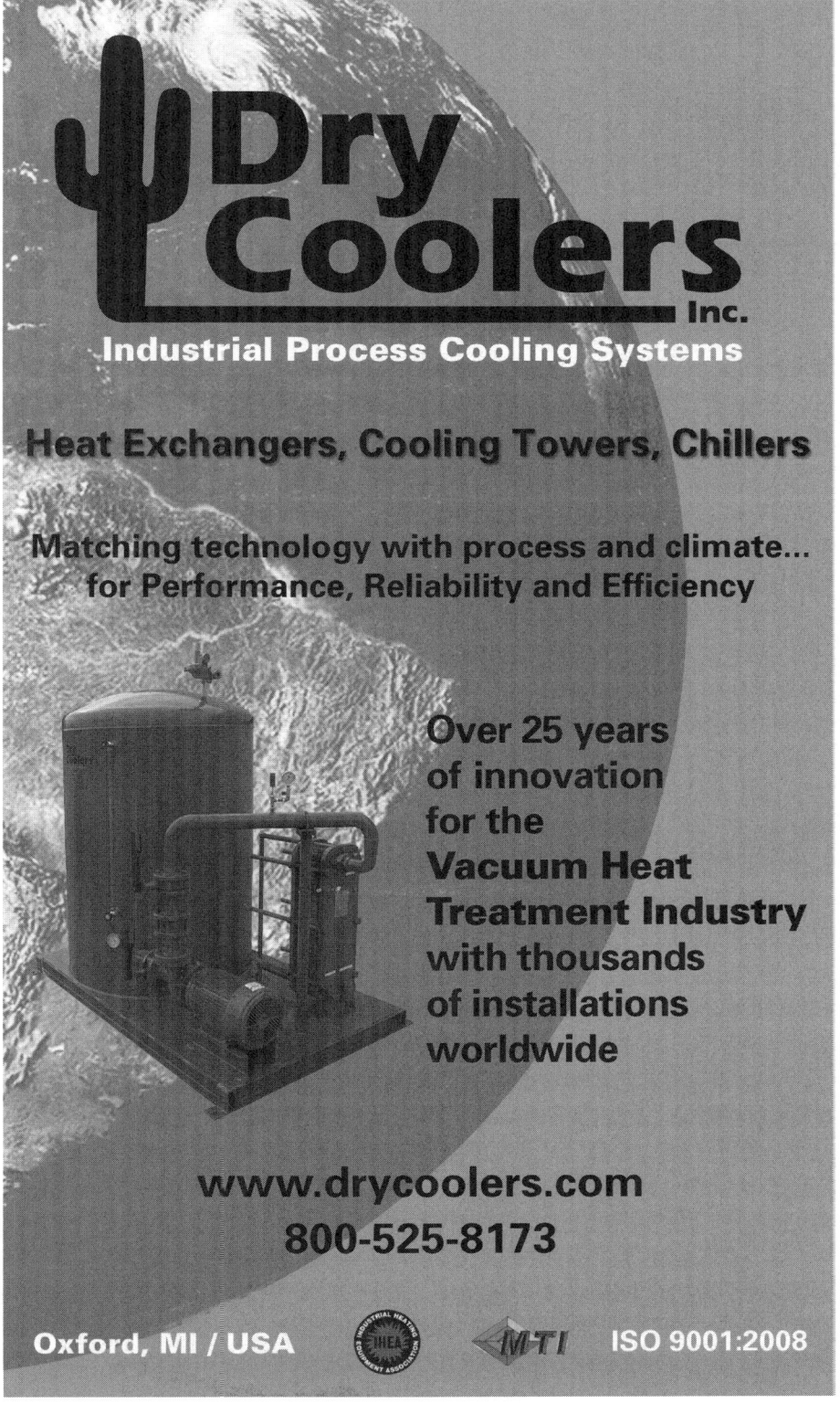

15.1 THE ROLE OF MANAGERS, SUPERVISORS AND OPERATORS

It is critically important to understand the role and responsibilities of management, supervisors and operators in support of the heat-treatment operation (Fig. 15.1.1). It is also everyone's responsibility to ask the right questions and emphasize the importance of safety in all that we do.

FIGURE 15.1.1 | A typical heat-treat shop (courtesy of Desert Fire Heat Treat)

The Manager's Role

People involved in the management of a heat-treatment operation must understand that the key to the operation's success is the control of process- and equipment-induced variability inherent in the heat-treat shop (Fig. 15.1.2). This includes the interactions and interrelations between the two. What surprises many managers is that the application of heat (or cold) causes the condition of the equipment to continuously change, so a series of checks and balances must be in place to detect, monitor and react appropriately. In addition, the manager is responsible for dealing with those external influences that affect the heat-treat department.

FIGURE 15.1.2 | Example of process variability in neutral hardening[3]

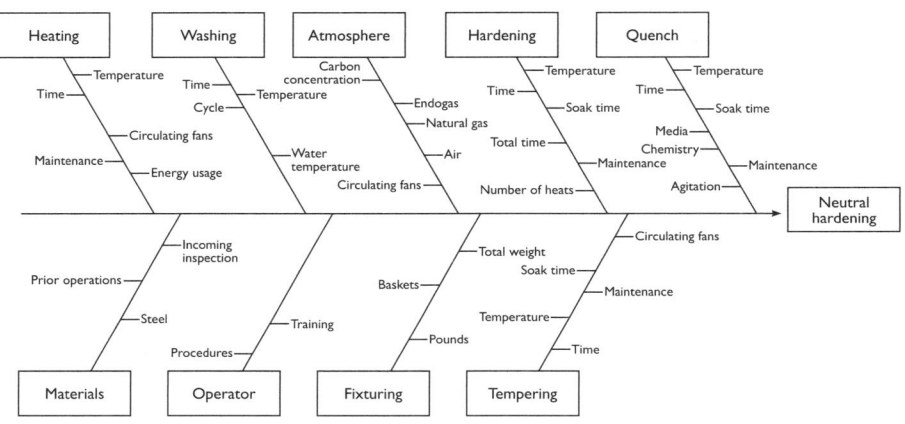

The following questions are important for managers to ask:
- ❖ What constitutes good heat-treating practice (Section 1.2)? Are the practices and policies being used effectively?
- ❖ What is the impact of lost productivity on profitability? What checks and balances are in place?
- ❖ What are the risks and rewards of our current (and future) practices? Are they controllable, repeatable and predictable?
- ❖ Are we proactive or reactive? Which management philosophy do we adhere to? Which management philosophy do we want?
- ❖ Does a true synergy exist with manufacturing? Does it exist in the real world or just on paper?
- ❖ Are we avoiding the "Pete & Repeat" syndrome? Do we learn from our mistakes?

15 | ANCILLARY TOPICS OF INTEREST

The Supervisor's Role

The supervisor is the person responsible for overseeing the day-to-day operation of the heat-treatment department. As such, the supervisor must ask the obvious questions (who, what, when, where, why and how), provide clarification of instructions, teach and mentor, and monitor the progress of the individuals doing the actual work. He/she is also responsible for defining the operator's responsibilities, coordinating maintenance efforts within the department and reporting to management.

A good supervisor anticipates problems (based on his/her experience) and makes sure they have the necessary materials and manpower available to deal with any situation that arises.

The Operator's Role

The operator represents the first line of defense to mitigate problems if and when they occur. The operator observes what is happening firsthand, loads and unloads furnaces, tests parts, and is often asked to make changes to the process or perform routine maintenance. One of the operator's key responsibilities is to look, listen and smell; that is, pay attention to the sights, sounds and smells present in the heat-treat shop.

Most furnaces do not fail catastrophically without warning (they deteriorate slowly or quickly), but there are often warning signs that an experienced operator should be able to detect and report to his supervisor as opposed to acting on his own (except in emergency situations). The operator is also responsible for ensuring that necessary actions and changes are being implemented.

The operator is the person most responsible for ensuring that a safe working environment is maintained and that unauthorized personnel (including visitors) are not put into dangerous situations.

Who is responsible for safety?

Safe operation must never be compromised. Safety (Section 14.2) is a multifaceted subject. What is important for this discussion is that managers and supervisors must ask the right questions and provide positive direction to their employees at all times. Some of the safety-related questions most pertinent to the heat-treat shop are:

- ❖ How safe is safe?
- ❖ Who controls the situation should a safety situation arise?
- ❖ Are the inherent dangers from furnace atmospheres (e.g., flammability, toxicity, asphyxiants) being dealt with in the most responsible manner?

Cost of Quality

Managers, supervisors, operators, maintenance personnel, quality control and really everyone who interfaces with the heat-treat department (including OEMs and suppliers) are responsible for quality. The total cost of ownership (Section 15.3) depends on our ability to manage all facets of the heat-treatment operation, not merely the ones that are the most obvious or the easiest to do. Part of the secret to success is to focus on three areas:

- ❖ Planned preventive maintenance (Section 13.1)
  - Is it effective?
  - Is it saving time and money?
- ❖ Determining the root cause of a problem (Section 12.5)
  - When a problem arises, do we always determine the root causes or are we content with Band-Aid fixes?
- ❖ Heat-treat audits (Section 11.2)
  - Do we perform self-audits and independent audits to check for major or minor deficiencies (nonconformance)?
  - Are our procedures living documents or just paperwork to satisfy auditors?
  - Do we have a real quality program?

Role of Continuous Improvement

One of the other responsibilities of those in charge of a heat-treatment operation is to plan for the future in addition to managing the present. In order to do this, one must focus on three fundamental areas.

- ❖ Improving the knowledge base (Chapter 1)
  - Identify necessary skill sets
  - Match capability with need
- ❖ Training the workforce (Section 11.1)
  - Positive reinforcement
  - Expand the knowledge base
- ❖ Networking with industry
  - Interface with outside experts, suppliers and OEMs to share common problems and their solutions.
  - Attend industry trade shows (e.g., Furnaces North America, ASM Heat Treat Society Conference & Exposition) to gain perspective on the current state of technology and where the industry is headed.

Traits for Success

Finally, step back and consider some of the characteristics that all leaders share and managers and supervisors look for in the individuals who report to them. They should have great vision and be creative, inspired, bold when boldness is needed, resilient and persistent at all times.

A person who embodies all of these traits is:
- A tireless worker
- Curious and always willing to learn
- A role model for others
- Someone who refuses to quit
- Someone filled with creative energy
- Self-reliant and responsible
- Relaxed and balanced
- Someone with a vision of where they need to be
- A forward thinker
- An honest and ethical person
- A doer, motivator and changer

Summary

We end this discussion with a simple thought:

> "Humankind is the sum total of what has come before, what is now and what will be. To be successful in life, one must contribute to the overall success of others."
> - Herring, 2011

REFERENCES

1. Herring, Daniel H., "Management Overview," white paper, 2009
2. Herring, Daniel H., "Traits of Greatness," Symposium in Honor of William R. Jones, ASM Heat Treat Conference & Exposition, 2011
3. *Heat Treater's Guide: Practices and Procedures for Irons and Steels*, 2nd Edition, ASM International, 1995, p. 17

Serving the Diemaking and Diecutting Industry Since 1953.

When only the best will do!

All AmeriKen tube and feed-thru diecutting punches are case hardened by vacuum carbonitriding, an exclusive heat treating process performed in a scientifically controlled vacuum environment, which prevents the formation of thin, weak spots that shorten the cutting life of ordinary punches. **Punch quality never deviates and uniformity is absolute!**

For all your die supply needs, contact us at 866.4.PUNCHES **(866.478.6243)** or **sales@ameriken.com** or visit us online at ameriken.com.

15.2
LEAN MANUFACTURING AND LEAN HEAT TREATMENT

Lean manufacturing strategies are designed to solve problems, save energy, reduce costs and extend environmental resources while addressing the challenge of global competitiveness.[2]

The long-standing practice of purchasing stand-alone pieces of equipment is starting to change. Manufacturers have become interested in purchasing complete systems, many of which are fully integrated into the manufacturing flow (Fig. 15.2.1).

FIGURE 15.2.1 | Fully integrated heat-treat system (courtesy of Ipsen)

Lean Manufacturing Strategies

The reality of global competition has placed new emphasis on improving the overall efficiency of manufacturing by using lean strategies (Fig. 15.2.2). The steps involved in a lean manufacturing strategy have extended the 5S strategy to one of 5S + 1. These steps are as follows.

- **Sift:** Decide what things you need to accomplish your work, and remove from the workplace all items that are not needed for current production (or clerical) operations.
- **Sweep:** Decide what things you do not need to do your job and remove them from your workplace.
- **Sort:** Ensure that people can do other people's work when they are not at their own workstation.
- **Standardize:** Create a standard for everything and try to use the same methodology and style throughout the manufacturing operation.
- **Sustain:** Perform the 5S + 1 methodology on an ongoing basis.

In addition, guarantee safety in everything that's done. Safe practices and methods ensuring worker safety are mandatory.

FIGURE 15.2.2 | Lean manufacturing strategy[1]

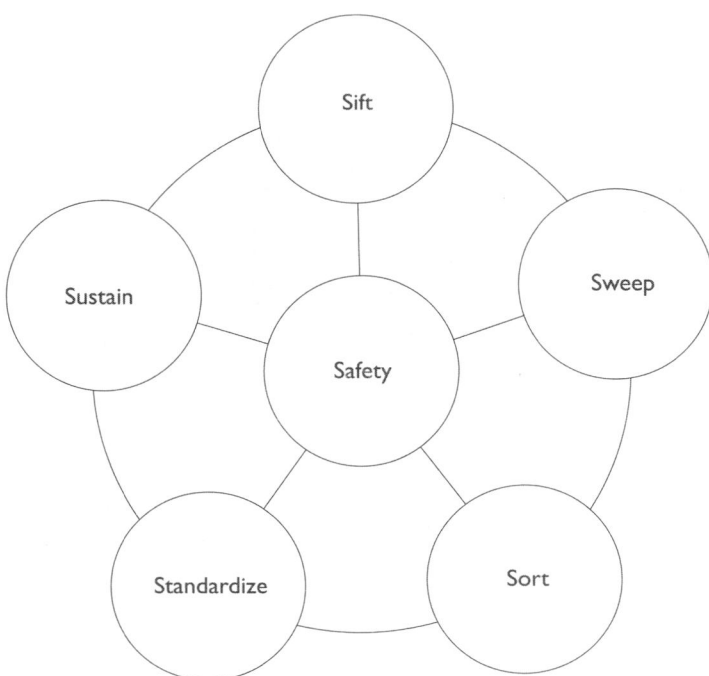

Lean Heat-Treatment Strategies

One of the changes that has taken place in manufacturing is to bring lean strategies into heat treatment given that it is a core manufacturing competency (Section 1.1) and, as such, an integral part of the success of any manufacturing strategy. Many equipment manufacturers are adapting their products to this on-demand philosophy. The 5S + 1 strategy for lean heat treatment (Fig. 15.2.3) involves the following items.

- ❖ **Strategize:** Optimize part and workload configurations to minimize distortion, reduce or eliminate post-heat-treatment manufacturing steps and achieve the best possible quality.
- ❖ **Synergize:** Organize and coordinate processes to reduce cycle times and increase throughput.
- ❖ **Service:** Perform the necessary actions to ensure uninterrupted operation and to maintain production throughput.
- ❖ **Satisfy:** Maintain absolute quality assurance.
- ❖ **Support:** Anticipate equipment and/or process problems to avoid unscheduled downtime.
- ❖ **Synchronize:** Create a safe and mistake-proof working environment within the heat-treatment operation and extend it to the entire manufacturing operation.

FIGURE 15.2.3 | Lean heat-treating strategy[1]

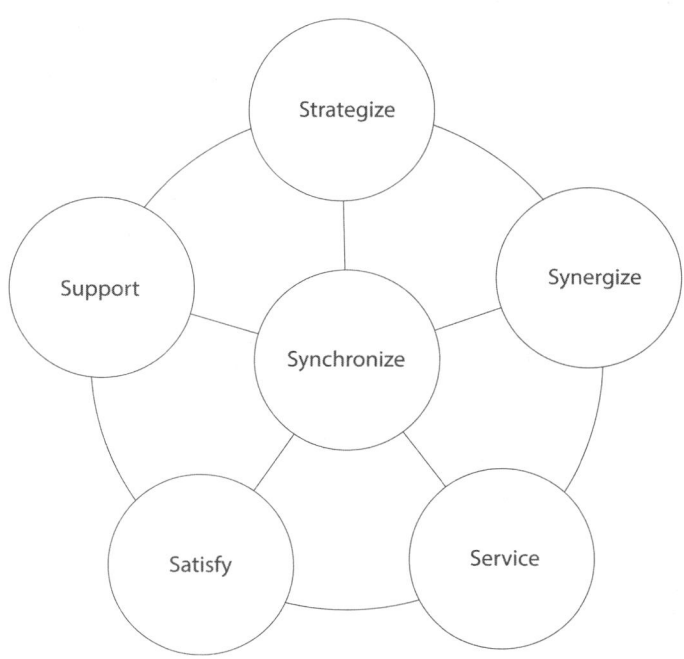

Aluminum Industry Example[2,3]

The processing of aluminum requires a combination of rapid heating and close temperature uniformity throughout the entire load. Many components are safety-critical and must be heat treated with high precision and repeatability. Often, conservation of plant floor space means compact designs are highly desirable. Integration with SCADA systems for real-time data acquisition with upstream and downstream processes is highly beneficial as well.

Batch designs (Fig. 15.2.4) are typically used for aluminum applications where throughput is not predictable or consistent, where the process varies from load to load or where a ramp-up in production favors sequential implementation. Batch units often run a variety of load configurations, so it is critical to have versatile airflow and tight temperature control. Systems typically operate in temperature ranges of 175-675°C (350-1250°F). Heating systems can be provided in electric, gas or indirect gas. Loading can be accomplished by truck, rack/shelf, car, overhead trolley or monorail. Single- and multi-chamber units provide maximum process flexibility. Common applications include aging, annealing, homogenizing and stress relief.

Continuous designs (Fig. 15.2.5) are typically used for automated production and include mesh-belt conveyors, roller hearths, roller rails, rotary hearths, pushers and walking-beam units. Capacities ranging from several hundred to several thousand units per hour are typical. Parts are loaded consistently and uniformly for continuous process flow, so minimum labor is required. Continuous furnace systems significantly reduce operating costs and typically operate in temperature ranges up to 675°C (1250°F). Heating systems can be provided in electric, direct gas or indirect gas. If desired, loading can be automated in a number of different ways. Common applications include aging, annealing and solution heat treating for products as diverse as castings, forgings, plate, bar and tube.

FIGURE 15.2.4 | Batch-style gas-fired drop-bottom furnace with common mobile quench tank for heat solution treatment of aluminum castings[2] (courtesy of Consolidated Engineering Co.)

FIGURE 15.2.5 | Continuous-style mesh-belt aging ovens (courtesy of Wisconsin Oven Corp.)

Many aluminum foundries look at their heat-treatment operations from a lean perspective. For example, high-pressure die-casting (HPDC) for the automotive industry is said to be one of the most cost-effective processes for making large quantities of complex aluminum components in near net shape or final form. In addition, the process offers the potential for die-casters to make complex HPDC components using up to 30% less alloy to achieve the same performance.

While other cast and wrought aluminum alloy parts can be heat treated to improve their mechanical properties (e.g., strength, hardness, toughness, fatigue and ductility), this has not been possible for HPDCs until recently. The reason for this is that HPDCs contain porosity (i.e., trapped gas pockets), which will expand, blister and distort the casting during heat treatment. Previous efforts to remove porosity from cast parts have proved time-consuming and costly.

Available technology permits an inline heat-treatment system to be compatible with a die-casting cell (Fig. 15.2.6).

FIGURE 15.2.6 | Integrated heat-treatment cell for high-pressure die-casting (courtesy of Consolidated Engineering Co.)

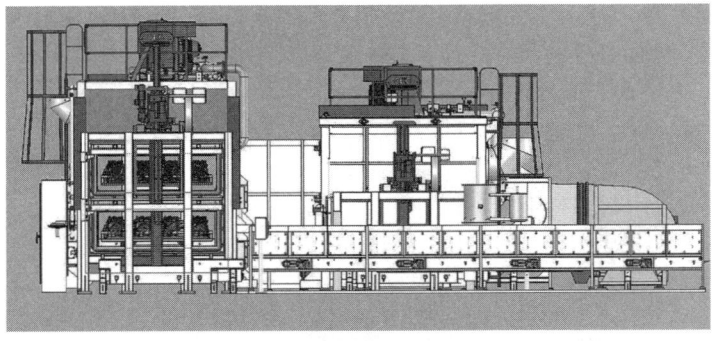

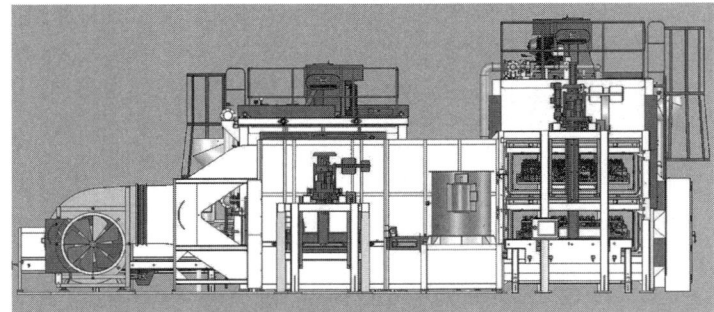

Engine blocks, transmission housings and other complex HPDC components are candidates for this technology. The resulting lighter vehicle weight directly translates into fuel savings and lower greenhouse-gas emissions.

Two aluminum alloys (A380 and A360) that account for more than 80% of the world's aluminum die-castings have been targeted, but a broad range of alloys respond to the treatment (Table 15.2.1).

TABLE 15.2.1 | HPDC alloys[4,a]

Alloy wt.%	Si	Fe	Cu	Mn	Mg	Ni	Zn	Pb	Sn	Ti	Other
A380	7.5-9.5	1.3	3.0-4.0	0.5	0.1	0.5	3.0		0.35		0.5
C380	7.5-9.5	1.3	3.0-4.0	0.5	0.1-0.3	0.5	3.0		0.35		0.5
A383	9.5-11.5	1.3	2.0-3.0	0.5	0.1-0.3	0.5	3.0		0.15		0.5
383	9.5-11.5	1.3	2.0-3.0	0.5	0.1	0.5	3.0		0.15		0.5
A384	10.5-12.0	1.3	3.0-4.5	0.5	0.1	0.5	1.0		0.35		0.5
B384	10.5-12.0	1.3	3.0-4.5	0.5	0.1-0.3	0.5	1.0		0.35		0.5
390	16.0-18.0	1.3	4.0-5.0	0.5	0.45-0.65	0.1	1.5			0.1	0.2
A360	9.0-10.0	1.3	0.6	0.35	0.4-0.6	0.5	0.5		0.15		0.25

Note:

[a] All values shown are maximum unless stated as a range.

TABLE 15.2.2 | Heat-treatment parameters[4,a]

Solution heat-treatment temperature, °C (°F)[b,c]	Quench	Aging (-T6), °C (°F)[d,e]	Aging (-T4), °C (°F)[f]
480-550 (900-1020)	Hot or cold water	150 (300) for 6-24 hours	≥ 20 (68) for four days minimum

Notes:

[a] Applicable to all aluminum alloys shown in Table 15.2.1.
[b] Optimal solution-treating temperature range.
[c] Time between 420°C (790°F) and maximum temperature ideally should not exceed 10 minutes.
[d] Recommended aging treatment for best results.
[e] Alternate temperature of 180°C (355°F) for 2.5-4.0 hours may be preferred for some alloys.
[f] Stabilized condition.
[g] -T7 temper involves over-aging – for example, 180°C (355°F) for 16 hours or 200°C (390°F) for 2-4 hours – and may be used to produce thermal stability at elevated temperature or for improved thermal conductivity.
[h] -T64 temper (e.g., 150°C/300°F for 6 hours) reduces peak hardness and strength (compared to -T6) but improves fracture resistance.

Properties achieved in a particular casting (Table 15.2.3) are dependent on the casting quality, which in turn depends on the casting process, the design and other manufacturing factors. It is important to note that this technology should not be used in an attempt to remedy defective castings.

TABLE 15.2.3 | Selected examples of mechanical property enhancement[4]

Alloy	Hardness, VHN Brinell	Tensile strength, MPa (ksi)[a]	Fatigue limit, R = 0.1 MPa (ksi)[b,c]	Fracture toughness, K_c MPa√m[d,e]	Thermal conductivity @ 23°C (73°F), W/m-K
A380/B380 (as cast)	90-110	320-370 (46.4-53.7)	205 (29.7)	31.3	111
A380/B380 (-T4)	120-130	380-410 (55.1-59.5)	240 (34.8)	41.0	120
A380/B380 (-T6)	145-160	430-460 (62.4-66.7)	260 (37.7)	21.0	129
A380/B380 (-T7)	120-130	360-400 (52.2-58.0)			136
A380/B380 (-T64)	125-135	340-370 (49.3-53.7)		37.1	
C380 – D380 (as cast)	100-115	330-380 (47.9-55.1)	215 (31.2)	36.3	
C380 – D380 (-T4)	130-150	400-420 (58.0-60.9)	230 (33.4)	49.0	
C380 – D380 (-T6)	160-180	450-480 (65.3-69.6)	260 (37.7)	21.6	
C380 – D380 (-T64)	125-145	410-430 (59.5-62.4)		40.8	

Notes:
[a] Results from cast test bars.
[b] Cast axial fatigue-test bars in tension-tension.
[c] Estimated from run-out data at 107 cycles.
[d] Recommended aging treatment for best results.
[e] Derived from ASTM B871 using specimens machined from cast plate coupons.
[f] Fracture toughness estimated from unit propagation energy.

Summary

Lean heat treatment goes hand-in-hand with lean manufacturing and is a cost and technology advantage to those who adopt its concepts and strategies.

REFERENCES

1. Herring, Daniel H., "The Role of Heat Treatment in Lean Manufacturing," *Industrial Heating*, April 2006
2. Howard, Robert D., "Aluminum Heat Treatment Processes, Applications, and Equipment," *Industrial Heating*, February 2007
3. Crafton, Paul, "Heat Treating Aluminum with Lean Strategies," *Foundry Management and Technology*, 2008
4. CSIRO, Australia (www.csiro.au)
5. Shanker Subramanian, Consolidated Engineering Co. (www.cec-intl.com), technical contributions and private correspondence

15 | ANCILLARY TOPICS OF INTEREST

Batch Furnace System: Single-Chain Model

ATLAS
Video Brochure

Ipsen's ATLAS atmosphere furnace line, which is manufactured and serviced in the USA, combines the achievements of past atmosphere furnaces with the evolutionary innovations of the future. This video showcases the single-chain model's features, benefits and technological advantages, including:

- Load size of 36" x 48" x 38" (W x L x H)
- Improved functionality and precision of the quenching system – TurboQuench™
- Intelligent controls with predictive process capabilities – Carb-o-Prof®

Scan the QR code to view the video:

IpsenUSA.com/ATLAS-Video-Brochure

15.3
TOTAL COST OF OWNERSHIP

Some view total cost of ownership (TCO) as merely a financial tool intended to estimate the direct and indirect costs of manufacturing a product. Alternately, TCO can be viewed as a full cost-accounting strategy used to understand product cost over time. Still others use TCO analyses to compare domestic and foreign manufacturing options. TCO has even been applied beyond the initial manufacturing cycle to consider costs such as tax credits and expedited delivery fees. For heat-treatment operations, a TCO analysis can help evaluate equipment purchases and rate suppliers as well as determine the best equipment and process value over time.

TCO as it applies to heat treatment includes life-cycle costs, operational costs and lost revenue due to downtime (Fig. 15.3.1). In particular, issues that create a downtime scenario if not corrected quickly and efficiently may also lead to lost business, which ultimately means loss of market share.

FIGURE 15.3.1 | Total-cost-of-ownership (TCO) model for heat treatment[1]

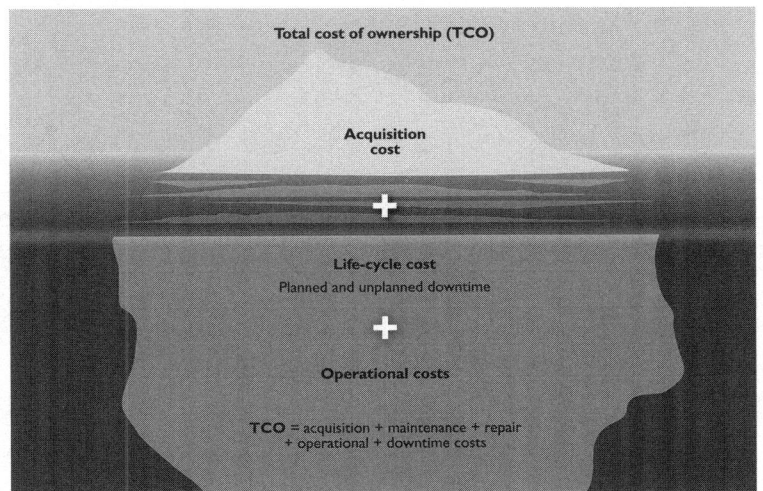

Maximizing Furnace Uptime Productivity

Maximizing furnace productivity requires a proactive approach that must continue throughout a unit's operational lifetime. This requires careful planning and anticipation of problems. The process should begin even before the purchase of a new or used piece of equipment by matching equipment and supplier capabilities with production and process needs. Buying good, well-built and high-quality equipment and operating and maintaining it properly will avoid hidden costs, saving both time and money.

OVERALL EQUIPMENT EFFICIENCY

The evaluative process should begin by calculating overall equipment efficiency (OEE), which is a measure of how effectively your equipment runs (Equation 15.3.1).

15.3.1)   OEE = availability x performance efficiency x rate of quality

Here, availability is the percentage of scheduled time that the unit is available to operate (available time divided by scheduled time).

For example, suppose a work center is scheduled to run for a 435-minute shift. However, the work center experiences 30 minutes of unscheduled downtime. The available time equals 435 minutes (scheduled time) less 30 minutes (downtime), or 405 minutes. The availability is 405 minutes divided by 435 minutes, or 93%. Not bad (or so you think).

Let's look at performance, which represents the speed at which the work center runs as a percentage of its designed speed. In other words, parts produced times ideal cycle time divided by available time.

Continuing our example, if the available time is 405 minutes and the standard rate for the part being produced is 40 units per hour (or 1.5 minutes per unit), then the work center produces 242 total units during the shift. If the time to produce the parts is 363 minutes (242 units times 1.5 minutes per unit), then the performance is 363 minutes divided by 405 minutes, or 90%. Again, not bad (or so you think).

Let's consider quality, which represents the good units produced as a percentage of the total units started, or good units divided by units started.

Continuing our example, if a given work center produces 230 good units of the 242 units that were started, the quality equals 230 good units divided by 242 units started, or 95%. Again, not bad (or so you think).

Now, let's look at the OEE. In our example, OEE = 93% (availability) x 90% (performance) x 95% (quality) = 79.5%. This tells us that our efficiency is not what we would have expected, and improvement is needed and warranted.

ANTICIPATORY MAINTENANCE PLANNING

As with any good maintenance operation, finding and eliminating problems even before they occur is an important proactive approach (Fig. 15.3.2). The trend is toward planned preventive-maintenance programs (Section 13.1) so that manufacturing output remains consistent. Some strategies use:

- ❖ Component usage time monitoring (via hour meters)
- ❖ Critical spare-parts identification (on-site or off-site inventories)
- ❖ Analysis of detailed operational and maintenance records
- ❖ Root-cause determination (when problems occur)
- ❖ Complete explanation of repairs (why, what, where, when and how)

FIGURE 15.3.2 | Hidden costs of unplanned maintenance[1]

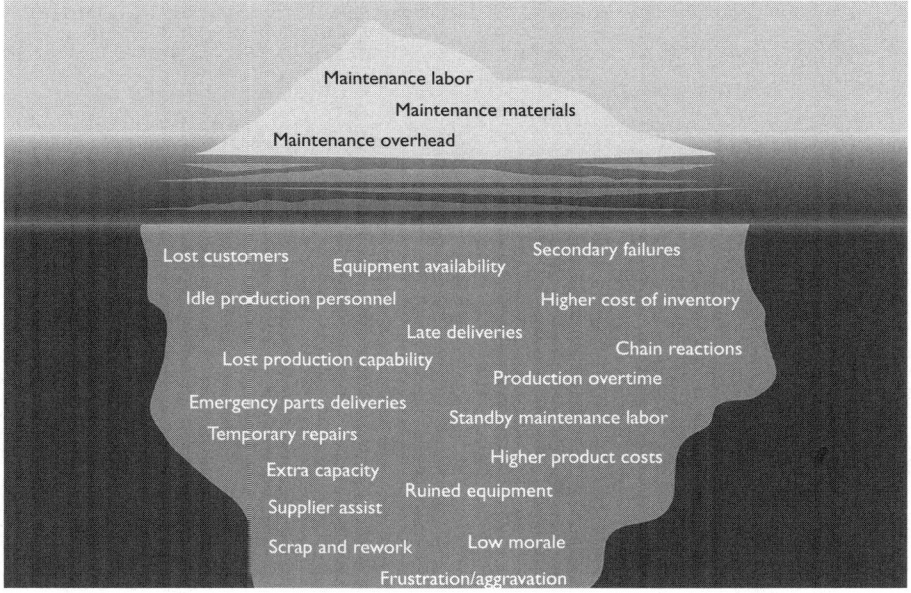

Planned preventive-maintenance programs also allow those involved with the equipment to:

- ❖ Understand the external constraints imposed by issues such as equipment usage, operator experience and budgetary constraints.
- ❖ Understand how the equipment must be serviced.
  - How should the equipment operate?
  - How is it working now?
- ❖ Tailor the plan to meet realistic expectations.
  - Identify necessary spares and have them in stock.
  - Understand which spares must come from an OEM provider and which spares can be purchased from third-party suppliers.

- ❖ Divide the work effort into small areas serviced by specific disciplines.
  - Focus on the components or assemblies (internal or external) that are critical to the functionality of the operation.
  - Do an exterior and interior review and observe how components function and interact.
- ❖ Put the repair information into a usable (i.e., searchable) and retrievable form in order to review needs with management, get feedback through team meetings and revise the plan as needed.
- ❖ Establish a mean time between failures for key components.
  - Conduct cause-and-effect analyses.
  - Determine the root cause of a failure (don't just fix the obvious).
- ❖ Be disciplined.
  - Realize the benefits of having a carefully structured, rigorously adhered-to program (this is not punishment but prevention). In other words, do the job right and on time.
  - Have the tools and supplies on hand to succeed.

Scheduled maintenance can also aid in the reduction of nonconforming production parts (i.e., scrap). Unacceptable parts are often the catalyst for unplanned downtime (Fig. 15.3.3). You should be able to minimize defective parts by being proactive in your maintenance plans.

FIGURE 15.3.3 | The effects of manufacturing downtime[1]

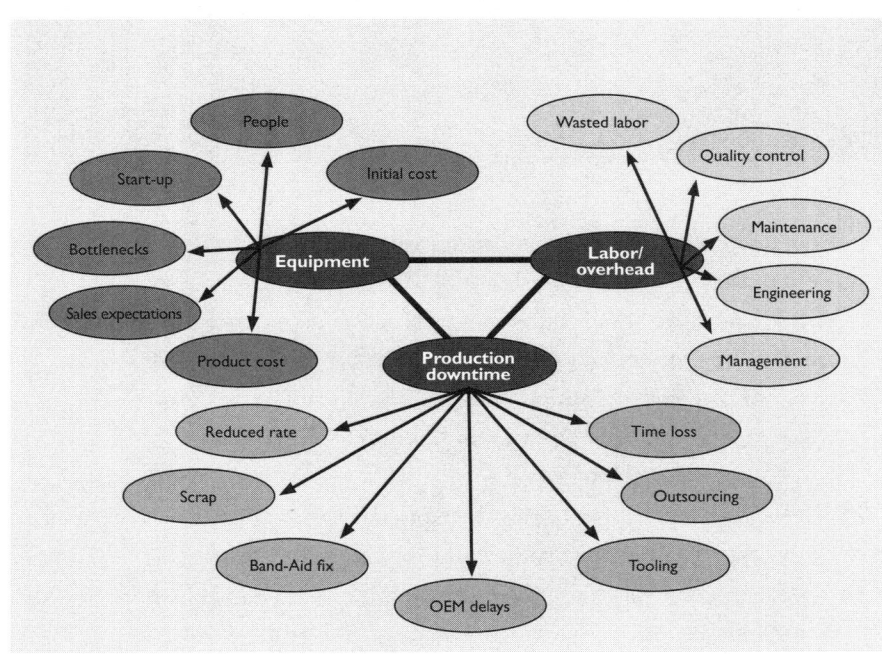

EXPOSING HIDDEN COSTS

Having reliable equipment is a huge factor in TCO. For example, a key consideration when looking to purchase a heat-treat furnace is understanding the True Equipment Cost (TEC), which equates to the initial investment plus the total cost to maintain (over time) divided by equipment life expectancy. In the heat-treat world, this is often 30-50 years!

When considering the purchase of new or used heat-treatment equipment, initial purchase cost should not be the only factor to consider. It is not necessarily the most important factor in determining what to buy either. In addition to the initial cost of the equipment, there are other criteria that should be considered.

1. **TEC.** When analyzing the cost to purchase, one must include not only the initial investment but also the costs to operate and maintain. This involves understanding the utilized capacity of the machine in terms of utilities, manpower, component parts, energy and environmental factors. Recognize that all equipment experiences downtime (planned or unplanned), but the amount of downtime is most often dictated by the robustness of the equipment and the responsiveness of those that service it.

    Dependability must be an integral part of the business decision. Ask questions such as: Will the equipment be reliable and available on a consistent basis? Will it produce quality parts safely while maintaining maximum uptime?

2. **Equipment and process compatibility.** Meeting process requirements is a necessity, and one that existing or new equipment must achieve. Lengthening cycles or compensating for design deficiencies is unacceptable. Equipment that fails prematurely or experiences processing problems or unplanned interruptions in production costs both time and money. Safety and environmental impact should also be considered because we can expect more regulations as time goes on.

    When equipment design reflects process requirements, optimal production capacity can be achieved. Understand if you are buying a heavy-duty or light-duty furnace and plan accordingly. Heavy-duty furnaces can be pushed up to (and often exceed) manufacturer's limits while light-duty equipment typically cannot.

    Uptime availability in the heat-treat industry is typically 85% and approaches over 90% in some (rare) instances. Do not expect more out of your system than you are willing to put into it. Costs associated with unplanned downtime may be 1-3% of asset value per year.

    Other factors to consider in the equipment purchase are how you believe your processes and capacity demands will change over time. Because we live in a customer-driven market, we can be assured that demands will change. Will the equipment be capable of meeting these changes in the future?

3. **Educate the workforce.** Training (Section 11.1) plays an important part in bottom-line profitability. Whether it is the operators, supervisors, engineers or maintenance staff, everyone needs to understand their product, how it is being heat treated and the equipment in which the process takes place. In this way, the support staff (including quality and purchasing) can participate more effectively when something goes wrong.

   A well-educated staff enables a company to achieve the fastest, most cost-effective solution to the problem. Training and experience help analyze the problem quicker, determine a solution and communicate a plan of action for service personnel to implement. This process takes time and commitment on the part of management.

4. **Apply automation and advanced controls.** Smart adaptation of automation, historical recordkeeping and the ability to adapt to the changing needs of heat treatment can increase profitability with only incremental investment costs. Automation takes the human element out of the operating equation, but humans are needed to select and design smart automation systems. This means that sensors and control devices must be part of the planning process.

5. **Avoidance of excess inventory and WIP.** To compensate for poor equipment performance, one often builds additional work-in-process (WIP) inventory. The costs associated with doing so (and often maintaining these increased levels for months or years) can be the difference between being profitable or incurring a loss. Knowing when scheduled maintenance will occur allows you to manufacture parts before your equipment goes down. One can build just enough inventory to make up for short-term interruptions and avoid long-term surprises. However, this assumes that maintenance can be done without discovering or creating additional problems. If these occur, furnaces can be down for extended periods while companies struggle to identify the source of a new problem or fix the result of a malfunction or catastrophe.

6. **Cost of emergency-parts purchases.** If an unplanned failure occurs, it is imperative to get the heat-treat equipment operational as quickly as possible. This will involve getting the necessary parts to repair the equipment, which leads to costs for installation and shipping. All of these increase the cost of manufacturing even though production is at a standstill. Having the right match of features and components on the furnace and having a primary and secondary (sometimes tertiary) supply chain can avoid huge cost overruns.

7. **Overtime and outsourced production.** During the time of an unplanned shutdown, you will still need parts. If other equipment can be used to

15 | ANCILLARY TOPICS OF INTEREST

produce those parts, overtime for manufacturing personnel will be required. This will also increase utility costs and may produce bottlenecks within the manufacturing process. If there is no other available equipment to use for production, it may be necessary to outsource your production to another party. This may jeopardize your promised delivery to your customer, and it will certainly increase your costs.

8. **Cost of product failure.** Ever wonder why we always have enough time to reprocess but never seem to have enough time to do the task right the first time? Product failures, especially those that escape the plant and make their way into the field, are devastating. These are true hidden costs that must be considered when selecting vendor partners.

9. **Impact of equipment deterioration over time.** Everything wears out, and all furnace components have a finite life and must be replaced. We know this and plan for it. When the same components repeatedly fail or the same systems break down time after time, only then do we become aware of the consequences of poor equipment choices. Premature replacement of component parts adversely affects the bottom line. Determining the root cause of these types of problems must be the top priority, and avoiding Band-Aid fixes is an absolute necessity.

10. **Making wrong assumptions.** The most overlooked areas of an equipment purchase are too often the service and support your supplier will provide. Given the complexity of today's heat-treat equipment, your supplier partner must be prepared to walk you through the discovery process to determine why something went wrong; help you understand what happened and what needs to be done to fix it; and help plan and execute its repair.

    Recordkeeping or maintenance logs, for example, should not be written with today in mind. They should tell a story that someone can understand and learn from years later when a similar problem reoccurs. Remember, the knowledge and expertise of your OEM is often the determining factor in the length of your downtime and the minimization of lost profits, so keeping them informed of your current equipment and lessons learned will not only benefit you but the industry as a whole.

11. **Look for the overlooked.** Uptime utilization is the goal in our cost-competitive and technology-driven world. To streamline the manufacturing process, flexibility is required. One must ask questions such as: Do furnace pits make sense, or is it wiser to buy small equipment and integrate heat treating into the manufacturing flow?

    Design and/or material defects can cause unnecessary direct costs (e.g., labor, materials), while fabrication, assembly or installation deficiencies can produce unwanted delays, all causing a hit to bottom-line profitability. Other potential threats that can cause diminished returns are operator and maintenance errors. When these factors are examined prior to equipment purchase and plans are created to supplement the skills of the workforce, there is a better chance of minimizing production costs and maximizing throughput.

12. **Be logical.** Having a blank check to purchase heat-treat equipment is a rare commodity. Every company wants to get the most value for the least expense, but not all equipment is created equal. It's important to understand that an equipment purchase is a complicated process that should be given adequate time to evaluate the hidden and obvious costs. For example, the life expectancy of heat-treat equipment varies, so you will need to partner with your supplier through its entire life.

    Ask yourself questions such as: Will the OEM be there to support you long term, and are they committed (and staffed) with proper engineering and service support? Do they subcontract? If so, how much and to whom?

    Be sure you and your supplier agree on costs (both initial and ongoing), level of service and support and mutual expectations. Finally, do not be overly impressed by technological glamour. Recognize that technology becomes obsolete quickly, and it is the adaptability of the system to new technology that is most important.

13. **Separate needs from wants.** It is important to examine what the company's needs are today and then project an estimate of future needs. The goal is to purchase equipment that is not only capable of meeting today's production capacity but is versatile enough to meet tomorrow's challenges as well. In many cases, the materials, processing requirements and load sizes change, but the equipment features do not. So, building in and paying for flexibility up front will result in long-term cost savings.

14. **Warranty.** The right supplier partner is not averse to extending the equipment warranty to 12-18 months provided certain restrictions and limitations apply. Be wary of the warranty period expiring while waiting for answers or repairs. The industry-standard 90-day warranty is unacceptable in most manufacturing projects.

Atmosphere Furnace Example

Atmosphere furnaces (Fig. 15.3.4) are a good example of sophisticated pieces of equipment whose TCO must be determined during the initial purchasing phase of the project. One of the distinct advantages of this technology is that most of the problems that can occur over time are well understood and solutions to them are well known. In addition, the technology is robust and mainstream, reducing risk and providing good quality.

FIGURE 15.3.4 | Typical atmosphere furnace installation (courtesy of Lawrence Industries Inc.)

Summary

OEMs and suppliers need to differentiate themselves by their support, service, speed of response and thoroughness. Many times, low-priced, light-duty furnaces with poor support ultimately cost more to own and operate. Supplier partners who lack "critical velocity" or who are unwilling or unable to assist, add cost to the user rather than save money in the long run. Doing the small things, such as managing and coordinating parts or having personnel and resources at the ready, can equate to a worthy vendor partner that will save you money by maximizing furnace uptime productivity.

REFERENCES

1. Herring, Daniel H. "Saving Money by Maximizing Uptime Productivity," *Industrial Heating*, January 2013
2. Fitchett, Don and Mike Sondalini, "True Downtime Cost Analysis," Business Industrial Network (www.bin95.com)
3. Drive Your Succe$$ (www.driveyoursuccess.com)
4. Bell, Donald R., "The Hidden Cost of Downtime: A Strategy for Improving Return on Assets," *Maintenance Technology*, July 2001
5. Downtime Central (www.downtimecentral.com)
6. Randall S. Wheeler, Pine Tree Castings, private correspondence
7. Alan Charky, VAC AERO International Inc. (www.vacaero.com), private correspondence

15.4
THE ROLE OF METALLURGY IN ENGINEERING AND MANUFACTURING

We live in a material world. It is the role of materials science (Fig. 15.4.1) to study, develop, design and operate processes that transform raw materials into useful engineering products intended to improve the quality of our lives. It is said by many that material science is the foundation on which today's technology is based, and real-world applications would not be possible without the materials scientist. If this is the case, why are we graduating so few metallurgical engineers? Part of the answer is the role the metallurgist plays in product design and manufacturing.

FIGURE 15.4.1 | Material-science disciplines

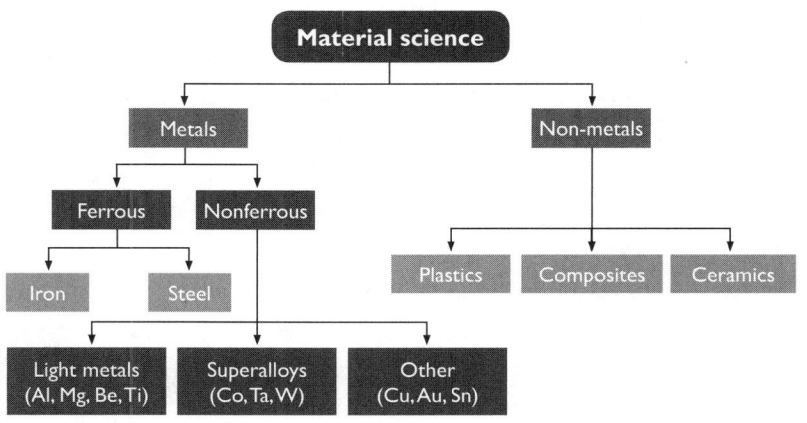

The industrial revolution thrust metals into the forefront of technology, and they have stayed there ever since, becoming the very foundation of modern society. One cannot envision a life where our transportation and communications systems, buildings and infrastructure, industrial machines and tools, and safety/convenience devices are not part of our daily lives. Other materials have emerged as complements or threats to metal's dominance. Composites are one such example.

How Metallurgy and Materials Science Differ

Metallurgy is that part of materials science and materials engineering that studies the physical and chemical behavior of metallic elements, intermetallic compounds and their alloys. This definition is all-encompassing and includes the study of processes run in furnaces and ovens, the forging and rolling of metals, foundry operations, electrolytic refining, creation and use of metal powders, welding, heat treatment and much more. One might say that metallurgy is the application of knowledge of materials, while materials science focuses on the theory of the structure, properties, processing and performance of engineering materials.

Metallurgy is also the technology of metals – the way in which science is applied to the production of metals (including heat treatment) and the engineering of metal components for use in consumer products and manufactured goods. The production of component parts made from metals is traditionally divided into several major categories:

* Mineral processing, which involves gathering mineral products from the Earth's crust.
* Extractive metallurgy, which is the study and application of the processes used in the separation and concentration of raw materials. Techniques include chemical processing to convert minerals from inorganic compounds to useful metals and other materials.
* Physical metallurgy, which links the structure of materials (primarily metals) with their properties. Concepts such as alloy design and microstructural engineering help link processing and thermodynamics to the structure and properties of metals. Through these efforts, goods and services are produced.

What is metallurgical engineering?

Metals and mineral products surround us everywhere – at home, on our way to and from work and in our offices or factories. They form the backbone of modern aircraft, automobiles, trains, ships and endless recreational vehicles; buildings; implantable devices; cutlery and cookware; coins and jewelry; firearms; and musical instruments. Their uses are endless. While threats abound from alternative material choices, metals continue to be at the forefront and are the only choice for many industrial applications.

Developing new materials, new processes to make them, and testing new theories and models to understand them are focal points for the metallurgist. We have the means to measure properties at the macro, micro, nano and atomic scales, giving us unprecedented access to fuel new developments. The strong dependence of our society on metals gives the profession of metallurgical engineering its sustained importance in the modern world.

It is believed by most that our economic and technical progress into the 21st century will depend in large part on further advances in metal and mineral technology. For example, advancements in energy technologies, such as the widespread use of nuclear fusion, will only be possible by material developments not yet in existence. The future is indeed bright for material scientists and those engineers who choose metallurgy for their career.

Why are there so few metallurgists?

The demand for careers in metallurgy is regrettably not at the forefront of our educational system. This is due in large part to the inability of the metallurgical community to communicate to management and academia our role in engineering and manufacturing. While metallurgists should be involved in all aspects of modern engineering, this is seldom the case. The reason for this is often centered on a misunderstanding of what we do, which is made more difficult by how metallurgists begin an answer to every question with "it depends…"

In many cases, this leads to management's belief that other engineering disciplines can replace our skill set. The failure of management to understand what we do is often a failure to understand the engineering life cycle and the interrelationship of engineering disciplines to each other.

Engineering Life Cycle

In the design of any engineered component, it is necessary to fully understand and address two key questions, which the metallurgist is best qualified to answer.

1. **What must the component endure during service (i.e., what are the product requirements)?** Questions such as the following must be addressed: What are the rigors of the application, and what is the design life? Must the component part provide premier service, or is there an adequate design life that is involved? What loading, lubricants, temperature and contaminants are involved? What other service/performance aspects specific to a particular product must also be factored into the selection process?

2. **How will the component part be made (i.e., what are the process/manufacturing requirements)?** Questions such as the following must be addressed: How will its basic form be generated, and how will it be heat treated (if at all)? Will it be important to introduce particular mechanical properties? If so, will they be introduced by heat treatment or mechanical means? Is geometry or surface finish important? Will special coatings be used? Is dimensional control (stability or stability at temperature) an issue? What other processing aspects specific to the particular product must be considered?

Obviously, product/process engineering, performance engineering and metallurgical engineering are not separate entities. They are highly interdependent, and all these disciplines must be considered in the product life cycle (Fig. 15.4.2). However, one must also recognize that cost demands often require compromises in material and manufacturing selection to meet logistical, supply-chain and inventory requirements. Fortunately, that does not mean that selection needs to be minimized. If done correctly, the needs of all parties can usually be met with excellent success while maintaining realistic economic, manufacturing and performance goals.

FIGURE 15.4.2 | Product life cycle[8]

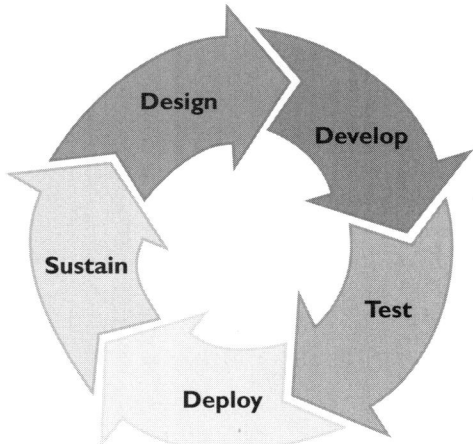

The role of the metallurgist is especially important during the engineering stage of product development. A metallurgist's participation enhances both the design and the capability of a manufacturing process to achieve the desired outcome. During this phase, there is a point at which manufacturing commences. The role of the metallurgist or metallurgical engineering group is to provide critical input in the following areas:

- ❖ Materials selection
- ❖ Manufacturing strategy
- ❖ Process development
- ❖ Equipment selection
- ❖ Controls development
- ❖ Variability assessment
- ❖ Testing criteria

Metallurgists and metallurgical engineers are also responsible for interfacing with manufacturing in an effort to meet production demands in an environmentally responsible way by designing processes and products that minimize waste, maximize energy efficiency, increase performance and facilitate recycling. Metallurgists have seldom been viewed as part of the manufacturing mainstream, however, which is another part of the problem. Gone are the days when every manufacturing plant had a chief metallurgist and multiple metallurgists on staff.

Summary

It is often said that mechanical, electrical, computer or software-related problems can always be solved if one dedicates enough time and money to the task. Solving a metallurgical problem, however, is not a function of money. The solution may be impossible to achieve, forcing one to revisit the very design of the product and its application end-use. It is for this reason, if no other, that the metallurgist exists and is the person who must be involved in every product design. As metallurgists, it is our responsibility to ensure that educators and executives understand the role we play.

REFERENCES

1. Herring, Daniel H., "The Relevance of Metallurgy in Engineering and Manufacturing," *Industrial Heating*, September 2014
2. The University of Utah (www.metallurgy.utah.edu)
3. The University of Queensland, Australia (www.uq.edu.au)
4. Wikipedia (www.wikipedia.com)
5. The Princeton Review (www.princetonreview.com)
6. New Mexico Tech (www.nmt.edu)
7. Illinois Institute of Technology (www.iit.edu)
8. Tech Mahindra (www.techmahindra.com)

15.5
THE ROLE OF HEAT TREATMENT IN REVERSE ENGINEERING

Reverse engineering has been called a journey of discovery, designed to uncover the fundamental principles of how a component was manufactured through analysis of its functionality, properties and microstructure. The goal of reverse engineering is to save cost by reducing analysis time and testing.

Reverse engineering often requires duplication or redesign of the component part, typically in a short time period using such tools as computer-aided-design modeling and rapid prototyping. This process is typically used for highly specialized components. Speed and flexibility over traditional design and manufacturing methods is the hallmark of reverse engineering.

One of the challenges that all engineers in this field face is how the components they are reverse engineering were heat treated and the determination and duplication of the heat-treated microstructure. It's time to understand more about this methodology and how our industry can contribute to its success.

Why use reverse engineering?
Corporations often benchmark their products against those of competition by systematic disassembly and evaluation of the function of each component. Reverse engineering is also commonly used in the aerospace industry to produce replacement parts when the original source no longer exists, the product has become obsolete or the original product design documentation has been lost (or never existed). Reverse engineering can also correct previous mistakes or make a good feature into a great one.

The Process

In general, reverse engineering includes the following:
- Form-fit-function analysis
- Application review
- Material identification
- Mechanical-property determination
- Heat-treatment determination
- Confirmation testing

Typical steps in the process involve:
1. Examine the product. Understand (predict) how the product should work or how the product works before disassembly.
2. Gather and analyze (interpret) all available information including customer requirements. With government projects, there may be overhaul manuals that define some or all of the critical dimensions and tolerances.
3. Acquire parts to use for dimensional data (e.g., size, shape, fits, interface considerations, etc.) and evaluate types of conditions to which the reverse-engineered component will be subjected (e.g., for wear issues, concerns should be part loading, seal surface condition, lubricant type, fluid erosion, etc.).
4. Inspect and record all dimensions of the assembly.
5. Document the assembly with digital photographs.
6. Create a plan. Disassemble and document each part. Pay special attention to areas of wear, coatings, etc.
7. Develop a detailed understanding of the actual product function (structure) based on the disassembly process.
8. Establish primary engineering data.
9. Inspect each part visually and optically (via stereomicroscope). If appropriate, use other NDT methods.
10. Create a materials/components list (bill of materials).
11. Apply engineering analysis, simulations and build models (prototypes). Create preliminary drawings of each part and the assembly.
12. Establish tolerances from statistical analysis of measurements (if there are sufficient samples to inspect) or preferred standards.
13. Perform a manufacturability review of the drawing.
14. Perform chemical analysis to determine type(s) of material(s) in use.
15. Perform metallurgical, mechanical and physical testing to establish mechanical properties, thermal treatment and coatings. Results should include surface treatments and hardness, case depth (if applicable) and core hardness.

16. Deduce heat-treatment processes including pre-treatments, localized effects, heat treating and mechanical working.
17. Deduce other manufacturing processes that will impact properties or end-use performance.
18. Examine metallurgical structure for grain size, shape and grain flow.
19. Revise drawing with material chemistry, heat-treat recipes and specifications.
20. Produce finished components and perform functionality and life tests.

The engineering drawings created will provide dimensions, tolerances, finishes and other critical requirements. Function, however, is not normally specified on drawings or even referenced in specifications, so it must be determined by performance testing that simulates the mechanical, physical, metallurgical and environmental requirements of the end-use application. This requires an understanding of the materials, processes and heat treatments as well as the gathering of information, supplier records and material callouts.

General notes on drawings may help if they specify manufacturing operations. For example, drawings may call out method of cleaning, type of heat treatment, type of material, surface treatment (e.g., coatings, shot peening), surface finish, ancillary processes (e.g., normalizing, annealing, stress relief), inspection (e.g., penetrant or magnetic particle), process sequence (e.g., heat treatment after welding, inspection after heat treatment), tooling (e.g., fixtures) and quality-control tests (e.g., magnetic particle inspection). Preferred suppliers may also be called out, indicating where to obtain specialized components and difficult-to-procure materials or where to go for specialized processing (e.g., casting, brazing, plating, straightening).

There is a simple cost equation (Equation 15.5.1) in the world of reverse engineering:

15.5.1)   COST = Material + Manufacturing + Inspection + Finishing + Rework

The Approach

Using a sample component known to be of good quality is a luxury seldom afforded the reverse-engineering team. The component delivered is usually worn, damaged or abused in some fashion. This makes the job much more difficult. Part dimensions determined by direct measurement must be altered to take these factors into account. Material must be determined by chemical analysis, but subtle issues (such as trace element chemistry, grain size and prior microstructure) must often be factored into the investigation. Selection of the proper mechanical-test methods is a critical component of the process. One must also realize that results may vary, and validation is an important aspect of the final determination.

Mechanical testing (Section 12.3) helps to determine and/or confirm the heat-treatment methods used. Drawings, if available, typically call out hardness and perhaps key strength or ductility values as a range or a minimum. Indirect mechanical-testing methods (preferred due to their cost) include hardness and conductivity testing, while direct methods include tensile testing, impact and torsion testing and corrosion testing.

Hardness alone cannot be relied on to provide the complete picture in determining heat-treatment details. For example, SAE 4340 bar can be specified in accordance to MIL-S-5000 (air melted) or MIL-S-8844 (vacuum melted). Either form produces identical tensile properties and hardness. However, MIL-S-8844 has superior toughness and low-temperature properties. Similarly, Inconel 718 sheet can be purchased per AMS 5596 or AMS 5597. While both have nearly identical tensile properties, they require different heat treatments and demonstrate different creep properties.

Heat-Treatment Examples

Heat treatment requires knowledge of both a process and a piece of equipment in order to be successful. Controlling both process and equipment variability is the key to ensuring that the design engineer gets the product response he/she is expecting. One of the dangers encountered when reverse engineering a product is to not take into account the service history. Many projects fail despite a correct heat treatment due to having an incorrect specification. Finished effective case depth, for example, is not the same as effective case depth. Also, international standards, global material sourcing and advanced heat-treatment methods and equipment in use may not have been in place when the component was originally manufactured.

EXAMPLE I: BALL SCREW

This B1 Lancer slat-actuator ball screw (Fig. 15.5.1) is made from SAE 4150 material. Examination of this item revealed that it was induction hardened after preliminary groove cutting. The scalloped pattern of the induction-hardened area resulted from material removal that improperly increased the distance to the coil.

FIGURE 15.5.1 | 4150 ball screw

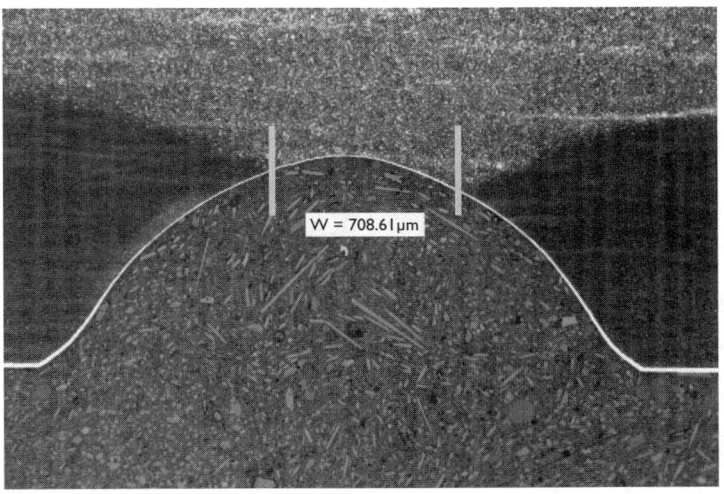

EXAMPLE 2: FLOATING PIN

This P3 Aileron control chain (Fig. 15.5.2) is a carburized component. It was case hardened to provide a wear surface. After installation, the ends of the pin were swaged to retain them in the assembly. Subsequent analysis determined that the ends of the pin were induction annealed to allow the swaging without breaking or cracking of the edges.

FIGURE 15.5.2 | 8620 pin

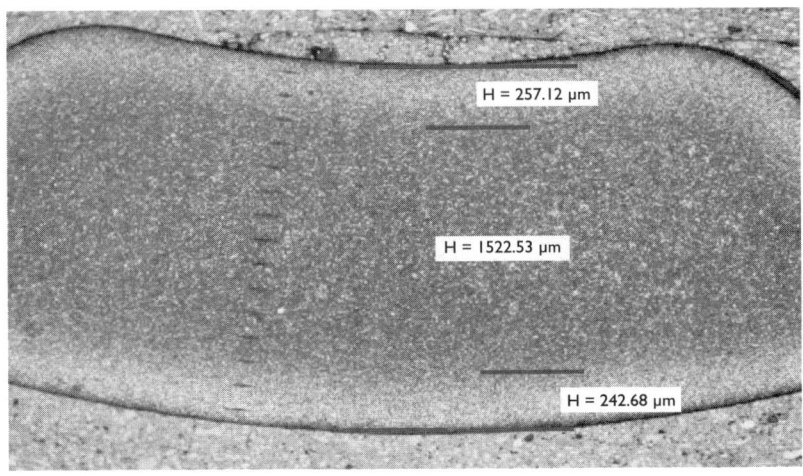

EXAMPLE 3: DEFLECTOR

The C5 Galaxy flap jack screw (Fig. 15.5.3) has carbon-enriched "tips" followed by an anneal from carburizing temperature. It was then induction hardened followed by edge tempering. The core was to remain very soft so that it conforms to a helical ball-screw groove in the nut and does not damage the groove in the screw.

FIGURE 15.5.3 | 8620 deflector

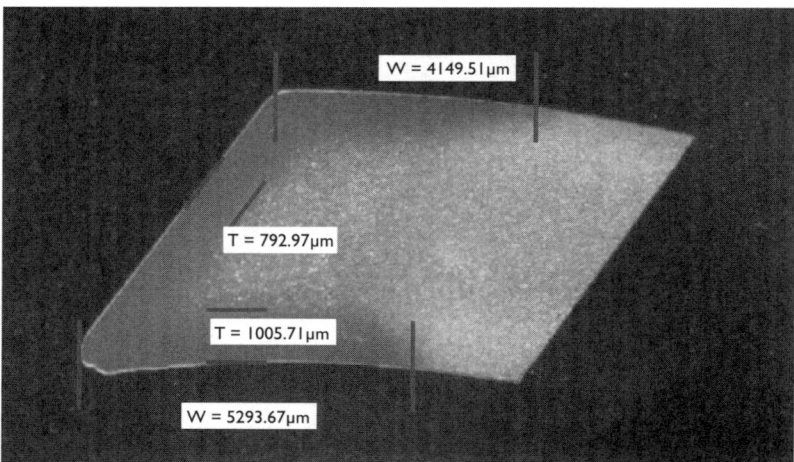

EXAMPLE 4: BALL SCREW

An unexpected process was discovered when examining a C5 Galaxy slat-actuator ball screw (Fig. 15.5.4), namely carburizing. The reverse-engineering assumption was induction hardening, but the chemistry and metallurgy supports carburizing per AMS-2759/7. Examination of this item also revealed that it was partially machined before carburizing.

FIGURE 15.5.4 | 8620 ball screw

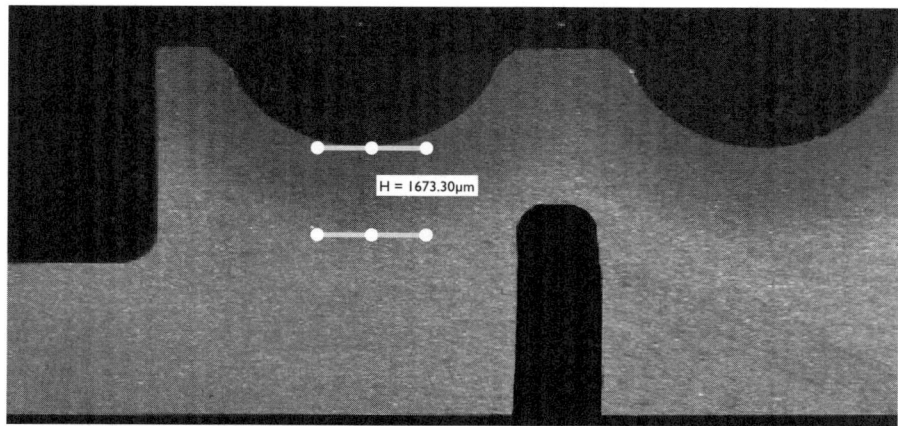

Confirmation Testing

A key question to ask is, "Are you absolutely sure that you perform relevant functional tests?" If not, the part will fail to perform its intended function. Material and process characterization may be incomplete, perhaps melting practices have been overlooked or further inspections are necessary. Information about auxiliary processes, specifications, tolerances and manufacturing sequences may need to be revisited.

Summary

Making an identical part can be extremely difficult or impossible, so extreme care must be taken to ensure that the reverse-engineered product performs to expectation (and beyond).

REFERENCES

1. Herring, Daniel H., "The Role of the Heat Treater in Reverse Engineering," *Industrial Heating*, March 2010
2. Khaled, Terry, "A Metallurgist Looks at Reverse Engineering," Federal Aviation Administration, 2005
3. *Reverse Engineering: An Industrial Perspective*, Vinesh Raja and Jude Fernandes Kiran (Eds.), Springer Series in Advanced Manufacturing, Springer, 2010
4. Bob Rainwater, Thomas Instrument Inc., private correspondence

15.6

EMBRITTLEMENT AND CORROSION

Failure of metals and alloys due to embrittlement and/or corrosion is a broad-based topic and the subject of many excellent books and organizations dedicated to the subject.[1-6] For their part, heat treaters need to be aware of how their processes might contribute to these phenomena.

Embrittlement Mechanisms

Simply stated, embrittlement is a loss of ductility in a material. It can occur in multiple forms, including:

- ❖ Hydrogen embrittlement
- ❖ Stress corrosion cracking
- ❖ Sigma-phase embrittlement
- ❖ Liquid-metal embrittlement
- ❖ Environmentally induced cracking

HYDROGEN EMBRITTLEMENT

Hydrogen embrittlement (Fig. 15.6.1), also known as hydrogen-induced cracking, involves the ingress of hydrogen into the metal, which causes reduced ductility and load-bearing capability with subsequent cracking and catastrophic (brittle) failure at stresses below the yield stress of the susceptible material. Steels, aluminum and titanium alloys are particularly vulnerable.

FIGURE 15.6.1 | Tin-plated electrolytic tough pitch (ETP) copper terminal lugs were embrittled during oxyacetylene brazing and subsequently cracked during forming.

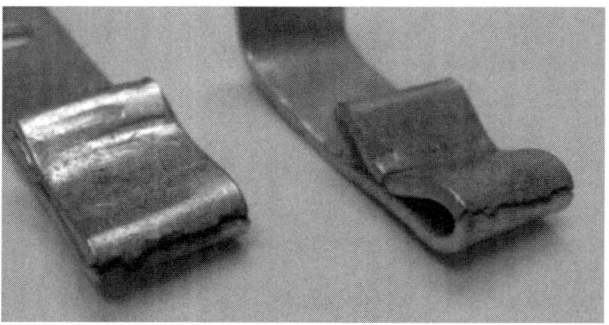

Proper heat treatment and surface-modification techniques offer many part performance benefits, but hydrogen can be lurking nearby and cause unexpected problems. Hydrogen attack in the form of embrittlement is responsible for a surprising number of delayed failures and problems with heat-treated parts, especially if they undergo secondary operations such as plating. Heat treaters need to better understand hydrogen embrittlement because it can affect all types of parts and many of our heat-treatment and brazing processes are conducted in hydrogen-bearing atmospheres.

How Hydrogen Gets In

Hydrogen in atomic form can enter and diffuse through a metal surface whether at elevated temperatures (Table 15.6.1) or ambient temperature. Once absorbed, dissolved hydrogen may be present as either atomic or molecular hydrogen or in combined molecular form (e.g., methane). As molecules, they are too large to diffuse out of the metal and pressure builds at crystallographic defects (dislocations and vacancies) or discontinuities (voids, inclusion/matrix interfaces), causing minute cracks to form. Whether this absorbed hydrogen causes cracking or not is a complex interaction of material strength, external stresses and temperature.

TABLE 15.6.1 | Carbon and hydrogen diffusion rates[7]

	Microstructural phase	Temperature, °C (°F)	Diffusion rate, cm²/second
Carbon	Austenite	925 (1700)	4.0×10^{-7}
Hydrogen	Austenite	925 (1700)	1.7×10^{-4}

Sources of hydrogen include heat-treating atmospheres, breakdown of organic lubricants, the steelmaking process (e.g., electric-arc melting of damp scrap), the working environment, arc welding (damp electrodes), dissociation of high-pressure hydrogen gas and grinding in a wet environment. Parts undergoing electrochemical surface treatments such as etching, pickling, phosphate coating, corrosion removal, paint stripping and electroplating are especially susceptible.

For example, large deep-carburized components can suffer corner exfoliation in the form of large slivers of the case (Fig. 15.6.2), which is characterized by a central initiation coincident with the case-core junction.[7] The initiation of the fracture is often near the center of the fracture surface (etched surface in Fig. 15.6.2) corresponding to the case-core junction. A scanning electron microscope reveals a region of intergranular fracture at the point of initiation.

In this particular study,[7] there was skepticism as to whether hydrogen absorbed during the carburizing process (not originally in the steel) could be the culprit. An accurate analysis of the sources of hydrogen (Fig. 15.6.3) was performed to prove it. A nitrogen purge introduced at the end of the carburizing cycle did not completely remove all the hydrogen. A significant proportion of the remaining hydrogen was removed during a subsequent spheroidize anneal at 620°C (1150°F). In this case, however, the nitrogen purge resolved the problem by lowering the hydrogen level at the critical period in the cycle to levels where no hydrogen-induced cracking was detected.

FIGURE 15.6.2 | Exfoliated steel sliver[7]

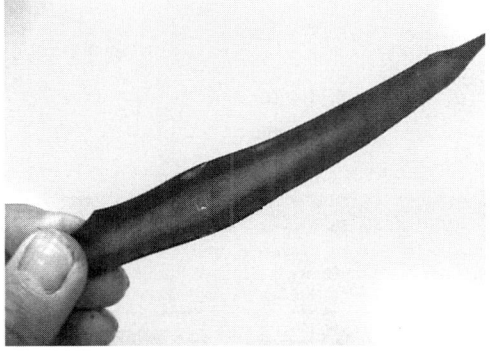

FIGURE 15.6.3 | Hydrogen content after various stages of processing[7]

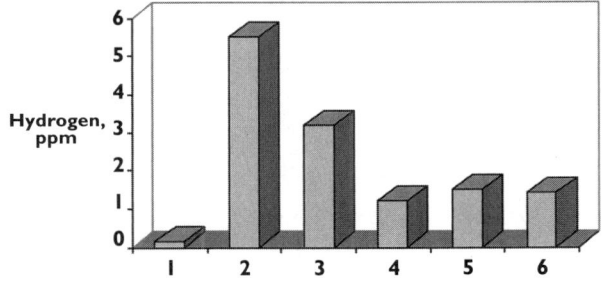

1. as machined
2. after 20 hours carburizing
3. after nitrogen purge
4. after spheroidize anneal
5. after hardening
6. after tempering

Nature and Effect of Hydrogen Attack

Although the precise mechanism(s) of hydrogen embrittlement is being actively investigated (Fig. 15.6.4), the reality is that components fail due to this phenomenon (Table 15.6.2). It is believed that all steels above 30 HRC hardness as well as copper (tough-pitch or oxygen-free), aluminum, titanium and nickel alloys are vulnerable.

FIGURE 15.6.4 | Proposed hydrogen-embrittlement mechanism[28]

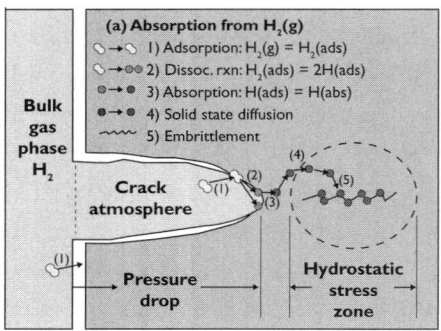
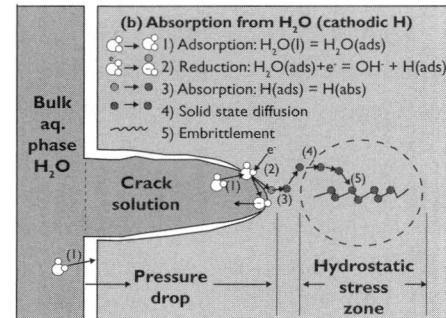

A metallurgical interaction occurs between atomic hydrogen and the atomic structure that inhibits the ability of the material to deform or stretch under load. This makes the material brittle under stress or load, resulting in fracture at a much lower load or stress than anticipated. This lower breaking strength makes hydrogen embrittlement so detrimental.

TABLE 15.6.2 | Examples of hydrogen damage and ways to avoid it

Symptom	Solution
Internal cracking or blistering	Use steel with low levels of impurities (S and P). Modify the environment to reduce hydrogen charging. Use surface coatings and effective inhibitors.
Loss of ductility	Use low-strength (hardness) or high-resistance alloys. Select the proper materials of construction and plating systems. Heat treat (bake-out) to remove absorbed hydrogen.
High-temperature hydrogen attack	Select the proper material (e.g., low- and high-alloy Cr-Mo steels). Limit the temperature and H_2 partial pressure.

In general, steel susceptibility to hydrogen embrittlement increases with increasing steel strength. High-strength steels, such as quenched-and-tempered and precipitation-hardened steels, are particularly vulnerable.

Nonferrous materials such as tough-pitch coppers and even oxygen-free coppers are subject to a loss of (tensile) ductility when exposed to reducing atmospheres. Bright annealing in hydrogen-bearing furnace atmospheres or torch/furnace brazing are typical processes that can induce embrittlement of these materials. Atomic hydrogen diffuses into the copper, subsequently reducing cuprous oxide (Cu_2O) to produce water vapor and pure copper. Embrittled copper can often be identified by a characteristic surface blistering, which results from expansion of water vapor in voids near the surface.

Purchasing oxygen-free copper is no guarantee against the occurrence of hydrogen embrittlement, but the degree of embrittlement depends on the amount of oxygen present. For example, CDA 101 (oxygen-free electronic) allows up to 5 ppm oxygen while CDA 102 (OFHC) permits up to 10 ppm. A simple bend test is often used to detect the presence of hydrogen embrittlement. Metallographic examination for the presence of voids at grain boundaries and at the near surface is also used.

How Hydrogen Gets Out

Hydrogen absorption need not be a permanent condition. If cracking does not occur and the environmental conditions are changed so very low levels of hydrogen is generated on the metal's surface, hydrogen can either diffuse back out or further inward into the material and ductility is restored. Performing an embrittlement relief, or hydrogen bake-out cycle, which involves both diffusion within the metal and outgassing, is a powerful method to eliminate hydrogen before damage can occur. Key variables are temperature, time at temperature and concentration gradient (atom movement).

For example, electroplating provides a source of hydrogen during cleaning and pickling cycles, but the most significant source is cathodic inefficiency. A simple hydrogen bake-out cycle can reduce the risk of hydrogen damage (Table 15.6.3). It should be noted that there is a great deal of confusion in the plating industry as to the proper length of time to use. In the author's opinion, a minimum bake-out cycle of 24 hours at temperature should be used in all cases. Caution: Over-tempering or softening of the steel may occur, especially on carburized or induction-hardened parts.

TABLE 15.6.3 | Hydrogen bake-out requirements for high-strength steels[a]

Tensile strength, MPa (ksi)	Hardness, HRC	Soak time, hours[b]
1,700-1,800 (247-261)	49-51	22+
1,600-1,700 (232-247)	47-49	20+
1,500-1,600 (218-232)	45-47	18+
1,400-1,500 (203-218)	43-45	16+
1,300-1,400 (189-203)	39-43	14+
1,200-1,300 (174-139)	36-39	12+
1,100-1,200 (160-174)	33-36	10+
1,000-1,100 (145-150)	31-33	8+

Notes:

[a] Per ASTM B850

[b] Post-plate bake-out at 190-220°C (375-430°F)

Influencing Factors

The severity and mode of the hydrogen damage depends on:
- Source of hydrogen – external (gaseous) or internal (dissolved)
- Exposure time
- Temperature and pressure
- Presence of solutions (e.g., acidic) or solvents that could react with metals
- Alloy type and production method
- Amount of discontinuities in the metal
- Treatment of exposed surfaces (e.g., barrier layers such as oxide layers to prevent hydrogen permeation)
- Final surface treatment (e.g., galvanic nickel plating)
- Heat-treatment method
- Level of residual and applied stresses

Low Hydrogen Concentrations Not Safe

Of concern is embrittlement from very small quantities of hydrogen where traditional loss-of-ductility bend tests cannot detect the condition. This atomic-level embrittlement manifests itself at levels as low as 10 ppm hydrogen, and there are many documented cases of embrittlement failures with hydrogen levels this low (1 ppm hydrogen reportedly has been a problem in certain critical applications).

This type of embrittlement occurs when hydrogen is concentrated or absorbed in certain areas of metallurgical instability (e.g., grain boundaries). Both residual and applied stress cause this concentrating action, which tends to sweep through the atomic structure and move the infiltrated hydrogen atoms along with it. These concentrated areas of atomic hydrogen can coalesce into molecular-type hydrogen, resulting in the formation of highly localized partial pressures of the actual gas.

Susceptible Part Types

Although almost any type of part is subject to hydrogen embrittlement, components such as fasteners, high-strength alloys and nuclear components are most susceptible.

Where to Go For Help

A good but relatively untapped source of information for heat treaters about the effects of hydrogen is NACE International, The Corrosion Society (www.nace.org). Local colleges and universities are another source.

Although many of the most severe problems associated with hydrogen embrittlement have occurred with aircraft/aerospace parts, a simple motto to remember is that the part doesn't have to "fly to die." The insidious nature of hydrogen embrittlement continues to cause product failures during both processing and service. These failures are often catastrophic, leading to injury or damage to adjacent structures. Hydrogen damage can and must be avoided.

STRESS CORROSION CRACKING (SCC)

Components fail for a variety of reasons, and among these is a corrosion phenomena characterized by the fact that stress (and/or deformation) is present to provide a trigger that leads to sudden crack formation, propagation and failure. Stress corrosion cracking (SCC) can take the form of:

- ❖ Sulfide stress cracking
- ❖ Chloride-induced SCC
- ❖ Caustic-induced SCC
- ❖ Hydrogen-induced cracking

SCC is a type of failure mechanism caused by a combination of environmental, material and stress conditions (Fig. 15.6.5). It is generally considered the most complex of the failure modes since it can attack soft or hard parts, ferrous or nonferrous materials, ferritic or austenitic structures, and materials in the unalloyed or alloyed state. Cracks may propagate in a transgranular or intergranular fashion or in a combination of the two. The stress, however, must be in the form of tensile stress above some minimum (i.e., threshold) value, usually below the yield stress of the material, and in the presence of a corrosive environment, which can include sulfides, chlorides, caustics and/or hydrogen.

Temperature is a significant environmental factor affecting crack formation, and pitting is commonly associated with SCC phenomena. In addition, catastrophic failure can occur without significant deformation or obvious (surface) deterioration of the component.

SCC phenomena can be affected by many factors in addition to stress level, including alloy composition, microstructure, concentration of corrosive species, surface finish, micro-environmental surface effects, temperature, electrochemical potential and the like. Further complications are initiation and propagation phases and the observation that cracks initiate at the base of corrosion pits in some cases.

FIGURE 15.6.5 | Factors contributing to stress corrosion cracking (SCC)[8]

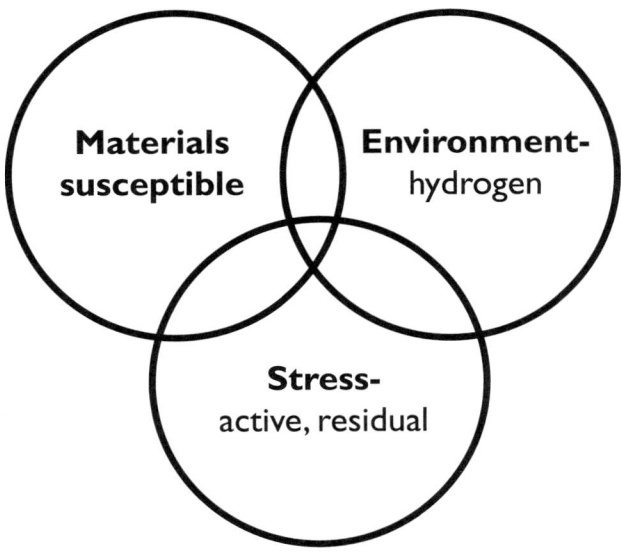

SCC Mechanisms

There is no identified single mechanism explaining SCC, but several theories have been proposed.

- Active-path propagation: localized preferential corrosion (aka dissolution) at the crack tip, along a susceptible path, with the bulk of the material remaining in a more passive state. The rate of metal dissolution can be several orders of magnitude higher when an alloy is in its active state compared to its passive condition.
- Hydrogen embrittlement: high hydrogen concentrates in highly stressed regions such as at the crack tip or other stress concentrators, leading to localized embrittlement.
- Brittle film-induced cleavage: cracks initiated in a brittle surface film may propagate (over a microscopic distance) into more ductile underlying material before being arrested by ductile blunting of the crack tip. If the brittle film re-forms over the blunted crack tip (under the influence of corrosion processes), the process can be repeated over and over again.

SCC Symptoms

SCC is a type of localized corrosion characterized by fine cracks (Fig. 15.6.6) that propagate quite rapidly, leading to failure of the component and potentially the associated structure. The effect is not limited to stainless steels. Steel components in sour-well applications and saltwater environments (Fig. 15.6.7) are other examples.

FIGURE 15.6.6 | SCC failure exhibiting branched cracks – 304 stainless steel fan in a chlorine environment (courtesy of Aliya Analytical Inc.)

FIGURE 15.6.7 | SCC of 4340 oil-rig fasteners in saltwater (courtesy of George Vander Voort Consulting)

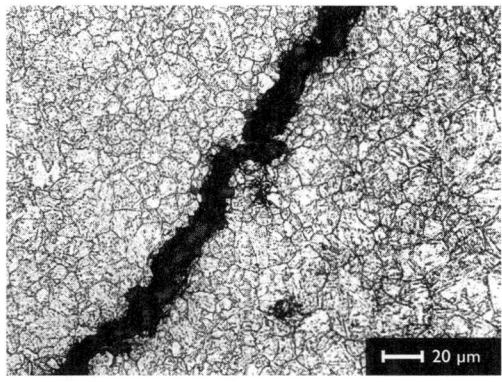

Negating the Effects of SCC

A combination of good design, correct selection of SCC-resistant materials, environmental management, maintenance and inspection can effectively control this type of corrosion. Stresses to consider include:

- ❖ Operational conditions
    - Applied (tensile) stresses
- ❖ Thermally induced factors
    - Temperature gradient
    - Differential thermal forces (expansion and contraction)
- ❖ Buildup of corrosion products
    - Volumetric-dependent
- ❖ Assembly issues
    - Poor fit-up (tolerance problems)
    - Tightening/torqueing

- Press and shrink fits
- Fastener interference
- Joining method
❖ Residual stresses (from the manufacturing processes)
  - Joining (welding, brazing, soldering)
  - Forging or casting
  - Surface treatment (plating, mechanical cleaning)
  - Heat treatment (quenching, phase changes)
  - Forming and shaping
  - Machining
  - Cutting and shearing

One of the most important considerations to negate the effects of SCC is choosing the proper alloy. It is relatively simple to choose a component with adequate strength and good (general) corrosion resistance. However, knowing the particular type of SCC issues that may be at work in the application is an important step in achieving a resistant material. In certain environments, it may be necessary to choose a material that will experience some general corrosion (since general corrosion is visually evident). On the other hand, SCC is rarely visually apparent and often occurs without warning (Fig. 15.6.8). When it does, a catastrophic failure often follows.

FIGURE 15.6.8 | Fastener failure due to SCC

a. Visual examination

b. Optical microscopy (500X, unetched) revealing cracking along grain boundaries (courtesy of Aston Metallurgical Services Co. Inc.)

c. Scanning electron microscopy detailing intergranular attack of plated fastener (courtesy of Aston Metallurgical Services Co. Inc.)

a.
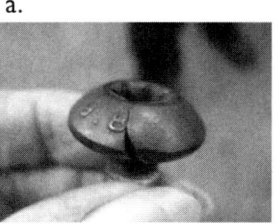

b.
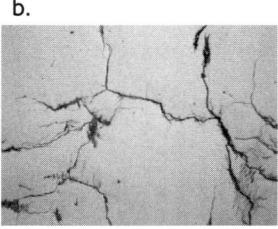

c.
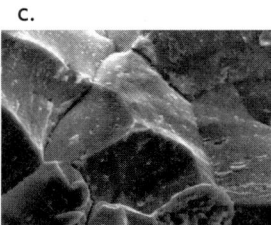

Other methods include removing the corrosive environment or changing the manufacturing process or design to reduce the tensile stresses. A combination of good design, careful selection of stress corrosion-resistant grades (e.g., stainless steel) and effective management, including maintenance and inspection, all can effectively control corrosion. Specific steps can be taken to prevent the onset of SCC and minimize its consequences when it does occur. These steps include:

- ❖ Consideration of the potential for SCC during the design and fabrication of components
- ❖ Selection of appropriate material grades
- ❖ Maintaining a chemical balance of the environment
- ❖ Ensuring that the potential for (organic or inorganic) contamination is minimized
- ❖ Maintaining proper environmental conditions (e.g., air quality)
- ❖ Regular inspections of components for signs of corrosion and SCC

Importance of Material Selection

In many applications, austenitic stainless steel fasteners (e.g., ASTM A193 grade B8) of 304 and 316 stainless steels provide good general corrosion resistance and are commonly requested. In marine environments where stainless steel would seem to be the logical choice, however, alloy-steel fasteners are preferred due to SCC concerns.

Chlorides, fluorides and other halogens are known catalysts for chloride SCC. In order to reduce their susceptibility to general corrosion, alloy-steel fasteners such as grade B7 are usually provided with some type of protective coating (e.g., zinc or cadmium plating). The designer, however, must still be aware that this can lead to another form of corrosion due to environmental stress cracking in the form of liquid-metal embrittlement (LME) or a related failure mode, solid-metal-induced embrittlement, so appropriate cautions must be taken.

In addition to SCC, other forms of embrittlement include environmentally induced cracking due to factors such as cold work (i.e., residual stress), welding, grinding, thermal treatment or service conditions; hydrogen embrittlement from plating, welding or cathodic protection; and as a byproduct of general corrosion or corrosion fatigue.

Finally, careful consideration of these factors as well as taking the time to understand how and where a component will be used in service can help minimize SCC in most applications.

SIGMA-PHASE EMBRITTLEMENT

Many stainless steels and other iron-chromium alloys are susceptible to a grain-boundary phenomenon known as sigma-phase (σ) embrittlement. This type of embrittlement has been shown to cause severe loss of ductility, toughness and corrosion resistance, which results in cracking (Fig. 15.6.9) and failure of components, especially those subjected to impact loads or excessive stress. As heat treaters, we need to know more about what sigma-phase embrittlement is and how to avoid its occurrence.

FIGURE 15.6.9 | Section of a cast HH (25% Cr, 12% Ni) stainless steel furnace load-lifting hook that failed due to sigma-phase embrittlement (courtesy of George Vander Voort)

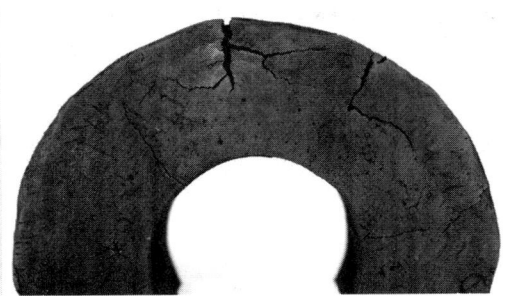

Prolonged exposure in the temperature range of 565-925°C (1050-1700°F) results in chromium depletion from the grain boundaries, which makes them susceptible to intergranular corrosion. The most rapid sigma-phase formation occurs in the range of 700-900°C (1290-1650°F). Alloy elements such as molybdenum, titanium and silicon promote the formation of sigma phase, while nitrogen and carbon reduce its tendency to form.

Sigma phase is an intermetallic compound consisting of chromium and iron, which is hard, brittle and nonmagnetic. Pure sigma forms between 42 and 50% chromium and is one of the equilibrium phases in the iron-chromium phase diagram (Fig. 15.6.10). A duplex structure (sigma and alpha phases) has been found to form in alloys with as little as 20% chromium and as much as 70% chromium when exposed to the critical temperature range noted previously. Sigma phase is difficult to form at chromium contents of less than 20%, but the presence of molybdenum, silicon, manganese or nickel has a tendency to shift the lower limit down.

Molybdenum reportedly promotes sigma-phase formation much more effectively than chromium, particularly at temperatures around 900°C (1650°F). This is why the molybdenum content of the alloy is deliberately kept around 0.5% in the HH cast-stainless example. Austenite-forming elements such as nickel or nitrogen can also accelerate the nucleation and growth of the sigma phase, although these elements may reduce the total amount formed because of the smaller volume fraction of ferrite. Sigma typically nucleates in the austenite-ferrite grain boundaries and grows into the adjacent ferrite. Additional austenite often forms in the areas of chromium depletion adjacent to the sigma phase.

FIGURE 15.6.10 | Iron-chromium phase diagram[9]

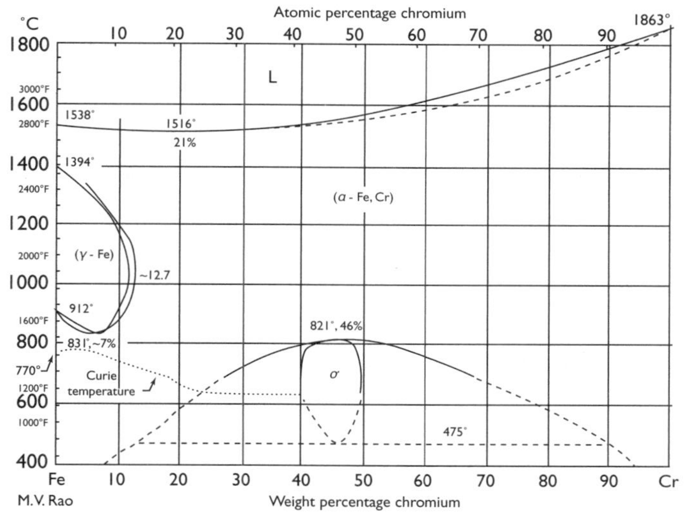

Although the formation of sigma phase is sluggish, cold working enhances the precipitation rate considerably. Sigma phase has even been found in air-cooled, as-cast structures in very high chromium-content alloys, usually appearing as a continuous network in the microstructure. Since sigma has a significantly lower corrosion resistance compared to the ferrite matrix, its presence can be detected by etching in a metallographic examination (Fig. 15.6.11).

FIGURE 15.6.11 | Sigma phase (dark areas; 1000X, KOH) in cast stainless steel (courtesy of Aston Metallurgical Services Co. Inc.)

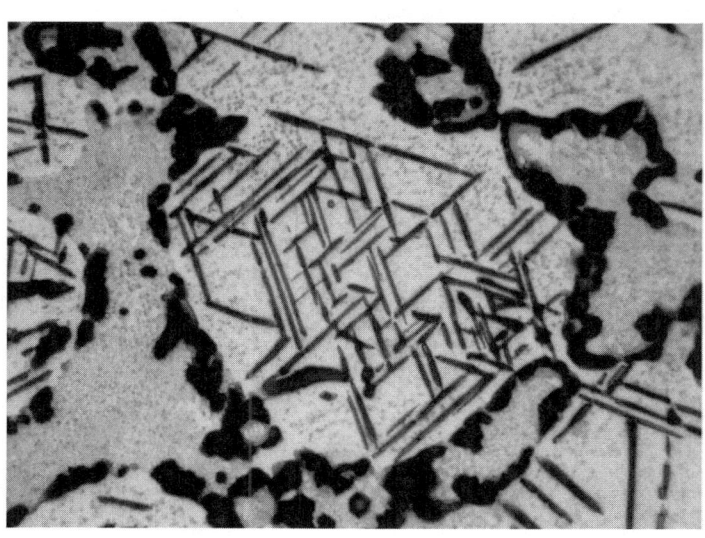

The temperature range of rapid sigma formation coincides with the normal temperatures used for annealing ferritic stainless steels. Consequently, highly alloyed ferritic stainless steels must be annealed and rapidly cooled through their critical range to avoid sigma-phase embrittlement. Any sigma phase already formed can be dissolved again by a solution-annealing process performed above 800-850°C (1475-1560°F) for relatively short times – approximately an hour (once the entire part has reached temperature) – followed by air cooling.

LIQUID-METAL EMBRITTLEMENT

It is well known that certain metals are susceptible to a phenomenon known as liquid-metal embrittlement (LME), aka liquid-metal-induced embrittlement or liquid-metal cracking, when exposed to other metals in the liquid (or solid) state. The embrittlement of aluminum in contact with liquid mercury is a classic example, which is why any mercury-filled device is prohibited on aircraft due to concerns over loss of structural integrity of the aircraft in flight.

Other examples include carbon steels and stainless steels, which are susceptible to LME by zinc and lithium; copper and copper alloys, which are susceptible to liquid-metal cracking by mercury and lithium; and aluminum and aluminum alloys, which are susceptible to LME from mercury and zinc. While the exact mechanisms of embrittlement are complicated, the penetration by the embrittling agent is normally intergranular and the requirements for embrittlement tend to vary depending on the materials involved.

TABLE 15.6.4 | Melting temperature of various liquid metals known to embrittle high-strength steels[10]

Liquid-metal species	Melting temperature, °C (°F)
Mercury	-38.9 (-38)
Gallium	29.4 (85)
Indium	156 (313)
Lithium	180 (356)
Tin	231.7 (449)
Cadmium	321.1 (610)
Lead	326.7 (620)
Zinc	419.4 (787)
Tellurium	449.4 (841)
Antimony	630.5 (1167)
Copper	1082.8 (1981)

The minimal conditions required for LME of steels are: the alloy must be in a state of tension (applied or internal); the surface must be clean and free of oxides, which act as a barrier; and the embrittling species (liquid) must intimately wet the metal surface (Table 15.6.4).

FIGURE 15.6.12 | Effect of tensile stress on embrittlement[10]

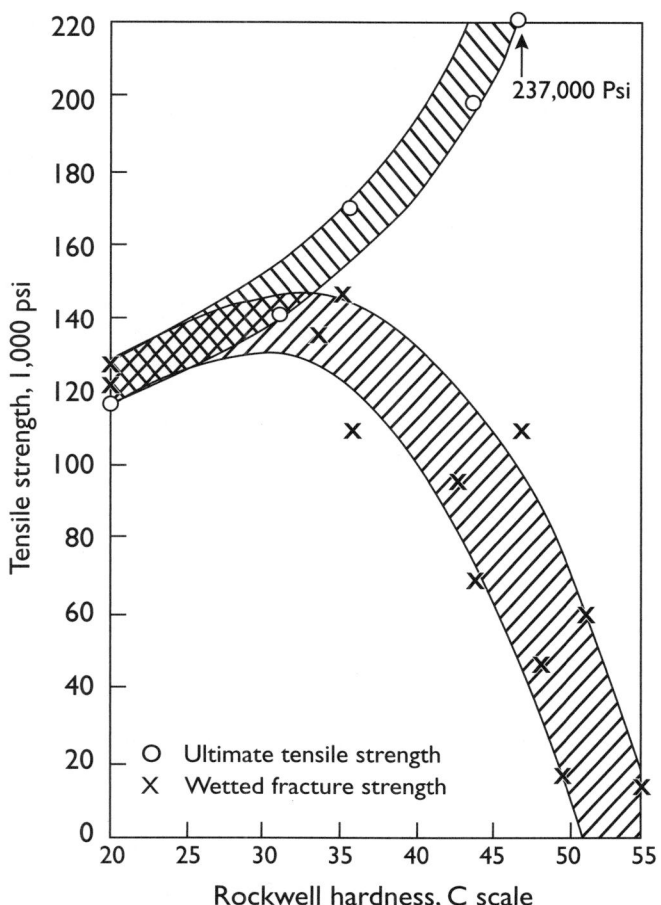

The embrittling effect is most severe when high tensile strength (Fig. 15.6.12) and associated high hardness exist (above 40 HRC). How fast a material will fail due to LME depends on many factors. Under certain conditions, fracture can take place in seconds (Fig. 15.6.13). Crack growth and propagation rates in the range of 0.25-1.0 m/second (0.82-3.3 feet/second) have been measured. An incubation period and a slow pre-critical crack-propagation stage generally precede final fracture.

FIGURE 15.6.13 | Effect of applied stress on time to failure – SAE 4130 exposed to molten lithium[10]

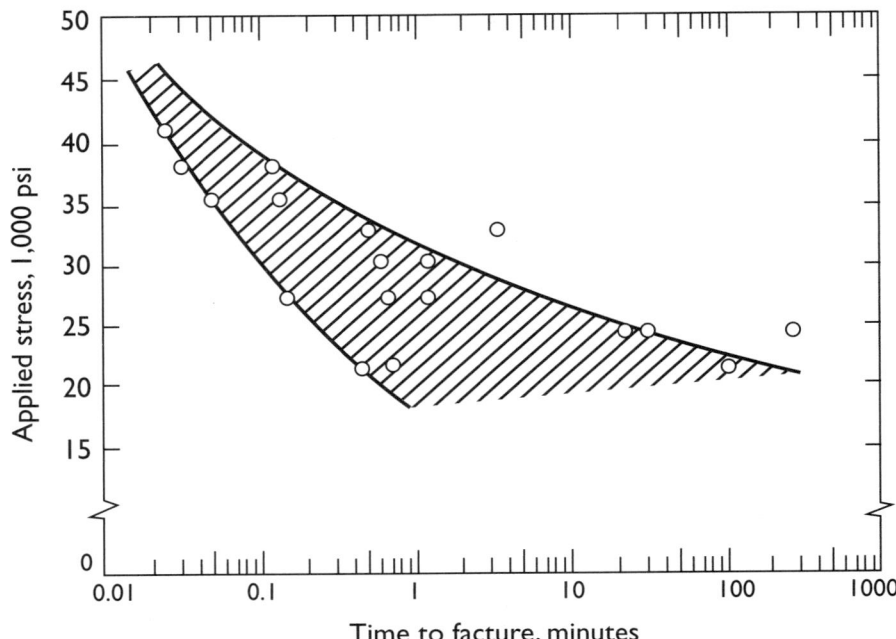

It is not uncommon for certain steels to experience ductility losses and cracking during manufacturing processes such as hot-dip galvanizing or during subsequent fabrication.

LME effects can be observed even in the solid state when one of the metals is brought close to its melting point. This type of phenomenon is also known as solid-metal-induced embrittlement.

Embrittlement Mechanism[11]

Many theories have been proposed for LME.[11,12] All of the following models utilize the concept of an adsorption-induced surface energy.

- ❖ The dissolution-diffusion model of Robertson and Glickman says that adsorption of the liquid metal on the solid metal induces dissolution and inward diffusion. Under stress, these processes lead to crack nucleation and propagation.
- ❖ The brittle fracture theory of Stoloff and Johnson, Westwood and Kamdar proposed that the adsorption of the liquid-metal atoms at the crack tip weakens interatomic bonds and propagates the crack.
- ❖ Gordon postulated a model based on diffusion penetration of liquid-metal atoms to nucleate cracks, which grow under stress to cause failure.

❖ The ductile failure model of Lynch and Popovich predicted that adsorption of the liquid metal leads to weakening of atomic bonds and nucleation of dislocations, which move under stress, pile up and work-harden the solid. Dissolution also helps in the nucleation of voids, which grow under stress and cause ductile failure.

ENVIRONMENTALLY INDUCED CRACKING

Environmentally induced cracking is the result of residual (tensile) stresses created during manufacturing processes. Examples include:
- Cold working
- Welding
- Grinding
- Heat treatment
- In-service conditions

OTHER FORMS OF EMBRITTLEMENT[13]

475°C (885°F) Embrittlement

Iron-chromium alloys containing 15-70% chromium may exhibit a pronounced increase in hardness accompanied by severe loss of ductility and corrosion resistance if exposed to the temperature range of 400-540°C (750-1005°F) for significantly shorter time periods than is required for sigma-phase formation. The name of this phenomenon comes from the fact that the peak hardness usually occurs at 475°C (885°F). In fact, it can occur during slow cooling from an elevated temperature as well as during elevated-temperature service. For alloys containing 18% Cr, the onset of embrittlement is fast enough to require rapid cooling from the annealing temperature to extend below 400°C (750°F) in order to ensure optimal ductility.

In service, alloys containing greater than 16% Cr should not be used at 375-540°C (707-1005°F) for extended periods of time or cycled from room temperature through this critical range. This embrittlement phenomenon is believed to be caused by the formation of a submicroscopic, coherent precipitate that is induced by the presence of a solubility gap below approximately 550°C (1020°F) in a chromium range where sigma phase forms at higher temperatures. Cold work intensifies the rate of 475°C (885°F) embrittlement, especially for the higher-chromium alloys. Reheating the alloy to above 550°C (1020°F) for a few minutes completely removes this form of embrittlement.

High-Temperature Embrittlement

Medium- and high-chromium ferritic alloys containing moderate amounts of carbon and/or nitrogen develop high-temperature brittleness if cooled slowly from above 950°C (1740°F). The mechanism is similar to that of sensitization and leads to severe intergranular corrosion. Work on two wrought ferritic stainless steels containing 18% Cr and 25% Cr has shown that the maximum amount of carbon plus nitrogen tolerable for good room-temperature toughness is 0.055% for the 18% Cr-containing alloy and 0.035% for the 25% Cr-containing alloy.

Duplex Steels Not Immune

In general, the presence of a high percentage of sigma phase is undesirable in duplex stainless steels due to its detrimental influence on corrosion (e.g., pitting) and mechanical properties.[14] Duplex and super-duplex stainless steels are ferrous alloys with up to 26% chromium, 8% nickel, 5% molybdenum and 0.3% nitrogen. They are intended for service in corrosive applications.[15] The metallurgy of duplex and super-duplex stainless steels, especially castings, is complex due to high sensitivity to sigma-phase precipitation on cooling from solidification temperature as well as from heat treatment.

The hardness of these materials is a strong indication of the presence of sigma phase in the microstructure. It has been found that the material hardness is inversely proportional to the heat-treatment temperature. When the heat-treatment temperature during solution treatment increases, the sigma-phase content in the microstructure decreases and, consequently, the material hardness diminishes. When the sigma phase is completely dissolved by heat treatment, the material hardness is influenced only due to ferrite and austenite contents in the microstructure.

The soak temperature also influences the percentage of sigma phase present in solution (as well as in the volumetric concentrations of the ferrite and austenite phases). The ferrite percentage increases with the increasing heat-treating temperature. From 1060°C (1940°F) and up, the sigma-phase quantity is eliminated and the volume fractions of ferrite and austenite each approach 50%.

The presence of sigma phase in stainless steels and iron-chromium alloys should be cause for concern among heat treaters, but awareness of what can trigger this form of embrittlement and what can be done to negate its effects is worth our time and effort.

Corrosion Mechanisms

Corrosion, in contrast to embrittlement, is the gradual destruction of a material by chemical reaction with its environment.

HOT GASEOUS CORROSION

Data and information on the effects of corrosion on engineered materials is available in many forms and from many sources.[1-6, 16-23] The focus for most corrosion engineers is on aqueous corrosion, which is an important topic in and of itself. As heat treaters, however, the effects of hot gaseous corrosion in our heat-treat furnaces are of more immediate concern.

Hot gaseous corrosion should be an area of focus in an effort to extend the life of alloy components, reduce downtime and save money.

Corrosion Basics

We begin with the realization that all materials are chemically unstable in some environments and corrosive attack will occur. It can often be predicted or modeled by studying thermodynamic data and knowing which of the many corrosion-related chemical states are active. In the real world, however, it is important to recognize the various forms of corrosion, namely:

- Uniform (or general) attack
- Intergranular attack
- Galvanic (or two-metal) action
- Erosion
- Dezincification (or parting)
- Pitting
- Stress corrosion
- Electrolytic (or concentration) cells

When it comes to corrosion, the greater the metal's solubility, the greater the degree and severity. There are many important variations of this, two of which are localized corrosive attack (e.g., pits, intergranular attack, crevices, galvanic action) and interaction with mechanical influences (e.g., stress, fatigue, fretting). These actions are frequently very rapid and have catastrophic effects.

There are also a number of ways to combat corrosion. These include alloying to produce better corrosion resistance, cathodic protection (via sacrificial anodes), coatings (metallic or inorganic), organic coatings (e.g., paints), metal purification, alteration of the environment, nonmetallic conversion and design (i.e., physical) changes.

Heat-Resistant Alloys

Furnace interiors contain numerous examples of heat-resistant iron-nickel-chromium (Fe-Ni-Cr) alloys for such items as radiant tubes, fans, heating elements, roller rails and rollers, chain guides and atmosphere inlet tubes. Baskets, grids and fixtures are other examples. These alloys are normally selected based on their strength (at temperature) and resistance to corrosive attack.

Since these heat-resistant alloy parts are often the most expensive furnace components, heat treaters should have an understanding of how they can be attacked and what can be done to extend their life by minimizing or preventing it.

Gas-Solid Reactions

A chemical reaction involving a solid and a non-equilibrium gas or gas mixture can be classified as a gas-solid reaction. Examples of intermediate and high-temperature reactions of this type include oxidation, sulfidation, carburization, catastrophic carburization (Section 13.7), nitriding and chloridation. The principles are the same for all these types; only the details differ. As heat treaters, our interest is in controlling, retarding or suppressing these reactions to prevent unwanted corrosion, gasification or embrittlement of the furnace alloy or materials being processed.

Pourbaix-Ellingham Diagrams

Thermodynamics can be applied to gas-solid reactions such as oxidation, carburization, sulfidation and nitriding to obtain equilibrium dissociation pressures below which no reactions occur. Diagrams are available of the free energies of formation versus temperature for most metallic compounds.

An interesting use of Pourbaix Diagrams (generally reserved for mapping out possible stable equilibrium phases of an aqueous electrochemical system) as a predictor of stable alloy systems is found by superimposing the various elemental constituents. These diagrams are read much like a standard phase diagram (with a different set of axes).

Summary

All materials are susceptible to the various forms of embrittlement and corrosion-related failure mechanisms in one way or another. The key for heat treaters is to avoid creating environments in which these phenomena can act.

REFERENCES

1. NACE International (www.nace.org)
2. ASM International (www.asminternational.org)
3. Uhlig, Hubert H., *Corrosion and Corrosion Control*, Wiley-Interscience, 2008
4. Fontana, Mars G. and Norbert D. Greene, *Corrosion Engineering*, McGraw-Hill, 2008
5. Nateson, K., *Corrosion-Erosion Behavior in Metals*, Metallurgical Society of AIME, 1980
6. *Gas Corrosion of Metals*, National Bureau of Standards, 1978
7. H. Walton, "Ubiquitous Hydrogen," Proceedings ASM Heat Treating Conference & Exposition, 1999

8. Herring, Daniel H., "Fastener Failures Due to Stress Corrosion Cracking," *Fastener Technology International*, August 2010
9. *Metals Handbook, Vol. 8: Metallography, Structures and Phase Diagrams*, ASM International, 1973, p. 291
10. de Rosset, William S., "Use of Liquid Metal Embrittlement (LME) for Controlled Fracture," Army Research Laboratory, ARL-TR-4976, September 2009
11. Wikipedia (www.wikipedia.com)
12. Kolman, D.G., "Environmentally Induced Cracking, Liquid Metal Embrittlement," *ASM Handbook Volume 13A, Corrosion: Fundamentals, Testing and Protection*, ASM International, 2003, pp. 381-392
13. McMahon Jr., C.J., "Brittle Fracture of Grain Boundaries," *Interface Science 12*, 2004, pp. 141-146
14. Reilly, Peter, "Swimming in the Dangerous Waters of Stress Corrosion," *Roof Consultant* (www.roofconsultant.co.uk)
15. Spence, Thomas, "Selecting the Right Fastener," *Materials Newsletter*, Flowserve (www.flowserve.com)
16. Staehle, R.W., "Engineering with Advanced and New Materials," *Materials Science and Engineering A*, Volume 198, Issues 1-2, 1995, pp. 245-256
17. *ASM Handbook Volumes 13A (Corrosion Fundamentals, Testing, Protection), 13B (Corrosion: Materials), and 13C (Corrosion: Environments and Industries)*, ASM International, 2003, 2005, 2006
18. *Oxidation of Metals and Alloys*, ASM International, 1971
19. Javaheradashti, Raza, *Microbiologically Induced Corrosion*, Springer-Verlag, 2008
20. Pourbaix, Marcel, *Atlas of Chemical and Electrochemical Equilibria in the Presence of a Gaseous Phase*, NACE International, 1998
21. Pourbaix, Marcel, *Atlas of Chemical and Electrochemical Equilibria in Aqueous Solutions*, NACE International, 1974
22. Schweitzer, Philip A., *Corrosion Engineering Handbook*, Marcel Dekker, 1996
23. Stempo, Michael J., "The Ellingham Diagram: How to Use it in Heat-Treat-Process Atmosphere Troubleshooting," *Industrial Heating*, April 2011
24. Herring, Daniel H., "What to Do About Metal Dusting," *Heat Treating Progress*, August 2003
25. Naumann, Friedrich Karl, *Failure Analysis: Case Histories and Methodology*, Riederer-Verlag GmbH, p. 183
26. "Hydrogen Embrittlement: A Guide to the Metal Finisher," Omega Research Inc. (www.omegaresearch.com)

27. "Hydrogen Embrittlement, Plating Systems and Technologies Inc." (www.mechanicalplating.com)
28. Kumar, Sidheshwar, "Hydrogen Embrittlement: Causes, Effects & Prevention," Department of Metallurgical and Materials Engineering, NIT – Rourkela
29. Kot, R.," Hydrogen Attack, Detection, Assessment and Evaluation," 10th NDT Conference, Brisbane, Australia
30. Spence, Thomas, "Selecting the Right Fastener," Materials Newsletter, Flowserve (www.flowserve.com)
31. Computational Thermodynamics (www.calphad.com)
32. Klar, Erhard and Parsan Samal, "Powder Metallurgy Stainless Steels: Processing, Microstructures and Properties," ASM International, 2007
33. Mathiesena, T. and J.V. Hansen, "Consequences of Sigma Phase on Pitting Corrosion Resistance of Duplex Stainless Steel," Conference Proceedings, Duplex World 2010, Beaune, France
34. Martins, Marcelo and Luiz Carlos Casteletti, "Heat Treatment Temperature Influence on ASTM A890 GR 6A Super Duplex Stainless Steel Microstructure," *Journal of ASTM International*, January 2005, Vol. 2, No. 1
35. *Hydrogen in Metals,* G. Alefeld and J. Volkl (Eds.), Springer-Verlag, Berlin, Vols. 1 and 2, 1978
36. *Metals-Hydrogen Systems,* R. Kirchheim, E. Fromm, E. Wicke, (Eds.), Verlag, Munchen, 1989
37. Brass, A.M. and J Chêne, "Hydrogen Uptake in 316L Stainless Steel: Consequences on Tensile Properties," *Corrosion Science*, 48 (2006), pp. 3,222-3,242
38. *Embrittlement of Engineering Alloys*, C.L. Briant and S.K. Banerji (Eds.), Academic Press, New York, 1983
39. *Hydrogen in Metals*, I.M. Bernstein and A.W. Thompson (Eds.), ASM International, 1974
40. Kamoutsi, H., G.N. Haidemenopoulos, V. Bontozoglou and S. Pantelakis, "Corrosion Induced Hydrogen Embrittlement in Aluminum Alloy 2024," *Corrosion Science*, 48 (2006), pp. 1,209-1,224
41. Siddiqui, R.A. and H.A. Abdullah, "Hydrogen Embrittlement in 0.31% Carbon Steel Used for Petrochemical Applications," *Journal of Materials Processing Technology*, 170 (2005), pp. 430-435
42. Pan, C., Y.J. Su, W.Y. Chu, et al., "Hydrogen Embrittlement of Weld Metal of Austenitic Stainless Steels," *Corrosion Science*, 44 (2002), p. 1,983
43. Sofronis, P. and I.M. Robertson, "Viable Mechanisms of Hydrogen Embrittlement – A Review," American Institute of Physics, 2006, p. 837
44. Key to Metals (www.key-to-metals.com)

45. Bogner, B., G. Rorvik and L. Marken, "Bolt Failures – Case Histories from the Norwegian Petroleum Industry," *Microscopy and Microanalysis* Volume 11 Supplement S02, August 2005
46. Corrosion Doctors (corrosion-doctors.org)
47. ASTM International (www.astm.org)
48. Alif A. Odeh, ATRONA Test Labs Inc. (www.atrona.com), private correspondence
49. Alan Stone, Aston Metallurgical Services Co. Inc. (www.astonmet.com), private correspondence
50. Fastenal (www.fastenal.com)
51. www.corrosion-club.com

15.7
APPLICATIONS FOR HEAT-RESISTANT MATERIALS

Heat-resistant alloys, ceramics and carbon/carbon composite materials (Section 4.6) are used in both furnace construction and for the transport of product through each step in various heat-treatment processes. Several examples of their use and factors that lead to proper material selection are detailed.

Grids

Grids (aka trays) are used to transport parts or baskets of parts through a heat-treatment furnace and are typically designed to support a specific load weight at a given temperature. There are two widely used designs: cast-alloy trays and wrought-alloy serpentine grids. A newer honeycomb design (Fig. 15.7.1), however, is showing promise. There are no sharp corners because the intersections have a generous radius. They have a tendency not to crack, but the sacrifice is that the legs are quite thick, up to 13 mm (0.5 inch), and they are heavier than conventional cast trays.

FIGURE 15.7.1 | Honeycomb grids (courtesy of Duraloy Technologies Inc.)

Cast grids are the most common type of work transport, with the design consisting of a series of webs with interconnected pockets. They are either single-piece construction (size-dependent) or multi-piece, which are bolted together allowing a certain degree of flexibility. Cast grids have higher creep strength than serpentine grids but may be subject to cracking at the intersections of the various-shaped components, especially in carburizing applications. Welding of cast grids is discouraged. Typical alloys include HT, HK and HX, with many other proprietary chemistries available.

Serpentine grids are made of wrought-alloy strips with alternating straight and bent legs in the shape of a snake, thus the name. Depending on the load to be supported, thin plate or sheet can be used. This design also has drill holes through each section so that a threaded rod is passed through all the strips. A nut is screwed onto the threads and welded to the end of the threaded bar to keep the assembly together. These are the only welds on the fixture. In addition, there is a gap between the nuts and the strips to allow for free expansion and contraction of the entire fixture without constraint, thus preventing cracking. The alloy-selection matrix is the same as for rod baskets.

Baskets

Rod (aka wire bar) and cast (Fig. 15.7.2) alloy baskets are designed to hold small- to medium-sized parts for transport through the furnace during heating and quenching (liquid or air) operations. Major considerations for alloy selection include operating temperature, thermal fatigue resistance and furnace atmosphere.

For high-temperature hardening operations above 980°C (1800°F), such as tool-steel hardening or hardening of stainless steel, the process is done in a protective atmosphere. If a rapid transfer to a liquid quench will be performed at the end of the process, 330 alloy, RA 602 CA® or cast HT alloy are good choices. RA 602 CA alloy is reported to have the best creep strength. With 2% aluminum, it will resist oxidation to 1230°C (2250°F) or slightly higher.

FIGURE 15.7.2 | Cast carburizing basket (courtesy of Duraloy Technologies Inc.)

For heat-treat facilities with multiple operations that include hardening and case hardening, 330 alloy (35% Ni, 15% Cr) is the de-facto industry standard. There is enough nickel to resist carburization, nitriding or a combination of the two. Basket life depends on many factors, but carburizing baskets reportedly last 1-2 years and nitriding baskets even longer in medium-duty service. Manufacturers indicate that over five years should be expected in neutral hardening. As long as the alloy has not completely embrittled from carburization, it can be straightened and reused. Many facilities use Inconel® 600 for nitriding.

For tempering and aluminum solution treat and age, temperatures are generally less than 538-565°C (1000-1050°F). In facilities where baskets can be dedicated, 304 or 316 are sufficient for such operations and are effective from a cost/life standpoint.

There was a trend toward hybrid trays with a cast bottom and rod-frame sides. The castings have higher creep strength and built-in stiffness, hence the tray bottom will not deform. However, one of the major causes of breakage in these baskets is impact from forklifts, and the cast portion is brittle and crack-susceptible, which is why the sides are wrought alloy.

Fixtures

In addition to baskets and trays, some heat-treatment facilities have the need for customized fixtures (Fig. 15.7.3) to support individual parts through the process, often in dedicated lines. Most fixturing is cast alloy and stackable.

FIGURE 15.7.3 | Ferritic nitrocarburizing fixture for brake rotors (courtesy of Rolled Alloys)

Retorts

Retorts isolate the component parts from the heating source and are used to contain the furnace atmosphere. They are typically manufactured from heat-resistant wrought or cast alloys (Fig. 15.7.4). Maximum operating temperature and type of atmosphere are of primary design importance. Thermal fatigue resistance is not much of a factor because they are not subject to quenching and are not normally subject to rapid temperature changes. Creep strength is a factor. Its importance depends on how the load is supported. Materials for construction of wrought-alloy retorts (Table 15.7.1) vary by application.

TABLE 15.7.1 | Wrought-alloy materials for construction of retorts

Process	Alloy selection
Neutral hardening above 1010°C (1850°F)	RA 602 CA®, Inconel® 600, 601
Neutral hardening below 1010°C (1850°F)	330
Carburizing	330
Nitriding	Inconel® 600, 601
Ferritic nitrocarburizing	Inconel® 600, 601, 330
Tempering	304, 316

Retorts are aluminized in certain instances. Recently, there is a trend away from Inconel 600 to RA 602 CA because of the superior creep strength and a multi-fold increase in usable life. This process is run toward the high end of the material's capability.

FIGURE 15.7.4 | Cast-alloy retorts (courtesy of Duraloy Technologies Inc.)

Muffles

Muffles are often manufactured from heat-resistant alloys and ceramics (e.g., silicon carbide, mullite). The standard material of construction for an alloy muffle is a 300-series stainless steel, such as 330. Designs include "D" shaped, rectangular and corrugated muffles (Fig. 15.7.5), some with internal or external stiffeners (or both).

When temperatures start to exceed 1010°C (1850°F), different manufacturers use different paths to achieve resistance to the high heat. Some of the solutions include higher alloy such as Inconel 600 or 601. Each has distinct advantages and disadvantages. Alloy or silicon-carbide hearth plates help protect muffle floors from excessive distortion. The use of silicon carbide is limited in hydrogen-bearing atmospheres because there is a risk of eutectic melting. Normally, muffles are continuously welded for use with a protective atmosphere.

Ceramic muffles (Fig. 15.7.6) are typically supplied in segments (sections) from 1-3 meters (3-10 feet) long. Fiber-blanket joints seal these individual sections, so it is necessary to run an inert atmosphere (e.g., argon, nitrogen) on the outside of the retort since it is not truly gas-tight.

FIGURE 15.7.5 | Alloy muffle (courtesy of Sinterite, a Gasbarre Furnace Group Company)

FIGURE 15.7.6 | Segmented silicon-carbide ceramic muffle in a continuous mesh-belt furnace (courtesy of Sinterite, a Gasbarre Furnace Group Company)

Radiant Tubes

Radiant tubes can be made from cast or wrought alloys and silicon carbide. Traditional alloys are HT and 330. A number of other alloys, such as Hastelloy® X and RA 602 CA, can be used to extend life. Wrought tubes are weldable, and damaged sections can be cut out and replaced without having to scrap an entire assembly.

Cautions

One should be aware of issues that can result during the heating of retorts, muffles and radiant tubes in gas-fired and electrically heated furnaces. In gas-fired units, aligning the burners firing in tubes or directing flame fronts to swirl around retorts extends the life of the alloy and prevents deformation or melting. In electrically heated units, the elements must not be allowed to come in contact with the retorts, muffles or tubes.

Summary

The decision of which heat-resistant material to use and in what design configuration is most often a life-versus-cost decision. Experience often dictates, but new alloys and designs offer attractive alternatives.

REFERENCES

1. Herring, Daniel H., *Atmosphere Heat Treatment, Volume I*, BNP Media, 2014
2. Marc Glasser, Rolled Alloys (www.rolledalloys.com), technical contribution and private correspondence
3. Roman Pankiw, Duraloy Technologies Inc. (www.duraloy.com), technical contribution and private correspondence
4. Marco Möser, Safe Cronite (www.safe-cronite.us), technical contribution and private correspondence

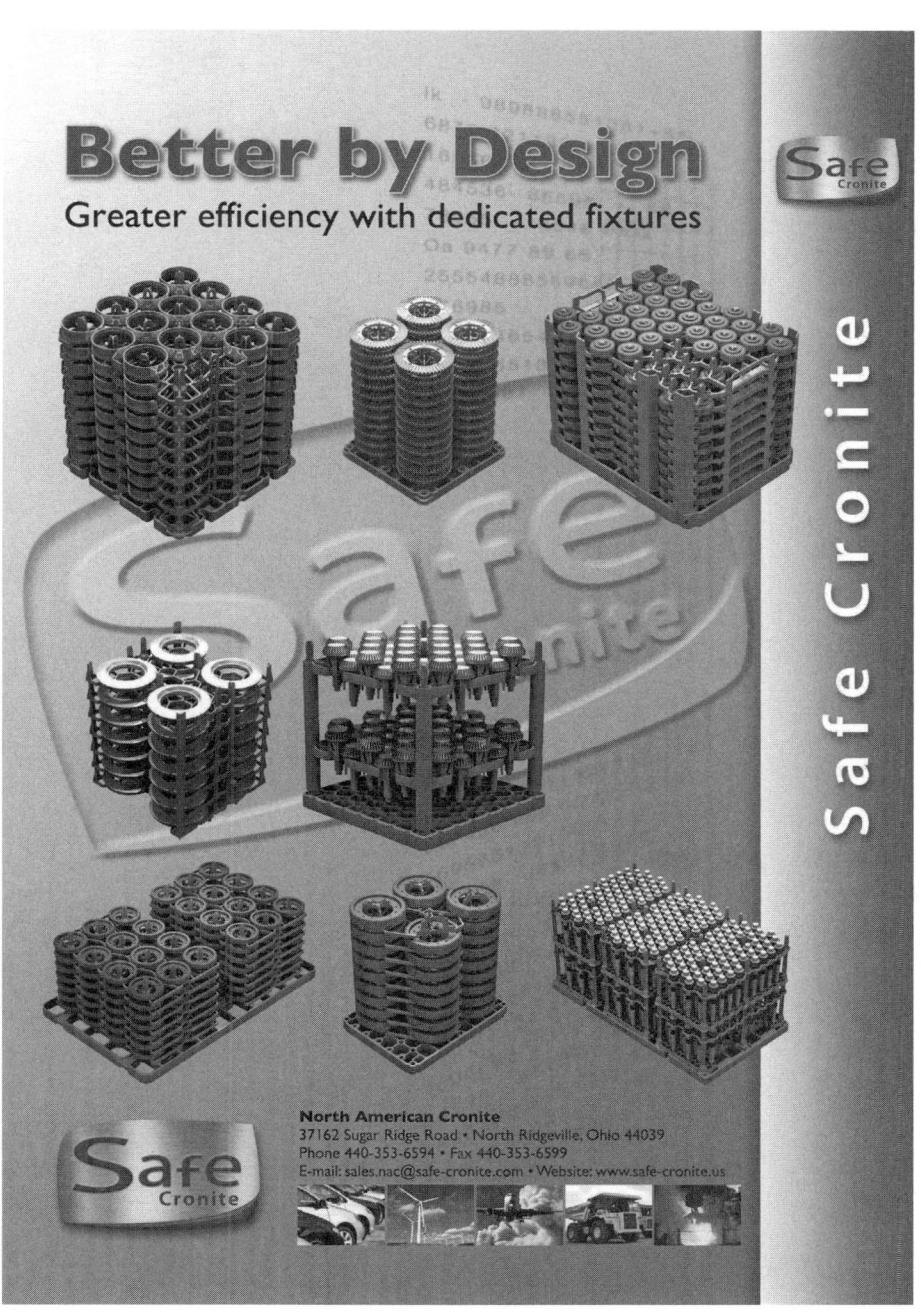

15.8

HEAT-TREATMENT REQUIREMENTS FOR THE FASTENER INDUSTRY

Fastener applications are as diverse as the industries they serve, driven primary by the aerospace, construction and automotive industries with petrochemical, nuclear, medical, marine and mining being important niche markets. It is estimated that some 350 manufacturing plants produce more than 200 billion fasteners per year in the U.S. alone.[2]

Fastener materials also cover the gamut, from ferrous to nonferrous materials including steel, stainless steel, tool steel, aluminum, titanium and exotic specialty grades. Coatings are also in common use and include zinc, cadmium, nickel, galvanized (hot-dip, sherardizing), phosphate and PTFE (polytetrafluoroethylene) to name a few.

Heat treatment (Fig. 15.8.1) plays a critical role in the manufacture of fasteners in order for them to achieve the desired performance properties. Heat treatment is typically conducted after the forming processes and before any coating or finishing process in both captive and commercial shops. The equipment necessary to effectively heat treat steel fasteners includes well-controlled atmosphere furnaces with temperature control, atmosphere control, quenching tanks and cleaning equipment. This equipment typically requires a large capital expense as well as continuing maintenance costs.

FIGURE 15.8.1 | Typical continuous mesh-belt furnace for case hardening of fasteners

Heat-treating systems for fasteners come in a variety of shapes and sizes, but all have one thing in common: flexibility, or the capability to produce large or small quantities of fasteners on demand. It is not uncommon to see production lots as small as 4.5 kg/hour (10 pounds/hour) and as large as 4,500 kg/hour (10,000 pounds/hour).

The most common heat treatments for steel fasteners involve annealing (Section 5.2) of the incoming material, through or selective hardening (Section 5.4) and case hardening (Section 5.5). Case depths typically fall between 0.0038-0.038 mm (0.0015-0.015 inch), and various ranges are specified as 0.076-0.13 mm, 0.13-0.25 mm, etc. (0.003-0.005 inch, 0.005-0.010 inch, etc.), depending on the fastener size. Quench media runs the gamut from brine, water, polymer, oil and molten salt, depending on engineering requirements. Stainless steels (Section 5.13) and many nonferrous fasteners are solution heat treated and age hardened (Sections 6.3-6.5).

Fastener Market

The world market for industrial fasteners represents a $67 billion industry divided between Asia (38.2%), Europe (25.3%), North America (21.9%) and the rest of the world (14.6%).[3] Of this total, the Asian market is dominated by China (14.8%) and Japan (10.1%). Strong growth prevails in the automotive, aerospace and energy sectors, and the industry is expected to show continued growth based on rapid global industrialization and rising demand for durable goods.

GLOBAL INDUSTRY GROWTH

The demand for fasteners is strong. Currently, the aerospace industry is the largest user of fasteners, accounting for over 30% of the total fastener market. Going forward, the use of fasteners by the construction industry is poised to overtake aerospace based on recent activity worldwide. The construction industry's use of fasteners is expected to increase at a compound annual growth rate of just over 9% through 2018.[4] Other OEM segments (e.g., fabricated metal products, electronic/electrical) have also exhibited strong growth in the last few years. The automotive sector is expected to continue to show above average short-term growth.

The Asia Pacific region is expected to account for over 45% of the market by 2018. This trend is due to factors such as rapid industrialization and favorable economic conditions, which are expected to boost the demand for durable goods and other manufacturing and development activities.

Fastener Applications

The heat treatment of fasteners is also highly application- and industry-specific. Highlights of major markets include the following.

AEROSPACE INDUSTRY

Aerospace applications (Fig. 15.8.2) include aircraft (manned and unmanned, fixed and flex wing), rotorcraft (helicopters, gyrocopters) and space vehicles (shuttles, space stations, satellites). Fasteners are one of the most critical components used in all of these applications and are required to meet the most demanding performance characteristics. The types of fasteners in the aerospace industry include screws, rivets, bolts, nuts, pins, collars and washers.

FIGURE 15.8.2 | Critical aircraft performance application for fasteners – jet engine (courtesy of Performance Review Institute)

Fasteners account for a significant number of parts in aircraft and directly affect strength characteristics and weight of structural assemblies. According to The Boeing Company, the 747 includes over six million parts, half of which are fasteners. On average, for example, 2.4 million fasteners are used to assemble a Boeing 787 aircraft. On average, 22% are structural bolts (mostly titanium), and the rest are aluminum rivets.

As the industry evolves to incorporate newer and more exotic materials, fasteners continue to figure prominently in the manufacturing and assembly processes. Fasteners play a critical role in defining the longevity, structural integrity and design philosophy of most metallic aircraft structures.

Typical aerospace fastener materials include aluminum, steel (e.g., A286, H-11) superalloys (e.g., Waspaloy, Hastalloy, Inconel 718), nickel alloys (e.g., Monel, K-Monel) and titanium.

AUTOMOTIVE INDUSTRY

The automotive industry is estimated to use between 25-35 billion fasteners alone. This manufacturing segment continues to see sales of greater than 14.5-15.5 million vehicles in North America. The automotive market share (as of 2012) is divided as follows: GM 17.9%, Ford 15.5%, Toyota 14.4%, Chrysler 11.4%, Honda 9.8% and Hyundai/KIA at 8.7%.

Steel and stainless steels, including duplex and austenitic grades (Tables 15.8.1-15.8.4), as well as plastic fasteners dominate the automotive landscape. Many fasteners are plated or coated for increased corrosion protection.

15 | ANCILLARY TOPICS OF INTEREST

TABLE 15.8.1 | Materials for automotive products (part 1a)

	Carbon steels for cold heading			Alloy steels[a]					Boron steels
	Rimmed steels	Al-killed steels	Killed steels	Ni-Cr-Mo steels	Cr steels	Cr-Mo steels	Mn steels[a]	Alloy steels[b]	Designated by customers
Engine bolts									
Cylinder head bolts				●	●	●			●
Connecting rod bolts					●	●			
Bearing cap bolts						●			●
Flywheel bolts					●	●			●
Tips						●			
Crank pulley bolts							●		
Drive plate bolts					●				
Metal cap bolts						●			●
Rocker stay bolts									●
Adjusting screws						●			
Crankshaft bolts						●	●		●
Brake parts									
Guide pins						●			
Lock pins						●			✗
Wedges						●			
Holder pins			●			●			
Sleeve bolts	●					●			
Adjusting screws	●		●	●		●			
Adjusting nuts	●	●							
Adjusting sockets	●								
Adjusting assemblies	●	●	●						
Roller tappets									

Notes:

[a] For machine structural use

[b] For bolts of high-temperature service

TABLE 15.8.2 | Materials for automotive products (part 1b)

	Steels for special use			Bearing steels				
	Stainless steels, austenitic	Stainless steels, martensitic	Heat-resistant steels	High C-Cr bearing steels	Non-HT steels	Ti alloys	Co alloys	High-strength steels
Engine bolts								
Cylinder head bolts								●
Connecting rod bolts							●	●
Bearing cap bolts								
Flywheel bolts								
Tips								
Crank pulley bolts								
Drive plate bolts								
Metal cap bolts								
Rocker stay bolts								
Adjusting screws								
Crankshaft bolts								
Brake parts								
Guide pins	●				●			
Lock pins					●			
Wedges								
Holder pins								
Sleeve bolts								
Adjusting screws				●				
Adjusting nuts								
Adjusting sockets								
Adjusting assemblies				●				
Roller tappets				●				

15 | ANCILLARY TOPICS OF INTEREST

TABLE 15.8.3 | Materials for automotive products (part 2a)

	Carbon steels for cold heading			Alloy steels[a]					Boron steels
	Rimmed steels	Al-killed steels	Killed steels	Ni-Cr-Mo steels	Cr steels	Cr-Mo steels	Mn steels[a]	Alloy steels[b]	Designated by customers
Special-shape parts									
Stud bolts	●		●	●		●		●	●
Line head bolts	●					●			●
Joint bolts	●		●			●			●
Hub bolts						●			
Serration bolts			●						●
Square head bolts									
Socket bolts	●		●			●			●
Plungers			●			●			
Weld bolts	●		●						
Caulking bolts			●			●			●
Cap square neck bolts		●							
Shoe bolts									●
Seal bolts			●			●			
Rivets	●								
Stepped bolts	●		●						●
Bolts			●						
Adjusting bolts	●								
Nuts									
Weld nuts	●								
Nylon nuts	●								
Multi-point nuts	●					●			
Other Parts									
Part by part-former	●		●						
Automatic joint	●					●			●
FT, FTN, FTY bolts	●	●	●			●			
Standard bolts									
Hexagon head bolts	●	●	●			●		●	●
Hexagon head bolts with flange	●	●	●		●	●			●
Bolts with captive washer	●		●			●			●
Hexagon head cap bolts	●		●			●			

Notes:

[a] For machine structural use

[b] For bolts of high-temperature service

TABLE 15.8.4 | Materials for automotive products (part 2b)

	Steels for special use			Bearing steels				
	Stainless steels, austenitic	Stainless steels, martensitic	Heat-resistant steels	High C-Cr bearing steels	Non-HT steels	Ti alloys	Co alloys	High-strength steels
Special-shape parts								
Stud bolts	●		●			●		
Line head bolts								
Joint bolts								
Hub bolts								
Serration bolts								
Square head bolts					●			
Socket bolts								
Plungers		●						
Weld bolts					●			
Caulking bolts								
Cap square neck bolts								
Shoe bolts								
Seal bolts								
Rivets								
Stepped bolts					●			
Bolts					●			
Adjusting bolts	●							
Nuts								
Weld nuts								
Nylon nuts								
Multi-point nuts								
Other Parts								
Part by part-former	●							
Automatic joint		●						
FT, FTN, FTY bolts			●					
Standard bolts	●							
Hexagon head bolts			●			●		
Hexagon head bolts with flange	●		●			●		
Bolts with captive washer					●			
Hexagon head cap bolts						●		

CONSTRUCTION INDUSTRY

Fasteners are the critical link in the load path of a building structure. They provide structural integrity and are a major point of energy dissipation under seismic and wind loads. Construction fasteners are generally classified as those that are used to secure building materials and can be classified as commodity fasteners or task-specific fasteners. Typical applications include:

- ❖ Roofing (flat, sloped)
- ❖ Decks (steel, wood, composite)
- ❖ Bridges (ladder deck, multi-girder, suspension)
- ❖ Buildings (residential, commercial, skyscraper)

Typical examples include cement-board screws, drywall screws, needlepoint screws, outdoor screws, pole-gripper screws, self-drilling screws and woodworking screws. Steel (Table 15.8.5) and stainless steel fasteners are commonly used in the construction industry.

TABLE 15.8.5 | Selected properties of steel construction-grade fasteners

ASTM designation no.	Size range (inclusive), mm (inch)	Minimum proof strength, ksi (MPa)	Minimum tensile strength, ksi (MPa)	Minimum yield strength, ksi (MPa)	Material
A307	6-38 (0.25-1.5)	33 (227)	60 (413)	36 (248)	Low carbon
A325, Type 1	12-25 (0.5-1.0), 29-36 (1.125-1.4)	85 (586) 74 (510)	120 (827) 105 (723)	92 (634) 81 (558)	Medium carbon, Q&T
A325, Type 2	12-25 (0.5-1.0), 29-38 (1.125-1.5)	85 (586) 75 (517)	120 (827) 106 (730)	92 (634) 82 (565)	Low carbon, Q&T
A325, Type 3	12-25 (0.5-1.0), 32-41 (1.25-1.6)	85 (586) 76 (524)	120 (827) 107 (737)	92 (634) 83 (572)	Alloy, Q&T
A354, Grade BD	6-101 (0.25-4.0)	120 (827)	150 (1034)	130 (896)	Alloy, Q&T
A449	6-25 (0.25-1.0), 29-38 (1.14-1.5), 45-76 (1.75-3.0)	85 (586) 74 (510) 55 (379)	120 (827) 105 (723) 90 (620)	92 (634) 81 (558) 58 (400)	Medium carbon, Q&T
ASTM 490, Type 1	12-38 (0.5-1.5)	120 (827)	150 (1034)[a] 173 (1192)[b]	130 (896)	Alloy steel, Q&T
ASTM 490, Type 2	12-25 (0.5-1.0)	120 (827)	150 (1034)[a] 173 (1192)[b]	130 (896)	
ASTM 490, Type 3	12-38 (0.5-1.5)	120 (827)	150 (1034)[a] 173 (1192)[b]	130 (896)	Low carbon, Q&T
ASTM A490M, Type 1 or 3[c]	12-36 (0.5-1.4)	17 (117)	150 (1034)	136 (937)	

Notes:

[a] Minimum value

[b] Maximum value

[c] Type 2 withdrawn

MINING INDUSTRY

A key requirement for fasteners in the mining and excavation industry is their resistance to vibration and corrosion. Vibrating conveyors, crushing and pulverizing equipment, mining machinery, railroad cars and material-handling devices (Fig. 15.8.3) are examples of applications where vibration is a significant issue that requires the fastener to be designed so that it won't "back off" in service. Furthermore, corrosion and/or cracking is frequently involved in mining fasteners related to hydrogen-induced cracking (aka hydrogen embrittlement), often in the form of stress corrosion cracking (SCC) or sulfide stress cracking.

FIGURE 15.8.3 | Fasteners for high-vibration material-handling applications (courtesy of Security Locknut LLC)

The types of fasteners used in the mining industry are quite diverse and include bolts (e.g., anchor, eye, structural, swing), rods (e.g., elevator, rock, tie, sag), washers, nuts and plates. Materials include carbon and alloy steels, construction steels and stainless steels.

MEDICAL INDUSTRY

The medical industry relies heavily on the use of fasteners. For example, medical devices (e.g., dental and orthopedic implants, instruments) employ literally hundreds of different types of fasteners to hold their assemblies together. Even though the components in the medical devices are small or even tiny, the device will almost always fail when a fastener fails.

Medical devices fall into two broad categories: surgical/non-implant devices and implantable devices. The correct fastener ensures that the device goes together and stays together for the intended life of the assembly and that it performs as desired. Fasteners can overcome challenges in assembly, solve quality problems and significantly reduce the total cost of the device.

Surgical and Non-implant Medical Devices

Surgical and dental instruments are examples of non-implant medical devices typically manufactured from austenitic stainless steels where good corrosion resistance and moderate strength are required. Examples include cannula, den-

15 ANCILLARY TOPICS OF INTEREST

tal impression trays, guide pins, hollowware, hypodermic needles, steam sterilizers, storage cabinets and work surfaces, and thoracic retractors. These applications often use a variety of stainless steels that can be easily formed into complex shapes. Their heat treatment is covered in Section 5.13.

Implantable Medical Devices
Specific grades of austenitic stainless steel and high-nitrogen austenitic stainless steels are used for some surgical implants. Examples include aneurysm clips; bone plates and screws; femoral fixation devices; intramedullary nails and pins; and joints for ankles, elbows, fingers, knees, hips, shoulders and wrists.

The vast majority of orthopedic implants worldwide, however, are manufactured from titanium (e.g., Ti-6Al-4V alloy) or cobalt-based alloys (e.g., ASTM F75 or cobalt-chromium-molybdenum alloys). They are manufactured from castings, forgings or bar stock. The heat treatment of these devices is typically performed in argon atmosphere furnaces or in vacuum.

Medical application examples include pins, bone plates, screws, bars, rods, wires, posts, expandable rib cages, spinal fusion cages, finger and toe replacements, hip and knee replacements, and maxillofacial prosthetics.

MARINE INDUSTRY
The marine industry is literally held together by fasteners designed to survive the extremely harsh and corrosive environments to which they are exposed. Bolts, screws, washers, locking washers, wing nuts, split rings and slating nails are just some of the many marine construction fasteners used on docks, ramps, ships, tanks, winches and for underwater construction projects.

Because fasteners are generally much smaller than the components they support, they must be more corrosion-resistant. As such, heat-treatment processes need to be carefully chosen and controlled in order to not have a negative impact on corrosion resistance or other important physical or mechanical properties.

In addition, since the fastener material is often dissimilar to the components being joined, galvanic-induced corrosion is a principal concern. Should the fastener be anodic to the remainder of the structure, the relative size effect could cause severe corrosion and degradation of the fastener in a short period of time. Failure of a nut or bolt can have catastrophic results.

Materials for Marine Fasteners
Many copper alloys are suitable for marine service, including coppers; copper-nickels; bronzes including aluminum and silicon bronze; brasses; and copper-beryllium. For seawater systems, copper-nickel and aluminum bronze are often preferred, although other copper alloys are used and have specific advantages.

Copper alloys differ from other metals in that they have an inherent high resistance to biofouling, particularly macrofouling, which can eliminate the need for antifouling coatings or water treatment.

Aluminum bronze refers to a range of copper-aluminum alloys that combine high strength and corrosion resistance and are widely used in both cast and wrought forms. These alloys are basically copper with 4-12% Al. They have a thin, adherent surface film of copper oxides and aluminum oxides, which will heal very rapidly if damaged. Furthermore, they have good resistance to erosion and wear as well as good corrosion-fatigue properties. Generally, the corrosion resistance of the aluminum bronzes increases as the aluminum and other alloying additions increase. At 9-10% Al, additions of 4-5% iron and nickel produce alloys with both high strength and corrosion resistance in unpolluted waters.

Aluminum-bronze alloys also resist ammonia SCC. Although there have been some instances of failure in high-pressure steam, extremely high resistance is found in marine environments.

Silicon bronzes, especially those containing about 3% Si and 1% Mn, have very good corrosion resistance in seawater and resistance to stress corrosion by ammonia. They have been widely used for screws, nuts, bolts, washers, pins, lag bolts and staples in marine environments, as well as screws used in wooden sailing vessels. Silicon bronzes have an alpha-phase metallurgical structure, and the silicon provides solid-solution strengthening. Strength and hardness can be increased by cold work. They generally have the same corrosion resistance as copper but with higher mechanical properties and weldability. They are tough with good shock and galling resistance.

Brasses (copper-zinc alloys) often have small additions of elements such as tin or arsenic for inhibition of dezincification or additions of lead to aid pressure tightness or machining. Their strength increases with zinc content and also with additional alloying elements. There are two distinct groups of brasses used for marine service, and they can be distinguished by their metallurgical structure. Alpha brasses have a single-phase structure containing up to 37% Zn. Alpha-beta (duplex) brasses have two-phase structures; the second phase – beta – starts to form above 37.5% Zn.

Many other materials can be used provided they demonstrate superior corrosion resistance. One such example is Inconel 686, a high-performance nickel-based alloy that exhibits high tensile strength and fracture toughness as well as resistance to corrosion, especially crevice corrosion, superior to that of other nickel-based alloys (such as K500 Monel).

Different combinations of properties can be produced in each of these materials by varying the heat treatment, which influences strength, hardness, ductility, conductivity, impact resistance and inelasticity.

15 | ANCILLARY TOPICS OF INTEREST

PETROCHEMICAL INDUSTRY

Petrochemical products are those derived from petroleum. They are mainly used in the production of petrochemical derivatives (e.g., ethylene, methanol, butadiene, propylene, formaldehyde, polyvinyl chloride, acetic acid and epoxy resins among many others).

The growing demand of petrochemicals from major end-use industries, including transportation, construction and packaging, drives the market. The demand for industrial fasteners is expected to grow at a rate of 5.4% through 2018[1] and is expected to fall just shy of $100 billion by 2018. In addition, the petrochemical industry is incredibly diverse, encompassing regional markets in North America, Latin America, Europe, China, Asia Pacific, Africa and the Middle East.

This diversity is also evident in the products and materials from which fasteners are constructed. Examples include threaded rod, studs (full-thread, double-end), stud bolts, hex bolts and nuts, gaskets (spiral wound, full face, rings), setscrews, sockets, standard fasteners and specialty products such as tie bars, socket screws and machined parts.

The range of chemically resistant materials includes petrochemical grades of stainless steel, duplex and super-duplex steels, nickel and cupro-nickel alloys, carbon and alloy steels, titanium and other nonferrous materials and superalloys. Coatings are common and include zinc, cadmium, nickel, galvanized (hot-dip, sherardizing), phosphate and PTFE.

NUCLEAR INDUSTRY

A typical nuclear power plant contains some 40,000 fasteners, including bolts, washers, nuts, studs, cap screws, pins, rivets and machine screws.[1] These fasteners are subjected to very specific design rules. Problems can develop, however, either immediately or over time.

The most common cause of fastener failure in the nuclear industry is reportedly due to SCC of certain stainless steel alloys.[1] Other contributory factors include improper material selection, poor installation practices (e.g., improper torque), bolts that were not properly sized and failures induced by heat-treatment issues such as overheating, decarburization and quench cracks.

Fasteners for the nuclear industry are made from a wide variety of materials, including carbon and alloy steels, stainless steel, Nitronic® and Monel® alloys, Inconel® and titanium as well as brass, bronze and copper. This diversity means that required properties and heat treatments are always specified, often in ASME and SAE standards (e.g., ASTM SA193/SA194/SA276/SA307 and SAE J429/J995). A variety of coatings can be applied after heat treatment. These include zinc, cadmium, phosphate, ceramic, chromium and black oxide.

SPECIALIZED APPLICATIONS

Critical applications require the use of materials whose performance encompasses both normal-duty and extreme-duty demands. It is the latter that differentiates specialty fasteners and components from standard ones.

What are specialty fasteners?

Specialty fasteners are those whose applications demand performance over cost. Mechanical, physical and metallurgical properties are more stringent than those for standard fasteners. Examples of equipment and industries that rely on specialty fasteners include power generation (e.g., gas turbines, offshore performance platforms), pulp and paper mills and electronic devices.

In aerospace applications (aircraft, rotorcraft, space), specialty fasteners are used on exteriors, interiors, avionics and flight systems (e.g., landing gear). Product examples include captive screws, rivets, gas springs, clamshell flexible couplings, quick-release pins, tension latches and telescopic slides. Fasteners and clamping devices are commonplace throughout the automotive industry (motor sports, cars, off-road, heavy truck), including the engine, body and subsystems.

Alloy Fastener Materials

There are some unique material selection challenges in the design of very high-strength, high-performance fasteners. These types of fasteners are often exposed to high stress concentration in the thread roots (caused by the tensile stresses produced from extremely high clamping loads), on top of which any fatigue loads are superimposed. To meet these challenges, the designer often selects alloys classified as "exotic" primarily due to their chemistry and ability to perform at elevated temperatures (Table 15.8.6).

TABLE 15.8.6 | Classification of exotic alloys for specialty fasteners

Class	Alloy	Examples
0	Steels[a]	A286
1A	Stainless steels[b]	Alloy 20, Alloy 50, Alloy 60, Custom 450, Carpenter 21, Carpenter CB3, 904L, AL6XN, Avesta 254SMo
1B	Stainless steels[a]	13-8, 17-4, 17-7
1C	Stainless steels[c]	2304, 2205, 2507, 2707, 3207, Ferralium 255-SD50
1D	Stainless steels[d]	50, 60
2A	Superalloys[e]	B, C-276, C-2000, C-22, C-4, G, X
2B	Superalloys[f]	230, HR120, HR160, ULTIMET 1233
2C	Superalloys[g]	600, 601, 625, 718, 800/800H, 825, 925, X750, 25-6Mo
2D	Superalloys[h]	Waspaloy
3A	Nickel alloys	B2, B3, G30
3B	Monel	400, 405, 500
4	Other	Co, Ta, Ti, Zr

Notes:

[a] Precipitation-hardening grades
[b] Chromium-nickel-molybdenum grades (with selected alloy additions)
[c] Duplex and super-duplex grades
[d] Nictronic alloys
[e] Hastalloy grades
[f] Haynes International alloys
[g] Inconel grades
[h] Other superalloy grades

Heat-Treat Equipment for the Fastener Industry

Fastener heat treatment can be performed in a wide variety of furnaces and ovens, and fastener designers should understand the variety of choices available to them. There are several examples of atmosphere furnaces that can be used.

MESH-BELT CONVEYOR FURNACES

Mesh-belt conveyor furnaces (Fig. 15.8.4) are the dominant technology for the heat treatment of fasteners. These units are often part of a completely automated heat-treating system that includes loaders, pre- and post-washers, a hardening furnace with quench tank and a tempering furnace. Soluble oil tanks and endothermic-atmosphere generators or nitrogen/methanol systems are common ancillary items.

FIGURE 15.8.4 | Typical atmosphere heat-treating system for fasteners (courtesy of Surface Combustion Inc.)

Heat-treating systems for fasteners must be flexible enough to handle the demands of both large and small quantities. Standard production capacities in mesh-belt furnaces typically range from 100 kg/hour (250 pounds/hour) to 3,000 kg/hour (7,000 pounds/hour). It is not uncommon to see fasteners loaded on the belt between 12.7-63.5 mm (0.5-2.5 inches) deep (Fig. 15.8.5). For heavier product (e.g., bolts) and higher production rates, cast link-belt furnaces are popular.

FIGURE 15.8.5 | Load of fasteners on a mesh belt for carbonitriding

15 | ANCILLARY TOPICS OF INTEREST

ROTARY-RETORT (ROTARY-DRUM) FURNACES

Rotary-retort furnaces (Fig. 15.8.6) are the most common alternative to mesh-belt conveyors. These systems are typically automated after loading parts into the hopper at the front end. Vibratory hoppers and weight-actuated skip loaders deposit precisely measured charges into the furnace to ensure the uniform loading of parts.

FIGURE 15.8.6 | Typical rotary-retort system with oil, polymer or water-quench capability (courtesy of SECO/WARWICK Corp.)

Typical standard system capacities for rotary-retort furnaces vary from around 225 kg/hour (500 pounds/hour) to 450 kg/hour (1,000 pounds/hour), but they can be manufactured to handle 1,800 kg/hour (4,000 pounds/hour) or more. The retorts are either cast or fabricated from high-temperature alloys. In general, cast retorts have superior mechanical-strength characteristics compared to wrought fabricated designs but often come at a cost premium. Auger flights with the retort convey the fasteners through the furnace. Variable-speed rotation of the retort provides flexibility of time-based processing cycles.

The actual quenching process of the part is typically completed within 2 to 5 seconds from the time fasteners drop into the quenchant, with 10 minutes being a typical residence time in the quench for continued cooling. Austempering processes, however, require extended times in the order of 20 minutes or more to produce a bainitic structure.

It is critical that an adequate amount of fluid flow is delivered to the active quench area because the overall quench-tank capacity is not enough to ensure individual quenching of each part. The quench-tank design must eliminate any clumps and clogs to ensure that both proper technique and adequate quenching is obtained.

Tempering furnaces, whether mesh-belt or rotary-retort type, use a recirculation system designed to keep heated gases in contact with the fasteners as they progress through the oven. Convection heating is used to full advantage for efficient heat transfer in minimum floor space. Operating temperature ranges vary from 150-650°C (300-1200°F).

Summary

Fastener heat treatment in atmosphere furnaces is widely used to meet both the quality and productivity demanded of them in a broad spectrum of industries. Atmosphere heat treatment provides a highly cost-competitive technology with high throughput and consistent, repeatable results.

REFERENCES
1. Herring, Daniel H., *Atmosphere Heat Treatment, Volume I*, BNP Media, 2014
2. Wikipedia (www.wikipedia.com)
3. Industrial Fasteners Institute Annual Report, 2012
4. Herring, Daniel H., "Overview of Fastener Heat Treatment: Present & Future Direction," *Fastener World International* (FW143), November/December 2013
5. Herring, Daniel H., "Heat Treatment of Fasteners for the Nuclear Industry," *Fastener Technology International*, August 2013
6. Herring, Daniel H., "Heat Treatment of Fasteners for the Petrochemical Industry," *Fastener Technology International*, October 2013
7. Herring, Daniel H., "Heat Treatment of Fasteners for the Medical Device Industry," *Fastener Technology International*, December 2013
8. Herring, Daniel H., "Heat Treatment of Fasteners for the Marine Industry," *Fastener Technology International*, February 2014
9. Herring, Daniel H., "Heat Treatment of Fasteners for the Mining Industry," *Fastener Technology International*, April 2014
10. Herring, Daniel H., "Heat Treatment of Fasteners for the Automotive Industry," *Fastener Technology International*, April 2015
11. Herring, Daniel H., "Heat Treatment of Specialty Alloy Fasteners," *Fastener World International* (FW148), September/October 2014
12. Herring, Daniel H., "The Heat Treatment of Aerospace Fasteners," *Fastener Technology International*, October 2014

16 | REFERENCE MATERIALS

16.1
HEAT-TREATMENT TERMINOLOGY

No matter the field of endeavor, an accurate knowledge of the vocabulary is essential to understanding the subject. The field of heat treatment, having a highly specialized vocabulary, is no different. As such, considerable emphasis is placed on the proper use, meaning and understanding of certain words or phrases. The reader is cautioned that it is important to recognize that these short descriptions cannot convey the full depth of understanding of each term, especially when used for a specific application. There is no substitution for a full scientific understanding of how the terms and their underlying science fit together.

These terms are common to a variety of manufacturing and heat-treatment technologies in the areas of:

- ❖ Brazing/welding/soldering
- ❖ Equipment (furnaces/ovens/applied energy/salt bath)
- ❖ Ferrous and nonferrous alloys
- ❖ Heat treatment/heat-treatment processes
- ❖ Manufacturing (related to heat treatment)
- ❖ Materials
- ❖ Metallurgical concepts
- ❖ Metallography
- ❖ Testing

The definitions for many terms used in these subject areas include:

Acid embrittlement – A form of embrittlement due to hydrogen that may be induced by acid treatment.

Aging – A hardening process caused by precipitation of a constituent from a supersaturated solid solution, which is typically performed on nonferrous alloys or precipitation-hardening steels after rapid cooling or cold working. Aging is a change in properties and generally occurs slowly at room temperature and more rapidly at slightly higher temperatures. Other common names and related subjects are: age hardening, artificial aging, interrupted aging, overaging, precipitation hardening, precipitation heat treatment, progressive aging, quench aging and strain aging.

Air hardening – A term usually referring to alloy steel, which will form martensite and develop a high hardness when cooled in air from its proper hardening temperature.

Alloy – A metallic compound consisting of one or more elements, one of which is typically a metal.

Aluminizing – A process resulting in the formation of a corrosion- and oxidation-resistant coating on a metal by diffusion of aluminum or an aluminum-rich alloy.

Annealing – Heating to and holding at a suitable temperature and then cooling at a suitable rate for such purposes as reducing hardness, improving machinability, facilitating cold working, producing a desired microstructure (for subsequent operations) or obtaining desired mechanical, physical or other properties. Other common names and related subjects are: black annealing, intermediate annealing, isothermal annealing, malleabilizing, process annealing, quench annealing, recrystallization annealing and spheroidizing.

Applied energy – The application of one of the following techniques: flame, induction or laser to perform a surface modification such as a hardening and/or tempering operation.

Austempering – A process involving rapid cooling from austenite to a temperature below that at which pearlite forms but above that at which martensite formation begins. This treatment results in a fully or primarily bainitic structure with a resulting combination of both hardness and ductility.

Austenite – A name given any solid solution in which gamma iron is the solvent. It has an FCC crystal structure. Austenite is a phase name and depends on but does not describe composition. Austenite is the phase from which most quenching operations start.

Austenite transformation – Austenite can transform by a nucleation and growth mechanism into ferrite, cementite and pearlite if the cooling rate is slow enough for full (atomic) diffusion. The iron-carbon phase diagram can be used to determine quantities of each type. Intermediate cooling rates form bainite, which is harder than pearlite but softer than martensite. Fast cooling forms martensite, which is very hard and occurs by an instantaneous diffusionless shear transformation.

Austenitic grain size – The size attained by the grains of steel when heated to the austenitic region; may be revealed by appropriate etching of cross sections after cooling to room temperature.

Austenitizing – Forming austenite by heating a ferrous alloy above the lower critical temperature of the material (partial austenitizing) or above the upper critical temperature of the material (complete austenitizing). When used without qualification, the term implies complete austenitizing.

Austenitizing temperature – The temperature region where partial or complete austenitizing takes place. Higher austenitizing temperatures allow more carbon to dissolve into the austenite, shifting CCT diagrams to the right. Higher temperatures promote larger grain sizes and increase time for transformation.

Bainite – A transformation product produced by cooling from austenite at a rate between that producing pearlite and martensite. Similar to pearlite, it is composed of ferrite and cementite, but unlike pearlite it is not lamellar. Basically, it is elongated cementite in a ferrite matrix. Upper bainite forms at higher temperatures closer to the pearlite formation temperature. Lower bainite forms at lower temperatures closer to martensite-forming temperatures. Properties also closely tied to each form of bainite.

Bake-out – See hydrogen bake-out.

Banded structure – A layer effect that is sometimes developed during the hot rolling of steel.

Bark – A term used to describe the oxidized (scaled), decarburized skin that develops on steel bars heated in a non-protective atmosphere.

Batch furnace – A furnace used to heat treat a single load. Batch-type furnaces are necessary for large parts such as heavy forgings and are preferred for processes or materials requiring long cycles.

Belt furnace – A continuous-type furnace that uses a belt (mesh or cast-link) to convey parts through the furnace.

Beta annealing – A process that produces a beta phase on heating certain titanium alloys in the temperature range in which this phase forms followed by cooling at an appropriate rate to prevent its decomposition.

Binary alloy – An alloy containing two alloying elements.

Black oxide – A black or grayish-black finish on a metal, which is produced by immersion in hot oxidizing salts or salt solutions.

Braze alloy – An alternative term for the brazing filler metal.

Brazing – Joining metals by flowing a thin capillary layer of nonferrous filler metal into the space between them. Bonding results from the intimate contact produced by dissolution of a small amount of base metal into the molten filler metal, without fusion of the base metal. The term brazing is used when the temperature exceeds some arbitrary value, typically 475°C (880°F).

Breaks – Creases or ridges usually in "untempered" steel or in age-hardened material where the yield point has been exceeded. Depending on the origin of the break, it may be termed a cross break, a coil break, an edge break or a sticker break.

Bright annealing – Annealing of work in a protective atmosphere so that there is no discoloration (oxidation) as the result of heating. Oxides may be reduced in some atmospheres. "Bright" is a highly subjective term.

Brinell hardness test – A test for determining the hardness of a material by forcing a carbide or hard steel ball of specified diameter into it under a specified load. The result is expressed as the Brinell hardness number, which is the value obtained by dividing the applied load in kilograms by the surface area of the resulting impression in square millimeters.

Brittle tempering range – Some hardened steels show an increase in brittleness when tempered within certain temperature ranges that depend on the characteristics of the steel.

Car furnace – A batch-type furnace utilizing a transfer car on rails to enter and remove a workload from the furnace. Car furnaces are typically used for stress relief, tempering or annealing processes.

16 | REFERENCE MATERIALS

Carbon potential – The carbon concentration expressed in weight percentage that a pure iron foil would have after being equilibrated with the atmosphere. A measure of the ability of a furnace atmosphere containing active carbon to alter or maintain the carbon level of the steel under prescribed conditions. Note: In any particular environment, the carbon level attained will depend on such factors as temperature, time and steel composition.

Carbon probe – See oxygen probe.

Carbon restoration – Heating and holding a metal above the critical temperature (Ac_1) in contact with a suitable carbonaceous atmosphere for the purpose of restoring carbon by adding it to the surface. The amount of carbon added is usually intended to return the material to its original carbon content (prior to a heat-treat operation that reduced this carbon level).

Carbonitriding – The introduction of carbon and nitrogen into a solid ferrous alloy by holding above the critical temperature (Ac_3) in an atmosphere that contains suitable gases such as hydrocarbons, carbon monoxide and ammonia. The carbonitrided alloy is usually rapidly cooled (quench hardened). Other common names and related subjects are case hardening, dry cyaniding, gas cyaniding, nicarbing and nitrocarburizing (obsolete).

Carburizing – The introduction of carbon into a solid ferrous alloy by holding above the critical temperature (Ac_3) while in contact with a suitable carbonaceous atmosphere. The carburized alloy is usually rapidly cooled (quench hardened). Other common names and related subjects are case hardening, gas carburizing and cyaniding.

Case – The surface layer of steel whose composition has been changed by the addition of carbon, nitrogen, boron or other elements at high temperature.

Case hardening – Hardening a ferrous alloy so that the outer portion, or case, is made substantially harder than the inner portion, or core. This is accomplished by increasing the surface carbon content and quenching or by selectively hardening the surface using applied energy techniques (flame, induction, laser).

Cast iron – A generic term for a large family of cast ferrous alloys in which the carbon content exceeds solubility of carbon in austenite at the eutectic temperature. Most cast irons contain at least 2% carbon plus silicon and sulfur and may or may not contain other alloying elements. For the various forms such as gray cast iron, white cast iron, malleable cast iron and ductile cast iron, the word "cast" is often left out.

Cementite – A compound of iron and carbon known chemically as iron carbide (Fe_3C). It is characterized by an orthorhombic crystal structure. When it occurs as a phase in steel, the chemical composition will be altered by the presence of manganese and other carbide-forming elements.

Charpy (impact) test – A pendulum-type single-blow impact test in which the specimen, usually notched, is supported at both ends as a simple beam and broken by a falling weight. The energy absorbed, as determined by the subsequent rise of the pendulum, is a measure of impact strength or notch toughness. Contrast with the Izod test.

Cold treatment – Exposure of a material to subzero temperatures for the purpose of obtaining the desired conditions or properties. It enhances dimensional or microstructural stability. When the treatment involves the transformation of retained austenite, it is usually followed by tempering.

Cold working – Plastic deformation of a metal at a temperature low enough so that re-crystallization does not occur during cooling.

Compounds – Described as materials whose components are present in atomic proportions, which are described by a chemical formula (e.g., Fe_3C). Many compounds in alloys are hard and therefore contribute a hardening effect to the alloy. Intermetallic compounds are compounds containing only metals.

Continuous cooling diagram – A phase diagram often used when heat treating steel to represent which types of phase changes will occur in a material as it is cooled at different rates. These diagrams are often more useful than time-temperature-transformation diagrams. There are two types: plot of transformation start, specific fraction of transformation produced and transformation finish temperature against transformation time on each cooling curve; and plot of transformation start, specific fraction of transformation produced and transformation finish temperature against cooling rate or bar diameter for each type of cooling medium.

Controlled cooling – A process in which a ferrous material is cooled from an elevated temperature in a predetermined manner to avoid cracking or other type of internal damage and produce a specific microstructure.

Cooling curve – A curve showing the relationship between time and temperature during the cooling of a material.

Cooling stress – Residual stresses resulting from non-uniform distribution of temperature during cooling.

Core – The interior part of steel whose composition has not been changed in a case-hardening operation.

Correction factor – The number of degrees, determined from the most recent calibration, that must be added to or subtracted from the temperature reading of a sensor, instrument or combination thereof (the system) to obtain NIST true temperature. When expressed as a percent, it means percent of reading. The correction factors of sensors and instruments are usually kept separately and added together algebraically when a combination is used.

Corrosion embrittlement – A severe loss of ductility of a metal resulting from corrosive attack, usually intergranular and often not visually apparent.

Creep strength – 1.) The constant nominal stress that will cause a specified quantity of creep in a given time at constant temperature. 2.) The constant nominal stress that will cause a specified rate of secondary creep at constant temperature.

Critical cooling rate – The rate of continuous cooling required to prevent undesirable transformation. For steel, it is the minimum rate at which austenite must be continuously cooled to suppress transformations above the M_s temperature.

Critical diameter – The diameter of a bar that can be fully hardened with 50% martensite at its center.

Critical point – A temperature point at which a structure change either starts, is completed or both when a material is being heated or cooled.

Critical range – The temperature range between an upper and lower critical point for a given material.

Cryogenic treatment – The cooling of a material to a temperature low enough to promote the transformation of phases to occur. Other common names and related subjects are deep-freezing, cold treatment and subzero treatment.

Crystalline – A material consisting of particles arranged in a repetitive pattern. In all metals and metal alloys, the atoms are arranged in fairly simple geometric patterns.

Crystallization – 1.) The separation, usually from a liquid phase on cooling, of a solid crystalline phase. 2.) Sometimes erroneously used to explain fracturing that actually has occurred by fatigue.

Dead soft – A temper of nonferrous alloys and some ferrous alloys corresponding to the condition of minimum hardness and tensile strength. Produced by full annealing.

Decarburizing – The process (usually unintentional) of removing carbon from the surface of steel, usually at high temperature, when in contact with certain types of atmosphere.

Deep freeze – See cold or cryogenic treatment.

Dendrite – Crystal that has a tree-like branching pattern most evident in cast metals slowly cooled through the solidification range.

Deoxidizing – 1.) The removal of oxygen from molten metals by use of suitable deoxidizers. 2.) Sometimes refers to the removal of undesirable elements other than oxygen by the introduction of elements or compounds that readily react with them. In metal finishing, the removal of oxide films from metal surfaces by chemical or electrochemical reactions.

Depth of hardening – The depth to which a specified hardness is obtained. Often determined by Jominy end-quench testing of round bars or flat plates.

Dew point – The temperature and pressure at which a gas begins to condense to a liquid.

Dew-point analyzer – An atmosphere monitoring device that measures the partial pressure of water vapor in a furnace atmosphere.

Diffusion – The degree to which the brazing filler metal penetrates the alloys with the base metal during brazing.

Double aging – Employment of two different aging treatments to control the type of precipitate formed from a supersaturated matrix in order to obtain the desired properties. The first aging treatment, sometimes referred to as intermediate or stabilization, is usually carried out at a higher temperature than the second.

Double tempering – A treatment in which a quench-hardened ferrous metal is subjected to two complete tempering cycles, usually at substantially the same temperature, for the purpose of ensuring completion of the tempering reaction and promoting stability of the resulting microstructure.

Draw (drawing) – A term used interchangeably with tempering.

Ductile fracture – Fracture characterized by tearing of metal accompanied by appreciable gross plastic deformation and expenditure of considerable energy. Contrast with brittle fracture.

Ductility – The ability of a material to deform plastically without fracturing. Measured by elongation or reduction of area in a tensile test, by height of cupping in an Erichsen test or by other means.

Effect of alloying element on hardening – Hardenability is affected by chemical composition and microstructure. For example, more carbon moves the nose of the TTT curve to the right. Carbon also lowers M_s and M_f temperatures. Steels greater than 0.6% C have M_f temperatures below room temperature. Equations and computer programs exist to describe the effects of major and minor alloying elements.

Elastic limit – The maximum stress to which a material may be subjected without any permanent strain remaining upon complete release of stress.

Embrittlement – The severe loss of ductility or toughness or both, of a material usually a metal or alloy.

Endurance limit – The maximum stress below which a material can presumably endure an infinite number of stress cycles. If the stress is not completely reversed, the value of the mean stress, the minimum stress or the stress ratio also should be stated. Compare with fatigue limit.

Eutectic – A mixture of two or more constituents that solidify simultaneously out of a liquid at a minimum freezing point.

Eutectoid – A mixture of two or more constituents that forms on cooling from a solid solution and transforms on heating at a constant minimum temperature. A eutectoid steel contains approximately 0.77-0.83% carbon.

Fatigue – The effect of repeated cycles of stress on metal. The insidious feature of fatigue failure is that there is no obvious warning. A crack forms without appreciable deformation of structure, making it difficult to detect the presence of growing cracks.

Ferrite – A solid solution of one or more elements in body centered cubic (BCC) iron. Unless otherwise designated, the solute is generally assumed to be carbon. In some cases two ferrite regions exist, separated by an austenite area. The lower area is alpha ferrite, and the upper area is delta ferrite.

Ferrous – A metal of, related to or containing iron as its principal addition.

Filler metal – The metal or alloy to be added in order to produce the braze joint.

Fillet – A region of brazing filler metal where the various parts of the assembly are joined.

Fixturing – See racking.

Flame hardening – A process of hardening steel by heating the surface layer above the transformation temperature by means of a high-temperature flame, followed by rapid cooling. Also see hardening.

Flow – The ability of molten filler metal to spread out over a surface.

Flow through – A term used to describe the motion of parts moving in a continuous furnace.

Flux – A chemical compound applied to the surface of a component to be brazed for the purpose of hindering or preventing the formation of oxides and other undesirable substances; used so as to allow the brazing filler metal and base metal surface to remain relatively clean while being heated to brazing temperature.

Forced-air quench – A quench utilizing blasts of air impinging on parts or a load.

Forming – A manufacturing operation that shapes a component part by deformation

Freezing – Solidification of the filler metal during cooling.

Gap – Another term for joint clearance.

Grain – An individual crystal in a polycrystalline metal or alloy; it may or may not contain twinned regions and subgrains.

Grain boundary – The surface boundary between crystals. When alloys yield new phases (as in cooling), grain boundaries are the preferred location for the appearance of the new phase. Certain deleterious phenomena (e.g., carbide precipitation on slow cooling of stainless steels) occur at grain boundaries.

Grain flow – Fiber-like lines that appear on polished and etched sections of forgings caused by orientation of the constituents of the metal in the forging direction. Grain flow produced by proper die design can improve the required mechanical properties of forgings.

Grain growth – Growth of some grains at the expense of others, resulting in an overall increase in average grain size. Usually a result of heating at elevated temperature.

Granular fracture – A type of irregular surface, produced when metal is broken, that is characterized by a rough, grain-like appearance as differentiated from a smooth, silky or fibrous type. It can be sub-classified into transgranular and intergranular forms. This type of fracture is frequently called crystalline fracture, but the inference that the metal broke because it "crystallized" is not justified because all metals are crystalline when in the solid state. Contrast with fibrous/silky fracture.

Hardenability – The measure of the depth of hardening. The standard methodology is the Jominy test, a graphical description of the depth of hardening obtained when a metal is quenched over a range of cooling rates. Used to predict the properties attainable for specific steel parts.

Hardening – An increase in the hardness of a material by suitable treatment usually involving heating and rapid cooling. Other common names and related subjects are neutral hardening, quench hardening, direct hardening, flame hardening, induction hardening, laser hardening and surface hardening.

Hardness – A measure of the resistance to penetration of a material by an applied load by a given indentation device. In general, hardness can be related to ultimate tensile strength.

Heat-affected zone – The portion of the base metal whose mechanical properties or microstructure have been altered by the heat created by a joining process (e.g., brazing, welding).

Heat treater – Someone who performs the heat-treatment process.

Heat treatment – The controlled application of time, temperature and atmosphere to produce a predictable change in the internal structure (microstructure) of a material. Heat treatment has the ability to vary the physical, mechanical and metallurgical properties of a material.

Heterogeneous – A material that is distinctly non-uniform in appearance. Heat-treat examples include terms such as globular or spheroidal (distributing particles in one component through another in the form of round particles); lamellar (where constituents appear in alternating layers); acicular (common distribution pattern of "needles" are arranged in a 60-degree pattern and gives the impression of equilateral triangles); and dendritic (often found in castings and appear as "trees" varying greatly depending on the rate of solidification of the material).

Holding – see soak.

Holding temperature – See soak temperature.

Holding time – See soak time.

Homogenizing – Holding at high temperature to eliminate or decrease chemical segregation by diffusion. The resultant composition is uniform, with coring and concentration gradients eliminated.

Homogeneous – A material that is distinctly uniform in appearance. As applied to alloys, it describes a material whose internal structure contains a uniform crystalline pattern.

Hot working – Shaping above the recrystallization temperature.

Hydrogen bake-out – A procedure used to remove hydrogen from inside a component by heating to 95-205°C (200-400°F) and soaking for an extended period until the material ductility is regained.

Ideal (critical) diameter – The maximum diameter of an alloy that will have 50% martensite at its center. Used as a guide to determine the largest part size that will harden properly.

Impact energy – The amount of energy required to fracture a material, usually measured by means of an Izod test or Charpy test. The type of specimen and test conditions affect the values and therefore should be specified.

Intergranular – Between crystals or grains. Also called intercrystalline.

Intergranular corrosion – Corrosion occurring preferentially at grain boundaries, usually with slight or negligible attack on the adjacent grains.

Intergranular cracking – Cracking or fracturing that occurs between the grains or crystals in a polycrystalline aggregate. Also called intercrystalline cracking.

Intergranular fracture – Brittle fracture of a metal in which the fracture is between the grains, or crystals, that form the metal. Also called intercrystalline fracture.

Intermediate annealing – An annealing process at one or more stages during manufacture and before final treatment.

Interrupted aging – Aging at two or more temperatures, by steps, and cooling to room temperature after each step. See aging, and compare with progressive aging and step aging.

Interrupted quenching – A two-stage quenching process for steel that involves heating to form austenite followed by an initial quench to a temperature above the start of martensite formation followed by a second cooling ramp down to room temperature.

Iron-carbon (iron-carbide) phase diagram – An equilibrium phase diagram generated by slow heating and cooling rates. Used to describe and interpret many metallurgical aspects of irons and steels. Specific details of the diagram are affected by composition, heating rate and cooling rate.

Isothermal (I-T) diagram – Also known as the time-temperature-transformation (TTT) curve. One version of the common graphical representation of transformation occurring under non-equilibrium conditions. Diagrams describe the phases present when a given steel is quenched to a specific temperature in a specific amount of time.

Isothermal transformation – A change in phase that takes place at a constant temperature. The time required for transformation to be completed and, in some instances, the time delay before transformation begins, depending on the amount of supercooling below (or superheating above) the equilibrium temperature for the same transformation.

Izod test – A pendulum-type single-blow impact test in which the specimen, usually notched, is fixed at one end and broken by a falling weight. The energy absorbed, as measured by the subsequent rise of the pendulum, is a measure of impact strength or notch toughness. Contrast with Charpy test.

Joint – The junction of members or the edges of members that are to be bonded or have been bonded.

Jominy end-quench test – Used to determine hardenability of steels. Data forms hardenability curves, which describe the hardness versus distance from quenched end. Each Jominy position represents a specific cooling rate.

Knoop hardness – Microhardness determined from the resistance of metal to indentation by an elongated pyramidal diamond indenter, having edge angles of 172 and 130 degrees, making a rhombohedral impression with one long and one short diagonal.

Latent heat – Thermal energy absorbed or released when a substance undergoes a phase change.

Ledeburite – The two-phase mixture obtained right below the eutectic point at 4.5% carbon concentration.

Liquid-penetrant inspection – A type of nondestructive inspection that locates discontinuities that are open to the surface of a metal by first allowing a penetrating dye or fluorescent liquid to infiltrate the discontinuity. The inspector then removes the excess penetrant and applies a developing agent that causes the penetrant to seep back out of the discontinuity and register as an indication. Liquid-penetrant inspection is suitable for both ferrous and nonferrous materials, but it is limited to the detection of open-surface discontinuities in nonporous solids.

Liquidus – The lowest temperature at which a metal or alloy is completely liquid.

Macroscopic stress – Residual stresses that vary from tension to compression in a distance that is comparable to the gauge length in ordinary strain measurements. Detectable by X-ray or other methods.

Macrostructure – The general crystalline structure of a metal and the distribution of impurities as seen on a polished or etched surface by either the naked eye or under low magnification of less than 50X.

Maraging – A precipitation-hardening treatment applied to a special group of iron-based alloys to precipitate one or more intermetallic compounds in a matrix of essentially carbon-free martensite.

Martempering (marquenching) – A heat treatment involving heating to austenitizing temperature followed by step quenching at a rate fast enough to avoid the formation of ferrite, pearlite or bainite to slightly above the M_s point. Martempering reduces thermal stresses (compared to conventional quenching).

Martensite – Rapidly cooled austenite in which little or no carbon diffusion takes place. Supersaturated solution of carbon in alpha iron (ferrite). Shear transformation is involved where trapped carbon induces internal stresses. Martensite is very hard and brittle and must be toughened by tempering. End result is still harder and stronger.

Materials science – The science and technology of materials. Metals and nonmetals are included in this field of study.

Mechanical properties – The properties of a material that reveal its elastic and inelastic behavior when force is applied, thereby indicating its suitability for mechanical applications (e.g., modulus of elasticity, tensile strength, elongation, hardness and fatigue limit). Compare with physical properties.

Metal – 1.) An opaque, lustrous element that is a good conductor of heat and electricity and when polished a good reflector of light. Most metals are malleable and ductile and, in general, heavier than other elemental substances. 2.)

Metals are distinguished from non-metals by their atomic binding and electron availability. Metallic atoms tend to lose electrons from the outer shells. 3.) An elemental structure whose hydroxide is alkaline. 4.) An alloy.

Metallizing – 1.) Forming a metallic coating by atomized spraying with molten metal or by vacuum deposition. Also called alloyed spray metallizing. 2.) Applying an electrically conductive metallic layer to the surface of a nonconductor.

Metallography – The study of the internal structure of metals and alloys via the use of a microscope or X-ray.

Metallurgy – The science and technology of metals. Metals are classified as ferrous and nonferrous and divided into classifications of wrought and powder metal. Process (chemical) metallurgy deals with extraction and refinement of ores, while physical metallurgy deals with the physical and mechanical properties of metals as they are affected by composition, mechanical working and heat treatment.

Microhardness – The hardness of a material as determined by forcing an indenter (e.g., Vickers or Knoop) into the surface of a material under very light load. The indentations are so small that they must be measured with an optical microscope. Capable of determining hardness of different microconstituents within a structure or of measuring steep hardness gradients such as those encountered in case hardening.

Microsegregation – Segregation within a grain, crystal or small particle.

Microstructure – The structure that is observed when a polished and etched specimen of metal is viewed in an optical microscope at magnifications in the range of approximately 50X to 1,500X.

Mill scale – The heavy oxide layer formed during hot fabrication or heat treatment of metals.

Modulus of elasticity – A measure of the rigidity of metal. Ratio of stress, below the proportional limit, to corresponding strain. Specifically, the modulus obtained in tension or compression is Young's modulus, stretch modulus or modulus of extensibility; the modulus obtained in torsion or shear is modulus of rigidity, shear modulus or modulus of torsion; the modulus covering the ratio of the mean normal stress to the change in volume per unit is the bulk modulus. The tangent modulus and secant modulus are not restricted within the proportional limit. The former is the slope of the stress-strain curve at a specified point; the latter is the slope of a line from the origin to a specified point on the stress-strain curve. Also called elastic modulus and coefficient of elasticity.

Modulus of rupture – Nominal stress at fracture in a bend test or torsion test. In bending, the modulus of rupture is the bending moment at fracture divided by the section modulus. In torsion, modulus of rupture is the torque at fracture divided by the polar section modulus.

Natural aging – Room-temperature aging of a supersaturated solid solution.

Nitriding – A case-hardening process that depends on the absorption of nitrogen into the steel. The nitrogen is introduced into a solid ferrous alloy by holding below the critical temperature (Ac_1) in contact with a suitable nitrogenous material. Quenching is not required to produce a hard case. Nitrided steels typically have a higher surface hardness when compared to carburized steels, are extremely resistant to abrasion and have a high fatigue strength. Other common names and related subjects are gas nitriding, ion/plasma nitriding or case hardening.

Nitrocarburizing – A case-hardening process similar to nitriding involving the introduction of nitrogen and carbon into a solid ferrous alloy by holding below the critical temperature (Ac_1) in contact with a suitable nitrogenous and carbonaceous material. A thin nitrogen and carbon-enriched layer possibly with accompanying carbonitrides and nitrides is produced. The compound layer (aka white layer) with an underlying diffusion zone contains dissolved nitrogen and iron (alloy) nitrides. Quenching is not required to produce a hard case. Other common names and related subjects are case hardening, ferritic nitrocarburizing (FNC) and austenitic nitrocarburizing.

Nonferrous – A metal in which iron is not the principal alloying element. Not ferrous.

Non-metals – A chemical element that lacks typical metallic properties and is able to form anions, acidic oxides and acids, and stable compounds with hydrogen.

Normalizing – Heating a ferrous alloy to a suitable temperature above the transformation range (and typically above the suitable hardening temperature) and then cooling in the equivalent of still air to a temperature substantially below the transformation range.

Oil hardening – A hardening treatment involving quenching in oil.

Oil quenching – Hardening of carbon steel by quenching in oil. Oils are categorized as fast, medium, slow, hot or martempering.

Oxidation (oxidizing) – 1.) A reaction in which there is an increase in valence resulting from a loss of electrons. 2.) A corrosion reaction in which the corroded metal forms an oxide; usually applied to reaction with a gas containing elemental oxygen, such as air.

Over-aging – Aging under conditions of time and temperature greater than those required to obtain maximum change in a certain property so that the property is altered in the direction of the initial value.

Overheating – Heating a metal or alloy to such a high temperature that its properties are impaired. When the original properties cannot be restored by another heat treatment, by mechanical working or by a combination of working and heat treatment, the overheating is known as burning.

Oxygen (carbon) probe – An atmosphere-monitoring device that is used to determine the percent carbon in the furnace atmosphere. An oxygen (carbon) probe electronically measures the difference between the partial pressure of oxygen in a furnace atmosphere and the external air. Oxygen potential can then be directly related to carbon potential for a given temperature.

Partial annealing – An imprecise term used to denote a treatment given to cold-worked material to reduce its strength to a controlled level or to affect stress relief. To be meaningful, the type of material, the degree of cold work and the time-temperature schedule must be stated.

Pearlite – A lamellar aggregate of ferrite and cementite, often occurring in steel and cast iron.

Physical metallurgy – The science that studies the processes that control, develop and modify the physical and mechanical properties of metals. Typically, the two most important types are mechanical operations (e.g., rolling, hammering or swaging) and heat treatment (e.g., annealing, hardening and carburizing).

Physical properties – Properties of a metal or alloy that are relatively insensitive to structure and can be measured without the application of force (e.g., density, electrical conductivity, coefficient of thermal expansion, magnetic permeability and lattice parameter). Does not include chemical reactivity. Compare with mechanical properties.

Plasma spray – A thermal-spraying process in which the coating material is melted with heat from a plasma torch that generates a non-transferred arc. Molten coating material is propelled against the basis metal by the hot, ionized gas issuing from the torch.

Poisson's ratio – The absolute value of the ratio of the transverse strain to the corresponding axial strain in a body subjected to uniaxial stress; usually applied to elastic conditions.

Powder metallurgy – The art of producing metal powders and of utilizing metal powders for the production of materials and shaped objects.

Precipitation heat treatment – See aging.

Preheating – Heating before a thermal or mechanical treatment. For tool steel, heating to one or more intermediate temperatures before final austenitizing. For some nonferrous alloys, heating to a high temperature for a long time to homogenize the structure before working. In welding and related processes, heating to an intermediate temperature for a short time before welding, brazing, soldering, cutting or thermal spraying to reduce stress in the material created by the subsequent operation.

Purpose for alloying – Alloying elements are used to create specific properties in metals (e.g., hardness, strength, toughness) and make it easier to transform to a specific microstructure.

Pusher furnace – A type of continuous furnace in which parts to be heated are periodically charged into the furnace on containers (i.e., grids, saggers, baskets) that are pushed along the hearth against a line of previously charged containers, thus advancing the containers toward the discharge end of the furnace where they are quenched then removed.

Quench aging – Aging induced by rapid cooling after solution heat treatment.

Quench annealing – Annealing an austenitic ferrous alloy by solution heat treatment followed by rapid quenching.

Quench cracking – Fracture of a component part during quenching from elevated temperature. Most frequently observed in hardened carbon steel, alloy steel or tool steel parts of high hardness and low toughness. Cracks often emanate from stress risers (e.g., fillets, holes, corners) and result from volume changes accompanying transformation to martensite.

Quenching – Cooling from high temperature, usually at a fast rate.

Racking – A term used to describe the placing of parts to be heat treated on a platform, such as a grid or fixture, to hold them in proper position and keep them separated during treatment.

Recarburize – See carbon restoration. The term is also used to describe a second carburizing process used on component parts that did not achieve desired results from the initial carburizing process.

Recrystallization – 1.) Formation of a new, strain-free grain structure from the existing cold-worked metal, usually accomplished by heating. 2.) The change from one crystal structure to another as occurs on heating or cooling through a critical temperature.

Recuperator – A device for recovering heat from gaseous products of combustion and transferring it to cold incoming air or fuel. The incoming gas passes through pipes surrounded by a chamber through which the outgoing gases pass.

Reduction of area – 1.) Commonly, the difference, expressed as a percentage of original area, between the cross-sectional area of a tensile-test specimen and the minimum cross-sectional area measured after complete separation. 2.) The difference, expressed as a percentage of original area, between the original cross-sectional area and that after straining of the specimen.

Refractory – 1.) A material of very high melting point with properties that make it suitable for such uses as furnace linings and kiln construction. 2.) The quality of resisting heat.

Residual stress – The stress that exists in an elastic solid body in the absence of, or in addition to, the stresses caused by an external load. Such stresses can arise from deformation during cold working, such as cold drawing or stamping; in welding from weld metal shrinkage; and in changes in volume due to thermal expansion.

Retained austenite – A room-temperature remnant of the transformation process from austenitizing temperature. Retained austenite is lower in hardness than other transformation products. Retained austenite is unstable at room temperature. It will transform to martensite (with a resulting volume expansion) upon the application of temperature, stress or over time. Most applications attempt to keep retained austenite under a prescribed percentage.

Rockwell hardness test – An indentation hardness test based on the depth of penetration of a specified penetrator into the specimen under certain arbitrarily fixed conditions.

Rotary-hearth furnace – A type of continuous furnace in which the work advances by means of an internal spiral or screw, providing better exposure to heat and the furnace atmosphere to small parts that tend to nest.

Scleroscope – A hardness test in which the loss in kinetic energy of a falling metal "tup," absorbed by indentation upon impact of the tup on the metal being tested, is indicated by the height of rebound.

Secondary hardness – An increase in hardness that can occur when hardened steel is reheated to an intermediate temperature (e.g., the conversion of retained austenite on tempering).

Selective heating – Intentionally heating only certain portions of a workpiece.

Selective quenching – Quenching only certain portions of a component part.

Setting the braze – A term used to describe an intermediate cooling step whereby the filler metal is allowed to cool to a semi-stable (liquid/solid) phase prior to the onset of rapid cooling.

Severity of quench – Ability of a quenching medium to extract heat from a hot steel workpiece expressed in terms of the H value. A measure of the "cooling power" of the quenching media. Air has the lowest severity; oil has medium severity; water has high severity; and brine has the highest severity. Agitation increases cooling power. More severe quenches cause deeper hardening.

Shaker-hearth furnace – A type of continuous furnace that uses a reciprocating shaker motion to move the parts along the hearth.

Shear – 1.) That type of force that causes or tends to cause two contiguous parts of the same body to slide relative to each other in a direction parallel to their plane of contact. 2.) A type of cutting tool with which a material in the form of wire, sheet, plate or rod is cut between two opposing blades. 3.) The type of cutting action produced by rake so that the direction of chip flow is other than at right angles to the cutting edge.

Shim – A thin piece of low-carbon metal that can be analyzed for carbon content after exposure and equilibrium with the furnace atmosphere. A measure of the furnace carbon potential. Alternately, a thin piece of material placed between two surfaces to obtain a proper fit, adjustment or alignment.

Sintering – The bonding of adjacent powder particle surfaces in a mass of metal powders or a compact by heating. Other common names and related subjects are cold/hot isostatic pressing, liquid-phase sintering and metal injection molding.

Skull – An unmelted residue of filler metal.

Slack quenching – The incomplete hardening of steel during quenching from the austenitizing temperature. The rate of cooling is slower than the critical cooling rate for the particular steel, resulting in the formation of one or more unwanted transformation products, usually in addition to martensite.

Slot furnace – A common batch furnace where stock is charged and removed through a slot or opening.

Soak – That portion of the thermal cycle during which the temperature of the part or load is maintained constant.

Soak temperature – The constant temperature at which the part or load is maintained.

Soak time – Time for which the temperature of the part or load is maintained constant.

Snap temper – A precautionary interim stress-relief treatment applied to high-hardenability steels immediately after quenching to prevent cracking due to a possible delay in tempering at the prescribed higher temperature.

Soaking – Prolonged holding at a selected temperature to effect homogenization of structure or composition.

Solution treating – Heating an alloy to a suitable temperature, holding at that temperature long enough to allow one or more of the constituents to enter into solid solution and then cooling rapidly enough to hold the constituents in solution. The alloy is left in a supersaturated, unstable state and may subsequently exhibit quench aging. Solution treating is most often followed by an aging treatment.

Solid solution – A single solid, homogenous crystalline phase containing two or more chemical species.

Solidus – The highest temperature at which a metal or alloy is completely solid.

Solute – A dissolved solid.

Solution – A homogeneous mixture composed of only one phase. In such a mixture, a solute is a substance dissolved in another substance, the solvent.

Solvent – A substance (typically a liquid) capable of dissolving or dispensing one or more substances within itself.

Solvus temperature – The temperature at which the solute atoms will dissolve or disperse completely within the material.

Soot – Free carbon present in a heat-treatment furnace due to an unstable atmosphere. Carbon deposits found on a workload typically resulting from when the load is not at temperature prior to or concurrent with the introduction of a carbonaceous species

Sorbite – A term referring to a dispersed variety of pearlite that was dropped from the metallographic lexicon because it inaccurately referred to microstructural constituents. Still found occasionally in the literature.

Spalling – A chipping or flaking of a surface due to any kind of improper heat treatment or material dissociation.

Spheroidization – The process of producing a microstructure consisting of coarse spherical (spheroidal) carbides in a ferrite matrix. Spheroidizing methods frequently used are: 1.) prolonged holding at a temperature just below Ae_1; 2.) heating and cooling alternately between temperatures that are just above and just below Ae_1; 3.) heating to a temperature above Ae_1 or Ae_3 and then cooling very slowly in the furnace or holding at a temperature just below Ae_1; 4.) cooling at a suitable rate from the minimum temperature at which all carbide is dissolved to prevent re-formation of a carbide network and then reheating in accordance with method 1 or 2 above. (Applicable to hypereutectoid steel containing a carbide network.)

Spray quenching – A quenching process using nozzles to spray water or other liquid on a part. The quench rate is controlled by the velocity and volume of liquid.

Spring temper – A temper of nonferrous alloys and some ferrous alloys characterized by tensile strength and hardness about two-thirds of the way from full-hard to extra spring temper (i.e., corresponding to approximately a fully cold-worked state).

Stabilizing treatment – 1.) Before finishing to final dimensions, repeated heating of a ferrous or nonferrous part to or slightly above its normal operating temperature and then cooling to room temperature to ensure dimensional stability service. 2.) Transforming retained austenite in quenched hardenable steels, usually by cold treatment. 3.) Heating a solution-treated stabilized grade of austenitic stainless steel to precipitate all carbon so that sensitization can be avoided on subsequent exposure to elevated temperature.

Statistical process control (SPC) – The application of statistical techniques for measuring and analyzing the variation in processes.

Statistical quality control (SQC) – The application of statistical techniques for measuring and improving the quality of processes and products (includes statistical process control, diagnostic tools, sampling plans and other statistical techniques).

Steel – A solid solution of iron and up to 2% carbon, usually containing manganese, silicon and various impurity elements. Other alloying elements are added to achieve specific properties and characteristics.

Strain – A measure of the relative change in the size or shape of a body. Linear strain is the change per unit length of a linear dimension. True strain (or natural strain) is the natural logarithm of the ratio of the length at the moment of observation to the original gauge length. Conventional strain is the linear strain over the original gauge length. Shearing strain (or shear strain) is the change in angle (expressed in radians) between two lines originally at right angles. When the term strain is used alone it usually refers to the linear strain in the direction of applied stress.

Strain hardening – An increase in hardness and strength caused by plastic deformation at temperatures below the recrystallization range.

Stress – Force per unit area, often thought of as force acting through a small area within a plane. It can be divided into components – normal and parallel to the plane – called normal stress and shear stress, respectively. True stress denotes the stress where force and area are measured at the same time. Conventional stress, as applied to tension and compression tests, is force divided by original area. For example, nominal stress in a notch bend test (ignoring stress risers and disregarding plastic flow) is determined by dividing the bending moment by the minimum section modulus.

Stress equalizing – A low-temperature heat treatment used to balance stresses in cold-worked material without an appreciable decrease in the mechanical strength produced by cold working.

Stress relief – A process consisting of heating to a suitable temperature, holding long enough to reduce residual stresses and then cooling slowly enough to minimize the development of new residual stresses. Other common names and related subjects are stress relieving or stress relaxation.

Structure – When applied to metals, structure refers to the arrangement of components that make the material. The common names given to particular steel structures are austenite, pearlite, ferrite and martensite. Physical properties of alloys are related directly to structure. The principal reason for heat treating of an alloy is to rearrange its structure to obtain certain desired characteristics.

Subcritical annealing – Process anneal performed on ferrous alloys at a temperature below Ac_1.

Surface hardening – A generic term covering several processes applicable to a suitable ferrous alloy that produces a surface layer that is harder or more wear resistant than the core on quenching. There is no significant alteration of the chemical composition of the surface layer. The use of the applicable process name is preferred.

Supercooling – Cooling below the temperature at which an equilibrium phase transformation can take place without actually obtaining the transformation.

Temper color – A thin, tightly adhering oxide skin that forms when steel is tempered at a low temperature, or for a short time, in air or in a mildly oxidizing atmosphere. The color, which ranges from straw to blue depending on the thickness of the oxide skin, varies with both temperature and exposure time.

Tempered martensite embrittlement (TME) – Embrittlement of high-strength steels caused by tempering in the temperature range of 205-400°C (400-750°F); also called 260°C (500°F) embrittlement. TME is thought to result from the combined effects of cementite precipitation on prior-austenite grain boundaries or interlath boundaries and the segregation of impurities at prior-austenite grain boundaries.

Tempering – Reheating a quench-hardened or normalized ferrous alloy to a temperature below the transformation temperature and then cooling at any rate desired. A reduction in strength and increase in ductility properties of the material generally results. Other common names and related subjects are draw, drawing and temper.

Thermal fatigue – Fracture resulting from the presence of temperature gradients that vary with time in such a manner as to produce cyclic stresses in a structure.

Thermal shock – The development of a steep temperature gradient and accompanying high stresses within a structure.

Thermal stress – Stresses in metal resulting from non-uniform temperature distribution.

Thermocouple – A device for measuring temperatures, consisting of lengths of two dissimilar metals or alloys that are electrically joined at one end (hot junction) and connected to a voltage-measuring instrument at the other end (cold junction). When one junction is hotter than the other, a thermal electromotive force is produced that is roughly proportional to the difference in temperature between the hot and cold junctions.

Total carbon – The sum of the free and combined carbon (including carbon in solution) in a ferrous alloy.

Toughness – The ability of a metal to absorb energy and deform plastically before fracturing.

Transformation diagrams – See time-temperature-transformation and continuous-cooling diagrams.

Transformation hardening – Heat treatment comprised of austenitization followed by quenching under conditions such that the austenite transforms more or less completely into martensite and possibly into bainite.

Transformation temperature – The temperature at which a change in phase occurs. This term is sometimes used to denote the limiting temperature of a transformation range. The following symbols are commonly used in the heat treatment of irons and steels:
- Ac_{cm} – In hypereutectoid steel, the temperature at which solution of cementite in austenite is completed during heating.
- Ac_1 – The temperature at which austenite begins to form during heating.
- Ac_3 – The temperature at which transformation of ferrite to austenite is completed during heating.
- Ac_4 – The temperature at which austenite transforms to delta ferrite during heating.
- Ae_{cm} – Ae_1, Ae_3, Ae_4. The temperatures of phase changes at equilibrium.
- Ar_{cm} – In hypereutectoid steel, the temperature at which precipitation of cementite starts during cooling.
- Ar_1 – The temperature at which transformation of austenite to ferrite or to ferrite plus cementite is completed during cooling.
- Ar_3 – The temperature at which austenite begins to transform to ferrite during cooling.
- M_f – The temperature at which transformation of austenite to martensite is completed during cooling.
- M_s – The temperature at which transformation of austenite to martensite starts during cooling.

Note: All these changes, except formation of martensite, occur at lower temperatures during cooling than during heating and depend on the rate of change in temperature.

Transgranular – Through (across) crystals or grains.

Transgranular cracking – Cracking or fracturing that occurs through or across crystals or grains. Also called transcrystalline cracking.

Transgranular fracture – Fracture through or across the crystals or grains of a metal. Also called transcrystalline fracture or intracrystalline fracture.

Troostite – A term referring to a very fine aggregate of ferrite and carbide that was dropped from the metallographic lexicon because it inaccurately referred to microstructural constituents. Still found occasionally in the literature.

Tunnel kiln – A continuous-type furnace where the work is conveyed through a tunnel-type heating zone, and the parts are hung on hooks or placed on fixtures to minimize distortion.

Ultimate (tensile) strength – The maximum conventional stress (tensile, compressive or shear) that a material can withstand.

Ultrasonic testing – A nondestructive testing method applied to sound-conductive materials having elastic properties for the purpose of locating inhomogeneities or structural discontinuities within a material.

Vapor plating – Deposition of a metal or compound on a heated surface by reduction or decomposition of a volatile compound at a temperature below the melting points of the deposit and the base material. The reduction is usually accomplished by a gaseous reducing agent such as hydrogen. The decomposition process may involve thermal dissociation or reaction with the base material. Occasionally used to designate deposition on cold surfaces by vacuum evaporation.

Vickers hardness – A microhardness technique for metals in which a 136-degree diamond pyramid is pressed into the surface of a metal being testing by a prescribed load weight (e.g., 5-120 kg).

Water quenching – A quench in which water is the quenching medium.

Wetting – The phenomenon in which a liquid filler metal or flux spreads or flows and adheres in a thin, continuous layer on a solid base metal.

White layer – Compound layer that forms as a result of the nitriding process.

Work hardening – Hardness developed in metal as a result of cold working.

Working zone – That portion of the enclosed volume of a piece of thermal-processing equipment occupied by parts or raw material during thermal treatment. It is usually, but not always, a high percentage of the total enclosed volume. It may include more than one (control) zone.

Wrought metallurgy – The art of producing metals by conventional steelmaking practices, either in blast furnaces or electric-arc furnaces. Iron-oxide reduction in the former and recycling of scrap steel in the latter are two main sources of raw materials.

Yield strength – The stress at which a material exhibits a specified deviation from proportionality of stress and strain. An offset of 0.2% is used for many metals. Compare with tensile strength.

REFERENCES
1. *Metals Handbook, 8th Edition, Volume 1: Properties and Selection of Metals*, (Definitions of Metals and Metalworking), ASM International, 1961, pp. 1-41
2. *Glossary of Terms*, Bodycote Inc. (www.bodycote.com)
3. *Manufacturing Engineering Handbook*, Hwaiyu Geng (Ed.) McGraw-Hill Book Company, 2004
4. Krauss, George, *Steels: Processing, Structure, and Performance*, ASM International, 2005
5. Wikipedia (www.wikipedia.org)
6. About Metals (www.metals.about.com)
7. Craig Darragh, AG Fox, LLC, technical review and private correspondence

16 | REFERENCE MATERIALS

16.2 REFERENCE LIBRARY FOR HEAT TREATERS

When it comes to understanding any subject, particularly heat treatment, having reference materials you can trust is invaluable. Establishing a good technical reference library, whether at work or at home (or both), is an absolute necessity because solving everyday problems requires an understanding of not only the root cause but the underlying science behind it. In addition, one must understand the technical issues and practical reasons why a particular phenomenon took place so that corrective action may be taken to avoid its reoccurrence in the future.

There are many ways for heat treaters to acquire useful information and valuable insights into their profession. They can use the Internet and interact with colleagues, mentors and supplier partners. Technical societies such as ASM International are also tremendous resources. Referencing industrial publications (e.g., *Industrial Heating*) and talking to universities, national laboratories and technical societies are still more ways. Finally, there are reference books.

Presented here is an alphabetical list of references by subject that heat treaters and metallurgists consider "must have" books. Newer editions may exist in some cases, but exercise caution to ensure their contents are equal to or better than the original. This is a reasonably comprehensive listing. It is typical for a company library or experienced consultant. Each reader can select and purchase the references that will be most pertinent to their products, processes, equipment and applications. They may or may not be readily available other than from used book suppliers.

Ferrous Metals (Iron and Steel)[1,2]

1. *ASM Handbook, Volume 4A: Steel Heat Treating: Fundamentals and Processes*, Jon L. Dossett and George E. Totten (Eds.), ASM

International, 2013, ISBN 978-1-62708-011-8, 768 pages
2. *ASM Handbook, Volume 4B: Steel Heat Treating Technologies*, Jon L. Dossett and George E. Totten (Eds.), ASM International, 2014, ISBN 978-1-62708-025-5, 528 pages
3. *ASM Handbook, Volume 4D: Heat Treating of Irons and Steels*, Jon L. Dossett and George E. Totten (Eds.), ASM International, 2014, ISBN 978-1-62708-066-8, 641 pages
4. *Modern Steels and Their Properties*, Bethlehem Steel Corp., 1980, ISBN 1-11441-106-X, 200 pages
5. Bain, Edgar Collins and Harold W. Paxton, *Alloying Elements of Steel*, 2nd Ed., ASM International, 1966, 301 pages
6. Atkins, M., *Atlas of Continuous Cooling Transformation Diagrams for Engineering Steels*, ASM International, 1977, ISBN 0-87170-093-X, 260 pages
7. *Atlas of Isothermal Transformation Diagrams*, United States Steel Corp., 1951, 143 pages
8. *Atlas of Isothermal Transformation Diagrams* (Supplement), United States Steel Corp., 1953, 529 pages
9. *Atlas of Isothermal Transformation and Cooling Transformation Diagrams*, ASM International, 1977, LOC 76-58536, 422 pages
10. *Atlas of Time-Temperature Diagrams for Irons and Steels*, George F. Vander Voort (Ed.), 1991, ASM International, ISBN 0-87170-415-3, 766 pages
11. *Boron in Steel*, S.K. Banerji, and J.E. Morral (Eds.), The Metallurgical Society of AIME, 1980, ISBN 0-89520-363-4, 215 pages
12. Bringas, John E. and Michael L. Wayman, *Casti Metals Black Book – European Ferrous Data*, 2nd Ed., Casti Publishing Inc., 2003, ISBN 1-894038-74-6, 785 pages
13. Bringas, John E. and Michael L. Wayman, *Casti Metals Black Book – North American Ferrous Data*, 5th Ed., Casti Publishing Inc., 2003, ISBN 1-894038-72-X, 764 pages
14. Hanson, Albert and J. Gordon Parr, *The Engineer's Guide to Steel*, Addison-Wesley Publishing Company, 1965, LOC 65-10407, 406 pages
15. *Engineering Properties of Steel*, Philip D. Harvey (Ed.), ASM International, 1982, ISBN 0-87170-144-8, 527 pages
16. Siebert, Clarence A., Douglas V. Doane and Dale H. Breen, *The Hardenability of Steel – Concepts, Metallurgical Influences, and Industrial Applications*, ASM International, 1977, LOC 77-22649, 218 pages
17. *Heat Treater's Guide: Practices and Procedures for Irons and Steels*, 2nd Ed., Harry E. Chandler (Ed.), 1995, ASM International, ISBN 0-87170-520-6, 903 pages
18. *Heat Treaters Guide: Standard Practices and Procedures for Steel*, Paul M.

Unterwieser, Howard E. Boyer and James J. Kubbs (Eds.), 1982, ASM International, ISBN 0-87170-141-3, 493 pages
19. *Steel Heat Treatment Handbook: Metallurgy and Technologies*, 2nd Ed., George E Totten (Ed.), CRC Press, 2007, ISBN 978-0-8493-8455-4, 831 pages
20. *Heat Treatment of Steel: Republic Alloy Steels Handbook*, Booklet Adv. 1009, 1961, 42 pages
21. *I–T Diagrams: Isothermal Transformation of Austenite in a Wide Variety of Steel*, 3rd Ed., United States Steel Corporation, 1963, 183 pages
22. Mangonon, P.L., *Iron & Steel*, Handbook Supplement HS30, 1982, SAE, ISBN 0-89883-404-X
23. *The Making, Shaping and Treating of Steel*, 9th Ed., Harold E. McGannon (Ed.), AISE (United States Steel Corp.), 1971, 1,420 pages
24. Duckworth, W.E. and T.F.J.N. Ryan, *Manganese in Ferrous Metallurgy*, Desforges, C.D., The Manganese Centre, 1976, ISBN 2-901-108-004, 87 pages
25. *Mechanical Properties of Alloy Steel: Republic Alloy Steels Handbook*, Booklet Adv. 1009, 1961, 84 pages
26. *Alloying Elements and Their Effect on Hardenability: Republic Alloy Steels Handbook*, Booklet Adv. 1009, 1961, 46 pages
27. *Nickel Alloy Steel*, International Nickel Co., 1949, 800 pages
28. Hall, A.M., *Nickel in Iron and Steel*, John Wiley & Sons, 1954, LOC 54-11982, 595 pages
29. Cias, Witold W., *Phase Transformation Kinetics and Hardenability of Medium-Carbon Alloy Steels*, Climax Molybdenum Co., 1972, 121 pages
30. *Republic Alloy Steels*, Republic Steel, 1961, 523 pages
31. Bullen D.K. and the metallurgical staff of Battelle Memorial Institute, *Steel and Its Heat Treatment, Vol. I: Principles, Processes, Control*, John Wiley & Sons, 1938, 445 pages
32. Bullen D.K., and the metallurgical staff of Battelle Memorial Institute, *Steel and Its Heat Treatment, Vol. II: Engineering and Special Purpose Steels*, John Wiley & Sons, 1939, 491 pages
33. *Steel and Its Heat Treatment: Bofors Handbook*, Karl-Erik Thelning (Ed.), Butterworths, 1973, ISBN 0-408-70651-1, 570 pages
34. *Steel Heat Treatment Handbook*, George E. Totten and Maurice A.H. Howes (Eds.), Marcel Dekker Inc., 1997, ISBN 0-8247-9750-7, 1,192 pages
35. *Steel Strengthening Mechanisms* (Symposium), Climax Molybdenum Co., 1969, 183 pages
36. Krauss, G., *Steels: Heat Treatment and Processing Principles*, ASM International, 1990, ISBN 0-87170-370-X, 497 pages
37. Llewellyn, D.T. and R.C. Hudd, *Steels, Metallurgy & Applications*, 3rd Ed., Butterworth Heinemann, 1998, ISBN 0-7506-3757-9, 389 pages

38. Krauss, George, *Steels: Processing, Structure, and Performance*, ASM International, 2005, ISBN 0-87170-817-5, 613 pages
39. *Steels for Elevated Temperature Service*, United States Steel, 1952, 87 pages
40. *Transformation and Hardenability in Steels* (Symposium), Climax Molybdenum Co., 1977, 212 pages
41. *Vanadium Steels and Irons*, Vanadium Corporation of America, 189 pages
42. Burgess C.O., *Heat Treating of Gray Iron*, Gray Iron Founders' Society Inc., 1984, 121 pages

Ferrous Metals (Stainless Steel)[1,2]

1. *ASM Specialty Handbook® Stainless Steels*, J.R. Davis (Ed.), 1994, ISBN 0-87170-503-6, 576 pages
2. *Stainless Steel Handbook*, Allegheny Ludlum Steel Corporation, 1959, 120 pages
3. Marshall, P., *Austenitic Stainless Steels: Microstructure and Mechanical Properties*, Elsevier Applied Science Publishing Ltd., 1984, ISBN 0-85334-277-6, 431 pages

Ferrous Metals (Tool Steel)[1,2]

1. Roberts, G.A., J.C. Hamaker Jr., and A.R. Johnson, *Tool Steels*, 3rd Ed., 1962, ASM International, 780 pages
2. Wilson, Robert, *Metallurgy and Heat Treatment of Tool Steels*, McGraw-Hill Book Company, 1975, ISBN 0-07-084453-4, 378 pages
3. Payson, Peter, *The Metallurgy of Tool Steels*, John Wiley & Sons, 1962, LOC 62-16155, 349 pages
4. Seabright, L., The *Selection and Hardening of Tool Steels*, Volumes 1 & 2, Seabright Texts, 1975, 235 pages
5. *Tool Steel Handbook*, Allegheny Ludlum Steel Corporation, 1959, 202 pages
6. Roberts, G., G. Krauss and R. Kennedy, *Tool Steels*, 5th Ed., ASM International, 1998, ISBN 0-87170-599-0, 364 pages
7. Palmer, Frank R. and George V. Luerssen, *Tool Steel Simplified*, Carpenter Technology Corp., 1977, ISBN 0-31055-805-6, 555 pages
8. *The Tool Steel Trouble Shooter: An Analysis of 107 Tool Failures*, Bethlehem Steel Corp., 1964, 125 pages
9. *Improving Production from Tools and Dies*, Bethlehem Steel Company, 1960, LOC 60-7475, 72 pages
10. Bryson, Bill, *Heat Treatment, Selection and Applications of Tool Steels*, Hansen Gardner Publications, 1997, ISBN 1-56990-238-0, 198 pages

16 REFERENCE MATERIALS

Furnace Atmospheres – Safety, Control and Generation[10,11]

1. Hotchkiss, A.G. and H.M. Webber, *Protective Atmospheres*, John Wiley & Sons, 1953, LOC 53-8521, 341 pages

General Reference[1,2]

1. *ASM Materials Engineering Dictionary*, J.R. Davis (Ed.), ASM International, 1992, ISBN 0-87170-447-1, 555 pages
2. Zimmerman, O.T. and Irvin Lavine, *Conversion Factors and Tables*, 3rd Ed., Industrial Research Services Inc., 1961, 680 pages
3. *Handbook of Mechanical Alloy Design*, George E. Totten, Lie Xie and Kiyoshi Funatani (Eds.), Marcel Dekker Inc., 2004, ISBN 0-8247-4308-3, 734 pages
4. Oberg, Erik, Franklin Day Jones and Henry H. Ryffel, *Machinery's Handbook*, 26th Ed., Industrial Press, 2000, ISBN 0-83112-625-6, 2,640 pages
5. *Mark's Standard Handbook for Mechanical Engineers*, 9th Ed., Eugene A. Avallone and Theodore Baumeister III (Eds.), McGraw Hill Book Co., 1978, ISBN 0-07-004127-X
6. *Metals Handbook, Vol. 1, Properties and Selection of Metals*, 8th Ed., ASM International, 1961, LOC 27-12046, 366 pages
7. *Metals Handbook*, 1948 Ed., ASM International, 1948, 1,332 pages
8. Wegst, C.W., *Stahlschlussel (Key to Steel)*, 20th Ed., Verlag, 2004, ISBN 3-922599-20-6, 720 pages
9. *Worldwide Guide to Equivalent Irons and Steels*, 4th Ed., ASM International, 2000, ISBN 0-87170-635-0, 1,132 pages
10. *Worldwide Guide to Equivalent Nonferrous Alloys*, 4th Ed., ASM International, 2001, ISBN 0-87170-741-1, 1,036 pages

Heating Methods (Electric and Gas)[1,2,5]

1. *Combustion Technology Manual*, 5th Ed., Industrial Heating Equipment Association, (IHEA)
2. *Efficient Use of Fuels in Metallurgical Industries Symposium*, Institute of Gas Technology, 1974, ISBN 0-91009-114-5
3. *IHEA Heat Processing Manual*, Industrial Heating Equipment Association
4. Reed, Richard J., *North American Combustion Handbook, Vol. 1*, 2nd Ed., North American Mfg. Co., 1978, ISBN 0-96015696-1-4, 501 pages
5. Reed, Richard J., *North American Combustion Handbook, Vol. 2*, 2nd Ed., North American Mfg. Co., 1986, ISBN 978-0960-159-63-5, 501 pages

6. *NFPA 86 Standard for Ovens and Furnaces,* National Fire Protection Agency, 2015 Ed.

Heat Treating (Processes and Equipment)[1,2]

1. Herring, Daniel H., *Atmosphere Heat Treatment, Volume I,* BNP Media, 2014, ISBN 978-0-692-28393-6, 714 pages
2. *Steel Heat Treatment Handbook: Equipment and Process Design,* 2nd Ed., George E Totten (Ed.), CRC Press, 2007, ISBN 978-0-8493-8454-7, 713 pages
3. Brooks, Charlie R., *Principles of the Heat Treatment of Plain Carbon and Low-Alloy Steels,* ASM International, 1996, ISBN 978-0-87170-538-0, 490 pages
4. Devis, C.E., *Ask Joe: Your Quick Reference to Fourteen Years of Practical Heat Treat Advice,* ASM International, 1983, ISBN 0-87170-167-3, 135 pages
5. *ASM Handbook, Vol. 4, Heat Treating,* ASM International, 1991, ISBN 0-87170-379-3, 1,012 pages
6. *Carburizing and Carbonitriding,* ASM International, 1977, LOC 76-055702, 215 pages
7. Parrish, Geoffrey, *Carburizing: Microstructure and Properties,* ASM International, 1999, ISBN 0-87170-666-0, 247 pages
8. *Handbook of Thermoprocessing Technologies,* Axel von Starck, Alfred Muhlbauer, and Carl Kramer (Eds.), Vulkan-Verlag GmbH, 2005, ISBN 3-8027-2933-1, 807 pages
9. *Heat Treating and Properties of Iron and Steel,* U. S. Department of Commerce National Bureau of Standards Monograph 88, 1966, LOC 66-61523, 46 pages
10. Trinks, W. and M.H. Mawhinney, *Industrial Furnaces,* 5th Ed., Vol. 1, John Wiley & Sons, 1967, 526 pages
11. Trinks, W. and M.H. Mawhinney, *Industrial Furnaces,* 5th Ed., Vol. 2, John Wiley & Sons, 1967, ISBN 0-31726-260-2, 486 pages
12. Parrish, Geoffrey, *Influence of Microstructure on the Properties of Case-Carburized Components,* 1980, LOC 810679, 236 pages
13. *Ion Nitriding* (Conference Proceedings), T. Spalvins (Ed.), ASM International, 1987, ISBN 0-87170-278-9, 202 pages
14. *Metals Handbook, Vol. 2, Heat Treating, Cleaning and Finishing,* 8th Ed., ASM International, 1964, LOC 27-12046, 708 pages
15. Haga, L.J., *Practical Heat Treating,* Metal Treating Institute
16. Haga, L.J., *Principles of Heat Treating,* Metal Treating Institute
17. Grossmann, M.A. and E.C. Bain, *Principles of Heat Treatment,* 5th Ed., ASM International, 1964, 302 pages
18. Brooks, Charlie R., *Principles of Heat Treatment of Plain Carbon and Low*

Alloy Steels, ASM International, 1996, ISBN 0-87170-538-9, 490 pages

19. *Source Book on Nitriding,* ASM International, 1977, LOC 77-23934, 320 pages

Induction Heating[1,2,7]

1. Haimbaugh, Richard E., *Practical Induction Heat Treating,* ASM International, 2001, ISBN 0-87170-743-8, 332 pages
2. Rudnev, V., D. Loveless, R. Cook and M. Black, *Handbook of Induction Heating,* Marcel Dekker, 2003
3. *ASM Handbook, Volume 4C: Induction Heating and Heat Treatment,* Valery Rudnev and George E. Totten (Eds.), ASM International, 2014, ISBN 978-1-62708-012-5, 820 pages
4. Tudbury, Chester A., *Basics of Induction Heating,* Vol. 1 and 2, 1960, LOC 60-8958, 132 pages. (Vol. 1) and 121 pages. (Vol. 2)
5. Zinn, S., and S.L. Semiatin, *Elements of Induction Heating,* 1988, ASM International, ISBN 0-87170-308-4, 335 pages
6. *Induction Heating for Forging,* FIA Plant Engineering Committee, Forging Industry Association
7. Davies, E.J., and P. Simpson, *Induction Heating Handbook,* McGraw-Hill, 1979, ISBN 978-0-07084-515-2, 426 pages
8. Semiatin, S.L. and D.E. Stutz, *Induction Heat Treatment of Steel,* ASM International, 1986, ISBN 978-0-087170-21101, 308 pages
9. Lozinskii, M.G., *Industrial Applications of Induction Heating,* Pergamon Press, 1969, ISBN 978-0-08011-586-3, 672 pages
10. Rowan, Henry M. and John Calhoun Smith, *The Fire Within: The Story of Inductotherm Industries Inc., and the Man Who Built It,* Penton Publishers, 1995, 408 pages
11. Muhlbauer, Alfred, *History of Induction Heating and Melting,* Vulkan-Verlag GmbH, 2008, ISBN 978-3-8027-2946-1, 147 pages
12. Davies, E.J., *Conduction and Induction Heating,* Peter Peregrinus Ltd., 1990, ISBN 0-86341-174-6, 397 pages
13. Lucas, Jean, *Electromagnetic Induction And Electric Conduction in Industry,* Centre Francais de L'Electricite 1997, ISBN 978-2-91066-826-6, 768 pages

Joining (Brazing, Soldering, Welding)[1,2,3]

1. *Brazing Handbook,* 5th Ed., American Welding Society, 2007, ISBN 978-08717-104-6-8, 740 pages
2. Peaslee, Robert L., *Brazing Footprints, Case Studies in High Temperature Braz-*

ing, Wall Colmonoy Corporation, 2003, ISBN 0-9724479-0-3, 299 pages

3. Schwartz, Mel M., *Brazing,* ASM International, 1967, ISBN 0-87170-246-0, 455 pages
4. Jacobson, David and Giles Humpston, *Principles of Brazing,* ASM International, 2005, ISBN 0-87170-812-4, 268 pages
5. *ASM Handbook, Volume 6: Welding, Brazing and Soldering,* ASM International, 1993, ISBN 978-0-87170-382-8, 1,299 pages
6. Roberts, Phillip M., *Industrial Brazing Practice,* 2nd Ed., CRC Press, 2013, ISBN 146-65677-4-0, 460 pages
7. *Aluminum Brazing Handbook,* 4th Ed., The Aluminum Association Inc., 1998, 84 pages
8. Schwartz, Mel M., *Ceramic Joining,* ASM International, 1990, ISBN 0-87170-373-4, 196 pages
9. Mohler, Rudy, *Practical Welding Technology,* Industrial Press Inc., 1983, ISBN 0-8311-1143-7, 220 pages
10. Manko, Howard H., *Solders and Soldering,* McGraw-Hill Book Company, 1964, ISBN 07-0-39895-X, 323 pages
11. Linnert, George E., *Welding Metallurgy; Carbon and Alloy Steels, Vol.1, Fundamentals,* 4th Ed., American Welding Society, 1995, 478 pages

Metallurgy[1,2]

1. Bauccio, Michael L., *ASM Metals Reference Book,* ASM International, 1983, ISBN 0-87170-478-1, 614 pages
2. Bhadeshia, H.K.D.H., *Bainite in Steels,* The Institute of Materials, 1992, ISBN 0-901465-95-0, 451 pages
3. Sinha, Anil Kumar, *Ferrous Physical Metallurgy,* Butterworths, 1989, ISBN 0-409-90139-3, 818 pages
4. Lipson, Charles and Robert C. Juvinall, *Handbook of Stress and Strength: Design and Material Applications,* Macmillan Co., 1963, 459 pages
5. Gaskell, David R., *Introduction to Metallurgical Thermodynamics,* 2nd Ed., McGraw-Hill Book Company, 1981, ISBN 0-07-022946-5, 611 pages
6. *Martensite,* G.B. Olson and W.S. Owen (Eds.), ASM International, 1992, ISBN 0-87170-434-X, 331 pages
7. Bamford, T.G. and H. Harris, *The Metallurgist's Manual,* (8 Volumes), Chapman & Hall Ltd., 1937, 256 pages
8. *Metallurgy for the Non-Metallurgist,* Harry E. Chandler (Ed.), ASM International, 1998, ISBN 0-87170-652-0, 284 pages
9. Brandt, Daniel A. and J.C. Warner, *Metallurgy Fundamentals,* 4th Ed., Goodheart-Willcox Co. Inc., 1999, ISBN 1-56637-543-6, 256 pages

10. Odeh, Atif A., *Metallurgy & Heat Treatment: The Pocket Book*, Atrona Metallurgical Services Inc., 1996, 153 pages
11. Allen, Dell K., *Metallurgy Theory and Practice*, American Technical Society, 1969, LOC 70-82378, 663 pages
12. Keffer, Robert, *Methods in Non-Ferrous Metallurgical Analysis*, McGraw-Hill Book Co., 1928, 335 pages
13. Reed-Hill, Robert E., *Physical Metallurgy*, D. Van Nostrand Co. Inc., 1964, 630 pages
14. Clark, Donald S. and Wilbur R. Varney, *Physical Metallurgy for Engineers*, 2nd Ed., Van Nostrand Reinhold, 1969, ISBN 0-44201-570-4, 629 pages
15. *Practical Data for Metallurgists*, 17th Ed., The Timken Co., 2014, 100 pages
16. *The Science and Engineering of Materials*, D. Askland (Ed.), PWS-KENT
17. Brick, Robert M., Robert B. Gordon and Arthur Phillips, *Structure and Properties of Alloys*, 3rd Ed., McGraw-Hill Book Company, 1965, LOC 64-18397, 503 pages

Metallography[1,2,4]

1. *Applied Metallography*, George F. Vander Voort (Ed.), Chapman & Hall, 1986, ISBN 0-042228-836-0, 301 pages
2. *ASM Handbook, Vol. 9, Metallography and Microstructures*, ASM International, 1985, ISBN 0-87170-007-7, 775 pages
3. Beraha, E. and B. Shpigler, *Color Metallography*, ASM International, 1977, ISBN 0-87170-045-X, 176 pages
4. Samuels, L.E., *Light Microscopy of Carbon Steels*, ASM International, 1999, ISBN 0-87170-655-5, 502 pages
5. Samuels, L.E., *Metallographic Polishing by Mechanical Methods*, 4th Ed., ASM International, 2003, ISBN 0-87170-135-9, 345 pages
6. Bramfitt, B.L. and A.O. Benscoter, *Metallographer's Guide: Practices and Procedures for Irons & Steels*, ASM International, 2002, ISBN 0-87170-748-9, 354 pages
7. Geels, K., et al., *Metallographic and Materialographic Specimen Preparation, Light Microscopy, Image Analysis and Hardness Testing*, ASTM International, 2007, ISBN 978-0-80314-265-7, 743 pages
8. Weck, E. and E. Leistner, *Metallographic Instructions for Colour Etching by Immersion, Part I: Klemm Colour Etching*, 1982, 60 pages; *Part II: Beraha Colour Etchants and their Different Variants*, 1983, 86 pages; and *Part III: Non-ferrous Metals, Cemented Carbides and Ferrous Metals, Nickel-Base and Cobalt-Base Alloys*, 1986, 84 pages; (German and English) Deutcher Verlag für Schweißtechnik GmbH, Düsseldorf

9. Richardson, R.H., *Optical Microscopy for the Materials Sciences*, Marcel Dekker, Inc., 1971, 692 pages
10. Huppman, W.J. and K. Dalal, *Metallographic Atlas of Powder Metallurgy*, Verlag Schmid, 1990, ISBN 978-9-99184-793-1, 190 pages
11. Petzow, Gunter, *Metallographic Etching*, 2nd Ed., ASM International, 1978, ISBN 0-87170-633-4, 143 pages
12. Mondolfo, Lucio, *Metallography of Aluminum Alloys*, John Wiley & Sons, 1943, 359 pages
13. Vander Voort, George F., *Metallography Principles and Practice*, ASM International, 1984, ISBN 0-87170-672-5, 752 pages
14. *Metals Handbook, Vol. 7, Atlas of Microstructures of Industrial Alloys*, 8th Ed., ASM International, 1972, LOC 27-12046, 366 pages
15. Brandon, D. and W.D. Kaplam, *Microstructural Characterization of Materials*, John Wiley & Sons, Chichester, England, 1999, 409 pages
16. Allmand, T.R., *Microscopic Identification of Inclusions in Steel*, British Iron and Steel Research Association, 1962, 72 pages
17. *Microstructure of Ductile Iron*, American Foundrymen's Society, 1965
18. Kehl, George L., *The Principles of Metallographic Laboratory Practice*, 3rd Ed., McGraw-Hill Publishing Co., 1949, ISBN 0-07033-479-X, 534 pages
19. *Typical Microstructures of Cast Metals*, G. Lambert (Ed.), The Institute of British Foundrymen, 1957, 224 pages
20. *Advanced Techniques for Characterizing Microstructures*, F.W. Wiffen and J.A. Spitznagel (Eds.), The Metallurgical Society of AIME, 1982, ISBN 0-89520-390, 1,507 pages

Nonferrous Materials (Aluminum)[1,2,6,12]

1. *Aluminum, Properties and Physical Metallurgy*, John E. Hatch (Ed.), ASM International, 1984, ISBN 0-87170-176-6, 424 pages
2. *Handbook of Aluminum, Vol. 1: Physical Metallurgy and Processes*, George E. Totten and D. Scott Mackenzie (Eds.), Marcel Dekker, Inc., 2003, ISBN 0-8247-0494-0, 1,296 pages
3. *Handbook of Aluminum, Vol. II: Alloy Production and Materials Manufacturing*, George E. Totten and D. Scott Mackenzie (Eds.), Marcel Dekker, Inc., 2003, ISBN 0-8247-0896-2, 1,296 pages
4. *Aluminum Standards and Data 2009*, The Aluminum Association, 2009
5. *Aluminium-Schlüssel: Key to Aluminium Alloys*, 7th Ed., Dr.-Ing. Werner Hesse (Ed.), Aluminium-Verlag Marketing & Kommunikation GmbH, Düsseldorf, 2011 (in German and English)
6. *ASM Specialty Handbook: Aluminum and Aluminum Alloys*, J.R. Davis (Ed.),

ASM International, 1993, ISBN 978-0-87170-496-X, 784 pages
7. ASTM B918 (Standard Practice for Heat Treatment of Wrought Aluminum Alloys)
8. *Heat Treater's Guide: Practices and Procedures for Nonferrous Alloys*, 2nd Ed., Harry Chandler (Ed.), ASM International, 1996, ISBN 978-0-87170-565-5, 669 pages
9. Campbell, F.C., *Manufacturing Technology for Aerospace Structural Materials*, Elsevier, 2006
10. *Atlas of Time-Temperature Diagrams for Nonferrous Alloys*, ASM International, 1991

Nonferrous Materials (Other)[1,2,12]

1. Polmear, I.J., *Light Alloys: Metallurgy of the Light Metals*, Arnold Publisher, 1990, ISBN 0-34049-175-2, 278 pages
2. Bringas, John E. and Michael L. Wayman, *Casti Metals Red Book – Non-Ferrous Data*, 4th Ed., Casti Publishing Inc., 2003, ISBN 1-894038-76-2, 760 pages
3. *Heat Treater's Guide: Practices and Procedures for Nonferrous Alloys*, 2nd Ed., Harry E. Chandler (Ed.), ASM International, 1996, ISBN 0-87170-565-6, 669 pages
4. Brooks, Charlie R., *Heat Treatment, Structure and Properties of Nonferrous Alloys*, ASM International, 1982, ISBN 0-87170-138-3, 420 pages
5. *Superalloys, A Technical Guide*, 2nd Ed., M.J. Donachie and S.J. Donachie (Eds.), ASM International, 2002, ISBN 0-87170-749-7, 439 pages
6. Sims, Chester T. and William C. Hagel, *The Superalloys*, John Wiley & Sons, 1972, ISBN 0-471-79207-1, 614 pages
7. *Titanium, A Technical Guide*, Matthew J. Donachie Jr. (Ed.), ASM International, 1988, ISBN 0-87170-309-2, 469 pages

Non-metals (Ceramics, Composites, Glasses, Plastics)[1,2]

1. *ASM Handbook, Vol. 21 : Composites*, ASM International, 2001, ISBN 0-87170-703-9, 1,201 pages
2. Chinn. Richard E., *Ceramography : Preparation and Analysis of Ceramic Microstructures*, The American Ceramics Society and ASM International, 2002, ISBN 0-87170-770-5, 214 pages
3. *Engineered Materials Handbook, Vol. 2: Engineering Plastics*, ASM International, 1988, ISBN 0-87170-280-0, 883 pages
4. *Engineered Materials Handbook, Vol. 4: Ceramics and Glasses*, ASM Interna-

tional, 1990, ISBN 0-87170-281-9, 893 pages
5. Van Vlack, Lawrence H., *Physical Ceramics for Engineers*, Addison-Wesley Publishing Co., 1964, LOC 64-11884, 342 pages
6. *Metallic Glasses*, Conference Proceedings, ASM International, 1978, LOC 77-24014, 348 pages

Other[1,2]

1. *Forging Industry Handbook*, Jon E. Jenson (Ed.), Forging Industry Association, 1970, 518 pages
2. *Forging Handbook*, Thomas G. Byrer (Ed.), ASM International, 1985, ISBN 978-087170-194-7
3. Davis, Joseph R. and S.L. Semiatin, *ASM Handbook, Volume 14: Forming and Forging*, ASM International, 1988, ISBN 10-087170-020-4, 978 pages

Plating/Finishing[1,2]

1. Durney, Lawrence J., *Trouble in Your Tank? Handbook for Solving Plating Problems*, 3rd Ed., Hanser Gardner Publications, 1996, ISBN 1-56990-200-3, 243 pages
2. *Metal Finishing Guidebook 2012/2013*, Volume 110 Number 9A, Elsevier, Inc., 888 pages
3. Dennis, J.K., and T.E. Such, *Nickel and Chromium Plating*, John Wiley & Sons, 1972, ISBN 0-470-20921-6, 325 pages

Quality Control (Failure Analysis)[1,2]

1. Uhlig, Herbert H., *Corrosion and Corrosion Control: An Introduction to Corrosion Engineering and Science*, 2nd Ed., John Wiley & Sons Inc., 1971, ISBN 0-471-89563-6, 419 pages
2. Fontana, Mars G. and Norbert D. Greene, *Corrosion Engineering*, 2nd Ed., McGraw-Hill Book Co., 1978, ISBN 0-07-021461-1, 465 pages
3. Wulpi, Donald J., *Understanding How Components Fail*, 3rd Ed., ASM International, 2013, ISBN 0-87170-189-8, 262 pages
4. Barer, R.D., and B.F. Peters, *Why Metals Fail*, Gordon and Breach Science Publishers, 1970, ISBN 0-677-02630-7, 345 pages
5. Schweitzer, Philip A., *Fundamentals of Metallic Corrosion: Atmospheric and Media Corrosion of Metals*, CRC Press, 2007, ISBN 0-8493-8243-2, 725 pages
6. Hertzberg, R.W., *Deformation and Fracture Mechanics of Engineering Materials*, 4th Ed., John Wiley & Sons, 1995, ISBN: 0-471-01214-9, 816 pages

7. *Ductility,* Conference Proceedings, ASM International, 1968, LOC 68-29168, 380 pages
8. *Fracture of Engineering Materials,* ASM International, 1964, LOC 64-8900, 248 pages
9. Scully, J.C., *The Fundamentals of Corrosion,* 2nd Ed., Pergamon Press, 1975, ISBN 0-08-018081-7, 235 pages
10. *Gas Corrosion of Metals,* Waldemar Bartoszewski (Trans.), U.S. Dept. of Commerce, National Technical Information Center, Report TT 76-54038, 1975, 528 pages
11. *IITRI Fracture Handbook: Failure Analysis of Metallic Metals by Scanning Electron Microscopy,* The Metals Research Division, IIT Research Institute, 1979, ISBN 0-915802-11-2
12. *Handbook of Case Histories in Failure Analysis, Vol. 1,* Khefa A. Esaklul (Ed.), ASM International, 1992, ISBN 0-87170-078-6, 484 pages
13. *Handbook of Case Histories in Failure Analysis, Vol. 2,* Khefa A. Esaklul (Ed.), ASM International, 1993, ISBN 0-87170-495-1, 581 pages
14. Dennies, Daniel P., *How to Organize and Run a Failure Investigation,* ASM International, 2005, ISBN 0-87170-811-6, 223 pages
15. *Hydrogen Embrittlement and Stress Corrosion Cracking,* R. Gibala and R.F. Hehemann (Eds.), ASM International, 1984, ISBN 0-87170-185-5, 324 pages
16. *Internal Stresses and Fatigue in Metals,* Gerald M. Rassmeiler and William L. Grubbe (Eds.), Elsevier Publishing Company, 1959, LOC 59-8945, 451 pages
17. *Basic Corrosion Course,* NACE International, 2014
18. Herro, Harvey M. and Robert D. Port, *The Nalco Guide to Cooling Water Systems Failure Analysis,* McGraw-Hill, Inc., 1993, ISBN 0-07-028400-8, 420 pages
19. Port, Robert D. and Harvey M. Herro, *The Nalco Guide to Boiler Failure Analysis,* McGraw-Hill, Inc., 1991, ISBN 0-07-045873-1, 293 pages
20. *Oxidation of Metals and Alloys,* ASM International, 1971, LOC 72-182305, 273 pages
21. Timmins, P.F., *Solutions to Hydrogen Attack in Steels,* ASM International, 1997, ISBN 0-87170-597-4, 198 pages

Quality Control (Inspection and Testing)[1,2]

1. *Analysis of Casting Defects,* American Foundrymen's Society, 2002, ISBN 978-0-87433-004-5, 140 pages
2. *Hardness Testing,* 2nd Ed., Harry E. Chandler (Ed.), ASM International,

1999, ISBN 0-87170-640-7, 192 pages
3. Lysaght, Vincent and Anthony DeBellis, *Hardness Testing Handbook*, American Chain & Cable Co., 1974, 140 pages
4. Anderson, Robert Clark, *Inspection of Metals: Vol. 1: Visual Examination*, ASM International, 1983, ISBN 0-87170-159-6, 149 pages
5. *Metallurgical Testing/Metallographic Atlas: Republic Alloy Steels Handbook*, Booklet Adv. 1009, 1961, 56 pages
6. *Practical Non-Destructive Testing*, 2nd Ed., B. Raj Jaykumar (Ed.), ASM International, ISBN 0-87170-763-2, 184 pages
7. Low, Samuel R., *Rockwell Hardness Measurement of Metallic Materials, Special Publication 960-5*, U. S. Department of Commerce, National Institute of Standards and Technology, 2001
8. *Tensile Testing*, J.R. Davis (Ed.), ASM International, 2004, ISBN 0-87170-806-X, 283 pages
9. Davis, Harmer E., George Earl Troxell, and Clement T. Wiskocil, *The Testing and Inspection of Engineering Materials*, 3rd Ed., McGraw-Hill Book Company, 1964, LOC 63-21885, 475 pages
10. ASTM Specifications (www.astm.org)

Quenching[1,2,6,8,12]

1. Totten, G.E., C.E. Bates and N.A. Clinton, *Handbook of Quenchants and Quenching Technology*, ASM International, 1992, ISBN 0-87170-448-X, 507 pages
2. Liščić, B., H.M. Tensi, L.C.F. Canale and G.E. Totten, *Quenching Theory and Technology* 2nd Ed,, CRC Press, 2010, ISBN 978-0-8493-9279-5, 725 pages
3. N.I. Kobasko, M.A. Aronov, J. Powell and G.E. Totten, *Intensive Quenching Systems: Engineering and Design*, ASTM International, 2010, ISBN 978-0-8031-7019-3, 253 pages
4. *Theory of Quenching and Quenching Technology: A Handbook*, B. Liščić, H.M Tensi, and W. Luty (Eds.), Springer-Verlag,1992, ISBN 978-36620-1598-8, 484 pages
5. Totten, G.E. and D.S. MacKenzie, , *Handbook of Aluminum: Vol. 1 - Physical Metallurgy and Processes*, Marcel Dekker Inc., 2003, ISBN 978-0-8247-0494-0 (Chapter 20 on Aluminum Quenching), 1,310 pages

Sintering[1,2,14]

1. German, Randall M., *Sintering Theory and Practice*, John Wiley & Sons, 1996, ISBN 0-471057-86-X, 550 pages
2. German, Randall M., *Powder Metallurgy of Iron and Steel*, John Wiley & Sons, 1998, ISBN 0-471-15739-2, 496 pages
3. Pease III, Leander F. and William G. West, *Fundamentals of Power Metallurgy*, Metal Powder Industries Federation, 2002, ISBN 1-87895-486-5, 452 pages
4. *High Temperature Sintering*, Howard I. Sanderow (Ed.), Metal Powder Industries Federation (MPIF), ISBN 0-918404-97-5, 400 pages
5. Hirschhorn, Joel S., *Introduction to Powder Metallurgy*, American Powder Metal Institute, 1969, LOC 76-83260, 341 pages
6. *MPIF Standard 35, Material Standards for PM Structural Parts*, Metal Powder Industries Federation, 2012 Ed.

Vacuum[1,2,13]

1. Herring, Daniel H., *Vacuum Heat Treating*, BNP Media, 2014, ISBN 978-0-9767565-0-7, 512 pages
2. *Vacuum Technology: Practical Heat Treating and Brazing*, Roger Fabian (Ed.), ASM International, 1993, ISBN 0-87170-477-3, 253 pages
3. Hablanian, Marsbed H., *High Vacuum Technology: A Practical Guide*, 2nd Ed., Varian Assoc. Inc., ISBN 0-8247-8197-X, 410 pages
4. *Handbook of Vacuum Science and Technology*, Dorothy M. Hoffman, Bawa Singh and John H. Thomas (Eds.), Academic Press, 1998, ISBN 81-8147-985-8, 835 pages
5. Brunner Jr., William F. and Thomas H. Batzer, *Practical Vacuum Techniques*, Robert E. Krieger Publishing Co., 1974, ISBN 0-88275-146-8, 198 pages
6. *The Vacuum Technology Book*, Pfeiffer Vacuum Inc., 2008
7. Dushman, Saul, *Scientific Foundations of Vacuum Technique*, 2nd Ed., J.M. Lafferty (Ed.), John Wiley & Sons, Inc., 1962, LOC 61-17361, 806 pages
8. *Critical Melting Points and Reference Data for Vacuum Heat Treating*, Virginia Osterman and Harry Antes Jr. (Eds.), Solar Atmospheres Inc., 2010, 40 pages
9. *Introduction to Helium Mass Spectrometer Leak Detection*, Varian Associates Inc., 1980, LOC 80-54419, 135 pages

REFERENCES

1. Herring, Daniel H., "A Heat Treaters Book Guide," *Industrial Heating*, July 2006
2. Herring, Daniel H., "A Reference Library for Heat Treaters," *Heat Treating Progress*, April/May 2003
3. Dan Kay, Kay & Associates, private correspondence
4. George Vander Voort, Vander Voort Consulting (www.georgevandervoort.com), private correspondence
5. Tom Bannos, TS Thermal, private correspondence
6. Dr. George E. Totten, Totten & Associates (www.getottenassociates.com), private correspondence
7. Fred Specht, Interpower Induction (www.interpowerinduction.com), private correspondence
8. Dr. D. Scott Mackenzie, Houghton International (www.houghtonintl.com), private correspondence
9. Craig Darragh, AgFox LLC, private correspondence
10. James Oakes, Super Systems Inc. (www.supersystems.com), private correspondence
11. Donald Bowe, Air Products & Chemicals (www.airproducts.com), private correspondence
12. Illinois Institute of Technology (http://tptc.iit.edu), private correspondence
13. William R. Jones, Solar Atmospheres (www.solaratm.com), private correspondence
14. Matthew J. Marzullo, C.I. Hayes, a Gasbarre Furnace Group Company (www.cihayes.com), private correspondence

16.3 METALLURGICAL SAMPLE PREPARATION

Over the years, a number of people have made suggestions and offered tips on the basic steps involved in the preparation of a sample for metallurgical examination, namely: sampling, sectioning, mounting, grinding, polishing and etching. These experienced-based insights are presented in outline form below.

A. Sampling

Deciding where and how to section a component part for an accurate representation of the condition of the microstructure (and mechanical properties) is a key aspect of the sampling process. Material specifications often dictate the sampling procedure. In failure analysis studies, specimens are not randomly sampled. They are instead examined to determine the origin of their failure, and sampling aids in determination of root cause.

(1) Specimen must be representative of the product being manufactured or the manufacturing/heat-treatment process used.
   (a) For example, wrought materials such as cold-rolled steels or forged parts will have a definite directional orientation and usually yield more significant information from a longitudinal as well as transverse section. Plate may require transverse, longitudinal and planar orientations for full characterization.

(2) The desired location and orientation of sections is best determined by collaboration between the metallographer and the technical contact requesting the analysis.

(3) Critical-area analysis is necessary.
   (a) Determine which area(s) are most critical to the investigation. Be aware that some specifications define the area and/or orientation of the specimen. This must be taken into consideration.
      (i) Be sure to compare these areas with other locations, looking for both specific and gross effects.
   (b) Determine the methods necessary to isolate these areas for further study.
   (c) Prepare samples (cut, mount, polish) in such a way as to yield the desired information.

(4) The number of samples is best determined by the project (e.g., carbonitriding of fasteners in a mesh-belt conveyor furnace can involve random sampling or taking samples from across the belt at select locations).

B. Sectioning

Samples are often sectioned from larger pieces using a variety of removal methods (such as core drilling, band or hack sawing, flame cutting or similar methods) to produce a sample that can be cut using a metallographic cutoff saw. Careful consideration must be given when using any sectioning method to avoid altering the microstructure of the sample.

(1) Process of removing a suitably sized sample to prepare and/or obtain a cross-sectional plane for examination.
   (a) Large sections often require multiple sectioning techniques (e.g., flame cutting, power or band saws, hack saws, metallurgical saw cuts) to obtain a representative sample.
   (b) Failures may require special techniques or planning in order to find, preserve and reveal the origin of the fracture.

(2) Purpose
   (a) To obtain a specimen from a defined location.
   (b) To produce a manageable-sized sample from a large bulk material.
   (c) To remove excessive surface or near-surface damage produced by shop cutting methods.

(3) Principle
   (a) Cut plane to be as near to the desired location as possible.
   (b) Surface produced should be as flat and damage-free as possible to minimize the severity of the steps that follow.

(c) Adequate coolant should be available to reduce surface damage.

(4) Sectioning methods vary by project (Table 16.3.1).
   (a) Use the least-aggressive method to obtain a satisfactory section from the sample at hand.

TABLE 16.3.1 | Typical sectioning methods

Sectioning method	Type of sample
Low-speed precision (diamond) saw (Fig. 16.3.1)	Small delicate pieces
Conventional (metallographic) cutoff saw (Fig. 16.3.2)	Common and advanced materials
Laser	Advanced materials[b]
EDM (spark) cutter	Complex shapes
Wire saw	Advanced materials[b]
Hack/band/power saw[a]	Softer materials, rough cutting
Torch[a,c]	Field cutting of bulk material samples

Notes:
[a] Severe damage will be produced, requiring corrective flattening and smoothing.
[b] Advanced materials are defined as non-metals such as ceramics, plastics, polymers and composites (metal matrix, carbon/carbon).
[c] Heat damage

FIGURE 16.3.1 | Precision diamond saw (courtesy of Struers Inc.)

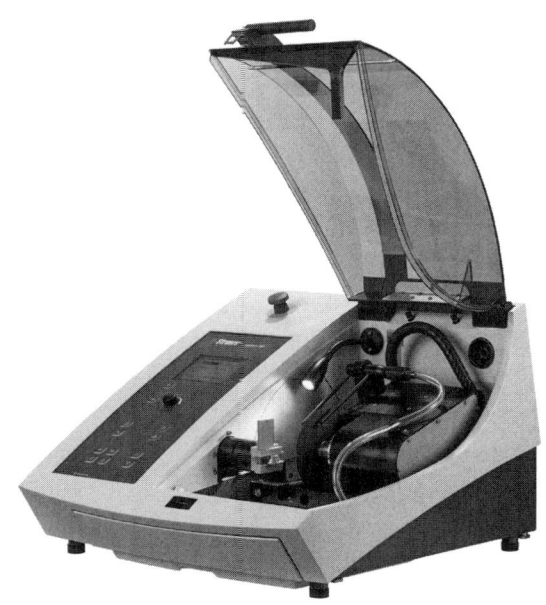

FIGURE 16.3.2 | Metallurgical sample cutoff saw (courtesy of Buehler, an ITW Company)

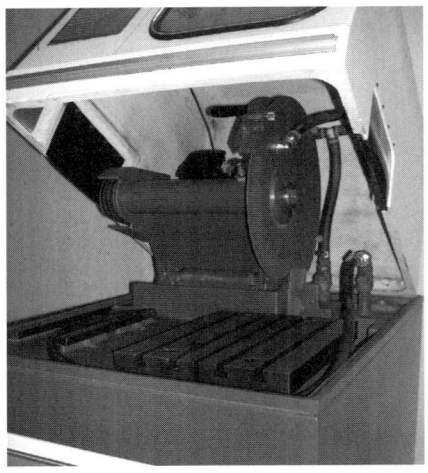

(5) Abrasive wheel selection
   (a) Effective free cutting occurs when the binder breaks down due to frictional heat and applied pressure.
   (b) Alumina abrasive wheels are best for ferrous materials, and silicon carbide wheels are best for nonferrous materials (e.g., aluminum).
   (c) Harder materials require a softer-bond wheel, while softer materials are best used with harder-bond wheels.

(6) Coolant
   (a) Adequate cooling of the sample material will prevent thermal damage.
      (i) Spray or flood cooling methods are acceptable.
      (ii) Dry cutting should be avoided if at all possible.
   (b) Uniform cooling will prevent excessive damage to the specimen surface and uneven wear of abrasive wheels that could cause breakage.

(7) Technique
   (a) Requires a moderate degree of skill to avoid cutting too rapidly.
      (i) Automatic cutoff saws are available to remove operator influence from the process and accomplish the same goals.
   (b) Clamp samples securely to prevent breaking the wheel, damaging the sample or injuring the operator.
   (c) Apply force at a reasonable rate to prevent overheating of sample or stalling of the cutter.

(8) Care and use of wheels
   (a) Proper care will ensure efficient cutting.
   (b) Store wheels flat on a smooth, rigid surface in a dry room.
   (c) Do not hang wheels on the wall.
   (d) Do not store wheels standing on edge or in a damp area.
   (e) Replace the wheel when it has reduced sufficiently in diameter so as not to cut through the sample in one smooth continuous motion.

(9) Guidelines for safe cutter operation
   (a) Read and follow the manufacturer's instructions.
   (b) Cutter should be in good operating condition and the door safety interlock confirmed to be working.
   (c) Check the coolant to see if there is enough liquid in the recirculating coolant tank. Add or replace if required.
   (d) Check the wheel flanges and remove any abrasive residues before installing a new wheel.
   (e) Check the wheel that is presently in place and replace it if it is too worn or damaged.
   (f) Be sure to return the wheel guard to the original operating position before cutting.
   (g) Wait for the wheel to stop before opening the cutting chamber door if cutter does not have a safety interlock or a wheel brake.

(10) Maintenance
   (a) Read and adhere to the recommendations of the manufacturer.
   (b) Dress the wheel as needed.
   (c) Clean debris from the cutting chamber frequently.
      (i) The use of magnets placed at strategic locations (near the drain area, on the floor of the machine, behind the cutting wheel, etc.) greatly simplifies this task.

(11) Troubleshooting
   (a) Cutting problems are easily diagnosed and corrected.
   (b) How the cutter responds, the appearance of the cut sample and the profile of the worn abrasive wheel may indicate the source of the cutting problem.

C. Mounting (Fig. 16.3.3)

The main purpose for mounting metallographic samples is for ease of handling of the different shapes and sizes received in the laboratory for metallographic analysis. Another important reason is to protect and preserve surfaces (e.g., defects) or to retain the edge during subsequent sample preparation (grinding and polishing). One must take care to ensure the mounting process is not detrimental to or does not change the microstructure of the sample. Mechanical deformation and heat are the most common problems faced when mounting samples. Examples of mounting methods include clamp mounting, compression mounting and castable mounting systems.

(1) Purpose
   (a) Facilitates handling of the specimen during later steps in the preparation process.
   (b) Standardizes both size and shape of the specimen.
   (c) Helps to maintain integrity of edges and intricate shapes.
   (d) Allows preparation of powders and small samples.
   (e) Protects technicians from sharp edges that could cause injury or discomfort during manual preparation and provides a sample of sufficient size to be easily grasped.

(2) Practice
   (a) Compression molding using a press.
      (i) Less expensive.
      (ii) Recommended for rigid materials that can withstand the heat and pressure of the mounting operation only.
      (iii) Avoid use for samples that can readily cold work (e.g., stainless steels).
   (b) Casting in a room-temperature curing resin.
      (i) More expensive.
      (ii) Ideal for delicate parts that are sensitive to heat and pressure.
      (iii) Sample cups or a special mounting ring made from a section of cut pipe, coated with grease and placed on a greased glass plate can be used.
      (iv) Casting should be done in a fume hood or well-ventilated area.
   (c) Clamping of samples in a specially designed polishing vice.
      (i) Useful for rapid analysis when the sample mounting process will consume too much time.
      (ii) Useful for odd-shaped parts or parts that have a size and shape that do not lend them to other techniques.
      (iii) Impregnation techniques can be used in special cases.
         1. Useful for examining the edges of pores, other internal void areas, fine through holes or porosity.

FIGURE 16.3.3 | Typical mounting press (courtesy of LECO Corp.)

(3) Sample considerations include
   (a) Shape
   (b) Size
   (c) Quantity
   (d) Hardness
   (e) Brittleness
   (f) Porosity
   (g) Heat sensitivity
   (h) Pressure sensitivity
   (i) Cross section or longitudinal section
   (j) Controlled material removal required
   (k) Edge retention

   Special note: Certain geometries pose edge-retention problems, and special care must be taken to avoid shrinkage gaps and subsequent staining when etched.

(4) Mounting material
   (a) Thermosetting resin powders
      (i) Fuse as temperature rises
      (ii) Becomes solid, cured mount after 6-9 minutes at temperature
      (iii) May be ejected while still hot
      (iv) Less shrinkage if cooled under pressure
   (b) Thermoplastic resins
      (i) Melt at a relatively low temperature
      (ii) Become soft, gummy masses until cooled to below the melting point
      (iii) Cannot be ejected hot

(c) Compression molding resins
  (i) Phenolic – least expensive, poor edge retention, opaque
  (ii) Filled epoxy – best edge retention, more expensive, opaque
  (iii) Diallyl phthalate – acid resistant, good edge retention, opaque
  (iv) Acrylic – highly transparent, longer curing cycle, sensitive to some solvents
  (v) Common problems (related to time, temperature or pressure in the mount)
    1. Radial splits
    2. Edge shrinkage
    3. Circumferential splits
    4. Bursting
    5. Unfused areas
(d) Room-temperature resins
  (i) Acrylics – rapid cure, translucent, low hardness, demountable, high exotherm, high shrinkage, strong odor
  (ii) Polyesters – transparent, low hardness, low shrinkage, slow curing, solvent sensitive, strong odor
  (iii) Epoxides – low shrinkage, semitransparency, moderate hardness, chemically resistant, slow to moderate cures
  (iv) Common problems
    1. Incorrect mixing; mixture not homogenous.
    2. Inaccurate proportioning of resin to hardener ratio.
    3. Shelf life is one year except for polyester, which is eight months.
    4. Samples not fully cured.

Special note:[2] Thermosetting epoxy resins provide the best edge retention of the hot-mounting resins and are degraded less by hot etchants than phenolic resins (melamine or methyl methacrylate), but they are more expensive. To reduce costs, the thermosetting epoxy resin can be used only for the bottom portion of the mount at the plane of polish. A less expensive resin, like phenolic, can be used to make up the top portion of the mount (for example, 20% thermosetting epoxy, 80% phenolic).

(5) Special mounting requirements
  (a) Edge retention
    (i) Use of foil (copper, aluminum, stainless steel) to wrap part contour prior to mounting.
      1. This technique is particularly useful when attempting to evaluate shallow case depths when microhardness traverses will be performed within 0.075 mm (0.003 inch) from the

edge of the material.
2. Prevents areas on a part sensitive to cracking from spalling off.
3. Protects thin layers (such as white layer on nitrided parts) or surface coatings from spalling off.
4. Allows for more accurate identification of surface defects and surface phenomena when using high (1000X or higher) optical microscopy techniques.

D. Grinding and polishing (Fig. 16.3.4)

Grinding and polishing are two independent yet interrelated related steps in the sample preparation process.

The goal of grinding is to remove the effects of sectioning and to produce extremely flat, scratch-free surfaces. Grinding uses progressively finer grit sizes to negate the damage from the prior grinding step. As with abrasive cutting, all grinding steps should be performed wet, provided water does not adversely affect the sample microstructure. Wet grinding keeps the heating of the specimen to a minimum.

Polishing follows grinding and is the final step in producing a deformation-free surface that is flat, scratch-free and mirror-like in appearance. Such a surface is necessary to observe the true microstructure during subsequent interpretation, testing or analysis (whether qualitative or quantitative). The polishing technique used should not introduce artifacts or other extraneous structures (e.g., disturbed metal, pitting, removal of inclusions, "comet-tails," staining or relief). Polishing is usually conducted in several stages.

(1) Grinding (manual or automatic)
  (a) Purpose
    (i) Remove oxide, paint dyes or other gross surface contamination.
    (ii) Flatten surface.
    (iii) Reach a specific plane that is too close to permit cutting.
    (iv) Key tip: round edges of specimen mount to prevent catching or tearing of the fine polishing cloths and to facilitate handling if hand polishing techniques are used.

(b) Goals
- (i) Freedom from deformation
- (ii) Flatness
- (iii) Freedom from scratches
- (iv) Absence of thermal damage

(c) Alternatives
- (i) Belt grinders – least expensive, easy to use but prone to damaging sample if care is not taken; long abrasive surface life; moderately high material removal rate.
- (ii) Disc grinders – more expensive than belt grinders; flattest surface (no belt seam); high material removal rate.

(d) Grinding safety
- (i) Sharp objects
    1. Grinding sharp-edged, unmounted samples presents risks of injury to the eyes and hands.
    2. Wear safety glasses.
    3. Do not use badly worn or torn abrasive belts or discs because tears can catch the edge of samples, causing them to be hurled away at high speeds.
- (ii) Dust and fumes
    1. Wet grind whenever possible to flush dust away before it becomes airborne.
    2. If grinding dry, use vacuum system and/or wear a dust mask or respirator.
    3. Know your sample material and its potential hazards.
        a. Magnesium alloy dust, for example, is flammable.
        b. Beryllium dust is highly toxic.

(2) Fine grinding
- (a) Older, traditional (manual) fine grinding uses grits of 240, 320, 400 and 600 paper.
    - (i) Adequately lubricate (typically water) for cooling.
    - (ii) Over-lubrication will cause mount to hydroplane on the abrasive surface.
- (b) Modern (automatic) fine grinding methods (Tables 16.3.2, 16.3.3).

TABLE 16.3.2 | Example of four-step method of fine grinding for iron-based alloys[2]

Abrasive and surface	Appropriate lubricant	rpm	Head/platen directions[c]	Load per specimen, N (pounds)	Time, minutes
120- to 220-grit SiC[a]; fixed diamond or diamond/phenolic	Water, oil, glycol[b]	240-300	Comp	27 (6)	Until planar
Napless medium fine cloth polishing (e.g., 9-μm diamond)	Water, oil, glycol[b]	120-150	Contra	27 (6)	5
Napless fine cloth polishing (e.g., 3-μm or 1-μm)	Water, oil, glycol[b]	120-150	Contra	27 (6)	4
Napped cloth on napless neoprene alumina or alumina/silica (non-drying)[d]	None	120-150	Contra	27 (6)	1-3

Notes:

[a] For steels ≥60 HRC, start with 120-grit SiC; for steels 35-60 HRC, start with 180-grit SiC; for steels ≤35 HRC, start with 220-grit SiC.

[b] Depends on diamond slurry used.

[c] Comp = head and platen both counter-clockwise (complementary); contra = head and platen in opposite directions (head speed should be ≤90 rpm).

[d] Optional but recommended for high-quality photomicrographs.

TABLE 16.3.3 | Example of three-step method for grinding steels harder than 200 HV[2]

Abrasive and surface	Appropriate lubricant	rpm	Head/platen directions[c]	Load per specimen, N (pounds)	Time, minutes
120- to 220-grit SiC	Water	240-300	Comp	27 (6)	Until planar
Napless medium fine cloth polishing (e.g., 9-μm diamond)	Water, oil, glycol[b]	120-150	Contra	27 (6)	5-8
Napped cloth on napless neoprene alumina or alumina/silica (non-drying)[d]	None	120-150	Contra	27 (6)	5

Notes:

[a] For steels ≥60 HRC, start with 120-grit SiC; for steels 35-60 HRC, start with 180-grit SiC; for steels ≤35 HRC, start with 220-grit SiC.

[b] Depends on diamond slurry used.

[c] Comp = head and platen both counter-clockwise ("complementary"); contra = head and platen in opposite directions (head speed should be ≤90 rpm).

[d] Optional but recommended for high-quality photomicrographs.

(3) Rough polishing
   (a) Purpose
      (i) To remove any remaining deformation and prepare the surface for final polishing.
   (b) Principle
      (i) Diamond abrasives are lodged in the hard but resilient surface of nap-less cloths.
      (ii) Material is removed by the planing action of suitably aligned grains.
   (c) Parameters
      (i) Select and apply a nap-less cloth. Discard the cloth when it becomes torn or contaminated (causing scratches) or when the cutting rate declines.
      (ii) Apply diamond abrasives for most sample materials for efficient, rapid material removal. Abrasives in the 15 to 1 micron grain size are most common.
      (iii) Apply liquid extender to moisten the cloth for lubrication and to help redistribute the abrasive. Do not apply so much liquid as to produce a visible wetness.
   (d) Practice (for manual methods)
      (i) Grasp the mounted or unmounted sample firmly.
      (ii) Apply moderate to heavy pressure while rotating the sample counter to the wheel rotation.
      (iii) Select 120 rpm or the nearest alternative speed.
      (iv) Some delicate materials may require complimentary rotation.

(4) Final polishing
   (a) Purpose
      (i) Removes any remaining superficial deformation (smearing) from the surface.
      (ii) Produces a lustrous, scratch-free surface.
   (b) Principle
      (i) Abrasive grains are loosely held in the compressed nap of a polishing cloth.
      (ii) Material is removed at a slow rate while producing a scratch-free surface.
   (c) Techniques
      (i) Automatic polishing equipment
      (ii) Manual (hand) polishing
   (d) Parameters

(i) Premoisten (but do not soak) a new cloth or add distilled water to a used cloth.
(ii) With polisher wheel rotating, apply a spiral pattern of fine abrasive such as 0.05 micron alumina.
(e) Practice (for manual methods)
(i) Grasp the mounted or unmounted sample firmly and apply a moderate pressure while rotating the sample counter to the wheel rotation.
(ii) Select 120 rpm or the nearest alternative speed.
(iii) Final polishing should be accomplished in the shortest time possible (typically under 60 seconds) to avoid pitting, excessive edge rounding and surface relief.

FIGURE 16.3.4 | Typical grinding/polishing setup (courtesy of Struers Inc.)

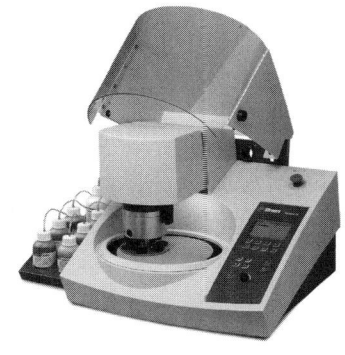

(5) Electrolytic grinding and polishing
(a) Advantages
(i) Produces a highly lustrous deformation-free surface without the use of abrasives.
1. No deformation and smeared surface layers.
2. Edges can be severely attacked.
(ii) Short time and good reproducibility; applies only when using a suitable recipe.
1. Once the polisher parameters are established for a particular material, the polishing results may be repeated over and over again.
(iii) No or little heat generation of the sample.
(iv) Possibility for subsequent etching with the same equipment.
(v) Easy removal of deformation or smeared surface layer from

mechanical grinding or polishing with a short electrolytic polish.
   (vi) Ideal for special investigations where any amount of deformation would influence the results.
(b) Limitations
   (i) Not a universal polishing method that replaces abrasive polishing or rotating wheels.
   (ii) Some abrasive sample preparation is required prior to electrolytic polishing.
   (iii) Samples are best polished unmounted, and edges as well as the bulk material cannot be prepared.
   (iv) Edges are especially heavily attacked. Conductive mounting materials with their good adhesion properties may help to support the edges. However, good edge retention is very difficult to obtain.
   (v) Only slight surface roughness is being leveled; large, uneven areas are not well polished.
   (vi) A surface layer is formed on the polished surface.
   (vii) Electrical contact can be lost with samples that oxidize easily. Special provisions have to be made for nonconductive mounting materials.
   (viii) Coarse-grained materials are not well suited for electrolytic polishing.
   (ix) At the interface of nonmetallic inclusions, the matrix metal is preferentially attacked, resulting in heavy relief and gap formation.
(c) Principles
   (i) The principles of electrolytic polishing are generally not well understood.
   (ii) A previously ground sample is made the anode in an electrolytic cell.
   (iii) The pump is activated, and the electrolyte flows across the surface of the sample.
   (iv) When DC current is applied, the micro-elevations are selectively removed by the electrolytic dissolutions.
(d) Practice
   (i) Samples are ground to a 600 grit or better finish and are inverted onto the aperture mask of the electrolytic cell.
   (ii) A suitable electrolyte is pumped against the ground surface of the sample.

- (iii) The power source is set to produce a current density that will remove material without pitting or etching the sample.
- (iv) Pure metals and solid-solution alloys electrolytically polish well.
- (v) Complex alloys having heterogeneous microstructures do not polish well and tend to become etched rather than polished.

(e) Parameters
- (i) Electrolyte composition
  1. Select according to the composition of the material to be polished.
  2. Some electrolytes may be used for a wide range of materials, while others have a much more limited range of usefulness.
  3. Electrolytes should be selected from approved lists such as appear in ASTM and other standards.
- (ii) Mask size
  1. Choose a mask that gives the needed area for examination.
  2. It is better to choose the smallest-diameter mask that will suit your needs.
- (iii) Electrolyte flow rate
  1. The electrolyte must cover the surface of the sample completely without turbulence.
     a. Preset the pump speed so that the electrolyte fills the mask aperture completely and rises slightly above the opening.
  2. Voltage setting
     a. The voltage control should be set to the range of 25-50 volts, depending on the materials, the electrolyte and the aperture size.
     b. Current density is the factor that actually controls the polishing process. Some references list the recommended current density for a particular material.
- (iv) Time
  1. The time selected depends on the electrolyte, the material composition and the condition of the sample surface before polishing.
  2. The finer the finish, the less time will be required to polish the sample.
- (v) Electrolyte temperature
  1. The temperature is important for several reasons.
     a. Some polishing recipes call for polishing at an elevated temperature. Most call for polishing at room temperature.

b. If the temperature rises due to frequent use or prolonged "on" times, the properties of the electrolyte will change.
c. In some cases, an explosive condition may result.
d. Use the cooling coil to control the electrolyte temperature.

(6) Evaluation of polishing methods
    (a) Reflectivity of the polished surface is a sensitive criterion for judging quality.
        (i) Alumina polishing produces the lowest reflectivity.
        (ii) Electrolytic polishing produces results approaching the ideal.
    (b) Varies from material to material.
    (c) Severe grinding and polishing stresses have a noticeable effect on microhardness.
    (d) When reporting surface condition, include information concerning:
        (i) Grinding technique
        (ii) Polishing technique
    (e) Differences in actual results compared with expected results may stem from differences in preparation.
    (f) Alumina and other oxides are widely used because they produce acceptable results at a reasonable cost.
    (g) Evaluation of a sample under low (10-50X) to medium (100X) magnification can reveal extent and orientation of surface damage (scratches, pits) on the part after polishing and may indicate if re-polishing is required.
    (h) Tip: To remove fine scratches present after grinding and after examining the sample mount for surface imperfections, which might negate this approach, over-etch the sample and repolish. In other words, use the etchant as a final polishing step to remove fine scratches.

(7) Cleaning
    (a) Remove all remnant grinding and polishing compounds after each step to avoid cross-contamination, which would reduce the efficiency of the next preparation step.
    (b) Often necessary to clean the sample several times within one preparation sequence.
    (c) Practice
        (i) Clean the sample under running water or using ultrasonic cleaning.

1. Overuse may remove inclusions or other microconstituents.
2. Ultrasonic cleaning is considered the most effective cleaning technique.
   a. Remnant polishing compounds can be completely removed.
   b. Usually needs only 10-30 seconds. Caution: Overly long times can cause severe damage.
(ii) Avoid touching sample or wiping the surface with tissues or cloth.
(iii) Rinse with alcohol.
(iv) Dry with hot-air dryer.
(d) A clean surface, free of oil, is a prerequisite for chemical or electrolytic treatment, especially for etching.
   (i) Traces of dirt can affect the overall evaluation.

(8) Troubleshooting
   (a) Relief
   (b) Pull-outs
   (c) Scratches
   (d) Comet tails
   (e) Smear
   (f) Stain

E. Etching

Metallurgical etching encompasses all processes used to reveal desired structural characteristics of a metal that are not evident in the as-polished conditions. Evaluation of an unetched sample should be performed because it may reveal items such as porosity, cracks, nonmetallic inclusions and, in some instances, grain size. Etching involves the use of selected chemicals to attack the polished surface of the sample and reveal microstructural features (e.g., tempered martensite, retained austenite, etc.).

(1) Etching develops fine microstructural detail by selectively attacking and/or staining (coloring) susceptible areas that can be observed by microscopic examination.
   (a) Chemical etching
      (i) Principle
         1. Sample is swabbed, immersed or has its metallic surface coated with a suitable etchant.

2. Etching proceeds without an externally applied electrical current or electrochemical reduction and oxidation process.
3. Selective attack results from the different dissolution rates of the different grain orientations or chemistry.

(b) Selected types of etchants (Tables 16.3.4-16.3.6) provide examples of the most common type and usage.

TABLE 16.3.4 | Selected etchants for aluminum and aluminum alloys[a]

Etchant	Usage
Flick (macro etchant)	Pure Al and Al alloys
Dix-Keller (micro etchant)	Pure Al and Al alloys
Keller's reagent	Outlines common constituents, reveals grain boundaries in certain alloys
Graff & Sargent etch	Reveals grain structure
Boss's etch	Reveals grain contrast

Note:
[a] Etchant compositions can be found in the literature.

TABLE 16.3.5 | Selected etchants for iron and steels[2]

Etchant	Usage
Combinations of nital and picral	Differentiate between austenite, martensite and tempered martensite and to determine the depth of nitrided layers.
Carapella (macro etchant)	Ni-Fe, Ni-Cu and Ni-Ag alloys; Ni-based superalloys; Monel
Fry (macro etchant)	Flow lines in low-carbon nitrogen-bearing steels
Fe M2 (macro etchant)	Unalloyed and low-alloy steels, carburized and decarburized zones, segregations
Fe M8	Austenite grain boundaries in case-hardened steels and steels for hardening and tempering
Nital 2-3% (micro etchant)	Pure iron, carbon steels
Nital, 5-10% (macro etchant)	Pure iron, carbon steels
Picral (micro etchant)	Pure iron, carbon steels
Villela's Reagent (micro etchant)	Ferritic Cr steels, austenitic CrNi steels; use fresh

(c) Etchant concentrations
   (i) Many etchants can be used for macroscopic or microscopic sample evaluation. The strength of an etchant must be carefully controlled (and monitored over time) so that the sample

being chemically attacked is done so under predicted (and controlled) conditions.
(ii) For example, alcohol will evaporate from etchants over time (especially in warm weather), increasing the etchant's concentration and interfering with predetermined etchant times.
(iii) Repolishing is required on every over-etched sample.

TABLE 16.3.6 | Additional etchants for steel[a]

Number	Composition	Comments
1	99-90 mL ethanol 1-10 mL HNO_3	Nital, the most commonly used etchant for steels. Do not stock nital with >3% nitric acid in ethanol. Use by immersion or swabbing (light pressure).
2	100 mL ethanol 4 g picric acid	Picral, better than nital for annealed microstructures. Does not reveal ferrite grain boundaries. Etch by immersion or swabbing.
3	100 mL ethanol 5 mL HCl 1 g picric acid	Vilella's reagent is good for higher-alloyed steels, tool steels and martensitic stainless steels. Etch by immersion or swabbing.
4	85 mL ethanol 15 mL HCl	Etch for duplex stainless steels developed by Carpenter Technology. Immerse specimens 15-45 minutes (time is not critical) to reveal the grain and phase boundaries in duplex stainless steels.
5	50 mL stock solution* 1 g $K_2S_2O_5$ (* water saturated with $Na_2S_2O_3$)	Klemm's I tint etch. It colors ferrite strongly; also colors martensite and bainite but not carbides or retained austenite. Use by immersion only until the surface is colored.
6	100 mL water 25 g NaOH 2 g picric acid	Alkaline sodium picrate, used 80-100°C by immersion only. Colors cementite (Fe_3C) and M_6C carbides.
7	100 mL water 20 g NaOH	Electrolytic etch for stainless steels. Use at 3 V dc, 10 seconds to color ferrite (usually tan or light blue) and sigma (orange) but not austenite. Mix fresh.
8	100 mL water 10 g NaOH (or KOH) 10 g $K_3Fe(CN)_6$	Murakami's reagent. Used to color ferrite and sigma (80-100°C for up to 3 minutes) in stainless steels. It will not color ferrite but will color certain carbides at room temperature. It colors ferrite, sigma and carbides at high temperature but not austenite.
9	100 mL water 3 g $K_2S_2O_5$ 2 g sulfamic acid 0.5 - 1 g $NH_4F \cdot HF$	Beraha's sulfamic acid reagent no. 4. Colors phases in highly alloyed tool steels and martensitic stainless steels. Use by immersion only, usually 30-180 seconds. Mix fresh. Best to use a plastic beaker and plastic tongs.
10	85 mL water 15 mL HCl 1 g $K_2S_2O_5$	Beraha-type etch for duplex stainless steels (similar to BI). Mix fresh. Use by immersion until the surface is colored. Colors ferrite but not austenite.

Note:
[a] See Reference 2 for additional etchants.

(2) Application of etchants
   (a) Total immersion and surface contact methods are the most common techniques in use.
      (i) Total immersion technique
         1. A container (such as a watch glass) can be used to immerse just the surface of the mount.
         2. Precise control of immersion time is required for best results.
      (ii) Surface covering technique
         1. A squirt bottle or eye dropper can be used to place the etchant on the surface of the sample. Note: Tongs should be used to hold the sample in position and avoid contact between the etchant and the skin.
         2. Swabbing is a preferred method by many metallographers.
      (iii) Etching residues are constantly flushed away.
      (iv) Operator has a better view of the specimen during etching.
      (v) Practice – use swabs that are larger than the familiar drug-store type.
         1. Apply for 2-3 seconds, then rinse immediately in running water.
         2. Rinse with alcohol.
         3. Dry using warm forced air.
         4. Examine the specimen to determine the degree of attack.
         5. Determine if the specimen needs further etching.
         6. Before re-etching, redo the finishing step and polish only long enough to remove the effects of the initial application.
   (b) Etching time
      (i) Etching time is usually determined empirically, although in well-defined situations (as in electrolytic etching) the specified times generally produce acceptable results.
      (ii) Etching times range from a few seconds to many minutes.
      (iii) Progress is judged by the surface appearance, which usually becomes less reflective (duller) as etching proceeds.
      (iv) With a little experience, one can gauge the proper degree of etching by the surface dulling or coloration produced.
      (v) Etching times and etching temperature are closely related
      (vi) Increasing the temperature will increase the rate of etching.
         1. This may not be advisable because the contrast could become uneven when the rate of attack is too rapid.
         2. Most etching is performed at room temperature.

(vii) A milder concentration of an etchant may be preferred over a stronger concentration.
1. A strong concentration produces rapid etching (an etching time of less than a few seconds), but good results can be difficult to obtain consistently.
(viii) Etchants that require 20 seconds or longer to develop the structure usually provide the needed control for etching.
(ix) Etching errors may lead to microstructural misinterpretation.
(x) Under-etching fails to reveal all the detail.
(xi) Many times, sample mount can be etched longer without repolishing.
(xii) It may be necessary to repolish the sample mount before re-etching.
(xiii) Over-etching obscures details.
1. The sample mount will need to be repolished and etched for a shorter time.
(c) Safety precautions
(i) Do not inhale concentrated or mixed etchant fumes. Use a fume hood to mix and apply etchants.
(ii) Do not allow exposed skin to contact etchant solutions. Use thin plastic gloves and hold the sample or mount with suitable tongs.
(iii) Identify the contents of all chemical containers including stock chemicals and mixed solutions.
(iv) Dispose of expended etchants according to locally approved procedures.

(3) Electrolytic etching
(a) Practice
(i) Sample is placed in an electrolytic cell as the anode.
(ii) Cathode made of compatible materials is also placed into the cell.
(iii) Low voltage is applied by a controlled source for a period determined experimentally or from previous experience.
(iv) Anode is selectively dissolved.
1. Used for austenitic stainless steels, Inconel® and many other corrosion-resistant alloys.
(b) Color etchants, also known as "tint" etchants, can be used to color specific constituents in steels. These can be quite useful for identifying constituents, studying grain size and detecting segregation

and residual deformation. There are etchants that will color either ferrite or austenite. Unlike standard etchants that reveal only a portion of the grain boundaries, a color etchant colors all grains. If the grains have a random crystal orientation, a wide range of colors is observed. If there is preferred orientation, on the other hand, then a narrow range of colors is obtained. Because color etchants are selective, they are very useful for image-analysis work where the contrast between what you want to measure and what you do not want to measure must be maximized. Sensitization and sigma-phase resolution are examples.[2]

Summary

Metallographic sample preparation is the key to our ability to evaluate the microstructure of a heat-treated component. By following long-established procedures, not cutting corners and with a great deal of practice, the metallurgist or metallurgical laboratory technician can reveal information about the heat-treatment process that will be invaluable to achieving our goal of producing the perfect part each and every time.

REFERENCES

1. Alan Stone, Aston Metallurgical Services Co. Inc. (www.astonmet.com), technical and editorial contributions, private correspondence
2. Rebecca Heckle, metallographer (deceased)
3. Vander Voort, George F., *Metallography Principles and Practices*, ASM International, 1984
4. George Vander Voort, George Vander Voort Consulting (www.georgevandervoort.com), private correspondence
5. William Durako, Materials Engineering Inc. (www.materials-engr.com), private correspondence
6. Debbie Aliya, Aliya Analytical Inc. (www.itothen.com), private correspondence
7. Craig Darragh, AgFox LLC, private correspondence

16 | REFERENCE MATERIALS

16.4
USEFUL PROPERTIES, TABLES AND CHARTS

The technical information contained in this section is intended to enhance the understanding of heat treaters, metallurgists and engineers who work with furnace atmospheres and atmosphere furnaces. While it is impossible to include all pertinent engineering data, the information selected is intended to aid those who design heat-treatment processes and operate equipment. In every instance the company providing the information retains copyright, and the material is used with permission.

FIGURE 16.4.1 | Equilibrium curves for furnace gases on iron, carbon and steel (source: Surface Combustion Inc.)

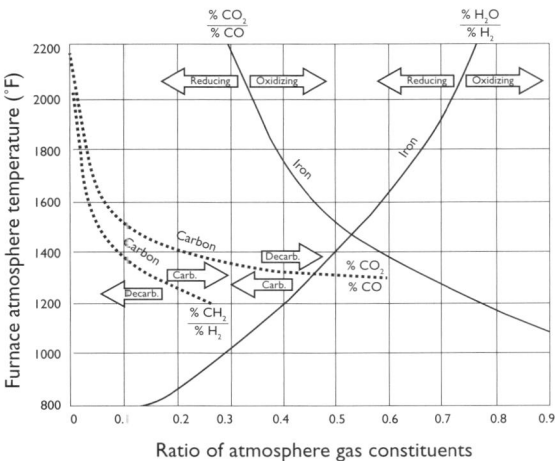
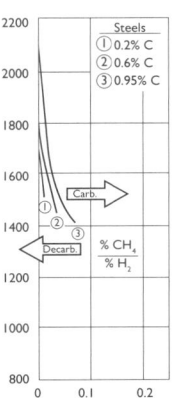

FIGURE 16.4.2 | Composition of exothermic and endothermic atmospheres (source: CMI Industry Americas)

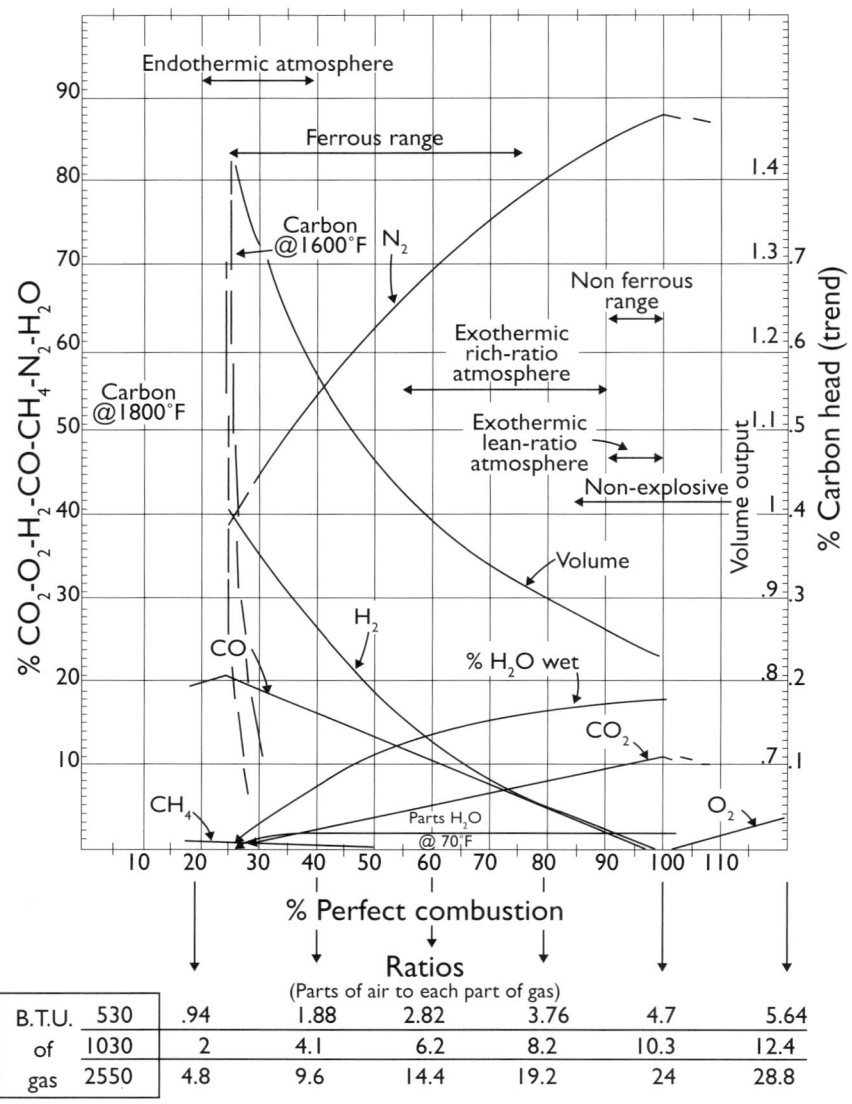

Curves shown are for gas analyzed at room temperature, except where noted.

16 | REFERENCE MATERIALS

FIGURE 16.4.3 | Carbon potential vs. temperature at a specified millivolt reading for typical endothermic gas from natural gas feedstock (source: Super Systems Inc.)

%CO = 20.0   %H$_2$ = 40.0

Temp °F	1400	1425	1450	1475	1500	1525	1550	1575	1600	1625	1650	1675	1700	1725	1750	1775	1800	1825	1850	1875	1900
Temp °C	760	774	788	802	816	829	843	857	871	885	899	913	927	941	954	968	982	996	1010	1024	1038
%C																					
0.05	957	959	961	963	965	967	968	970	972	974	976	978	979	981	983	985	987	989	991	992	994
0.10	989	991	993	996	998	1000	1002	1005	1007	1009	1011	1014	1016	1018	1020	1023	1025	1027	1030	1032	1034
0.15	1007	1010	1012	1015	1018	1020	1023	1025	1028	1030	1033	1035	1038	1040	1043	1045	1048	1050	1053	1055	1058
0.20	1021	1024	1026	1029	1032	1034	1037	1040	1042	1045	1048	1050	1053	1056	1059	1061	1064	1067	1069	1072	1075
0.25	1031	1034	1037	1040	1043	1046	1048	1051	1054	1057	1060	1063	1065	1068	1071	1074	1077	1080	1082	1085	1088
0.30	1040	1043	1046	1049	1052	1055	1058	1061	1064	1067	1070	1073	1076	1078	1081	1084	1087	1090	1093	1096	1099
0.35	1048	1051	1054	1057	1060	1063	1066	1069	1072	1075	1078	1081	1084	1087	1090	1093	1096	1100	1103	1106	1109
0.40	1054	1057	1061	1064	1067	1070	1073	1076	1079	1082	1086	1089	1092	1095	1098	1101	1104	1108	1111	1114	1117
0.45	1060	1063	1067	1070	1073	1076	1079	1083	1086	1089	1092	1096	1099	1102	1105	1108	1112	1115	1118	1121	1124
0.50	1065	1069	1072	1075	1079	1082	1085	1089	1092	1095	1098	1102	1105	1108	1112	1115	1118	1121	1125	1128	1131
0.55	1070	1074	1077	1080	1084	1087	1091	1094	1097	1101	1104	1107	1111	1114	1117	1121	1124	1127	1131	1134	1138
0.60	1075	1078	1082	1085	1089	1092	1095	1099	1102	1106	1109	1113	1116	1119	1123	1126	1130	1133	1136	1140	1143
0.65	1079	1083	1086	1090	1093	1097	1100	1104	1107	1111	1114	1117	1121	1124	1128	1131	1135	1138	1142	1145	1149
0.70	1083	1087	1090	1094	1097	1101	1104	1108	1111	1115	1119	1122	1126	1129	1133	1136	1140	1143	1147	1150	1154
0.75	1087	1091	1094	1098	1101	1105	1108	1112	1116	1119	1123	1126	1130	1134	1137	1141	1144	1148	1151	1155	1159
0.80	1091	1094	1098	1102	1105	1109	1112	1116	1120	1123	1127	1131	1134	1138	1141	1145	1149	1152	1156	1160	1163
0.85	1094	1098	1101	1105	1109	1112	1116	1120	1123	1127	1131	1134	1138	1142	1146	1149	1153	1157	1160	1164	1168
0.90	1097	1101	1105	1109	1112	1116	1120	1123	1127	1131	1135	1138	1142	1146	1149	1153	1157	1161	1164	1168	1172
0.95	1101	1104	1108	1112	1115	1119	1123	1127	1131	1134	1138	1142	1146	1149	1153	1157	1161	1164	1168	1172	1176
1.00	1104	1107	1111	1115	1118	1122	1126	1130	1134	1138	1142	1145	1149	1153	1157	1161	1164	1168	1172	1176	1180
1.05	1107	1110	1114	1118	1121	1125	1129	1133	1137	1141	1145	1149	1153	1157	1160	1164	1168	1172	1176	1180	1183
1.10	1109	1113	1117	1121	1124	1128	1132	1136	1140	1144	1148	1152	1156	1160	1164	1168	1172	1175	1179	1183	1187
1.15	1112	1116	1120	1124	1127	1131	1135	1139	1143	1147	1151	1155	1159	1163	1167	1171	1175	1179	1183	1187	1191
1.20	1115	1119	1123	1127	1130	1134	1138	1142	1146	1150	1154	1158	1162	1166	1170	1174	1178	1182	1186	1190	1194
1.25	1118	1122	1126	1130	1133	1137	1141	1145	1149	1153	1157	1161	1165	1169	1173	1177	1181	1185	1189	1193	1198
1.30	1120	1124	1128	1132	1135	1139	1143	1147	1151	1155	1159	1163	1168	1172	1176	1180	1184	1188	1192	1197	1201
1.35	1123	1127	1131	1135	1138	1142	1146	1150	1154	1158	1162	1166	1171	1175	1179	1183	1187	1191	1195	1200	1204
1.40	1125	1130	1134	1138	1141	1145	1149	1153	1157	1161	1165	1169	1174	1178	1182	1186	1190	1194	1198	1203	1207
1.45	1128	1132	1136	1140	1144	1149	1153	1157	1161	1165	1169	1173	1178	1182	1186	1190	1194	1198	1202	1206	1211
1.50	1130	1135	1139	1143	1147	1151	1155	1160	1164	1168	1172	1176	1180	1185	1189	1193	1197	1201	1205	1210	1214

Note: The shaded portion of this table corresponds to saturation limits of carbon in steel.

ATMOSPHERE HEAT TREATMENT

FIGURE 16.4.4 | Oxygen-sensor millivolt output vs. percent oxygen as a function of temperature for processes (e.g., nitrogen or nitrogen/hydrogen atmospheres) where the oxygen percentage is higher than might typically be found in an endothermic atmosphere (source: Super Systems Inc.)

Note: values shown in table are %O₂

Furnace temp °C / °F	20	30	40	50	60	70	80	90	100	110	120	130	140	150	160	170	180	190	200	210	220
600 / 1112	7.235	4.252	2.4988	1.468	0.863	0.507	0.298	0.175	0.103	0.060	0.036	0.021	0.012	0.007	0.004	0.002	0.001	0.001	0.001	0.000	0.000
625 / 1157	7.4526	4.445	2.6512	1.5812	0.9431	0.5625	0.3355	0.2001	0.1193	0.0712	0.0425	0.0253	0.0151	0.009	0.0054	0.0032	0.0019	0.0011	0.0007	0.0004	0.0002
650 / 1202	7.664	4.636	2.8038	1.696	1.026	0.620	0.375	0.227	0.137	0.083	0.050	0.030	0.018	0.011	0.007	0.004	0.002	0.001	0.001	0.001	0.000
675 / 1247	7.870	4.824	2.9565	1.812	1.111	0.681	0.417	0.256	0.157	0.096	0.059	0.036	0.022	0.014	0.008	0.005	0.003	0.002	0.001	0.001	0.000
700 / 1292	8.071	5.009	3.1090	1.930	1.198	0.743	0.461	0.286	0.178	0.110	0.068	0.042	0.026	0.016	0.010	0.006	0.004	0.002	0.002	0.001	0.001
725 / 1337	8.266	5.192	3.2612	2.048	1.287	0.808	0.508	0.319	0.200	0.126	0.079	0.050	0.031	0.020	0.012	0.008	0.005	0.003	0.002	0.001	0.001
750 / 1382	8.456	5.372	3.4128	2.168	1.377	0.875	0.556	0.353	0.224	0.143	0.091	0.058	0.037	0.023	0.015	0.009	0.006	0.004	0.002	0.002	0.001
775 / 1427	8.641	5.549	3.5638	2.289	1.470	0.944	0.606	0.389	0.250	0.161	0.103	0.066	0.043	0.027	0.018	0.011	0.007	0.005	0.003	0.002	0.001
800 / 1472	8.821	5.724	3.7139	2.410	1.564	1.015	0.658	0.427	0.277	0.180	0.117	0.076	0.049	0.032	0.021	0.013	0.009	0.006	0.004	0.002	0.002
825 / 1517	8.996	5.895	3.8631	2.531	1.659	1.087	0.712	0.467	0.306	0.200	0.131	0.086	0.056	0.037	0.024	0.016	0.010	0.007	0.004	0.003	0.002
850 / 1562	9.167	6.064	4.0113	2.653	1.755	1.161	0.768	0.508	0.336	0.222	0.147	0.097	0.064	0.043	0.028	0.019	0.012	0.008	0.005	0.004	0.002
875 / 1607	9.334	6.230	4.1583	2.776	1.853	1.237	0.825	0.551	0.368	0.245	0.164	0.109	0.073	0.049	0.033	0.022	0.014	0.010	0.006	0.004	0.003
900 / 1652	9.496	6.393	4.304	2.898	1.951	1.313	0.884	0.595	0.401	0.270	0.182	0.122	0.082	0.055	0.037	0.025	0.017	0.011	0.008	0.005	0.003
925 / 1697	9.654	6.553	4.4485	3.020	2.050	1.392	0.945	0.641	0.435	0.295	0.201	0.136	0.092	0.063	0.043	0.029	0.020	0.013	0.009	0.006	0.004
950 / 1742	9.808	6.711	4.5917	3.142	2.150	1.471	1.006	0.689	0.471	0.322	0.221	0.151	0.103	0.071	0.048	0.033	0.023	0.015	0.011	0.007	0.005
975 / 1787	9.958	6.866	4.7334	3.263	2.250	1.551	1.069	0.737	0.508	0.350	0.242	0.167	0.115	0.079	0.055	0.038	0.026	0.018	0.012	0.009	0.006
1000 / 1832	10.105	7.018	4.8737	3.385	2.351	1.633	1.134	0.787	0.547	0.380	0.264	0.183	0.127	0.088	0.061	0.043	0.030	0.021	0.014	0.010	0.007
1025 / 1877	10.248	7.167	5.0125	3.506	2.452	1.715	1.199	0.839	0.587	0.410	0.287	0.201	0.140	0.098	0.069	0.048	0.034	0.023	0.016	0.011	0.008
1050 / 1922	10.387	7.314	5.1498	3.626	2.553	1.798	1.266	0.891	0.628	0.442	0.311	0.219	0.154	0.109	0.076	0.054	0.038	0.027	0.019	0.013	0.009
1075 / 1967	10.523	7.458	5.2856	3.746	2.655	1.882	1.334	0.945	0.670	0.475	0.336	0.238	0.169	0.120	0.085	0.060	0.043	0.030	0.021	0.015	0.011
1100 / 2012	10.656	7.599	5.4198	3.865	2.757	1.966	1.402	1.000	0.713	0.509	0.363	0.259	0.184	0.132	0.094	0.067	0.048	0.034	0.024	0.017	0.012
1125 / 2057	10.785	7.739	5.5524	3.984	2.858	2.051	1.472	1.056	0.758	0.544	0.390	0.280	0.201	0.144	0.103	0.074	0.053	0.038	0.027	0.020	0.014
1150 / 2102	10.912	7.875	5.6835	4.102	2.960	2.136	1.542	1.113	0.803	0.580	0.418	0.302	0.218	0.157	0.113	0.082	0.059	0.043	0.031	0.022	0.016
1175 / 2147	11.035	8.009	5.8129	4.219	3.062	2.222	1.613	1.171	0.850	0.617	0.448	0.325	0.236	0.171	0.124	0.090	0.065	0.047	0.034	0.025	0.018
1200 / 2192	11.156	8.141	5.9408	4.335	3.164	2.309	1.685	1.229	0.897	0.655	0.478	0.349	0.254	0.186	0.135	0.099	0.072	0.053	0.038	0.028	0.020
1225 / 2237	11.274	8.270	6.067	4.451	3.265	2.395	1.757	1.289	0.946	0.694	0.509	0.373	0.274	0.201	0.147	0.108	0.079	0.058	0.043	0.031	0.023
1250 / 2282	11.389	8.398	6.1917	4.565	3.366	2.482	1.830	1.349	0.995	0.734	0.541	0.399	0.294	0.217	0.160	0.118	0.087	0.064	0.047	0.035	0.026
1275 / 2327	11.502	8.522	6.3148	4.679	3.467	2.569	1.903	1.410	1.045	0.774	0.574	0.425	0.315	0.233	0.173	0.128	0.095	0.070	0.052	0.039	0.029
1300 / 2372	11.612	8.645	6.4363	4.792	3.567	2.656	1.977	1.472	1.096	0.816	0.607	0.452	0.337	0.251	0.187	0.139	0.103	0.077	0.057	0.043	0.032
1325 / 2417	11.720	8.766	6.5562	4.904	3.668	2.743	2.052	1.535	1.148	0.858	0.642	0.480	0.359	0.269	0.201	0.150	0.112	0.084	0.063	0.047	0.035
1350 / 2462	11.825	8.884	6.6746	5.015	3.767	2.830	2.126	1.598	1.200	0.902	0.677	0.509	0.382	0.287	0.216	0.162	0.122	0.092	0.069	0.052	0.039
1375 / 2507	11.928	9.000	6.7914	5.125	3.867	2.918	2.202	1.661	1.253	0.946	0.714	0.539	0.406	0.307	0.231	0.175	0.132	0.099	0.075	0.057	0.043
1400 / 2552	12.029	9.115	6.9067	5.233	3.966	3.005	2.277	1.725	1.307	0.991	0.751	0.569	0.431	0.327	0.247	0.188	0.142	0.108	0.082	0.062	0.047
1425 / 2597	12.128	9.227	7.0204	5.341	4.064	3.092	2.353	1.790	1.362	1.036	0.788	0.600	0.456	0.347	0.264	0.201	0.153	0.116	0.089	0.067	0.051
1450 / 2642	12.224	9.338	7.1327	5.448	4.162	3.179	2.428	1.855	1.417	1.082	0.827	0.632	0.482	0.369	0.281	0.215	0.164	0.125	0.096	0.073	0.056
1475 / 2687	12.319	9.446	7.2434	5.554	4.259	3.266	2.504	1.920	1.473	1.129	0.866	0.664	0.509	0.390	0.299	0.230	0.176	0.135	0.104	0.079	0.061
1500 / 2732	12.411	9.553	7.3527	5.659	4.356	3.353	2.581	1.986	1.529	1.177	0.906	0.697	0.537	0.413	0.318	0.245	0.188	0.145	0.112	0.086	0.066
1525 / 2777	12.502	9.658	7.4605	5.763	4.452	3.439	2.657	2.052	1.585	1.225	0.946	0.731	0.565	0.436	0.337	0.260	0.201	0.155	0.120	0.093	0.072
1550 / 2822	12.591	9.761	7.5669	5.866	4.548	3.525	2.733	2.119	1.643	1.273	0.987	0.765	0.593	0.460	0.357	0.276	0.214	0.166	0.129	0.100	0.077
1575 / 2867	12.678	9.862	7.6719	5.968	4.643	3.612	2.809	2.185	1.700	1.323	1.029	0.800	0.623	0.484	0.377	0.293	0.228	0.177	0.138	0.107	0.083
1600 / 2912	12.763	9.962	7.7754	6.069	4.737	3.697	2.886	2.252	1.758	1.372	1.071	0.836	0.652	0.509	0.398	0.310	0.242	0.189	0.148	0.115	0.090

16 | REFERENCE MATERIALS

FIGURE 16.4.5 | Surface carbon potential vs. oxygen-probe millivolt output as a function of temperature for endothermic gas (source: Super Systems Inc.)

%CO = 20.0

TEMP °C	800	810	820	830	840	850	860	870	880	890	900	910	920	930	940	950	960	970	980	990	1000
TEMP °F	1472	1490	1508	1526	1544	1562	1580	1598	1616	1634	1652	1670	1688	1706	1724	1742	1760	1778	1796	1814	1832
%C									Note: Oxygen-probe millivolt values are shown below.												
0.05	963	964	965	967	968	969	971	972	973	975	976	977	979	980	981	983	984	985	987	988	989
0.10	995	997	999	1000	1002	1003	1005	1007	1008	1010	1012	1013	1015	1016	1018	1020	1021	1023	1025	1026	1028
0.15	1015	1017	1018	1020	1022	1024	1026	1027	1029	1031	1033	1035	1036	1038	1040	1042	1044	1045	1047	1049	1051
0.20	1029	1031	1032	1034	1036	1038	1040	1042	1044	1046	1048	1050	1052	1054	1056	1058	1060	1062	1063	1065	1067
0.25	1040	1042	1044	1046	1048	1050	1052	1054	1056	1058	1060	1062	1064	1066	1068	1070	1072	1074	1076	1078	1080
0.30	1049	1051	1053	1055	1057	1059	1061	1064	1066	1068	1070	1072	1074	1076	1078	1081	1083	1085	1087	1089	1091
0.35	1056	1059	1061	1063	1065	1067	1070	1072	1074	1076	1078	1081	1083	1085	1087	1089	1092	1094	1096	1098	1100
0.40	1063	1066	1068	1070	1072	1075	1077	1079	1081	1084	1086	1088	1090	1093	1095	1097	1099	1102	1104	1106	1108
0.45	1069	1072	1074	1076	1079	1081	1083	1086	1088	1090	1093	1095	1097	1100	1102	1104	1106	1109	1111	1113	1116
0.50	1075	1077	1080	1082	1084	1087	1089	1092	1094	1096	1099	1101	1103	1106	1108	1110	1113	1115	1118	1120	1122
0.55	1080	1082	1085	1087	1090	1092	1095	1097	1099	1102	1104	1107	1109	1111	1114	1116	1119	1121	1124	1126	1128
0.60	1085	1087	1090	1092	1095	1097	1100	1102	1104	1107	1109	1112	1114	1117	1119	1122	1124	1127	1129	1132	1134
0.65	1089	1092	1094	1097	1099	1102	1104	1107	1109	1112	1114	1117	1119	1122	1124	1127	1129	1132	1134	1137	1139
0.70	1093	1096	1098	1101	1104	1106	1109	1111	1114	1116	1119	1121	1124	1126	1129	1131	1134	1137	1139	1142	1144
0.75	1097	1100	1102	1105	1108	1110	1113	1115	1118	1121	1123	1126	1128	1131	1133	1136	1139	1141	1144	1146	1149
0.80	1101	1104	1106	1109	1112	1114	1117	1119	1122	1125	1127	1130	1132	1135	1138	1140	1143	1145	1148	1151	1153
0.85	1105	1107	1110	1113	1115	1118	1121	1123	1126	1128	1131	1134	1136	1139	1142	1144	1147	1150	1152	1155	1158
0.90	1108	1111	1113	1116	1119	1121	1124	1127	1129	1132	1135	1138	1140	1143	1146	1148	1151	1154	1156	1159	1162
0.95	1111	1114	1117	1120	1122	1125	1128	1130	1133	1136	1138	1141	1144	1147	1149	1152	1155	1157	1160	1163	1166
1.00	1115	1117	1120	1123	1126	1128	1131	1134	1136	1139	1142	1145	1147	1150	1153	1156	1158	1161	1164	1167	1169
1.05	1118	1120	1123	1126	1129	1131	1134	1137	1140	1143	1145	1148	1151	1154	1156	1159	1162	1165	1167	1170	1173
1.10	1121	1123	1126	1129	1132	1135	1137	1140	1143	1146	1149	1151	1154	1157	1160	1163	1165	1168	1171	1174	1177
1.15	1124	1126	1129	1132	1135	1138	1141	1143	1146	1149	1152	1155	1157	1160	1163	1166	1169	1172	1174	1177	1180
1.20	1126	1129	1132	1135	1138	1141	1144	1146	1149	1152	1155	1158	1161	1163	1166	1169	1172	1175	1178	1181	1183
1.25	1129	1132	1135	1138	1141	1144	1146	1149	1152	1155	1158	1161	1164	1167	1169	1172	1175	1178	1181	1184	1187
1.30	1132	1135	1138	1141	1144	1146	1149	1152	1155	1158	1161	1164	1167	1170	1173	1175	1178	1181	1184	1187	1190
1.35	1135	1138	1140	1143	1146	1149	1152	1155	1158	1161	1164	1167	1170	1173	1176	1178	1181	1184	1187	1190	1193
1.40	1137	1140	1143	1146	1149	1152	1155	1158	1161	1164	1167	1170	1173	1176	1179	1181	1184	1187	1190	1193	1196
1.45	1140	1143	1146	1149	1152	1155	1158	1161	1164	1167	1170	1173	1176	1178	1181	1184	1187	1190	1193	1196	1199
1.50	1142	1145	1148	1151	1154	1157	1160	1163	1166	1169	1172	1175	1178	1181	1184	1187	1190	1193	1196	1199	1202

FIGURE 16.4.6 | Surface carbon potential vs. dew point as a function of temperature for endothermic gas (source: Super Systems Inc.)

%CO = 20.0 %H$_2$ = 40.0 A$_f$ = 1.00

Note: Dew-point values are shown below in °C.

Temp °C	760	775	790	805	820	835	850	865	880	895	910	925	940	955	970	985	1000	1015	1030	1045	1060
Temp °F	1400	1427	1454	1481	1508	1535	1562	1589	1616	1643	1670	1697	1724	1751	1778	1805	1832	1859	1886	1913	1940
% C																					
0.05	67	64	61	58	55	53	50	48	46	44	41	39	38	36	34	32	31	29	27	26	24
0.10	52	49	47	44	42	39	37	35	33	31	29	27	25	24	22	21	19	17	16	15	13
0.15	44	42	39	37	34	32	30	28	26	24	22	20	19	17	16	14	13	11	10	8	7
0.20	39	36	34	31	29	27	25	23	21	19	17	16	14	12	11	9	8	7	5	4	3
0.25	35	32	30	27	25	23	21	19	17	16	14	12	11	9	7	6	5	3	2	1	0
0.30	31	29	26	24	22	20	18	16	14	13	11	9	8	6	5	3	2	1		-2	-3
0.35	28	26	24	21	19	17	15	14	12	10	8	7	5	4	2	1	0	-2	-3	-4	-5
0.40	26	23	21	19	17	15	13	11	10	8	6	5	3	2	0	-1	-2	-4	-5	-6	-7
0.45	24	21	19	17	15	13	11	9	8	6	4	3	1	0	-2	-3	-4	-6	-7	-8	-9
0.50	22	19	17	15	13	11	9	8	6	4	3	1	0	-2	-3	-5	-6	-7	-8	-9	-11
0.55	20	18	16	13	11	10	8	6	4	3	1	0	-2	-3	-5	-6	-7	-9	-10	-11	-12
0.60	18	16	14	12	10	8	6	4	3	1	0	-2	-3	-5	-6	-7	-9	-10	-11	-12	-13
0.65	17	15	13	10	9	7	5	3	2	0	-2	-3	-5	-6	-7	-9	-10	-11	-12	-13	-14
0.70	15	13	11	9	7	5	4	2	0	-1	-3	-4	-6	-7	-8	-10	-11	-12	-13	-14	-16
0.75	14	12	10	8	6	4	2	1	-1	-3	-4	-5	-7	-8	-10	-11	-12	-13	-14	-16	-17
0.80	13	11	9	7	5	3	1	0	-2	-4	-5	-7	-8	-9	-11	-12	-13	-14	-15	-17	-18
0.85	12	10	8	6	4	2	0	-1	-3	-5	-6	-8	-9	-10	-12	-13	-14	-15	-16	-17	-18
0.90	11	9	7	5	3	1	-1	-2	-4	-6	-7	-9	-10	-11	-13	-14	-15	-16	-17	-18	-19
0.95	10	8	6	4	2	0	-2	-3	-5	-7	-8	-9	-11	-12	-13	-15	-16	-17	-18	-19	-20
1.00	9	7	5	3	1	-1	-3	-4	-6	-7	-9	-10	-12	-13	-14	-15	-17	-18	-19	-20	-21
1.05	8	6	4	2	0	-2	-4	-5	-7	-8	-10	-11	-13	-14	-15	-16	-17	-19	-20	-21	-22
1.10	7	5	3	1	-1	-3	-5	-6	-8	-9	-11	-12	-13	-15	-16	-17	-18	-19	-21	-22	-23
1.15	6	4	2	0	-2	-4	-5	-7	-8	-10	-11	-13	-14	-15	-17	-18	-19	-20	-21	-22	-23
1.20	5	3	1	-1	-3	-5	-6	-8	-9	-11	-12	-14	-15	-16	-17	-19	-20	-21	-22	-23	-24
1.25	4	2	0	-2	-4	-5	-7	-9	-10	-12	-13	-14	-16	-17	-18	-19	-20	-22	-23	-24	-25
1.30	3	1	-1	-3	-4	-6	-8	-9	-11	-12	-14	-15	-16	-18	-19	-20	-21	-22	-23	-24	-25
1.35	2	0	-2	-3	-5	-7	-8	-10	-12	-13	-14	-16	-17	-18	-20	-21	-22	-23	-24	-25	-26
1.40	2	0	-2	-4	-6	-8	-9	-11	-12	-14	-15	-16	-18	-19	-20	-21	-23	-24	-25	-26	-27
1.45	1	-1	-3	-5	-7	-8	-10	-11	-13	-14	-16	-17	-18	-20	-21	-22	-23	-24	-25	-26	-27
1.50	0	-2	-4	-6	-7	-9	-11	-12	-14	-15	-16	-18	-19	-20	-22	-23	-24	-25	-26	-27	-28

FIGURE 16.4.7 | Surface carbon potential vs. CO_2 (via infrared analysis) as a function of temperature for endothermic gas (source: Super Systems Inc.)

%CO = 20   Process factor (PF) = 150

Temp °C	760	788	816	843	871	899	927	954	982	1010	1038	1066	1093	1121	1149
Temp °F	1400	1450	1500	1550	1600	1650	1700	1750	1800	1850	1900	1950	2000	2050	2100
%C					Note: Percent CO_2 values are shown below.										
0.2	5.644	3.788	2.595	1.812	1.287	0.929	0.681	0.506	0.381	0.291	0.224	0.175	0.138	0.109	0.088
0.3	3.686	2.474	1.695	1.183	0.840	0.607	0.445	0.331	0.249	0.190	0.146	0.114	0.090	0.071	0.057
0.4	2.707	1.817	1.245	0.869	0.617	0.446	0.327	0.243	0.183	0.139	0.107	0.084	0.066	0.052	0.042
0.5	2.119	1.423	0.975	0.680	0.483	0.349	0.256	0.190	0.143	0.109	0.084	0.066	0.052	0.041	0.033
0.6	1.728	1.160	0.794	0.555	0.394	0.284	0.208	0.155	0.117	0.089	0.069	0.053	0.042	0.033	0.027
0.7	1.448	0.972	0.666	0.465	0.330	0.238	0.175	0.130	0.098	0.075	0.058	0.045	0.035	0.028	0.023
0.8	1.238	0.831	0.569	0.397	0.282	0.204	0.149	0.111	0.084	0.064	0.049	0.038	0.030	0.024	0.019
0.9	1.075	0.722	0.494	0.345	0.245	0.177	0.130	0.096	0.073	0.055	0.043	0.033	0.026	0.021	0.017
1	0.945	0.634	0.434	0.303	0.215	0.155	0.114	0.085	0.064	0.049	0.038	0.029	0.023	0.018	0.015
1.1	0.838	0.562	0.385	0.269	0.191	0.138	0.101	0.075	0.057	0.043	0.033	0.026	0.020	0.016	0.013

FIGURE 16.4.8 | Dew point vs. oxygen-probe millivolts as a function of temperature for endothermic gas (source: Super Systems Inc.)

Temp °C	704	732	760	788	816	843	871	899	927	954	982	1010	1038	1066	1093
Temp °F	1300	1350	1400	1450	1500	1550	1600	1650	1700	1750	1800	1850	1900	1950	2000
Dew PT °C(°F)						Note: Oxygen-sensor millivolt values are shown below.									
-18(0)	1196	1194	1192	1190	1188	1186	1184	1182	1180	1178	1176	1173	1171	1169	1167
-17(2)	1192	1190	1188	1186	1183	1181	1179	1177	1175	1173	1171	1168	1166	1164	1162
-16(3)	1188	1186	1184	1181	1179	1177	1175	1172	1170	1168	1166	1163	1161	1159	1157
-14(7)	1184	1182	1180	1177	1175	1173	1170	1168	1165	1163	1161	1158	1156	1154	1151
-13(9)	1181	1178	1176	1173	1171	1168	1166	1163	1161	1158	1156	1153	1151	1148	1146
-12(10)	1177	1174	1172	1169	1167	1164	1161	1159	1156	1154	1151	1149	1146	1143	1141
-11(12)	1173	1171	1168	1165	1162	1160	1157	1154	1152	1149	1146	1144	1141	1138	1136
-10(14)	1170	1167	1164	1161	1158	1156	1153	1150	1147	1144	1142	1139	1136	1133	1130
-9(16)	1166	1163	1160	1157	1154	1151	1149	1146	1143	1140	1137	1134	1131	1128	1125
-8(18)	1162	1159	1156	1153	1150	1147	1144	1141	1138	1135	1132	1129	1126	1123	1120
-7(20)	1159	1156	1152	1149	1146	1143	1140	1137	1134	1131	1128	1125	1122	1118	1115
-6(21)	1155	1152	1149	1146	1142	1139	1136	1133	1130	1126	1123	1120	1117	1114	1110
-4(25)	1152	1148	1145	1142	1138	1135	1132	1129	1125	1122	1119	1115	1112	1109	1105
-3(27)	1148	1145	1141	1138	1135	1131	1128	1124	1121	1118	1114	1111	1107	1104	1101
-2(28)	1145	1141	1138	1134	1131	1127	1124	1120	1117	1113	1110	1106	1103	1099	1096
-1(30)	1141	1138	1134	1130	1127	1123	1120	1116	1112	1109	1105	1102	1098	1094	1091
0(32)	1138	1134	1130	1127	1123	1119	1116	1112	1108	1105	1101	1097	1093	1090	1086
1(34)	1134	1131	1127	1123	1119	1115	1112	1108	1104	1100	1097	1093	1089	1085	1081
2(36)	1131	1127	1123	1119	1116	1112	1108	1104	1100	1096	1092	1088	1084	1081	1077
3(38)	1128	1124	1120	1116	1112	1108	1104	1100	1096	1092	1088	1084	1080	1076	1072
4(40)	1124	1120	1116	1112	1108	1104	1100	1096	1092	1088	1084	1080	1076	1071	1067
6(43)	1121	1117	1113	1109	1104	1100	1096	1092	1088	1084	1079	1075	1071	1067	1063
7(45)	1118	1114	1109	1105	1101	1097	1092	1088	1084	1080	1075	1071	1067	1063	1058
8(46)	1115	1110	1106	1102	1097	1093	1089	1084	1080	1076	1071	1067	1062	1058	1054
9(48)	1111	1107	1103	1098	1094	1089	1085	1080	1076	1072	1067	1063	1058	1054	1049
10(50)	1108	1104	1099	1095	1090	1086	1081	1077	1072	1068	1063	1058	1054	1049	1045
11(52)	1105	1101	1096	1091	1087	1082	1077	1073	1068	1064	1059	1054	1050	1045	1040
12(54)	1102	1097	1093	1088	1083	1078	1074	1069	1064	1060	1055	1050	1046	1041	1036
13(55)	1099	1094	1089	1085	1080	1075	1070	1065	1061	1056	1051	1046	1041	1037	1032
14(57)	1096	1091	1086	1081	1076	1071	1067	1062	1057	1052	1047	1042	1037	1032	1027
16(61)	1093	1088	1083	1078	1073	1068	1063	1058	1053	1048	1043	1038	1033	1028	1023

FIGURE 16.4.9 | Equilibrium relationships between dew point and tool steels at various furnace atmospheres when using endothermic gas. Data (Koebel) was developed by direct measurements of dew-point temperature and percent carbon for these tool steels. (source: Lindberg/MPH)

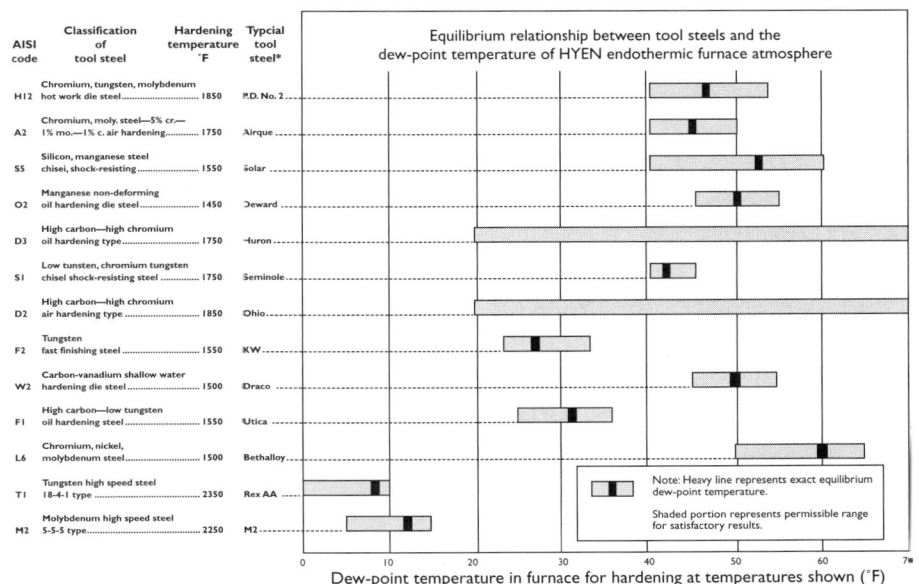

FIGURE 16.4.10 | Equilibrium relationship between dew point and percent carbon in plain-carbon steels at various furnace temperatures when using endothermic gas (source: Lindberg/MPH)

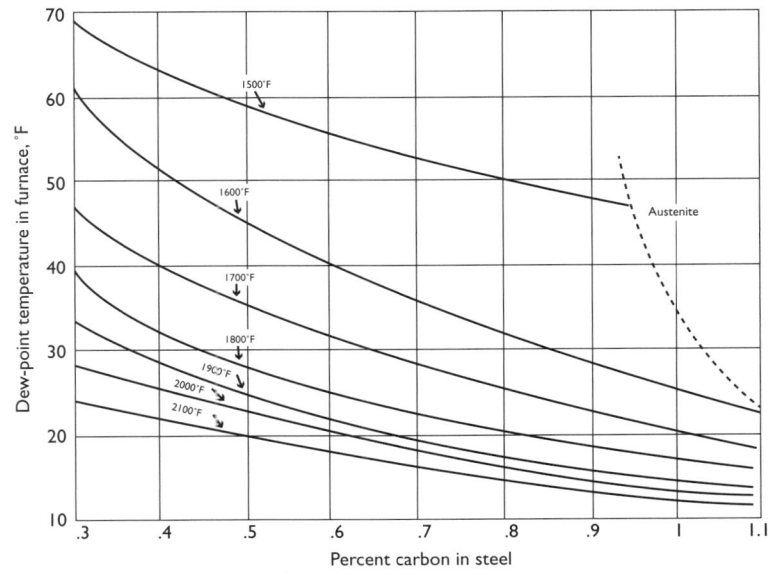

FIGURE 16.4.11 | Dew-point vs. surface-carbon concentration at temperatures indicated. Solid lines show calculated theoretical relationships for atmosphere with 40% hydrogen and 20% total carbon monoxide. Shaded areas indicate correlation with experimental results of two investigators with atmospheres that approximated the same conditions. (source: Lindberg/MPH)

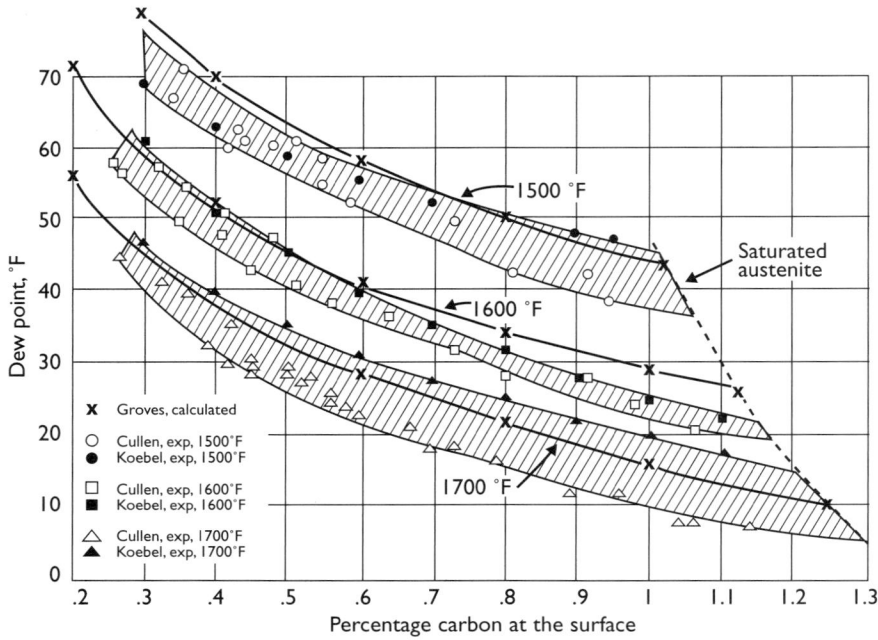

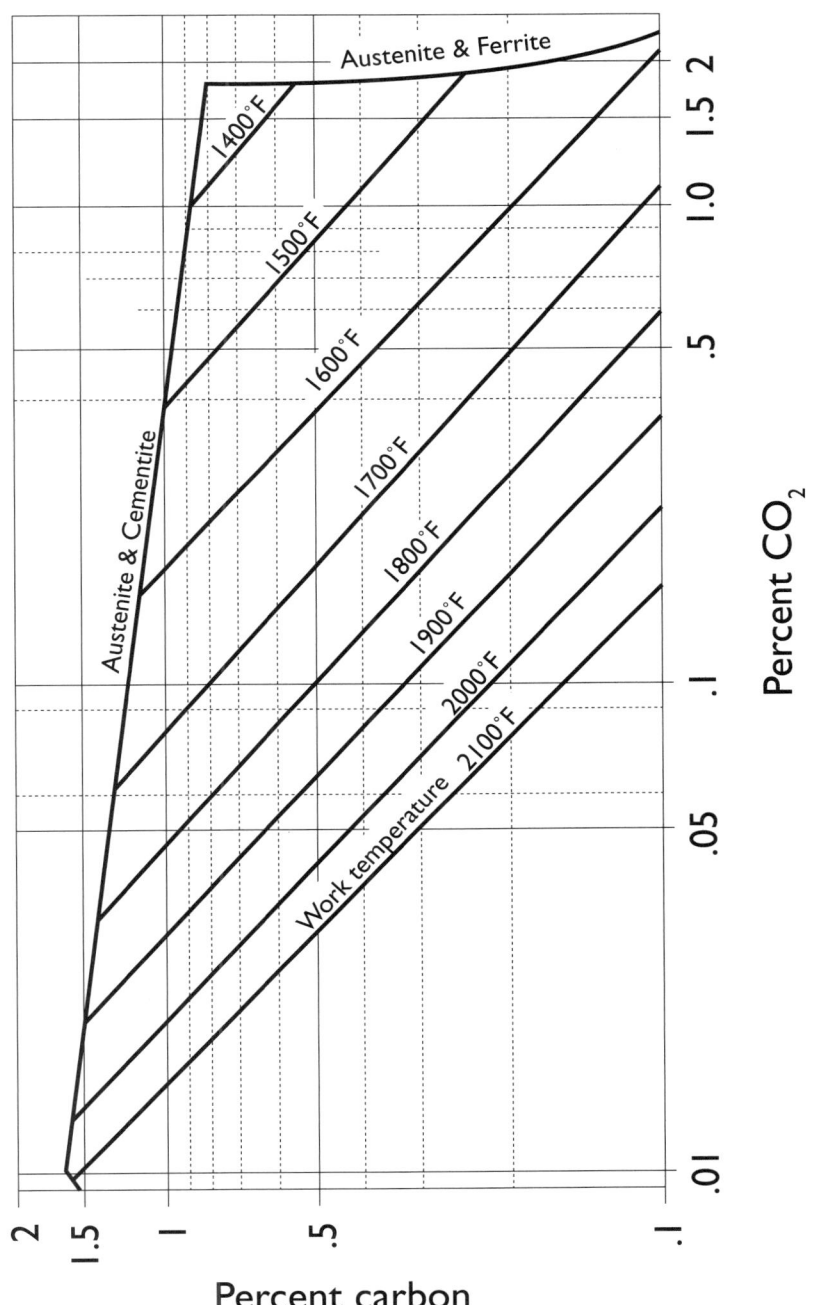

FIGURE 16.4.12 | Percentage carbon dioxide vs. carbon in austenite for endothermic (RX®) gas with 20% CO (source: Surface Combustion Inc.)

FIGURE 16.4.13 | Relationships between H_2O/H_2 and CO_2/CO at different temperatures (1500-1750°F) in accordance with the reaction $CO + H_2 = CO_2 + H_2$ (source: Lindberg/MPH)

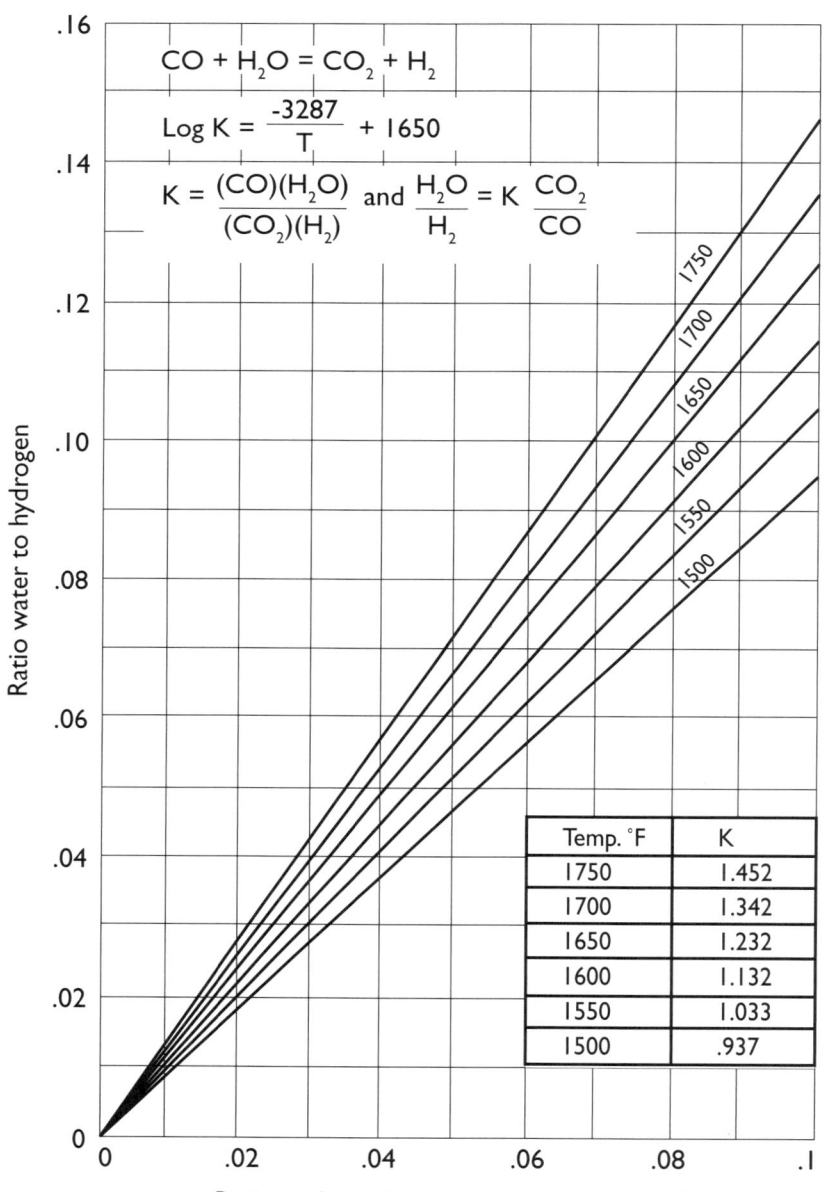

FIGURE 16.4.14 | Equilibrium constants (Austin and Day) for the reaction $Fe_3C + CO_2 = 3Fe + 2CO$ as a function of temperature and carbon content (source: Lindberg/MPH)

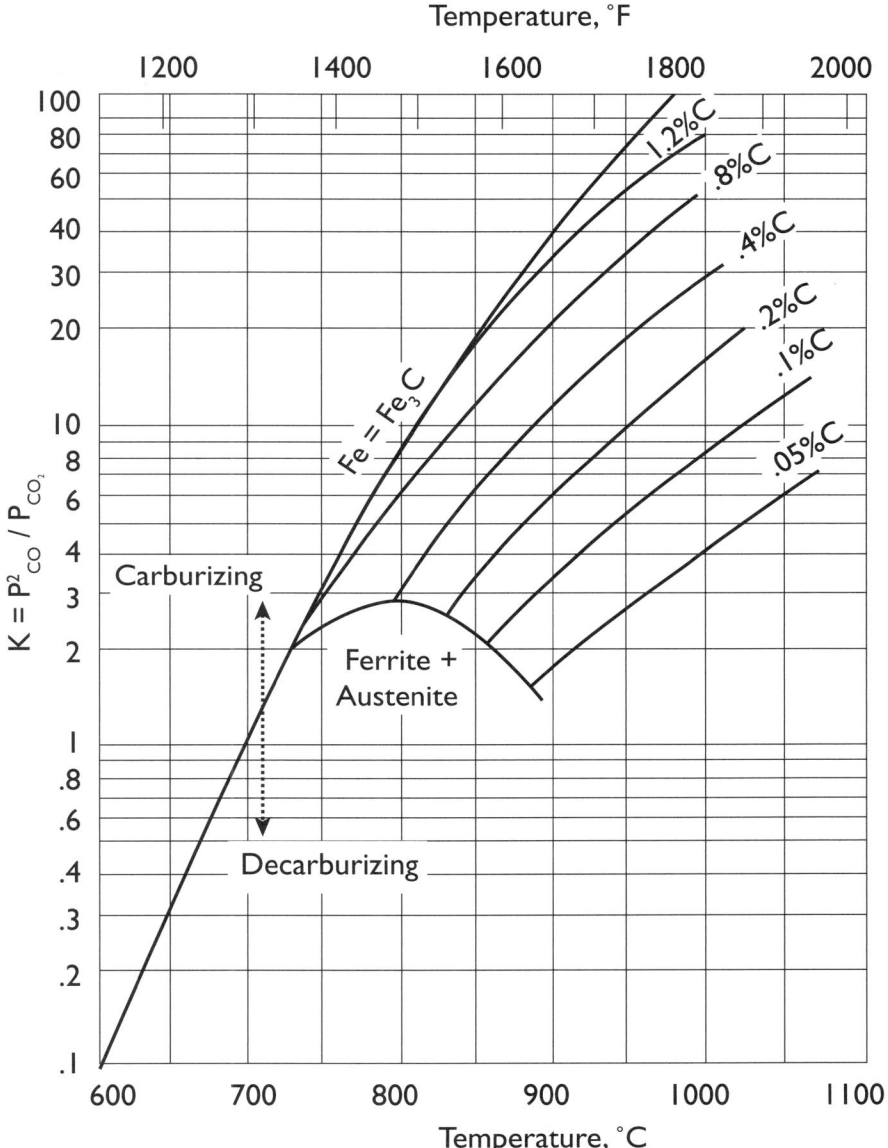

FIGURE 16.4.15 | The equilibrium constant (Austin and Day) for the reaction $3Fe + 2CO = Fe_3C + CO_2$ as a function of temperature and carbon content (source: Lindberg/MPH)

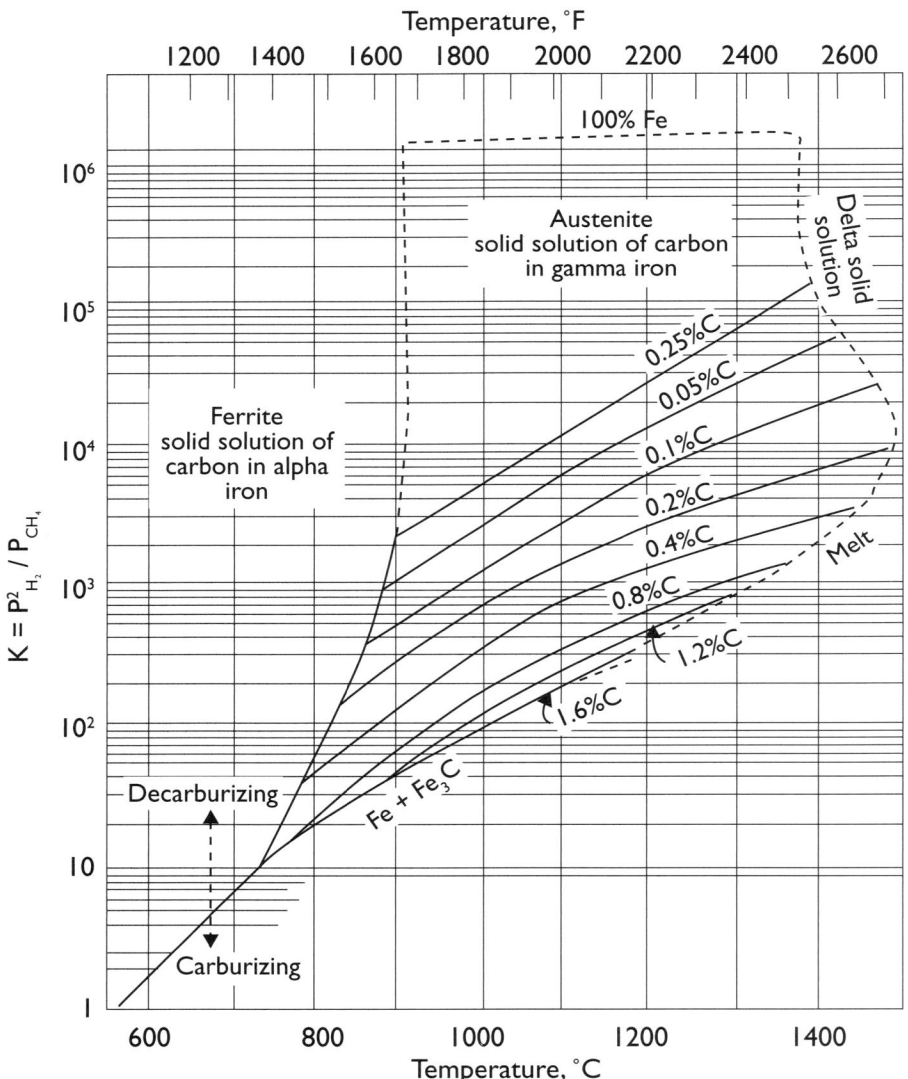

16 | REFERENCE MATERIALS

TABLE 16.4.1 | Properties and characteristics of liquid methanol (methyl alcohol) (source: Surface Combustion Inc.)

Technical Data			
Chemical Formula	CH_3OH	Calorific Value	9,646.19 Btu/lbs
Molecular Weight (lb. Mol)	32.0 lbs	Purify (commercial grade)	>99.9%
Approximate Liquid Density	49.6 lbs/ft³ 6.63 lbs/gal	Physiological Properties	Toxic
Specific Density,		Maximum permissible concentration in working enviornment	200.0 ppm
at 32°F	6.72 lbs/gal	Heat of Evaporation	
at 68°F	6.61 lbs/gal	at 149°F	475.11 Btu/lb
Boiling Point	149°F	Heat of Decomposition	1,220.17 Btu/lb
Miscibility with Water	Unlimited		
Reaction when heated: ...to 150°F Vaporizes and generates approx. 380ft³ per lb. Mol. ...to 600°F Starts to dissociate and precipitate soot. ...to 1650°F Completely dissociates and generates appoximately 1140ft³ per lb Mol.		The volume of gases produced per pound Mol, upon complete dissociation: 380 ft³ of CO weighing 28 lbs + 760 ft³ of H_2 weighing 4 lbs 1,140 ft³ CO + H_2 weighing 32 lbs 32 lbs of dissociated methanol is generated from 4.83 gallons of liquid methanol.	
Chemical Reaction (when heated above 1380° F) $CH_3OH \rightarrow CO + 2H_2$			

FIGURE 16.4.16 | Thermal cracking of methanol (source: Surface Combustion Inc.)

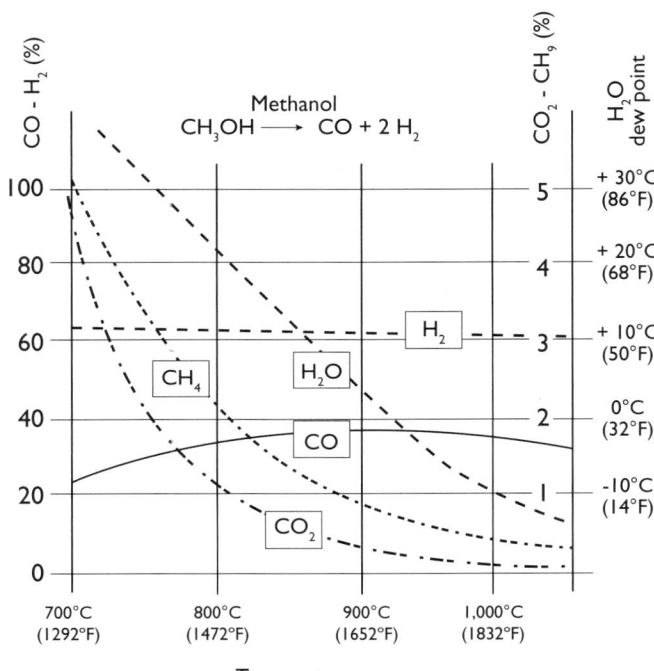

FIGURE 16.4.17 | CO values as a function of temperature using a nitrogen/methanol atmosphere (source: Surface Combustion Inc.)

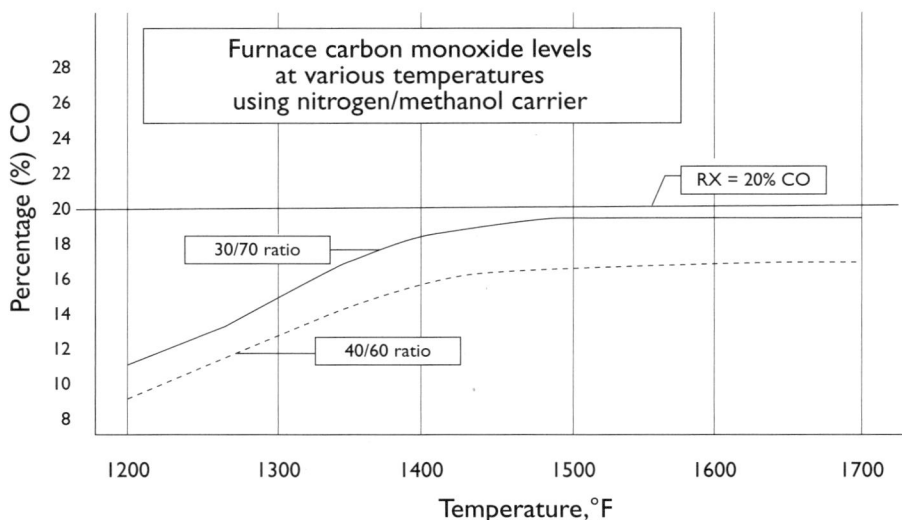

FIGURE 16.4.18 | CH_4 values as a function of temperature using a nitrogen/methanol atmosphere (source: Surface Combustion Inc.)

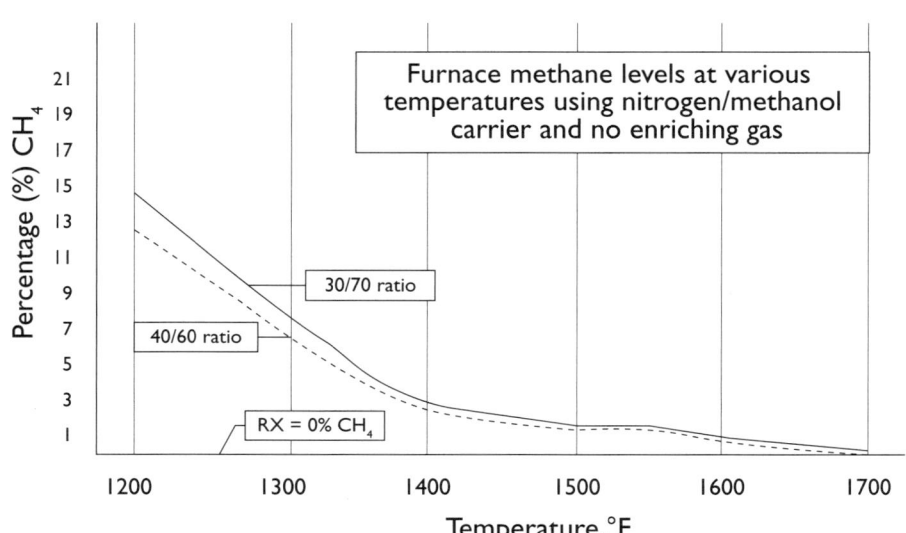

TABLE 16.4.2 | Relationship of ammonia dissociation to nitriding potential, K_N value (source: Super Systems Inc.)

% Dissociation	K_N
1.33	986.67
2.67	344.13
6.67	83.48
13.33	27.41
20.00	13.77
26.67	8.20
33.33	5.33
40.00	3.65
46.67	2.58
53.33	1.84
60.00	1.33
66.67	0.94
73.33	0.65
80.00	0.43
86.67	0.25
93.33	0.11
100.00	0.00

TABLE 16.4.3 | Effect of agitation on Grossman "H" values (quench severity) for various quench mediums (source: Gulf Super Quench 70 product brochure)

Method of cooling	Oil	Water	Brine
No agitation	0.25-0.30	0.90-1.0	2.0
Mild agitation	0.30-0.35	1.0-1.1	2.0-2.2
Moderate agitation	0.35-0.40	1.2-1.3	
Good agitation	0.40-0.50	1.4-1.5	
Strong agitation	0.50-0.80	1.6-2.0	
Violent agitation	0.80-1.00	4.0	5.0

FIGURE 16.4.19 | Effect of bath temperature on quench speed (source: Petrofer product brochure: Quenchants for the Heat-Treatment of Steel, Cast Iron and Aluminium Alloys)

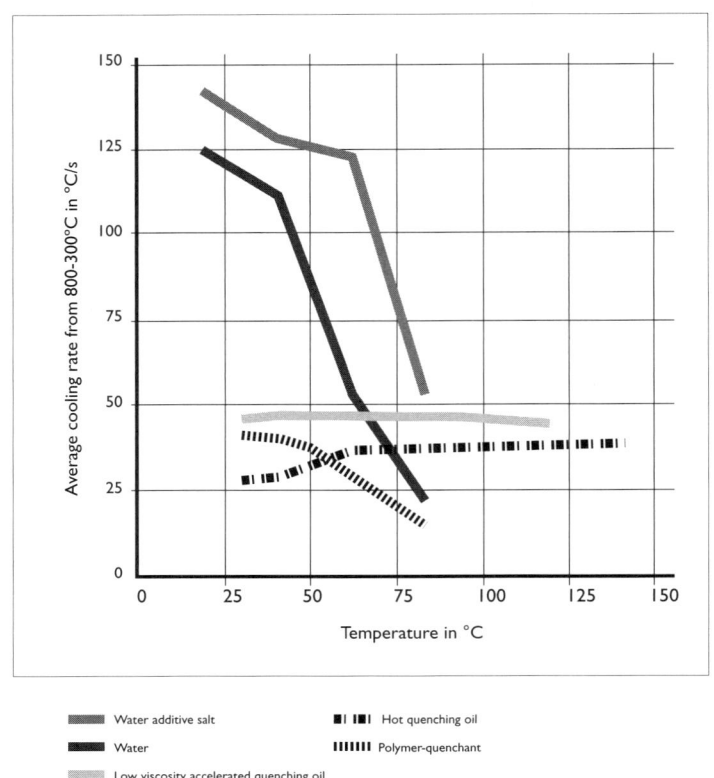

FIGURE 16.4.20 | Comparison of cooling curves for various quench media (source: Reference 1, page 604)

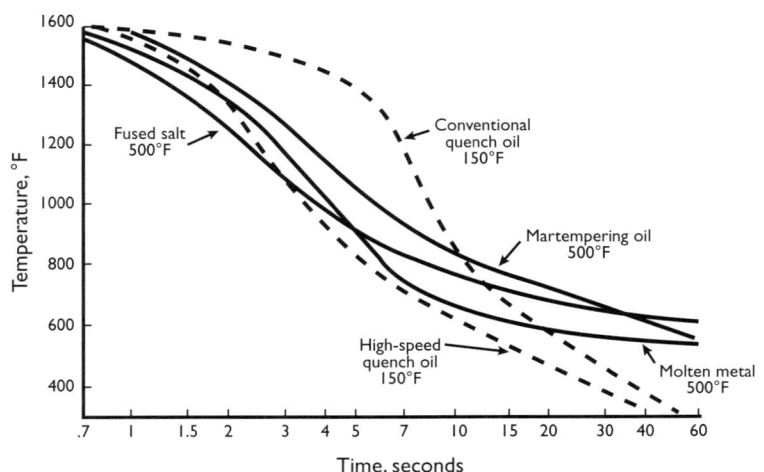

FIGURE 16.4.21 | Cooling time-temperature curves for water, petroleum oil and an aqueous polymer (PAG) quenchant superimposed on the continuous-cooling transformation diagram for SAE 1045 steel (source: ASM Handbook, Volume 4A, ASM International, 2014)

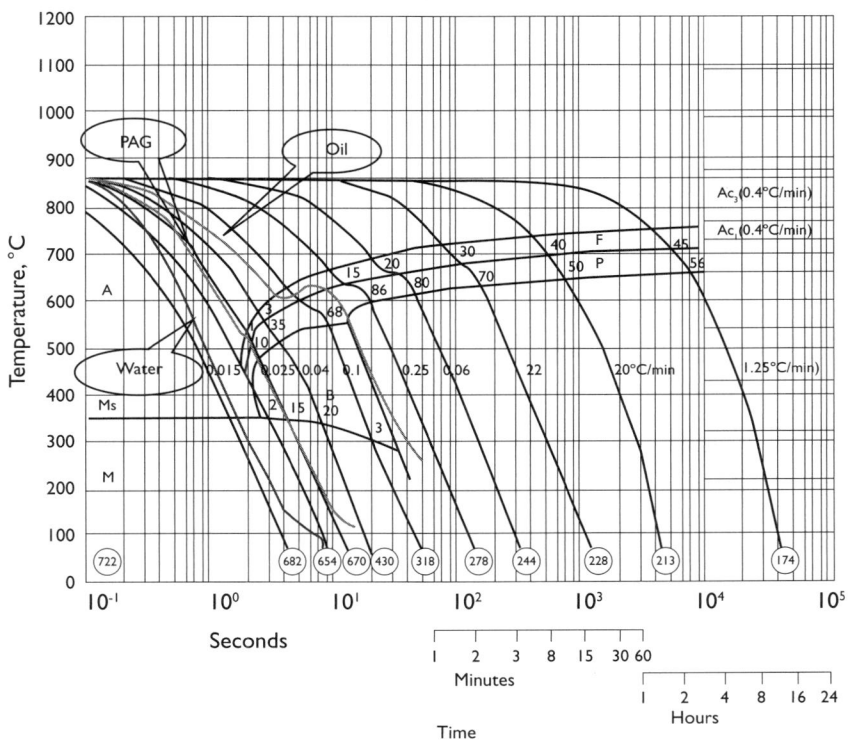

FIGURE 16.4.22 | Effect of time to wet a metal surface (wetting time) on surface hardness of SAE 1045 steel as a function of distance from the end of the specimen (source: Reference 1, page 602)

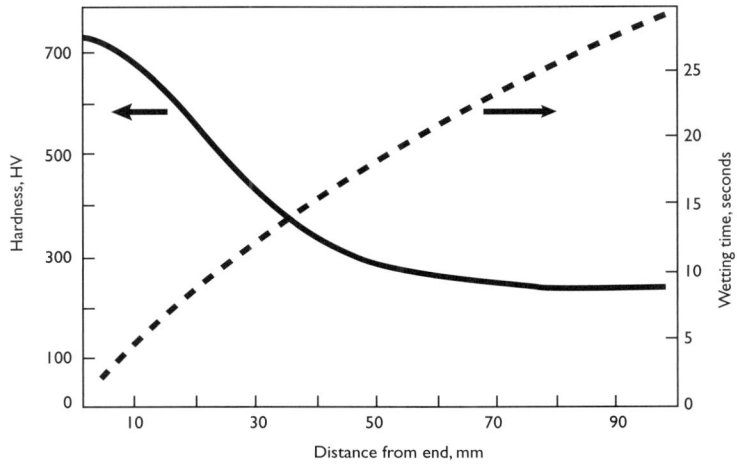

FIGURE 16.4.23 | Quenchant viscosity as a function of concentration and temperature for an aqueous solution of a quenchant (source: Reference 1, page 618)

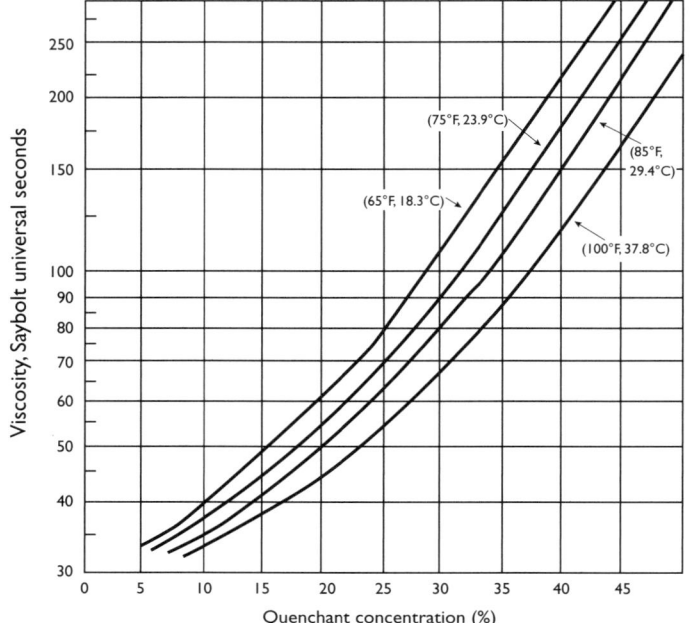

FIGURE 16.4.24 | Illustration of cooling-curve performance of a severely degraded aqueous polymer quenchant compared to water and a fresh solution at the same concentration, bath temperature and agitation (source: Reference 1, page 622)

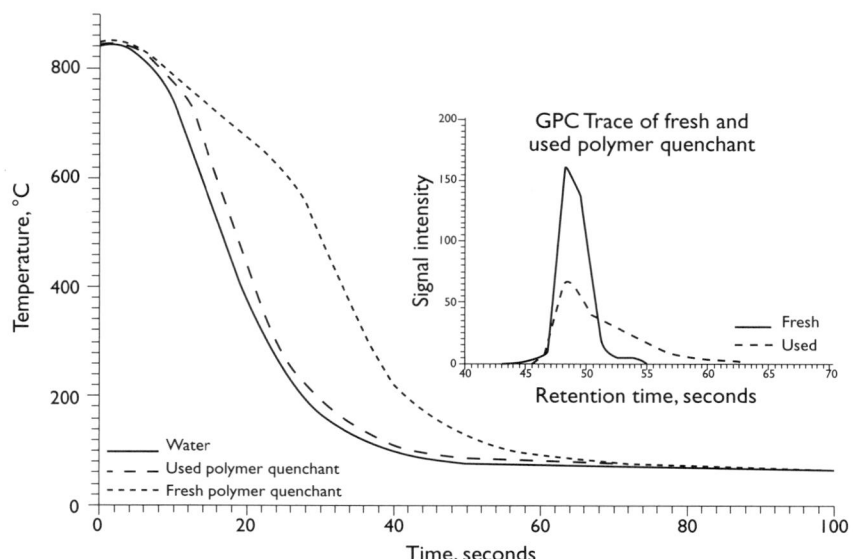

16 | REFERENCE MATERIALS

FIGURE 16.4.25 | Continuous furnace production nomograph (source: C.I. Hayes, a Gasbarre Furnace Group Company)

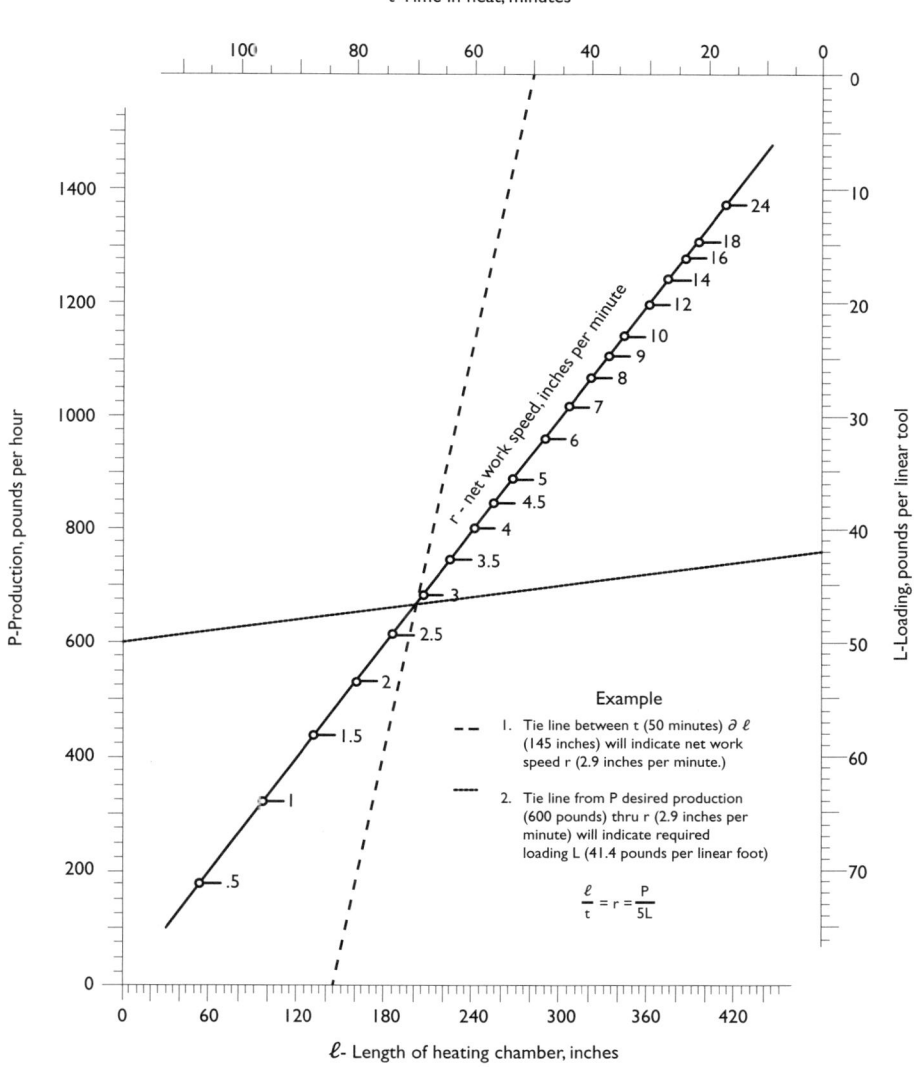

ATMOSPHERE HEAT TREATMENT

TABLE 16.4.4 | Properties of some common fuels and hydrocarbons (source: Reference 2, Appendix A-1, Table A-27)

Fuel (phase)	Formula	Molar mass, kg/kmol	Density,[1] kg/L	Enthalpy of vaporization,[2] kJ/kg	Specific heat,[1] c_p kJ/kg·K	Higher heating value,[3] kJ/kg	Lower heating value,[3] kJ/kg
Carbon (s)	C	12.011	2	—	0.708	32,800	32,800
Hydrogen (g)	H_2	2.016	—	—	14.4	141,800	120,000
Carbon monoxide (g)	CO	28.013	—	—	1.05	10,100	10,100
Methane (g)	CH_4	16.043	—	509	2.20	55,530	50,050
Methanol (l)	CH_4O	32.042	0.790	1168	2.53	22,660	19,920
Acetylene (g)	C_2H_2	26.038	—	—	1.69	49,970	48,280
Ethane (g)	C_2H_6	30.070	—	172	1.75	51,900	47,520
Ethanol (g)	C_2H_6O	46.069	0.790	919	2.44	29,670	26,810
Propane (l)	C_3H_8	44.097	0.500	335	2.77	50,330	46,340
Butane (l)	C_4H_{10}	58.123	0.579	362	2.42	49,150	45,370
1-Pentene (l)	C_5H_{10}	70.134	0.641	363	2.20	47,760	44,630
Isopentane (l)	C_5H_{12}	72.150	0.626	—	2.32	48,570	44,910
Benzene (l)	C_6H_6	78.114	0.877	433	1.72	41,800	40,100
Hexene (l)	C_6H_{12}	84.161	0.673	392	1.84	47,500	44,400
Hexane (l)	C_6H_{14}	86.177	0.660	366	2.27	48,310	44,740
Toluene (l)	C_7H_8	92.141	0.867	412	1.71	42,400	40,500
Heptane (l)	C_7H_{16}	100.204	0.684	365	2.24	48,100	44,600
Octane (l)	C_8H_{18}	114.231	0.703	363	2.23	47,890	44,430
Decane (l)	$C_{10}H_{22}$	142.285	0.730	361	2.21	47,640	44,240
Gasoline (l)	$C_nH_{1.87n}$	100–110	0.72–0.78	350	2.4	47,300	44,000
Light diesel (l)	$C_nH_{1.8n}$	170	0.78–0.84	270	2.2	46,100	43,200
Heavy diesel (l)	$C_nH_{1.7n}$	200	0.82–0.88	230	1.9	45,500	42,800
Natural gas (g)	$C_nH_{3.8n}N_{0.1n}$	18	—	—	2	50,000	45,000

[1] At 1 atm and 20°C.
[2] At 25°C for liquid fuels, and 1 atm and normal boiling temperature for gaseous fuels.
[3] At 25°C. Multiply by molar mass to obtain heating values in kJ/kmol.

TABLE 16.4.5 | Pressure relationships (source: Reprinted with permission from "Guide to Burners/Combustion Data" presented by Hauck Manufacturing)

Conversion Table of Pressure Equivalents

Lbs. per sq. In.	Oz. per sq. in.	Ft. H_2O	Inches H_2O	mm H_2O	mbar	kPa	Inches Hg.	kg per cm²
1.000	16.00	2.309	27.710	703.834	68.970	6.897	2.040	.070
.063	**1.000**	.144	1.732	43.993	4.345	.431	.127	.004
.433	6.928	**1.000**	12.000	304.800	29.864	2.985	.888	.030
.036	.572	.083	**1.000**	25.400	2.464	.2464	.074	.003
.001	.023	.003	.039	**1.000**	.098	.010	.003	.0001
.015	.232	.033	.402	10.208	**1.000**	.100	.030	.001
.145	2.320	.334	4.020	102.00	10.183	**1.000**	.295	.010
.491	7.858	1.134	13.610	345.694	33.864	3.386	**1.000**	.035
14.220	227.600	32.842	394.100	10010.140	980.700	98.070	28.960	**1.000**

Various relationships for flow and pressure

$$Q_2 = Q_1\sqrt{\frac{\Delta p_2}{\Delta p_1}} \qquad \Delta p_2 = \Delta p_1\left(\frac{Q_2}{Q_1}\right)^2$$

$$Q_2 = Q_1\left(\frac{A_2}{A_1}\right) \qquad \Delta p_2 = \Delta p_1\left(\frac{A_1}{A_2}\right)^2$$

$$Q_2 = Q_1\sqrt{\frac{Sg_1}{Sg_2}} \qquad \Delta p_2 = \Delta p_1\left(\frac{Sg_2}{Sg_1}\right)$$

$$Q_2 = Q_1\sqrt{\frac{F_1+460}{F_2+460}} \qquad \Delta p_2 = \Delta p_1\left(\frac{F_2+460}{F_1+460}\right)$$

$$Q_2 = Q_1\sqrt{\frac{p_2+14.7}{p_1+14.7}} \qquad \Delta p_2 = \Delta p_1\left(\frac{p_1+14.7}{p_2+14.7}\right)$$

Where:
Δp = pressure drop   p = pressure in PSI
Q = STP flow   F = degrees °F
A = area   Sg = specific gravity

FIGURE 16.4.26 | Flue-gas (hot-mix) temperature as a function of excess air (source: Reprinted with permission from "Guide to Burners/Combustion Data" presented by Hauck Manufacturing)

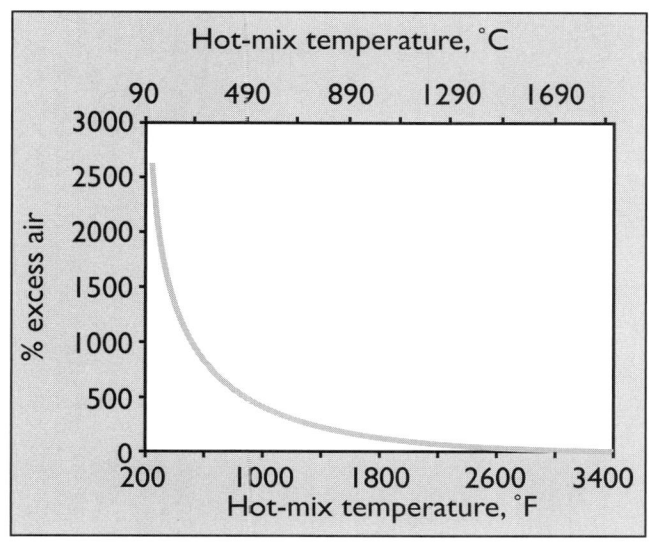

FIGURE 16.4.27 | Fuels and flue data (source: Reprinted with permission from "Guide to Burners/Combustion Data" presented by Hauck Manufacturing)

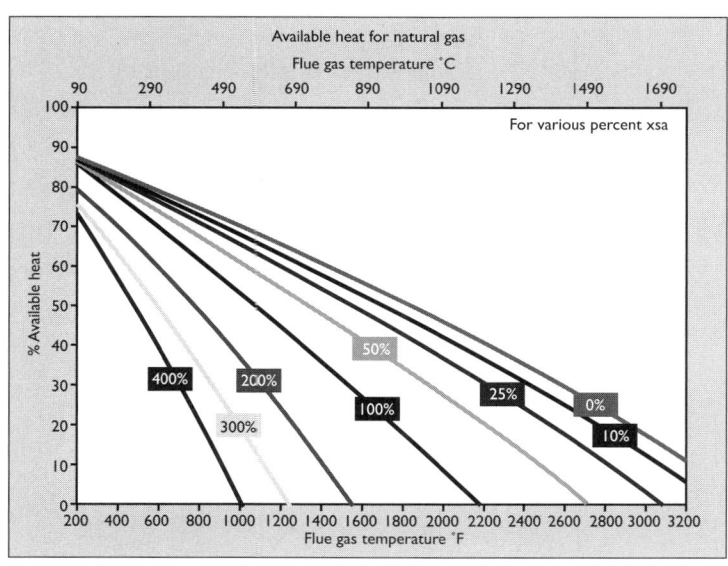

FIGURE 16.4.28 | Emissions conversion chart (source: Reprinted with permission from "Guide to Burners/Combustion Data" presented by Hauck Manufacturing)

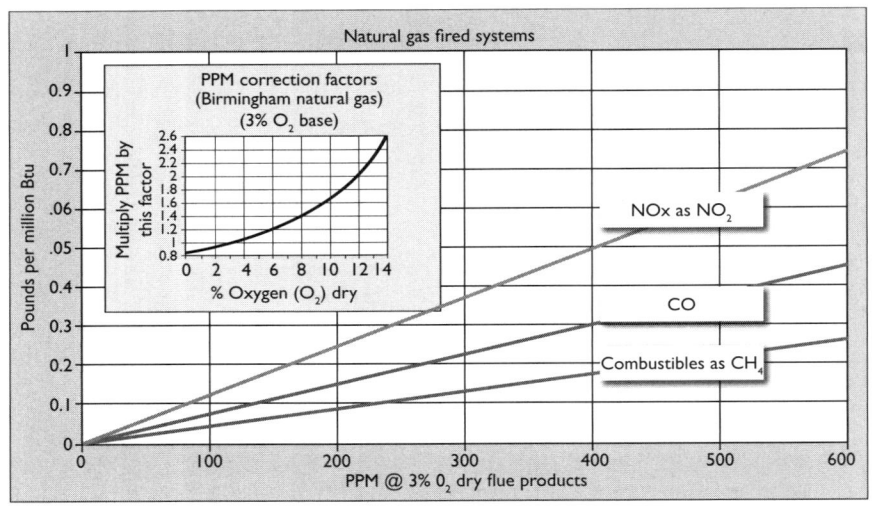

TABLE 16.4.6 | Heat recovery and fuel savings (source: Combustion Technology Manual, 5th Edition, Industrial Heating Equipment Association, 1994)

Percent fuel savings for various combustion air preheat temperatures											
Furnace outlet temperature, °F	Preheat temperature, °F										
	400	500	600	700	800	900	1000	1100	1200	1300	1400
2600	22	26	39	34	37	40	43	46	48	50	52
2500	20	24	28	32	35	38	41	43	45	48	50
2400	18	22	26	30	33	36	38	41	43	45	47
2300	17	21	24	28	31	34	36	49	31	43	45
2200	16	20	23	26	29	32	34	37	39	41	43
2100	15	18	22	25	27	30	33	35	38	39	41
2000	14	17	20	23	26	29	31	33	36	38	40
1900	13	16	19	22	25	27	30	32	34	36	38
1800	13	16	19	21	24	26	29	31	33	35	37
1700	12	15	18	20	23	25	27	30	32	33	35
1600	11	14	17	19	22	24	26	28	30	32	34
1500	11	14	16	19	21	23	25	27	29	31	33
1400	10	13	16	18	20	22	25	27	28	30	—
Fuel is natural gas at 10% excess air.											

TABLE 16.4.7 | Common thermocouple types (source: ASTM E230 (Standard Specification and Temperature-Electromotive Force Tables for Standardized Thermocouples), ASTM International)

ANSI code	Positive (+) connector	Negative (-) connector	Temperature range	Limits of error Range	Limits of error Standard	Limits of error Special
J	Iron (magnetic) Color: white	Constantan Color: red	0-760°C (32-1400°F)	-18°C to 293°C (0-560°F) 293-760°C (560-1400°F)	± 2.2°C (± 4°F) ± 0.75%[a]	± 1.1°C (± 2°F) ± 0.4%[a]
K	Chromel Color: yellow	Alumel (magnetic) Color: red	-200°C to 1260°C[c] (-328°F to 2300°F)	-18°C to 293°C (0-560°F) -200°C to 0°C (-328°F to 32°F) 293-1260°C (560-2300°F)	± 2.2°C (± 4°F) ± 2.2°C (± 4°F) or ± 2%[a] ± 0.75%[a]	± 1.1°C (± 2°F) ± 0.4%[a]
T	Copper Color: blue	Constantan Color: red	-200°C to 370°C (-328°F to 700°F)	-200°C to 0°C (-328°F to 32°F) 0-370°C (32-700°F)	± 3.5°C (± 1.8°F) or ± 1.5%[a] ± 3.5°C (± 1.8°F) or ± 0.75%[a]	±0.5°C (0.9°F) or ± 0.4%[a]
E	Chromel Color: purple	Constantan Color: red	-200°C to 870°C (-328°F to 1600°F)	-200°C to 340°C (-328 to 644°F) 340-870°C (644-1600°F)	± 1.7°C (± 3.1°F) ± 3.1°F or ± 1%[a] ± 0.50%[a]	± 1°C (± 1.8°F) ± 0.4%[a]
S or R	Platinum/10% Rhodium Platinum/13% Rhodium Color: black[b]	Platinum/13% Rhodium Platinum Color: red	0-1480°C (32-2700°F)	0-600°C (32-1112°F) 600-1480°C (1112-2700°F)	± 1.5°C (± 3°F) ± 0.25%[a]	± 0.6°C (± 1.1°F) ± 0.1%[a]
B	Platinum/30% Rhodium Color: gray[b]	Platinum/6% Rhodium Color: red	870-1700°C (1600-3100°F)	870-1700°C (1600-3100°F)	± 0.50%[a]	± 0.25%
N	Nicrosil Color: orange	Nisil Color: red	0-1260°C (32-2300°F)	0-293°C (32-560°F) 293-1260°C (560-2300°F)	± 2.2°C (± 4°F) ± 0.75%[a]	± 1.1°C (± 2°F) ± 0.40%[a]
C	Tungsten/ 5% Rhenium	Tungsten/26% Rhenium	0-2330°C (32-4200°F)	0-450°C (32-842°F) 450-2330°C (842-4200°F)	± 4.5°C (± 8°F) ± 1.00%[a]	

Notes:

[a] When determining the accuracy of thermocouples using a percent of accuracy, the temperature scale used for these calculations is Celsius.

[b] The color codes black and gray are used on extension grade wire. Type R and S use the same extension wire.

[c] Limited in vacuum use to 1150°C (2100°F).

TABLE 16.4.8 | Common thermocouple applications (source: "The Right Thermocouple Makes A World of Difference," Cleveland Electric Laboratories, white paper)

ANSI code	Application usage
J	Comparatively inexpensive. Suitable for continuous service to 760°C (1400°F) in vacuum, reducing or inert atmospheres. Reduced life in oxidizing atmosphere; iron oxidizes rapidly above 538°C (1000°F) so only heavy-gauge wire is recommended for high temperature. Bare elements should not be exposed to sulfur-bearing atmospheres above 538°C (1000°F). Protection tubes should always be used in a contaminating atmosphere and above 480°C (900°F).
K	Recommended for continuous oxidizing or neutral atmospheres. Mostly used above 540°C (1000°F) since it provides a more mechanically and thermally rugged unit than platinum/rhodium (Type R, S or B) and has longer life than iron/constantan (Type J). Subject to failure if exposed to sulfur-bearing atmospheres. Preferential oxidation of chromium (positive leg) at certain low-oxygen concentrations causes a phenomenon called "green rot" and large (negative) calibration drifts, most serious in the 815-1150°C (1500-2100°F) range. Applications up to 1260°C (2300°F) are possible using special precautions. Requires protection tubes when used in these temperature ranges. Excellent performance when supplied in mineral-insulated metal-sheathed cable form.
T	Usable in oxidizing, reducing or inert atmospheres, as well as vacuum. Its stability makes it useful for subzero temperatures and has high conformity to published calibration data. Not subject to corrosion in moist atmospheres. Copper oxides above 315°C (600°F).
E	Recommended for continuously oxidizing or inert atmospheres since both elements are highly corrosion resistant. Does not corrode at subzero temperatures. Highest thermoelectric output of common calibration. Stability is unsuitable in reducing atmospheres.
S or R	Recommended for higher-temperature environments. Must be protected with non-metallic protection tubes and ceramic insulators since it is easily contaminated in other than oxidizing atmospheres. Continued high-temperature usage causes grain growth, which can lead to mechanical failure. Calibration drift (negative) caused by rhodium diffusion to pure leg as well as from rhodium vaporization.
B	Better stability than Types S or R since it has increased mechanical strength. Recommended for higher temperature environments. Lower output values. Must be protected with non-metallic protection tubes and ceramic insulators. Easily contaminated in non-oxidizing atmospheres. Continued high-temperature usage causes grain growth, which can lead to mechanical failure. Calibration drift (negative) caused by rhodium diffusion to pure leg as well as from rhodium vaporization.
N	Nickel-based thermocouple alloy used primarily at high temperatures. While not a direct replacement for Types K or N, provides better resistance to oxidation at high temperature and longer life in applications where sulfur is present. Used in temperature range of 315-1260°C (600-2300°F), this material is less susceptible to preferential oxidation than type K. It provides excellent performance when supplied in mineral-insulated metal-sheathed cable form.
C	Material has no oxidation resistance, so its use is restricted to vacuum, hydrogen or (truly) inert atmospheres.

16 | REFERENCE MATERIALS

TABLE 16.4.9 | Other less-common thermocouple designations (source: Wang. T.P., "Thermocouples for Special Applications," Proceedings of the International Conference: Equipment and Processes, 1994, ASM International)

Designation	Positive (+) connector	Negative (-) connector	Maximum use temperature	Most common atmosphere type
Nickel/Nickel Molybdenum	Ni/18% Mo (20 Alloy)	Ni/1% Co (19 Alloy)	1315°C (2400°F)	Reducing
Platinel II	55% Pd/ 31% Pt/14% Au	65% Au/35% Pd	1260°C (2300°F)	Oxidizing
W	W	W/26% Re	2300°C (4200°F)	Reducing
W-3	W/3% Re	W/26% Re	2300°C (4200°F)	Reducing

TABLE 16.4.10 | Metallurgical sample preparation defects, causes and remedies

Defect	Probable Cause	Suggested Remedy
Comet tails	Poorly bonded very hard phase in softer matrix, or pored in matrix results in unidirectional grooves emanating from particles of hole.	Use hard, napless cloths; reduce applied pressure; reduce step times; impregnate pores with epoxy or wax; use complementary rotation. In manual preparation, avoid unidirectional grinding.
Edge rounding	Shrinkage gaps between specimen and mount are main problem and, when present, avoid napped cloths and long polishing times.	Avoid gaps by cooling thermosetting mount under pressure after polymerization. Use EpoMet resin (least prone to shrinkage gaps); of castable resins, epoxy is best (adding Flat Edge Filler particles to epoxy improves edge retention); use central force rather than individual force; use Apex Hercules rigid grinding discs; use napless cloths.
Embedding	Hard abrasive tends to embed in soft metals; finer particles are more likely to embed then larger particles.	Coat SiC abrasive paper with parafrin wax (candle wax is best, soaps are also effective); beeswax is less effective; reduce applied pressure rpm and times; avoid finer SiC abrasive papers; avoid using diamond slurries with fine diamond sizes (for ≤3μm diamond, paste embeds less than slurries). SiC embedded in soft metals (e.g. Pb) can be removed by polishing with alumina slurries
Pull-outs	Excessive grinding time on worn SiC paper; excessive polishing time on napped cloths; excessive pressure; and inadequate lubrication lead to pull out of second-phase particles.	Grind no more than 60 seconds on a sheet of SiC paper; use napless cloths; reduce applied pressure; use proper degree of lubrication.
Relief	Use of napped cloths, long polishing times and low pressure create height differences between matrix and second-phase particles.	Use napless cloths, higher pressure and shorter time. If relief is observed using contra rotation in last step, repeat the last step using complementary rotation.
Scratches	Contamination of grinding or polishing surfaces by coarser abrasives; pull out of hard particles from matrix or broken pieces of brittle materials; or inadequate grinding or polishing times.	Maintain clean operating conditions, washing hands and equipment between steps; for cracked or porous specimens, use ultrasonic cleaning after each step (especially polishing); execute each step thoroughly (avoid short cutting methods); when using the Apex magnetic disc system, store the discs in a clean drawer or cabinet and scrub surfaces if contamination is believed to have occurred; and high pressures can ease a heavy pattern.
Smear	Flow of softer metals may be caused by inadequate lubrication, excessive pressure or rpms.	Use proper degree of lubrication, low pressures and rpms; use a medium napped cloth on final step; lightly etch the specimen after final polishing and repeat the last step; use the vibratory polisher to remove smeared surface metal.
Stain	Stains around second-phase particles can be induced by interactions between abrasive and particle, perhaps affected by water quality; or they can occur due to inadequate cleaning or improper solvents used after the abrasive step.	Some staining issues are unique to the specimen but most result from using impure tap water (switch to distilled water; sometimes de-ionized water in needed) or inadequate drying. When using colloidal silica, which can be hard to remove, direct the water jet onto the cloth and wash the specimens (and cloth) for the last 6-10 seconds of the polishing cycle. This make subsequent cleaning easy. Otherwise, scrub the surface with water containing a detergent solution using cotton. Then, rinse the alcohol and blow dry with hot air. Compressed air systems can contain impurities, such as oils, the can stain the surface.

REFERENCES

1. *Fuels and Lubricants Handbook: Technology, Properties, Performance and Testing*, George E. Totten (Ed.), 2003; "Chapter 22: Non-Lubricating Process Fluids: Steel Quenching Technology," by Bozidar Liscic, Hans M. Tensi, George E. Totten and Glenn M. Webster, ASTM International, 2003
2. Cengel, Yunus and Michael Boles, *Thermodynamics: An Engineering Approach*, 7th Edition, McGraw-Hill Book Company, 2011, Table A- 27E
3. Robert Brodeur, C.I. Hayes, a Gasbarre Furnace Group Company, private correspondence

EQUATION INDEX

DESCRIPTION	NUMBER
Both ammonia and (atomic) nitrogen can cause nitrogen pickup on the surface of a material	9.6.2
Brinell hardness if a non-standard ball or force is used	12.2.1
Calculating the actual flow when a change in specific gravity occurs	13.3.1
Calculating the actual flow when a change in pressure occurs	13.3.3
Calculating the actual flow when a change in temperature occurs	13.3.2
Calculating the actual flow to compensate for both temperature and pressure	13.3.4
Calculation of the carbon potential of the atmosphere	13.6.1
Carbon dioxide oxidation of iron reaction	9.3.2, 9.3.3
Carbon monoxide and hydrogen is the rate-determining reaction in carburizing atmospheres	9.3.11
Carbon monoxide (CO) is highly reducing to steel	9.3.5
Carbon potential (C_p), which is equal to the carbon content in weight percent in a binary Fe-C system, can be related to the molar fraction x_c, which in turn is related to the carbon activity	9.5.3, 9.5.4
Carbon reversal reaction	9.3.6
Changes to a metal's surface by oxidation	9.4.1a, 9.4.1b, 9.4.1c
Changes to a metal's surface by reduction	9.4.2a, 9.4.2b, 9.4.2c
Combustion reactions involving the partial or total burning of a hydrocarbon fuel and air	9.3.7, 9.3.8
Convective heat transfer	10.4.2
Critical heat-flux density	10.2.1
Decarburization reaction of carbon dioxide	9.3.4
Dissociation of ammonia (NH_3) reaction	9.3.1
Dissociated-ammonia gas reaction	9.6.1
"Endothermic equivalent" (aka synthetic Endo) gas atmosphere obtained by cracking liquid methanol (methyl alcohol) and combining it with nitrogen	9.5.2
Endothermic gas reaction	9.6.4, 9.6.5
Engineering measures for stress and strain	12.3.1, 12.3.2
Exothermic gas reaction – two steps	9.6.3a, 9.6.3b
First stage of metal dusting occurs when carbon monoxide decomposes to carbon dioxide and carbon on the surface of the alloy	13.7.1
Formula for vapor pressure (P) of zinc (in atmospheres) is a function of temperature (T) expressed in degrees Kelvin (K) over the range from melting (693K) to boiling (1177K)	9.5.1
Hardness used to approximate the tensile strength of the material	12.3.3a, 12.3.3b, 12.3.4a, 12.3.4b
Heat-transfer coefficient due to convective heat transfer	10.4.1

ATMOSPHERE HEAT TREATMENT

Hydrocarbons dissociate at elevated furnace temperatures	9.3.13
Hydrogen is highly reducing to steel as well as decarburizing when present in concentrations higher than equilibrium	9.3.14, 9.3.15
Hydrogen (H_2) is the byproduct when steel absorbs carbon (C)	9.3.12
Methanol/nitrogen mixtures or methanol only are used to produce either a neutral or carburizing atmosphere	9.3.16, 9.3.17
Nitriding potential equation	9.7.1
Overall equipment efficiency (OEE)	15.3.1
Oxygen is decarburizing to steel	9.3.21
Oxygen is oxidizing to steel	9.3.18, 9.3.19, 9.3.20
Oxygen reaction with carbon (soot) present in the furnace	9.3.22
Reaction between hydrogen or carbon monoxide present and oxygen	9.3.9, 9.3.10
Rockwell hardness calculations (diamond indenter, regular)	12.2.2
Rockwell hardness calculations (diamond indenter, superficial)	12.2.3
Rockwell hardness calculations (ball, regular)	12.2.4
Rockwell hardness calculations (ball, superficial)	12.2.5
Second stage of metal dusting of ferrous metals, iron-carbide particles form and grow	13.7.2
Simple cost equation for reverse engineering	15.5.1
Water-gas reaction	9.3.23
Water vapor as a strongly decarburizing gas	9.3.24
With the use of CO_2 (common in pit-type furnaces), there is a reaction with hydrogen that produces CO to create carbon activity, K_C	9.7.2

TABLE INDEX

DESCRIPTION	NUMBER
Additional etchants for steel	16.3.6
Anhydrous ammonia product specifications	9.6.1
Applications for endothermic gas	9.6.5
Atmosphere choices for annealing	9.5.1
Benefits and limitations of manual and automated control	9.7.1
Carbon and hydrogen diffusion rates	15.6.1
Characteristic temperature history, °C (°F)	13.4.3
Chemical reaction categories by gas type	9.3.1
Classification of exotic alloys for specialty fasteners	15.8.6
Classification of gas-quenching pressure ranges for atmosphere furnaces	10.4.2
Classification of generated furnace atmospheres	9.2.5
Classification of quench oils	10.2.3
Classification of quench oils	13.4.1
Common blended atmosphere substitutions for generated atmospheres	9.2.4
Common brazing temperatures	9.5.4
Common furnace atmospheres for brazing	9.5.3
Common heat-treatment atmosphere gas constituents	14.1.1
Common problems and troubleshooting steps	13.5.1
Common thermocouple applications	16.4.8
Common thermocouple construction techniques	13.3.5
Common thermocouple types	16.4.7
Common types of furnace atmospheres (listed alphabetically)	9.2.1
Comparison of airfoil-type draft tubes to pump horsepower for different volumes of water	10.6.1
Comparison of NDT methods	12.6.1
Compositional ranges for endothermic gas	9.5.5, 9.6.4
Composition of selected generated furnace atmospheres	9.2.6
Compositions of nitrocarburizing atmospheres	9.5.7
Cooling-rate history at 315°C (600°F), °C/s (°F/sec)	13.4.4
Dew point vs. surface carbon at selected temperatures	9.7.5
Differences between shear and cleavage fracture	12.5.1
Effect of agitation on Grossman "H" values (quench severity) for various quench mediums	16.4.3
Effect of water temperature on cooling rate	10.4.3
Energy equivalents	14.3.1
Example of a calculation for the maximum oil operating temperature	10.2.4
Example of common test methods for fasteners	12.3.1
Example of flow estimation from pipe size	14.5.2

Example of four-step method of fine grinding for iron-based alloys	16.3.2
Example of inspection criteria for carburized, quench-and-tempered steel	11.5.1
Example of process variables in material selection	11.5.2
Example of three-step method for grinding steels harder than 200 HV	16.3.3
Examples of blended and diluted atmosphere types and compositions	9.2.3
Examples of hydrogen damage and ways to avoid it	15.6.2
General categories of wear	12.5.3
Hardness ranges for testing files	12.2.1
Heat recovery and fuel savings	16.4.6
Heat-treatment parameters	15.2.2
HPDC alloys	15.2.1
H-value history	13.4.5
Hydrogen bake-out requirements for high-strength steels	15.6.3
K_N and K_C values	9.7.7
Limit of solubility of carbon in austenite	13.6.8
Materials for automotive products (part 1a)	15.8.1
Materials for automotive products (part 1b)	15.8.2
Materials for automotive products (part 2a)	15.8.3
Materials for automotive products (part 2b)	15.8.4
Maximum cooling-rate °C/s (°F/sec) history	13.4.2
Maximum flame curtain gas and air demands	13.3.4
Melting temperature of various liquid metals known to embrittle high-strength steels	15.6.4
Metallurgical sample preparation defects, causes and remedies	16.4.10
Minimum shim cooling time (weight-gain method)	13.6.7
Nitralloy 135M core hardness as a function of tempering temperature	9.5.8
Other less common thermocouple designations	16.4.9
Overview of process tables	11.3.1
Physiological effects of ammonia	14.1.3
Physiological effects of carbon monoxide	14.1.2
Physiological effects of oxygen deficiency	14.1.4
Polymer selection guide (part 1)	10.2.5
Polymer selection guide (part 2)	10.2.6
Pressure conversion factors	13.3.3
Pressure relationships	16.4.5
Properties and characteristics of liquid methanol (methyl alcohol)	16.4.1
Properties of molten salt	10.4.1
Properties of some common fuels and hydrocarbons	16.4.4
Quench-oil selection guide (part 1)	10.2.1
Quench-oil selection guide (part 2)	10.2.2
Quench-oil test standards	13.4.6

Recommended burnout frequency 9.7.2
Relationship between dew point, moisture content and percent
    dissociation 9.6.2
Relationship between dissociation and nitriding potential (K_N) 9.7.6
Relationship between part properties, steel selection and
    nitriding/nitrocarburizing parameters 9.5.6
Relationship of ammonia dissociation to nitriding potential, K_N value 16.4.2
Relative cooling rate as a function of quench medium 10.4.4
Results of copper-steel-stainless steel test 13.6.2
SAT differences by furnace class 11.4.1
Selected annealing temperatures 9.5.2
Selected etchants for aluminum and aluminum alloys 16.3.4
Selected etchants for iron and steels 16.3.5
Selected examples of mechanical property enhancement 15.2.3
Selected properties of gases used for furnace atmospheres 9.4.1
Selected properties of steel construction-grade fasteners 15.8.5
Shim dwell time 13.6.6
Shim insertion depth 13.6.5
Shim-stock size recommendations 13.6.4
Sintering temperature and equipment type by material 9.5.9
Smoke-bomb burning times and volumes 13.6.1
Specific-gravity conversion factors 13.3.1
Troubleshooting guide for oxygen analyzers 13.5.2
Typical characteristics of ductile and brittle fractures 12.5.2
Temperature conversion factors 13.3.2
Temperature duration for in-furnace data-logging devices 13.6.3
Types of protection tubes used in the heat-treating industry 13.3.6
Typical design influences 12.5.6
Typical dew-point levels 9.7.4
Typical environmental influences 12.5.9
Typical exothermic-gas composition and specifications 9.6.3
Typical field data for an operating endothermic-gas generator 9.7.3
Typical heat-treatment influences 12.5.8
Typical industry standards for thermocouples 13.3.7
Typical material influences 12.5.5
Typical mechanical tests 12.1.1
Typical processing (manufacturing) influences 12.5.7
Typical Rockwell hardness scales by applications 12.2.2
Typical sectioning methods 16.3.1
Typical service/application influences 12.5.4
Typical water requirements for open systems 14.3.2
Volume changes required for safe purging of furnaces 9.2.2
Water-quality requirements for open cooling towers 14.5.1
Water requirements for closed hydronic systems 14.3.3
Wrought-alloy materials for construction of retorts 15.7.1

FIGURE INDEX

DESCRIPTION	NUMBER
200-mm-diameter (8-inch-diameter) bar of SAE 4150 entering spray water quench for hardening	9.5.27
330 stainless steel alloy illustrating catastrophic carburization (Kallings, 1000X) – radiant-tube failure in the area where the tube passed through the insulation	12.1.7
440C stainless steel (composite topographical image, 30X) – example of thread damage from machining	12.1.3
4150 ball screw	15.5.1
8620 ball screw	15.5.4
8620 deflector	15.5.3
8620 pin	15.5.2
52100 bearing race (3680X, 2% nital) – SEM reveals an acceptable microstructure of fine carbide distributed in a matrix of tempered martensite	12.1.8
A356.0 aluminum casting (unetched, 125X) – blistering due to hydrogen embrittlement	12.1.4
A typical heat-treat shop	15.1.1
Air cooling after stress relief in an outdoor environment	10.4.3
Alloy muffle	15.7.5
Ammonia tank installation	9.6.5
Anatomy of a high-performance, high-velocity burner	13.2.4
Application uses for polymer quenchants	10.2.19
Area around heating elements after cleaning	13.4.9
Atmosphere control based on gas analysis – nitrogen and endothermic gas	9.5.2
Atmosphere integral-quench furnace adapted for IntensiQuench®	10.4.13
Austempered stampings exiting the salt quench tank	10.4.5
Balance between productivity and quality	11.2.1
Basket loaded with rectifying tablets for loading into high-temperature tank	13.4.8
Batch-style gas-fired drop-bottom furnace with common mobile quench tank for heat solution treatment of aluminum castings	15.2.4
C356.0 aluminum castings ready for solution heat treatment and water quenching	10.4.10
Calcium-carbonate level as a function of temperature	14.5.7
Carbon determinator for shim-stock analysis by the combustion method	13.6.14
Carbon-potential analyzer setup	13.6.12

    a. Test instrument
    b. Test coil in end of insertion rod
    c. Test coil being inserted in furnace

Carbon potential vs. temperature at a specified millivolt reading
  for typical endothermic gas from natural gas feedstock 16.4.3
Carbon-rich subsurface layer inside a refractory wall..13.7.1
Carburized gears 9.5.20
Carburized shafts 9.5.21
Carburizing/decarburizing equilibrium of CH_4 for various carbon steels......9.4.14
Carburizing or decarburizing reactions of CO and CO_2 and their
  relation to carbon in the steel at different steel temperatures........................9.4.13
Carburizing recipe-control scheme 9.7.9
Case-hardened bearing races.. 9.1.2
Case hardening of steel by austenitizing and quenching 9.5.16
Cast-alloy retorts 15.7.4
Cast carburizing basket..15.7.2
Catastrophic carburization (aka metal dusting) of a furnace
  alloy fan shaft... 13.3.2
CH_4 values as a function of temperature using a nitrogen/methanol
  atmosphere..16.4.18
Charpy impact test setup 12.3.10
Circulation pattern of a radial-blade fan in a pit-style furnace....................... 13.3.1
  a. Circulation pattern
  b. Fan detail
CO detector in a manufacturing shop located outside the
  heat-treat department 14.1.2
  a. CO detector and emergency instructions
  b. Close-up of CO emergency instructions
CO values as a function of temperature using a nitrogen/methanol
  atmosphere..16.4.17
Cold-drawing dies with sharp corners... 11.5.8
Combination airflow oven design..11.4.7
Combination furnace oxygen probe and infrared control 9.7.2
Combustibles analyzers in use.. 13.5.4
  a. Portable combustion analyzer
  b. Panel-mounted combustion analyzer
Common atmosphere furnace maintenance items... 13.2.3
  a. Alloy (mesh) belt
  b. Burner
  c. Gas train
  d. Furnace door
  e. Furnace fan
  f. Flame curtain
  g. Flowmeter and mass-flow control devices
  h. Furnace alloy
  i. Furnace-atmosphere controls
  j. Quench solenoid and related components
  k. Water-to-oil heat exchanger
  l. Electric heating elements
  m. Instruments and controls

      n. Furnace insulation
      o. Oil agitators
      p. Oxygen probes
      q. Seals and gaskets
      r. Limit switches, electric eyes and load movement sensors

Common quenchant choices and their effect 10.1.1
Comparison of cooling curves for various quench media 16.4.20
Comparison of intensive quenching to other quenching technologies 10.4.12
Comparison of nitrided 135M with nitrided SAE 4140 and SAE 4340 .... 9.5.31
Comparison of oil-quenching methods 10.5.4
      a. Martempering
      b. Hot-oil quenching
Composition of exothermic and endothermic atmospheres 16.4.2
Consequence of excessive vibration 13.6.15
Continuous furnace production nomograph 16.4.25
Continuous-style mesh-belt aging ovens 15.2.5
Cooling time-temperature curves for water, petroleum oil and an aqueous
    polymer (PAG) quenchant superimposed on the continuous-cooling
    transformation diagram for SAE 1045 steel 16.4.21
Copper-steel-stainless steel coupons 13.6.5-13.6.8
Crack surface illustrating multiple initiations, each with their own
    fatigue propagation 12.5.3
Critical aircraft performance application for fasteners – jet engine 15.8.2
Critical heat-flux densities 10.2.3b
Cross-sectional view of an exothermic-gas generator 9.6.11
Cross-sectional view of a self-recuperative burner 14.3.3
      a. Single-ended radiant tube burner
      b. Cross-sectional view
Cross-sectional view of an ultra-low emission burner capable of
    limiting NOx emssisons to less than 30 ppm at 3% oxygen 14.3.9
Dew-point analyzers 9.7.7
      a. Portable, Model DP-2000
      b. Panel-mounted, Model DPC 2530
Dew-point sample port on a continuous mesh-belt furnace 9.7.8
Dew point vs. oxygen-probe millivolts as a function of temperature
    for endothermic gas 16.4.8
Dew-point vs. surface-carbon concentration at temperatures indicated 16.4.11
Difference in bubble formation due to concentration after 6.5 seconds 10.2.17
      a. 10%
      b. 22%
Dissociated-ammonia generator gas chemistry 9.6.1
Disturbed metal layers that sample preparation must remove 12.4.2
Double-shear testing apparatus 12.3.8
Draft-tube impeller system design 10.6.3
Drive-shaft assemblies for combination brazing and carbonitriding
    followed by oil quenching 10.2.14

Effect of agitation and water content on quench severity in molten salt 10.4.7
    a. Agitation
    b. Water content
Effect of applied stress on time to failure – SAE 4130 exposed to molten lithium 15.6.13
Effect of bath temperature on quench speed 16.4.19
Effect of belt tension on service life 13.3.6
Effect of crystals on vapor layer 10.4.14
    a. 1st stage
    b. 2nd stage
Effects of manufacturing downtime 15.3.3
Effect of tensile stress on embrittlement 15.6.12
Effect of time to wet a metal surface (wetting time) on surface hardness of SAE 1045 steel as a function of distance from the end of the specimen 16.4.22
Electronic flowmeter 13.3.4
    a. Electronic flowmeter
    b. Mass-flow controller
Ellingham-Richardson diagram 9.3.1
Emissions conversion chart 16.4.28
Endoinjector® fuel-injection gas mixing system mounted on an endothermic-gas generator 9.6.18
Endothermic-gas generator chemistry 9.6.14
Endothermic-gas generator schematic 9.5.17
Endothermic-gas generator schematic piping arrangement 9.6.17
Endothermic generator dew-point control scheme via oxygen probe 9.7.3
Energy-efficient gas-fired box furnaces 14.3.2
Energy-saving locations in a pusher furnace system 14.3.1
Equilibrium constant (Austin and Day) for the reaction $3Fe + 2CO = Fe_3C + CO_2$ as a function of temperature and carbon content 16.4.15
Equilibrium constants (Austin and Day) for the reaction $Fe_3C + CO_2 = 3Fe + 2CO$ as a function of temperature and carbon content 16.4.14
Equilibrium curves for furnace gases on iron, carbon and steel 16.4.1
Equilibrium relationship between dew point and percent carbon in plain-carbon steels at various furnace temperatures when using endothermic gas 16.4.10
Equilibrium relationships between dew point and tool steels at various furnace atmospheres when using endothermic gas 16.4.9
Example of a cleavage-type failure 12.5.5
Example of a decohesive rupture-type failure 12.5.6
Example of a equiaxed dimpled rupture-type failure 12.5.4
Example of a world-class heat-treat department 11.2.2
Example of bad and good gear design and a real-world example 10.5.3
    a. Poor design resulting in distortion
    b. Good design avoiding distortion
    c. Real-world example of improper hole placement leading to distortion of adjacent gear teeth
Example of balancing section size to minimize cracking 11.5.1

Examples of fracture loading/opening modes (Type I = tearing,
Type II = shear via edge sliding, Type III = shear via screw sliding) 12.5.2
Example of free quenching 10.2.6
Example of process variability in neutral hardening 15.1.2
Example of quenching parts in baskets 10.2.8
Example of quenching parts on fixtures 10.2.9
Example of reducing stress risers 11.5.2
Example of restricted (press) quenching 10.2.7
Example of the use of statistical data to evaluate an in-control process 12.1.1
    a. Process capability histogram
    b. Probability plot showing data is normally distributed
Example of the use of statistical data to evaluate an out-of-control process .. 12.1.2
    a. Process capability histogram
    b. Probability plot showing data is normally distributed
Examples of part surfaces after oil quenching 13.1.2
    a. Unacceptable part surface appearance after heat treatment – part
       staining, sludge buildup, decarburized surfaces
    b. Acceptable part surface appearance after heat treatment – parts are
       clean; typically gray/green in appearance in the as-quenched condition
       with no evidence of staining, soot, scale or other surface contamination
Exfoliated steel sliver 15.6.2
Exoinjector® fuel-injection gas mixing system 9.6.12
Exothermic-gas composition as a function of air/gas ratio 9.6.10
Exothermic-gas generator chemistry 9.6.7
Exothermic-gas generator with cogeneration capability 9.6.13
Factors contributing to stress corrosion cracking 15.6.5
Failed ASTM A490 bolts from structural beams in a power plant 12.5.1
Failure triangle 12.5.7
Family of smoke bombs 13.6.3
Fastener failure due to stress corrosion cracking 15.6.8
    a. Visual examination
    b. Optical microscopy (500X, unetched) revealing cracking along grain
       boundaries
    c. Scanning electron microscopy detailing intergranular attack of plated
       fastener
Fasteners for high-vibration material-handling applications 15.8.3
Fatigue testing equipment 12.3.7
Ferritic nitrocarburizing fixture for brake rotors 15.7.3
Firecheck on a endothermic-gas generator 9.6.19
Flow blockage in the top cool of an integral-quench furnace
   leading to sludge buildup 14.3.10
Flue-gas (hot-mix) temperature as a function of excess air 16.4.26
Fluid-flow patterns comparing open and draft-tube systems 10.6.2
Forged bolts 12.3.1
Four modes of cooling during quenching 10.2.3a
Fuels and flue data 16.4.27
Fully integrated heat-treat system 15.2.1

Furnace atmosphere analysis during heating of coil stock in a bell furnace.... 9.5.3
Furnace equipped with a fan cooling section...10.4.1
Furnace infrared (three-gas) control scheme 9.7.6
Gas cylinders quenched into an ACR 10.2.22
Gear teeth with stress-concentration points 11.5.6
GM Quench-O-Meter test shortcoming – cooling pathway A, B or C
 is not revealed 10.3.3
Gonser's curves showing slight carburizing action of purified
 exothermic gas (curve 5) compared to increasing decarburizing
 results (curves 4-1) with increasing CO_2 content.....................................9.4.11
Heat-transfer characteristics in different quench media................................10.1.2
    a. Liquid quenching
    b. Gaseous quenching
Hidden costs of unplanned maintenance 15.3.2
High-velocity self-recuperative burners firing inside a box-type furnace 14.3.8
Honeycomb grids 15.7.1
Hydrogen analyzer mounted on a circuit board 13.5.3
Hydrogen content after various stages of processing......................................15.6.3
Illustration of cooling-curve performance of a severely degraded
 aqueous polymer quenchant compared to water and a fresh
 solution at the same concentration, bath temperature and agitation 16.4.24
Impeller arrangements and resultant flow patterns 10.6.1
Importance of maintenance in total cost containment 13.1.1
Improper spray-quench coverage and spray parameters resulting in
 failure to uniformly remove heat, which lead to cracking of the part..........10.4.11
Improved design of a blanking die to avoid cracking.......................................11.5.7
Inconel 600 retort failure on the raw ammonia (inlet side) of an
 ammonia dissociator (Kallings, DIC, 25X) 12.1.5
Influence of hot-oil age on cooling-rate/temperature profile........................ 10.5.5
Influence of part geometry on cooling 10.2.4
Infrared gas analyzers..9.7.5
    a. Portable unit
    b. Panel-mounted unit
Infrared measurement of a gas 9.7.4
    a. Principle of operation
    b. Board-mounted IR sensor
Integral-quench-style furnace 9.5.25
Integrated heat-treatment cell for high-pressure die-casting......................... 15.2.6
In-process thermal-profiling tools 11.4.3
Iron-chromium phase diagram 15.6.10
IR spectra – new vs. used oil 10.3.2
    a. Moderately degraded oil
    b. Severely degraded oil
Ishikawa diagram for quenching 10.1.4
Ishikawa distortion diagram... 10.5.2
Jack arch repair...13.7.5
Keyways in shafts 11.5.4

Kinematic-viscosity instrument	13.4.2
Large pen indenter applying load into clay cube	12.2.11

    a. Initial force
    b. Intermediate force
    c. Excessive force (bulging or side distortion in evidence)

Large mill pinion being lowered into a polymer quench	10.2.16
Large polymer quench tank for rolled rings	10.2.20

    a. Typical rolled ring quenched in a PAG
    b. Polymer quench tank

Leakage field in the area of a crack	12.6.2
Lean heat-treating strategy	15.2.3
Lean manufacturing strategy	15.2.2
Endothermic gas igniting during load discharge from an atmosphere box-carburizing furnace	14.1.1
Load of 140-mm-diameter (5.5-inch-diameter) bearing races	10.2.10
Load of fasteners on a mesh belt for carbonitriding	15.8.5
Load of piston plates after ferritic nitrocarburizing	9.5.32
Load of pivot hinges for hardening and oil quenching	10.2.13
Load of SAE 4130 cylinder liners being brine quenched	10.4.15
Load of structural bolts	10.2.11
Long shafts with keyways	11.5.9
Manual survey method for an integral-quench furnace	11.4.2
Material-science disciplines	15.4.1
Mesh-belt conveyor sintering furnace with hypercooler for sinter hardening	13.6.4
Metallurgical sample cutoff saw	16.3.2
Microcrack fracture detection by acoustic emission	12.6.8
Microhardness testing	12.2.7
Microhardness values as a function of test load	12.2.10
Microstructure testing	12.6.5

    a. Annealed microstructure (ferrite and pearlite)
    b. Hardened microstructure (tempered martensite)

Milling cutter at risk of distortion and cracking	11.5.5
Minimum thickness charts	12.2.9

    a. Knoop
    b. Vickers

Modular multi-retort endothermic-gas generator	9.6.16
Neutral-hardened retaining rings	9.1.1
Nickel brazing of orthodontic braces	9.5.11
Nitriding control scheme	9.7.11
Nitrocarburizing control scheme	9.7.13
Nitrogen/methanol system	9.5.18
Nitrogen purge cylinder setup in a typical heat-treat shop	14.2.2
Normalizing large paper rolls using fan-assisted cooling due to part mass	10.4.2
Oil breakdown on quenching	13.7.6
Oil circulation pattern in a batch-style integral-quench furnace	10.6.8
Oil-saturated oven walls	14.3.7

Oily fumes present in the heat-treat shop 14.4.1
"Old Number One," the first controlled-atmosphere furnace sold
   to American industry – "Certain Curtain" furnace May 11, 1927 .... 9.1.3
Original bag-house smoke fume-abatement arrangement............................. 14.3.6
Oven temperature uniformity survey.. 11.4.8
Oxidation boundaries for various elements 9.3.2
Oxidation/reduction diagrams for 18% chromium stainless steel
   at 1095°C (2000°F) 9.6.6
Oxygen probe and its alloy protection tube destroyed by melting at the
   exact location of the carbon-rich subsurface layer...13.7.2
Oxygen (carbon) probes and recommended mounting arrangement............9.7.1
   a. Family of oxygen (carbon) probes
   b. Oxygen-probe mounting arrangement
Oxygen-sensor millivolt output vs. percent oxygen as a function of
   temperature for processes where the oxygen percentage is higher
   than might typically be found in an endothermic atmosphere 16.4.4
Paddle switch 14.5.9
Part/load thermocouples.. 13.6.9
Parts ready for loading into a pit nitriding furnace 9.7.10
Penetrant inspection revealing a circumferentially orientated heat-treat
   indication 12.6.1
Percentage carbon dioxide vs. carbon in austenite for endothermic
   (RX®) gas with 20% CO...16.4.12
Pipe scale..14.5.10
Pit furnace system for carburizing wind-energy gears......................................9.5.24
Pit nitriding furnaces..9.5.33
Pit quench-tank design: deflector vanes under workload 10.6.7
Pit quench-tank design: draft-tube assembly.. 10.6.6
Pit quench tank design: side view 10.6.5
Pit quench-tank design: top view 10.6.4
Plant-wide energy monitoring device 14.3.5
   a. Real-time power meter
   b. Power-meter status light
Portion of right side of tertiary diagram for hydrogen, oxygen and nitrogen
   at elevated temperatures.. 9.4.3
Power-transmission gears..10.2.12
Precision diamond saw... 16.3.1
Pressure-monitoring system for a two-chamber integral-quench furnace
   with oil quench 13.6.1
Product life cycle 15.4.2
Proper low-temperature bath appearance 13.4.10
Properly adjusted integral-quench furnace flame curtain................................ 13.3.7
Proposed hydrogen-embrittlement mechanism... 15.6.4
Pusher furnace 9.5.26
Quench-chute design for use with continuous furnaces 10.6.9
Quenchant circulation in a typical press quench... 10.5.7

Quenchant viscosity as a function of concentration and temperature
   for an aqueous solution of a quenchant 16.4.23
Radiant tube damaged by soot buildup in the refractory 13.7.3
Radiographic crack detection 12.6.3
Refractive index vs. % Brix... 13.4.3
Refractometers 13.4.1
Relationship between cooling rate, metallurgical transformation
   and internal stress 10.1.3
Relationship of the volume of oil and the weight of the work quenched
   in relation to the temperature rise 10.2.5
Relationship of the polymer volume and the weight of the work
   quenched in relation to the temperature rise 10.2.18
Relationships between H_2O/H_2 and CO_2/CO at different temperatures
   (1500-1750°F) in accordance with the reaction $CO + H_2 = CO_2 + H_2$ 16.4.13
Resulting gas composition upon cracking of methanol in an atmosphere
   containing 40% nitrogen and 60% cracked methanol................................... 9.5.19
Results of weight-change tests using an atmosphere of purified
   (CO_2-free, dried) and unpurified exothermic gas.. 9.4.9
Results of weight-change tests using an atmosphere of (dry)
   endothermic gas 9.4.10
Retort with catalyst.. 9.6.4
Ring-gear processing cycle 9.5.22
Role of a sintering furnace atmosphere by location inside the furnace 9.5.34
Rotary-hearth furnace for neutral hardening.. 9.5.28
SAE 4140 pipe moving into water-spray quench... 10.4.9
SAE 8620 (1250X, 2% nital) – pearlite and ferrite formed in the core of a
   carburized 8620 gear tooth, indicating improper quenching 12.1.6
SAE 8620H shaft processing cycle 9.5.23
Schematic of a dissociated-ammonia generator.. 9.6.3
Schematic of an exothermic-gas generator.. 9.6.9
Schematic of closed-loop air-cooled system 14.5.5
Schematic of closed-loop evaporative cooling-tower system 14.5.4
Schematic of closed-loop water-cooled system 14.5.6
Schematic representation of membrane technology 9.4.5
   a. Basics of membrane technology
   b. Cut-away view of a membrane unit
   c. Installed membrane unit
Schematic representation of PSA adsorption technology 9.4.6
   a. Sieve bed
   b. Working PSA unit
Section of a cast HH (25% Cr 12% Ni) stainless steel furnace load-lifting
   hook that failed due to sigma-phase embrittlement 15.6.9
Section of fastener thread showing decarburization, 50X 9.4.12
Segmented silicon-carbide ceramic muffle in a continuous
   mesh-belt furnace 15.7.6
Self-recuperative burner installed on a heat-treatment furnace...................... 14.3.4
Sight flow device 14.5.8

Sigma phase (dark areas) in cast stainless steel (1000X, KOH) 15.6.11
Silver brazing of HVAC components 9.5.10
Single-retort endothermic-gas generator 9.6.15
Sintering of valve-seat inserts 9.5.35
Small pen indenter applying load to clay cube with minimal
 bulging (side distortion) 12.2.12
Soot accumulation to a depth of 25 mm (1 inch) depositing on an external
 furnace load table in less than one hour due to a furnace
 atmosphere that is out of control 13.7.4
Sources of gear distortion 10.5.1
Stress corrosion cracking failure exhibiting branched cracks 15.6.6
Stress corrosion cracking of 4340 oil-rig fasteners in saltwater 15.6.7
Stress relief of 304 and 316 stainless steel plates – distortion improperly
 blamed on air quenching 10.4.4
Stress-rupture test 12.3.9
Surface and subsurface attack by metal dusting 13.7.7
Surface carbon potential vs. CO_2 (via infrared analysis) as a function
 of temperature for endothermic gas 16.4.7
Surface carbon potential vs. dew point as a function of temperature
 for endothermic gas 16.4.6
Surface carbon potential vs. oxygen-probe millivolt output
 as a function of temperature for endothermic gas 16.4.5
Surface decarburization as a function of hydrogen
 content of the atmosphere 9.4.8
Temperature measurement package in position to survey a batch furnace 13.6.11
TempTab® ceramic shrinkage devices 13.6.10
   a. Fixtured on a load
   b. Measurement tool
Tensile-testing apparatus 12.3.2
Tertiary diagram for hydrogen, oxygen and nitrogen
 (with flammability envelope for ambient conditions) 9.4.2
Thermal cracking of methanol 16.4.16
Thermographic analysis – furnace power leads showing potential hot spot .. 12.6.9
Theoretical equilibrium relations between iron and iron oxide
 when in contact with carbon monoxide and carbon dioxide
 or hydrogen and water vapor at heat-treating temperature 9.4.1
Theoretical purge-down curve 9.2.1
Thermal-conductivity cell for measuring hydrogen 9.7.12
Thin and thick sections on the same component part 11.5.3
Three stages of liquid quenching 10.2.2
Tin-plated electrolytic tough pitch (ETP) copper terminal lugs embrittled
 during oxyacetylene brazing and subsequently cracked during forming 15.6.1
Torsion testing apparatus 12.3.5
Total-cost-of-ownership model for heat treatment 15.3.1
Training options 11.1.1
TUS rack of carbon/carbon-fiber material 11.4.5
Two-chamber integral-quench furnace pressure chart 13.6.2

Typical atmosphere box furnace... 9.5.8
Typical atmosphere furnace installation... 15.3.4
Typical atmosphere heat-treating system for fasteners 15.8.4
Typical batch-style atmosphere furnaces 13.2.1
    a. Box furnace
    b. Box furnace with retort
    c. Pit furnace
    d. Bell urnace
    e. Integral-quench furnace
    f. Car-bottom furnace
Typical bell furnace.. 9.5.7
Typical box furnace with retort.. 9.5.15
Typical brazing cycle 9.5.12
Typical carbon restoration cycle with carbon (shaded area) added................9.4.15
Typical combustion-control scenario.. 13.5.5
Typical continuous mesh-belt furnace for case hardening of fasteners 15.8.1
Typical continuous-style atmosphere furnaces.. 13.2.2
    a. Mesh-belt furnace
    b. Roller-hearth furnace
    c. Pusher furnace
    d. Screw conveyor furnace
    e. Rotary-drum furnace
    f. Rotary-hearth furnace
Typical cooling curves and cooling-rate curves for new oils 10.2.1/13.4.6
Typical cooling curves and cooling-rate curves for new vs. used
    quench oil... 13.4.5
Typical cooling and cooling-rate curves for salt quenching............................ 10.4.6
Typical cutoff saw 12.4.3
Typical cryogenic nitrogen storage system.. 9.4.7
Typical depth-hardness curves for Nitralloy 135M nitrided at 525°C
    (975°F) with 30% dissociation rate by the single-stage process................... 9.5.29
Typical depth-hardness curves for Nitralloy 135M nitrided by the
    two-stage process 9.5.30
Typical dissociated-ammonia generator... 9.6.2
Typical eddy-current setup for materials property evaluation 12.6.4
Typical etchant station 12.4.6
Typical exothermic-gas generator... 9.6.8
Typical flow control panel (multiple-zone continuous furnace) 13.3.3
Typical grinder/polisher.. 12.4.5/16.3.4
Typical horizontal mesh-belt conveyor furnace for brazing............................ 9.5.13
Typical horizontal mesh-belt conveyor sintering furnace................................9.5.36
Typical inclined (humpback) conveyor furnace...9.5.14
Typical indentation hardness testing arrangement 12.2.2
Typical in-plant ammonia tank hookup..14.1.3
Typical load of bars quenched into a PVP 10.2.21
Typical mesh-belt conveyor furnace... 9.5.4

Typical mesh-belt sintering furnace 13.3.5
    a. Front end
    b. Back end
Typical metallurgical laboratory setup 12.4.1
    a. Sample preparation area
    b. Sample evaluation area
Typical microhardness indenters 12.2.8
    a. Dimensions for Knoop indenter
    b. Dimensions for Vickers indenter
    c. Knoop and Vickers indenters and test block
Typical mounting press 12.4.4/16.3.3
Typical nitrogen generator system 9.4.4
    a. Typical mounting arrangements
    b. Nitrogen generator
Typical oxygen (carbon) probe 13.5.1
Typical polymer quenchant report 13.4.4
Typical press-quench operation 10.5.6
Typical pusher furnace with entrance vestibule to prevent atmosphere
    disruption 9.5.6
Typical quench-oil report 13.4.7
Typical rebound (dynamic) hardness testing arrangement 12.2.3
    a. Instrument
    b. Probe close-up
Typical Rockwell test 12.2.6
Typical roller-hearth furnace 9.5.5
Typical rotary-retort system with oil, polymer or water-quench capability 15.8.6
Typical S-N curve 12.3.6
Typical gas-atmosphere sampling system 13.5.2
Typical schematic of one-pass water-cooling system 14.5.1
Typical schematic of recirculating open cooling-tower system 14.5.2
Typical schematic of recirculating open cooling-tower system
    with intermediate heat exchanger 14.5.3
Typical scratch (file) hardness testing arrangement 12.2.1
Typical shim-puller arrangement 13.6.13
Typical single-chamber batch vacuum furnace 9.5.38
Typical stress-strain curves for various materials 12.3.4
Typical survey chart 11.4.4
Typical survey rack in an annealing furnace 11.4.1
Typical thermocouples and thermocouple assemblies 13.3.8
Typical tip-up furnace 9.5.9
Typical ultrasonic flaw detector 12.6.6
Typical walking-beam furnace 9.5.37
Ultrasonic detection process 12.6.7
Value of cooling-curve analysis 10.3.4
    a. Effect of quench-oil oxidation on cooling rate
    b. Effect of water contamination on cooling rate

Variables in high-pressure gas-quenching performance	10.4.8
Variables influencing choice of atmosphere	9.5.1
Variable-valve timing assemblies	10.2.15
Various regions and points on the stress-strain curve	12.3.3
Various styles of Rockwell hardness indenters	12.2.4
Various styles of support anvils	12.2.5
Viscosity change in martempering oil	10.3.1
Well-spaced and supported oven load of aluminum coils	11.4.6
Wire product quenched into a PEO	10.2.23
Worker pouring molten metal in a small foundry	14.2.1

SUBJECT/TERMINOLOGY INDEX

Note: Page numbers followed by f or t indicate figures or tables, respectively.

SUBJECT/TERM   PAGE NUMBER

A

Abrasive wear 358t
Abrasive wheels 702
Accreditation
    audits for 241–54
    by CQI-9 .... 241, 247–53, 281
    by Nadcap 167, 241–47, 281
Acid number 165
Acid rain 535
Acoustic emission testing 370t, 377–78, 377f
ACR (sodium polyacrylate), as quench media 156, 159, 159f
Adhesive wear 358t
Aerospace industry 244, 246, 281–82, 283, 595, 608, 637, 648
Aerospace Material Specification (AMS)
    2750 .... 244, 257, 260, 264, 265, 401, 429, 488
    5596 598
    5597 598
    International 683
Aging (precipitation hardening) 48t, 130
Agitation
    in austempering 177
    in quench tanks 143, 175, 176f, 195, 206–7, 207f, 208f, 209, 212, 214
Agitators 386f, 401, 495
Air atmosphere 8t
Air burnouts 499–500
Air cooling, ... 173–74, 174f
Air leaks 128f
    differentiated from water leaks 23–24, 482–86, 484t–486t
    tests for 116, 480–82, 481f, 482f
Air pollution 534–38
Air quenching 128f, 173–75, 453
Alarm systems 251
Alloy muffles 631, 631f
Alloys. *See also specific alloys*
    heat-resistant 621–22
    salt quenching of 175
Alumina polishing 714
Aluminum
    annealing of 93
    etchants for 716t
    liquid-metal embrittlement of 616–17

polymer quenching of 212–14
  reference sources about 692–93
Aluminum alloys
  air processing of 22
  die-casting 574t
  etchants for 716t
  flow rate sensitivity of 214
  liquid-metal embrittlement of 616–17
Aluminum-bronze alloys, as fastener material 646
Aluminum casting, blistering of 299
Aluminum heat treatment 250t
  training course in 224–26
Aluminum industry, lean manufacturing strategies in 572–76
American Society for Testing and Materials (ASTM)
  A434 64
  A490 .... 355f
  D 91 166
  D 92 165
  D 1744 164
  D 3520 140
  D 6200 .... 139, 167
  D-6200-01 450
  E3 ....
  321
  E10 320
  E18 .... 318, 321
  E92 320
  E103 320
  E140 .... 314, 320, 321
  E384 .... 319, 320, 321
  International 297
Ammonia 17t
  anhydrous 82, 82t, 86–87
  effect on oxygen (carbon) probes 110
  generated atmosphere-produced 12t
  metallurgical 83
  in steel carburization 55–56
  toxicity/hazards of 514t, 516, 517t, 522
Ammonia atmosphere
  for nitriding 65
  for nitrocarburizing 65, 66, 66t
Annealing
  bell 47, 48f
  bright/clean 85, 93, 487f
  decarburization-free 46
  dissociated ammonia in 85
  exothermic gas in 93

of fasteners 636
  in neutral (inert) atmosphere 4–5
  process tables for 250t
  temperatures for 48t, 252
Annealing furnaces 48–50, 262f
Antioxidants 195–96
Anvils, for hardness testing 306, 306f, 318
Aqua-Quench® 152t–153t
Argon 17t, 35
  hazards/toxicity of 514t, 522
Argon atmosphere 8t, 35, 51
AS9100 ..... 246, 281
Asphyxiants 517, 518t, 521–22
Atmosphere
  for annealing 45–50, 46t
  applications of 45–74
  blended 8t, 10, 10t–11t
  bright/clean 4
  for carburizing 39–42, 41f, 55–58
  for case hardening and carbon restoration 27, 39–42
  catalytic reactions in 17
  chemical reactions in 16–21
  choice of 45f
  constituents of 15
  CQI-9 specifications for 248
  for decarburization 39–42, 41f
  for decarburization prevention 27, 36–39, 37f
  flow rate in 16
  for oxidizing prevention and/or reducing oxides 27–36
  passive (chemically inert/neutral) 1–2, 4–5
  purpose of 1–5, 15
  reactive (chemically active) 1, 2–3, 5
  for special applications 27, 42
  toxicity/hazards of 512–14
  types of 7, 8t
  volume requirements for 7, 9, 9t
Atmosphere controls, monitoring of 251–52
Atmosphere furnaces 4
  batch 54, 391f–392f, 487f, 500–501, 501f, 572f
  batch, integral-quench 215, 215f
  belt 16
  box 50f, 53, 54f, 64, 391f, 498
  box-carburizing 511f
  car-bottom 392f
  components, maintenance of 391–407
  continuous 16, 392f–393f, 501–4
  continuous sampling 247

    direct-fired box 529, 529f
    energy efficiency of 527–33
    for fastener heat treatment 649–51, 650f
    history of 3, 3f
    impurities in 17
    in-situ monitoring of 247
    integral-quench 64, 184f, 212, 265f, 392f, 480f, 498, 500–501, 501f, 540f, 545
    maintenance of 391–407, 409–36
    mesh-belt conveyor 49, 49f, 53, 71, 392f, 394f, 422f, 482, 483f,
                                                                              502–3, 636f, 649–50, 650f
    mesh-belt conveyor, horizontal 49, 49f, 52, 53, 54f, 71, 72f
    mesh-belt conveyor, inclined (humpback) 53, 54t
    for neutral hardening 64, 64f
    for nitriding 68, 68f
    pit 62f, 64, 68, 68f, 392f
    pressure quenching in 179–80, 180t
    pusher 49, 50f, 53, 62, 63f, 64, 392f, 503–4, 504f, 528, 528f, 545
    roller-hearth 49, 53, 64, 64f, 392f
    rotary-drum 64, 393f
    rotary-hearth 64, 393f, 504, 527–28
    rotary-retort 247, 650–51, 650f
    safety procedures with 407
    screw conveyor 392f
    shaker-hearth 64
    for sintering 71, 72f
    surveys of 265–68, 265f, 266f, 268f
    for tempering 651
    typical installation of 587f
Atmosphere generation. *See also* Generated atmospheres
    monitoring of 251
Atmosphere requirements 385
Audits
    for accreditation 167, 241–54, 281
    for energy management 530–33
    for prime approval 281–82
    role of 566
Austempered ductile iron (ADI) 177
Austempered stamping, salt quenching of 175f
Austempering 177–78, 179
Austenite, carbon solubility in.. 493, 493t
Austenite-to-martensite transformation 145
Austenitizing ....55f, 251. *See also* Hardening
Automotive industry 69–70, 70f, 637–38, 638t–642t

B

B1 Lancer slat-actuator ball screws 598, 599f
Bainite, austempering-produced 177

Bainitic steel, austempering of ... 177
Ball screws, reverse-engineering analysis of ... 598, 598f, 600, 600f
Baskets ... 388, 628–29, 628f
Batch intensive water quenching process ... 184, 184f
Beachmarks ... 357, 357f
Bearing races ... 2–3, 2f, 147, 147f, 197, 301f
Belt drives ... 422–25
Belts ... 399
Bend test ... 296f
Beryllium copper, annealing of ... 85
Beryllium nickel, annealing of ... 85
Best practice, in the heat-treatment industry ... 283–85
Biocides ... 169
Blanketing ... 15, 93
Blanking dies, design of ... 287, 287f
Blended atmosphere ... 8t, 10, 10t–11t
Blistering ... 299f
Bluing ... 93
Boeing Company ... 637
Boiling point, effect on heat-transfer rates ... 141–42
Bonding ... 51
Brake rotors, fixtures for ... 629f
Brass, as fastener material ... 646
Brass alloys, annealing of ... 85
Brass HVAC components, silver brazing of ... 52
Brass powder steel, sintering of ... 86
Brazing ... 51–54
    applications of ... 51–52
    copper ... 93
    definition of ... 51
    dissociated ammonia in ... 85–86
    furnace atmospheres for ... 51, 51–52t
    furnaces for ... 53, 54f
    nickel ... 23, 52, 52f
    oxyacetylene ... 604f
    silver ... 52, 52f, 93
    temperatures for ... 52, 53t
Brazing joints, wetting and spreading of ... 85–86
"Breathing" ... 116
Brine, as quench media ... 128f
Brine/caustic quenching ... 185–87, 185f, 186f, 186t, 403
Brinell hardness testing ... 305, 307–8
    common problems in ... 316–17
    standards for ... 320
Brittle cleavage ... 610
Brittle fractures ... 336, 356–57, 357t
Brix measurements ... 169, 445, 445f

Bronze alloys, annealing of 85
Bronze powder steel, sintering of 86
Bubble testing 377
Burners 394f, 396, 397f
   energy efficiency of 527–28, 530f, 532f
   NOx emissions from 537–38, 538f
   self-recuperative, ... 528, 529, 530f, 532f, 536, 536f
Burnouts 110–11, 250, 499
   flash (improper) 498
   methods 499–500
   in specific types of furnaces 499–500, 501, 504
Butane 19, 90

C

C5 Galaxy screws, reverse-engineering analysis of 600, 600f
Calcium carbonate 556, 557f
Calibration .... 249, 401. *See also under specific types of equipment*
   accreditation audits of 245
   for atmosphere furnace surveys 266–67
   guidelines for 249–51
Carbomartempered parts 178–79
Carbon
   diffusion rates of 604t
   as furnace atmosphere component 15
   interaction with oxygen 20
   nascent (soot) 19
   as oxygen (carbon) probe contaminant 110
   penetration into refractory wall 498–500, 498f
   solubility in austenite 493, 493t
Carbon activity 63
Carbon controllers 111
Carbon dioxide 17t, 35, 36
   decarburizing effect of 18
   effect of radiant-tube leaks on 478–79
   oxidizing effect of 17
   toxicity/hazards of 514t
Carbon dioxide atmosphere 8t, 15, 35
Carbon dioxide detectors 512, 513f
Carbonitriding 55, 250t, 493
Carbon molecular sieve (CMS) 17t, 32
Carbon monoxide
   hazards/toxicity of 514–16, 514t, 515t, 522
   interaction with hydrogen 19
   reactions in furnace atmosphere 18
Carbon monoxide atmosphere 8t, 15
Carbon monoxide monitors 94
Carbon potential 23, 109, 461

equation	63–64, 493
measurement/monitoring of	114, 116, 248, 488, 489, 489f, 500
of neutral hardening atmosphere	63
in sooting conditions	502
tests for	488–94
Carbon-potential analyzers,	488, 489, 489f

Carbon probes. *See* Oxygen probes

Carbon restoration	39, 41–42, 42f
Carbon-reversal reaction	18, 502

Carbon rot. *See* Metal dusting

Carbon steel

austempering of	177
liquid-metal embrittlement of	616–17
martempering of	178
sintering of	86
Carbon-steel tubes, annealing of	47
Carburizing	17t, 55–62
applications of	59–61
baskets for	628f
burnouts in	110–11

catastrophic. *See* Metal dusting

definition of	1, 55
furnace atmosphere for	58–59
mechanism of	39
oxygen (carbon) probes in	109, 110–11, 110t
process control in	118
process tables for	250t
rate-determining reaction in	19
single-stage	55
two-stage	55
Case-depth uniformity tests	495
Case hardening	127
atmospheres for	39–42
baskets for	628
of bearing races	2–3, 2f
equipment for	61, 62f
of fasteners	636
furnaces for	64
of steel	55f
Case-hardening operations	277
Cast-alloy baskets	628, 628f
Cast-alloy fixtures	629
Cast-alloy retorts	629, 630f
Cast aluminum alloys, water quenching of	181, 181f
Cast grids	627
Cast stainless steel, sigma-phase embrittlement of	615
Catalytic reactions	17

Ceramic-fiber insulation 404
Ceramic muffles 631, 631f
Ceramic radiant tubes 404
Ceramics, reference sources about 693–94
Ceramic shrinkage devices 486, 487f
Certification, of heat treaters 281–82
Charcoal, generated atmosphere-produced 11t
Charpy impact test 296f, 325t, 336, 336f
Chart-recorder data...283
Chemical analysis tests..325t, 337
Chemical exposure..524
Chromium... 23, 413
Cleaning, of parts..387
   of oily parts 548–49
Cleveland Open Cup procedure 165
Closed-loop process...4
Coil stock, annealing of 48f
Color ("tint") etchants 719–20
Combustion analyzers.. 472–73, 474f
Combustion products, as furnace atmosphere...8t
Combustion reactions 18
Combustion systems, maintenance of 396, 397f
Commercial heat-treatment shops, specifications for... 279–82
Composites, reference books about 693–94
Computational fluid dynamics (CFD) 441
Conductivity meters 446
Conductivity testing 598
Confirmation testing 601
Constant carbon potential method 55
Construction industry, fastener use in 636, 642–43, 643t
Contact injuries 524–25
Contact stress fatigue 358t
Continuing education, in heat treatment 221
Continuous improvement (CI) .... 294, 566-567. *See also* CQI-9
Continuous improvement (CI) programs, audits of 237
Continuous strip charts 248, 250, 252
Continuous systems, ...572f
Control devices
   maintenance of... 401
   manual vs. automated 107–8, 108t
Controllable parameters, CQI-9 assessment of..247–54
Convective heat transfer 141, 141f, 142f, 174
Cooling
   effect of part geometry on 143f
   stages of 141–42, 142f
Cooling-curve analysis 139, 139f, 167, 168f
Cooling-curves, in oil quenching.. 448–50, 449f, 450–51t

Cooling rate 130
    effect of water temperature on 181, 181t
    in hot-oil quenching systems 195, 197f
    relationship to quenching method 186t
Cooling-rate curves, in oil quenching, ... 448–50, 449f, 450t–451t
Cooling-tower systems 551, 552–53, 552f, 553f, 556, 556t, 558
Copper
    annealing of 85, 93
    embrittlement of 606, 616–17
    melting temperature of 24, 483
    reduction temperature of 22, 23f
Copper alloys
    annealing of 84
    as fastener material 645–46
    liquid-metal embrittlement of 616–17
Copper HVAC components, silver brazing of 51–52, 52f
Copper steel, sintering of 86
Copper-steel-stainless steel test, ... 23–24, 483–84, 484t–486t
Copper-zinc alloys, as fastener material 646
Corrosion 358–59
    brine quenching-related 186–87
    definition of 358, 620
    failure analysis of 358–59
    fretting 358t
    of furnace fans 413
    galvanic-induced 645
    hot gaseous 620–21
    mechanisms of 620–22
    polymer quenching-related 151
Corrosion Society 608
Corrosion testing 337, 344
Corrosive attack 337
Corrosive materials 521
Cost-benefit analysis 345
Cost containment 381f
Cost equation 597
CQI-9 accreditation audits 167, 245, 247–54, 281
Cracking
    effect of part design on 285–87, 285–87f
    environmentally-induced 619
    of fasteners 643–44
    hydrogen-induced 603–8, 603f, 606t, 607t, 643–44
    liquid-metal 616–19, 617t, 618f, 619f
    nondestructive testing methods for 369–80
    oil quenching-related 212
    polymer quenching-related 151
    stress-corrosion 609–16, 610f, 611f, 612f, 643–44

sulfide stress 643–44
water quenching-related 182, 182f
Creep tests 335
Critical-area analysis 700
Critical-cooling rate, of steel 193
Cryogenic materials 521
Cryogenic nitrogen storage system 30, 34, 34f
Cusps 359
Custom blended atmosphere 8t
Cutting tools, design of 286–87, 287f
Cylinder liners, brine quenching of 186f
Cylinders, maintenance of 397–98

D

Data acquisition systems 4, 247
Data interpretation 117
Data loggers 248, 250, 252
Decarburization 17t
atmospheres for prevention of 36–39, 37f
carbon dioxide-based 18, 21
definition of 1
hydrogen-based 19
mechanism of 39–40
oxygen-based 20
water vapor-based 20–21
Deep-carburized components 605, 605f
Deflectors, reverse-engineering analysis of 599–600, 600f
Delubrication 68
Design factors, in product failure 364, 366t
Design/material/manufacturing analysi 283–85
Design-specific features 385
Dew point
in burnouts 499
calibration of 245, 247
as carbon potential indicator 115
definition of 115
in dissociated ammonia generators 83, 84t
effect on gas-sampling values 464
in endothermic-gas generators 112–13, 113f, 505
in stainless steel brazing 23
Dew-point analyzers and meters 21, 115–17, 115f, 116f, 116t, 250, 467–69, 500
Dezincification 51
Dies
cracking of 285, 285f
spalling and flaking of 288, 288f
Dissociated ammonia
applications of 84–86

  as brazing atmosphere 51
  gas reactions of 77–78, 79f
  in nitriding 66, 120t–121t
  as nitrogen source 86
  properties of 77
Dissociated ammonia atmosphere 36, 69
Dissociated-ammonia generators 77–87
  impurities in 82–84
  maintenance of 81–82
  retort failure on 300f
  safety of 86–87
Distortion
  air cooling-related 174, 174f
  air quenching-related 174, 175f
  factors influencing 191–92, 191f
  management of 191–202, 193f, 198f, 212
  polymer quenching-related 151
  types of 194–95
Distortion profile, in oil quenching 212
Documentation 282–83
Doors, on furnaces 394f, 397–98
Downtime 583
  effects of 582f
  unplanned 497
Draft tubes 207, 208f, 214–15
Drawings, in reverse engineering 597
Drive-shaft assembly, oil quenching of 149, 150f
Drive test 325t
Ductile fractures 356–57, 357t
Ductility 297
  definition 329
  loss of. *See* Embrittlement
  measures of 327
Duplex steel, embrittlement of 620
DX® gas. *See* Exothermic gas
Dye penetrant testing 370t, 371–72, 372f

E

Eddy-current testing 370t, 373–75, 374f
Eductors 214, 402
Elastic deformation 326, 330, 358
Elasticity 296
Elastic modulus 326, 329
Electrically-heated furnaces 498, 530, 632
Electric heating elements 395f
Electric heat treatment, reference sources about 687–88
Electrolyte etching 719

Electrolyte grinding and polishing	711–14
Electronic flow meter	415f
Electroplating	607
Ellingham-Richardson diagram	21–23, 22f, 28
Elongation	327
Embrittlement	603–20
475° C (885° F)	619
definition of	603
high-temperature	619–20
hydrogen embrittlement	299f, 337, 603–9, 604f, 605f, 606f, 606t
liquid-metal embrittlement (LME)	613, 616–19, 616t, 617f, 618f
sigma-phase embrittlement	614–16, 614f, 620
Endoinjection fuel-injecting gas mixing system	98f
Endo. *See* Endothermic gas	
Endothermic equivalent gas atmosphere	58
Endothermic gas	
characteristics and applications of	36, 56, 94, 95f
composition of	12t, 56, 56t, 94–95, 95t
ignition of	511f
Endothermic-gas atmosphere	8t, 11t, 12t, 112
burnouts in	499
case hardening in	39
effect on decarburization	37, 38f
sooting in	502–3
Endothermic-gas generators	56, 57f, 95–99
components of	95–99, 96f, 97f, 98f, 99f
dew-point control systems of	112–13, 113f
dew points in	505
gas chemistry in	94, 95f
infrared analyzers on	114, 115t
maintenance of	99–102
oxygen sensors on	112–13, 113f
problems with	57, 102–3
safety of	98–99, 99f, 103
soot removal from	405
Energy equivalents	531, 533t
Energy management programs	527–33
Energy monitoring devices	532f
Engineering	
life cycle of	591–93
metallurgical	590–93
reverse	595–601
Environmental factors, in product failure	365, 367t
Environmental issues, in heat treatment	533–38
Equilibrium reactions	16–21
Equipment. *See also* Furnaces	
deterioration of	585

hidden costs of 583–86
process compatibility of 583
stand-alone vs. fully integrated 569
suppliers' service and support for 585–86, 587
warranties for 586
Erosion 358t
Etchant station 352f
Etching, metallurgical 715–20, 716t, 717t
Exhaust stacks, fires in 546
Exothermic gas 11t, 12t
    characteristics and applications 36, 88, 93
    lean 89, 90t
    rich 89, 90t
Exothermic-gas atmosphere 36, 37, 38f, 39f, 87, 89, 90t
Exothermic-gas generators 88–93
    atmospheres for 89, 90t
    cogeneration technology system of 93, 93f
    components of 90–92, 90f
    gas chemistry in 87, 88f, 89f
    maintenance of 92
    mixed-gas technology system of 92, 92f
    sizes and types of 89–90
Exotic alloys, as fastener material 648, 649t
Explosions/explosive gases 212, 521, 522
Extension wires 431
Extractive metallurgy 590

F

Failure
    factors contributing to 364–67, 365t–367t
    mechanisms of 355
Failure analysis 346–48, 355–68
    failure classification in 364–67
    methods and tools in 359
    as preventive-maintenance plan component 386
Failure triangle 364f
Fan-assisted cooling 174, 174f
Fans 394f, 409–14
    excessive vibration from 495, 496f
    maintenance of 398, 409–14
    tests of 412–13
    types of 409–11, 410f
Fastener industry
    heat-treatment equipment in 649–52
    heat-treatment requirements for 635–52
Fasteners
    applications of 635–48

coatings for 635
decarburization of.. 40f
hardness test of... 325t
market for... 636–37
microhardness test of... 325t
specialty...648
stress-corrosion cracking of..611f, 612–13, 612f, 644, 647
tensile strength test of.. 325t
Fatigue, definition of 332
Fatigue analysis, equipment and techniques in...362–63
Fatigue failure.. 332, 333–34
Fatigue fractures... 357, 357f, 362
Fatigue testing 296f, 332–34, 333f
Ferrite 300f
Ferrous alloys, metal dusting of 506–9
Ferrous-induction treating 250t
Ferrous metals, reference sources about 684–87
Filters..465
    bag-house 534, 534f
    for hydrogen sulfide.. 19
Filtration
    in cooling-tower water systems 553, 555
    in oil quench systems 167, 195
    in polymer quench systems..169
    in salt quench systems 402
Finishing, reference sources about 694
Firebrick 404
Firecheck 98–99, 99f
Fires, oil quenching-related, ...165, 212, 524, 545–49
Fixtures..147f, 629, 629f
    inspection of 388
Flame curtains..394f, 398, 425–29, 426f, 429t
Flammable materials..30, 31, 521. *See also* Fires
Flash point, of quench oils 164–65
Floating pins, reverse-engineering analysis of.. 599, 599f
Flow control panel...414f
Flow indicators...557–58, 557f
Flowmeters 395f, 414–21, 464
    for burnouts..499
    definition of..414–15
    maintenance of..399, 417
    sooting of 405
    types of 415–16
Flue-gas recirculation 537
Forged bolts 324f
Fourier transform infrared (FTIR) spectroscopy 167
Fractography..359

Fracture point 331
Fractures
    brittle 336, 356–57, 357t
    cleavage 360–61, 361f
    decohesive rupture 361, 361f
    dimpled rupture 360, 360f
    ductile 356–57, 357t
    fatigue 357, 357f, 362
    macroscopic and microscopic 355–57, 357t
Fume abatement programs 534, 534f
Furnace alloy 395f
Furnace-atmosphere controls 395f
Furnace-atmosphere reactions 16–21
Furnace pressure measurement 479–80, 480f
Furnaces. *See also* Atmosphere furnaces; Electrically-heated furnaces; Gas-fired furnaces
    air purging in 9, 9f
    batch vacuum 72, 73f
    for brazing 53, 54f
    continuous 215, 216f
    energy efficiency of 530–33
    inspection of 387
    life span of 527
    for neutral hardening 64f
    for tempering 651
    tip-up 49, 51f
    walking-beam 72f

G

Gas-atmosphere sampling system 464f
Gas coolers 91
Gas cylinders, safe storage and use of 525, 526f
Gas equilibrium reactions, in furnace atmosphere 16–21
Gases. *See also specific gases*
    specific gravity of 31t
    thermal conductivity of 31t
    thermal content of 31t
    toxicity/hazards of 512–18, 521–23
Gas-fired furnaces 3
    batch-style drop-bottom 572f
    carbon penetration in 498
    energy efficiency of 530
    retorts, muffles, and radiant tubes heating in 632
    soot in 501, 501f
Gas flowmeters 416–17
Gas generators. *See also* Dissociated-ammonia generators; Endothermic-gas
    generators; Exothermic-gas generators
    carbon-reversal reaction in 18

CQI-9 specifications for 24
maintenance of 399–400
process control in 118
transmission lines from 406
Gas heat treatment, reference sources about 687–88
Gaskets 396f, 405
Gas measurements 419
Gas poisoning 521–23
Gas quenching 453–54
Gas ratio 16
Gas sampling systems 463–65
Gas-solid reactions 622
Gas trains 394f, 396
Gauges 400
Gears
    carburizing of 59, 59f, 62f
    distortion of 191f, 192, 192f
    press quenching of 197
Gear teeth 287, 287f, 300f
Generated atmosphere .... 8t, 36, 77-104. *See also* Gas generators
    classification of 11–13
Glass, reference sources about, ... 693–94
Glycols, as cooling medium 551, 554
GM Quench-O-Meter 140t, 167, 168f
Gonser's curves 37, 39f
Gouging 358t
Gray cast iron 177, 178
Grids (trays) 388, 627–28, 627f
Grinders 351f
Grinding 358t, 711–14

H

Hardenability, relationship to distortion 192
Hardening
    baskets for 628
    definition of 62–63, 277
    direct 62, 127
    in dissociated ammonia atmosphere 86
    of fasteners 636
    neutral 1–2, 2f, 62, 63–64, 65f, 564f, 628
    quench 63, 64f
    surface 63, 64f
    temperature uniformity surveys of 251
Hardening operations 277
Hardness
    definition of 303
    probability plot of 292f, 293f

process capability of 292f, 293f
  relationship to tensile strength 330
Hardness testing 253, 303–19, 495, 597–98
  basics of 305–7
  Brinell tests 305, 307–8, 316–17, 320
  common problems in 316–20
  equipment calibration for 305–6
  files for 251
  indentation hardness tests 304, 304f
  macro and micro 296t
  microhardness tests 305, 312–14, 312f, 313f, 315f, 319–20, 325t
  practical considerations in 315, 316f
  rebound hardness tests 304, 305f
  Rockwell and Rockwell superficial tests ... 282, 309–12, 309t, 310f, 311t, 317–19, 325t
  scratch hardness tests 304, 304f
  standards for 320–21
  types of tests 307–12
  value of 295
Hardness-testing equipment, calibration and verification of 251, 252
Hastelloy® X 632
Health issues, in the heat-treatment industry .... 511-518. *See also* Safety
Heat-exchanger efficiency 166–67
Heat exchangers
  in dissociated ammonia generators 82
  maintenance of 400
  in water-cooling systems 553, 555
  water-to-oil 395f
Heating elements
  maintenance of 400–401
  soot buildup on 498, 500–501
Heating methods, reference sources about 687–88
Heat-resistant materials, applications of 626–32
Heat treatment, categories of 127
Heat-treatment departments
  checklist for 386–88
  world-class 232f
Heat-treatment factors, in product failure 365, 366t
Heat-treatment processes, classification of 277
Heat-treatment systems, fully-integrated 569f
Heat-treat shops, typical 563f
Helium 17t
  hazards/toxicity of 514t, 522
Helium atmosphere 8t, 35
Hidden costs, in heat treatment 583–86
High-chromium ferritic alloys, embrittlement of 619–20
High-pressure die-casting (HPDC) 573–76, 573f, 573t–576t, 574
High-pressure gas quenching. *See* Pressure quenching

High-strength steel, hydrogen bake-out cycle of 607, 607t
Honeycomb grids 627, 627f
Hooke's law 329
HVAC components, silver brazing of..52, 52f
Hydrocarbon gas atmosphere 8t
Hydrocarbon reactions, in furnace atmosphere 19
Hydrogen
    auto-ignition temperature of 30, 32
    as brazing atmosphere 51
    characteristics of..30
    diffusion of.. 604, 604t
    flammability of 30, 30f
    interaction with carbon monoxide 18
    in nitriding..65, 119
    production of...30
    reactions in furnace atmosphere 19
    as a reducing gas 30
    thermal conductivity of.. 30, 30t
    toxicity/hazard of..514t, 522
Hydrogen analyzers...470, 472, 472f
Hydrogen atmosphere 8t, 15, 30–31, 31t
Hydrogen bake-out cycle 607, 607t
Hydrogen detectors..512
Hydrogen embrittlement................................ 299f, 337, 603–9, 604f, 605f, 606f, 606t
Hydrogen sensors 31, 33
Hydrogen sulfide, reactions in furnace atmosphere... 19

I

Impact tests 336, 336f
Impellers, in quench tanks 206–7, 207f, 208f
Inconel 600 ....168, 300f, 630, 630t, 631
Inconel 601 .... 600, 630t, 631
Inconel 686 646
Inconel 718..598
Indenters, for hardness testing 306, 306f, 307, 308, 309, 312, 313, 313f, 316f, 317, 318
Induction-coupled plasma (ICP) spectroscopy 167
Induction heating, reference sources about 689
Industrial Heating... 220, 683
Information resources, in heat treatment...683
    See also Reference library, for heat treaters
Infrared (NDIR) analyzers,...113–17, 114f
    calibration of 250, 457t, 465–66
    for endothermic atmosphere monitoring.................... 111–12, 111f, 114, 114f, 115f
    for ferritic nitriding process control.. 122–23
    maintenance of... 465–67
    three-gas... 114f, 115f, 247, 250, 251, 465, 500
Infrared pyrometers..251

Infrared spectroscopy, of total acid number (TAN) 165, 166f
Infrared thermal imaging 370t, 378–79, 379f
Inspection 325t
    criteria for 278, 279t
    of furnaces 387
    of parts 370–71, 370t
    reference sources about 695
Instruments, maintenance of 401
Insulation 395f, 404, 413, 414, 500, 523
Intensive water quenching (IWQ) technology 183–84, 184f
Internal furnace pressure measurement 421–22
Inventory, work-in-process (WIP) 584
IR compensation 118
Iron
    etchants for 716t
    as furnace atmosphere component 15
    oil quenching of 206–12
    oxidation of 29, 29f
    oxidation temperature of 29, 29f
    reference sources about 684–86
Iron carbide 15
Iron-chromium alloys, embrittlement of 619
Iron-chromium phase diagram 615f
Iron-copper steel, sintering of 86
Iron-nickel-chromium alloys, heat-resistant 505–6, 621–22
Iron-nickel steel, sintering of 86
Iron nitriding, process tables for 250t
Iron oxide 15
Iron steel, sintering of 86
Ishikawa (fishbone) diagrams 132f, 191f, 345
ISO 297
ISO 6506-1 320
ISO 6507 320
ISO 9001 281
ISO 9950 .... 139, 167, 450
Izod specimens 336

J
Job compliance 243
Joining, reference sources about 689–90

K
Karl Fisher analysis 164
Keyways, in shafts 286, 286f, 288, 288f
Kinematic-viscosity instrument 444, 445f
Knoop indenters 313, 313f

L

Landing-gear components, oil quenching of 150, 150f
Leaks. *See also* Air leaks; Water leaks
    magnetic-flux.. 372, 372f
Leak testing... 370t, 376–77, 462
Lean manufacturing strategies 569–76, 570f
    in the aluminum industry 572–76
Leidenfrost phenomenon 140
Liquid metals
    as embrittlement cause 613, 616–19, 616t, 617t, 618f, 619f
    melting temperature of... 616t
Lithium, liquid-metal embrittlement of... 616–17
Load, high surface-area 547
Load hang-up... 546
Load ratings... 385
Lockout/tagout procedures 407
Low-alloy steel... 46t, 86, 178
Low-carbon steel.. 93, 178–79
Low-iron brick 499
Low-pressure carburizing, process tables for....................................... 250t

M

Magnetic-particle inspection.. 370t, 372–73, 372f
Maintenance
    of atmosphere furnace components.. 391–407
    of atmosphere furnace systems 409–36
    cost-containment effects of.. 381f
    of flowmeters 414–21
    of furnace fans... 409–14
    heat-treat checklist for 386–88
    PEER system of 382
    planned preventive.. 381–89, 406–7, 566, 580–82
    processing issues in 497–509
    of quenching systems 439–58
    recordkeeping in 384
    safety in 407, 523
    scheduling of 497
    of sensors and controls 459–74
    shutdown periods for 497
    steps in... 383
    system accuracy tests after.. 259, 260
    unplanned... 581f
Maintenance logs 585
Managers, of heat-treatment operations 564–65, 567
Manometers 400, 479, 480f
Marine industry, fastener use in 645–46
Marquenching

definition of	193
hot-oil quenching modification	193–98, 193f, 198f
Martempering	178–79
combined with austempering	179
definition of	193
hot-oil quenching modification of	193–98, 193f, 198f
Martensitic stainless steel, hardening of	86
Mass-flow controllers	415f, 416
Mass spectrometry	377
Material certification	278
Material certification sheets	279, 280
Material data sheet (MDS)	279
Material factors, in product failure	364, 365t
Material selection, process variables in	284t
Material specification systems	279
Materials science	589f, 590
Materials testing	297
Measurement devices, in heat treatment	108–17
Mechanical testing	323–39, 325t, 344, 597
chemical analysis	337
creep and stress rupture tests	335
fatigue tests	332–34, 333f
impact tests	336, 336f
metallurgical (structure) analysis	337
residual-stress analysis	338
shear strength tests	334, 334f
statistical methods in	338
stress durability tests	337
torque/torque-tension tests	331–32, 332f
types of tests	323–25
typical tests in	296t
value of	295–97
vibration tests	337
Medical industry, fastener use in	644–45
Membrane technology	32, 33f
Mercury, aluminum-embrittling effect of	616
Mesh belts	394f, 422-425. *See also* Atmosphere furnaces, mesh-belt conveyor
Metal dusting	300f, 413f, 505–9, 508f
Metal-injection-molded (MIM) parts	69
Metallic oxides, reduction potential of	21–23, 22f, 23f
Metallographic testing	297–301
Metallography, comparative	298
Metallurgical (structure) analysis. *See* Metallurgical evaluation	
Metallurgical engineering	590–93
Metallurgical evaluation	337, 341–53
comparative analysis in	345
cost-benefit analysis in	345

failure analysis in	346–48
of field failures	346–48
laboratory procedures in	343–44
photography use in	342
processing history/background information component	342–43
raw material analysis component	343
root-cause analysis in	345
sample preparation for	348–51, 350f, 351f, 352f
Metallurgical laboratories	341f, 343–45
Metallurgists, roles and responsibilities of	278, 589–93
Metallurgy	589–93
reference sources about	690–92, 691–92
Metal powder, annealing of	85
Metal Treating Institute (MTI)	285
Metal tubes and pipes, annealing of	46–47
Methane	522
Methanol	20, 514t
Methanol atmosphere	39
Microbiological contamination, of polymer quench media	169
Microhardness testing	305, 312–14, 312f, 313f, 315f, 325t
Knoop	313, 313f, 314f, 321
problems in	319–20
standards for	321
Vickers	313, 313f, 314f, 321
Microstructure	297
analysis of	325t
eddy-current testing of	374, 374f
Mill pinion	151f
MIL-S-5000	598
MIL-S-8844	598
Mineral processing	590
Mining industry, fastener use in	643–44, 644f
Mixers, in quench tanks	206–7, 207f
Molecular sieve dryers	83
Molten salt, as quench medium	128f, 177t. *See also* Salt quenching
Molybdenum, embrittlement-promoting effect of	614
Motors, maintenance of	401
Mounting, of samples	703–7, 705f
Mounting presses	351f, 705f
Muffles	631, 631f, 632f

N

NACE (National Association of Corrosion Engineers) International	608
Nadcap accreditation audits	167, 241–46, 281
National Aerospace and Defense Contractors Accreditation Program. *See* Nadcap	
National Association of Corrosion Engineers. *See* NACE	
National Institute of Standards and Technology (NIST)	297

Natural gas 501, 504, 514t, 528
Neutral hardening 1–2, 2f, 62, 63–64, 65f, 564f, 628
Neutralization number, of quench oils 165
Neutral reactions 17t
NFPA 520
NFPA 86 .... 87, 136
Nichrome, melting point of 139
Nickel, annealing of 85
Nickel alloys 19, 85
Nickel pickup 84
Nickel steel, sintering of 86
NIST. *See* National Institute of Standards and Technology
Nitralloy 135M 66–68, 66t, 67f
Nitriding 64–68
   atmospheres for 65
   process control in 119–22, 119f, 120f
   single-stage 66
   two-stage 66, 67f
Nitrocarburizing 64–65
   atmospheres for 66, 66t
   process control in 119, 121–22, 122f
Nitrogen
   in ammonia dissociation 17
   atomic 20
   characteristics of 31–32
   combined with generated gas 10, 10t
   cryogenic storage systems for 30, 34, 34f
   hazards/toxicity of 514t, 522
   industrial-grade 34
   as nitriding dilution gas 65
   properties of 31–32
   reactions in furnace atmosphere 20
Nitrogen atmosphere 8t, 11t, 36, 51
Nitrogen dioxide 535
Nitrogen/dissociated ammonia atmosphere 86
Nitrogen generators 32, 32f–34f
Nitrogen/hydrocarbon atmosphere 10, 11t
Nitrogen/hydrogen atmosphere 10, 10t, 11t, 68
Nitrogen/methanol atmosphere 10, 11t, 39, 57–58, 58f, 112
Nitrogen monoxide 535
Nitrogen pickup 84
Nitrogen purge 605
Nitrogen purge cylinders 526f
Nitrogen purge systems 546
Nondestructive testing (NDT) methods 325t, 343–44, 369–80
   acoustic emission 370t, 377–78, 377f
   comparison of 370t

dye penetrant	370t, 371–72, 372t
eddy-current testing	370t, 373–75, 374f
infrared thermal imaging	370t, 378–79, 379f
leak testing	370t, 376–77
magnetic-particle inspection	370t, 372–73, 372f
Nadcap audits of	243
radiographic testing	370t, 373, 373f
ultrasonic testing	370t, 375–76, 375f, 376f
visual inspection	370–71, 370t
Nonferrous alloys	93
Nonferrous materials, reference sources about	692–93
Non-metals, reference sources about	693–94
Normalizing	93, 192
Notched-bar impact tests	336
NOx emissions	535–38
Nuclear industry, fastener use in	647

O

Occupational Safety and Health Administration (OSHA)	520
OEM (original equipment manufacturer)	384, 585, 587
Oil, as quench media	128f, 135, 163–68
classification of	139, 140t, 451t
cleanliness	209
cooling rates of	138–40, 139f, 208–9
degradation of	164–65, 166, 166f
as distortion and cracking cause	212
as fire/explosion cause	545–49
flash point of	164–65
fractionation of	505, 505f
neutralization number of	165
oxidation of	165, 166f, 168f
precipitation number of	166
properties of	143–44, 163–67, 164f, 166f
routine analysis of	440, 452, 452t, 453t
as sludge source	166–67
types of	147, 147f, 148f
volume required	144–45, 145f
water content of	164, 168f, 212, 545
Oil capacity	505
Oil drag-out	546
Oil-quench furnaces	136
Oil quenching	136–51, 163–70
accelerators of	167
agitation in	195, 209, 212, 386f
applications of	147–50
bath temperature in	208–9
cooling-curve data in	448–50, 449f, 450t–451t

cooling stages in	208–9
distortion management with	193–97, 193f, 198f, 212
hot-oil	193–97
of iron and steel	206–12
maintenance issues in	402–3, 440, 447–53
oil velocity in	210
part surfaces after	388f
quench tank design	205–12
quench tank maintenance in	447–53
temperature for	145, 145t
Oil-rig fasteners, stress corrosion cracking of	611f
On-the-job (OTJ) training	219, 221, 227
Operators, of heat-treatment operations	565–66
Optical microscopy	344
Original equipment manufacturer (OEM)	384, 585, 587
Orthodontic braces, nickel brazing of	52, 52f
Orthopedic implants	645
Outsourcing	585
Ovens	
airflow within	269, 270–71
combination airflow design of	270f
continuous-style mesh-belt aging	573f
damaged	269
exhausters on	271
leakage checks of	271
load arrangement in	272–73
oil-saturated	534, 535f
positive pressure within	270
sizes	271, 273
temperature ranges in	272
Overall equipment efficiency (OEE)	580
Overpressurizing test	481
Overtime	584
Oxidation	17t
atmospheres for prevention of	27–36
definition of	28
prevention of	27–36
of quench oils	165, 166f, 168f
water vapor-based	21
Oxidation potential	15
Ellingham-Richardson diagram of	21–23, 22f, 23
Oxidation/reduction (redox) reactions	27–37
Oxides	27–36, 93
Oxygen, reactions in furnace atmosphere	20
Oxygen analyzers	470, 471t
Oxygen atmosphere	8t, 15
Oxygen deprivation	517, 518t, 521

Oxygen monitors ... 35
Oxygen (carbon) probes ... 108, 109–13, 386f, 459f
    automatic burnouts of ... 500
    calibration of ... 250
    carbon potential verification of ... 248
    carbon-related damage to ... 498, 499f
    for endothermic atmosphere ... 112
    in endothermic environment ... 111
    insertion of ... 460–61
    lambda-style ... 112
    maintenance of ... 123, 401–2, 459–63
    relationship to setpoint ... 248
    troubleshooting procedures for ... 461–63
Oxygen-sensing devices ... 523

P

P3 Aileron control chain ... 599, 599f
Paddle switches ... 558, 558f
PAG (polyalkylene glycol), as quench media ... 156, 157–58, 212, 214
Part drawings (prints) ... 28
Parts
    design of ... 285–88, 285t–288f
    distortion of. *See* Distortion
    emergency purchase of ... 584
    hydrogen embrittlement of ... 608
    oily ... 548–49
    reverse-engineered ... 595–601
    unacceptable/defective ... 582
PEG (polyethylene glycol), as quench media ... 156, 157–58
PEO (polyethyl oxazoline), as quench media ... 156, 159, 160f
Percentage Brix ... 169
Performance Review Institute (PRI)
    Heat Treatment Pyrometry Guide ... 244
    Nadcap accreditation audit program of ... 167, 241–46, 281
Performance testing ... 596–97
Personal protective equipment (PPE) ... 512
Petrochemical industry, fastener use in, ... 647
Physical metallurgy ... 590
Physical tests ... 325t
Pilots ... 425, 426
Pipes, annealing of ... 46–47
Piston plates, ferritic nitrocarburizing of ... 67, 67f
Pitting ... 609
Plastic deformation ... 329, 330–31, 358
Plasticity. *See* Ductility
Plastics, reference sources about ... 693–94
Plating, reference sources about ... 694

Polishers 351f
Polishing, of metallurgical samples 709–14
Polyalkylene glycol (PAG), as quench media 156, 157–58, 212, 214
Polyethylene glycol (PEG), as quench media...156, 157–58
Polyethyl oxazoline (PEO), as quench media 156, 159, 160f
Polymer quenching 151–60
   of aluminum 212–14
   heat-removal mechanism in 154
   maintenance guidelines for...403, 444–46
   oil contamination in 444
   of steel 215, 216f
   types of polymer quenchants in 156–60
   volume of polymer required..155–56, 156f
Polymers, as quench media 128f, 156–60, 169, 446f
   viscosity of 444–46
Polyvinyl pyrrolidone (PVP), as quench media......................................156, 158, 158f
Potassium hydroxide test 165
"Potato test,"..186
Pourbaix-Ellingham diagrams 622
Powder-metallurgy (P/M) components...69–70, 71t, 548–49
Power failures..411
Power-transmission gears, oil quenching of..148, 149f
Precipitation hardening (aging) 48t, 130
Precipitation number, of quench oils 166
Precision diamond saws 702f
Press quenching...197–200, 198f, 200f
Pressure, effect on flowmeter readings 420–21, 421t
Pressure compensation/conversion calculations 420–21, 421t
Pressure gauges 400, 405, 558
Pressure measurement tests 479, 480f
Pressure quenching..179–80, 180t, 453
Pressure swing adsorption (PSA) system 32, 33f–34f
Prime approval 281–82
Probe impedance (resistance) test 462–63
Process-capability histogram...292f, 293f
Process certifications 283
Process compatibility..583
Process control..117–23
Process controllers...103
Processes, reference sources about 688–89
Processing factors, in product failure...365, 366t
Processing operations, maintenance of...497–509
Process-monitoring frequencies..251–52, 253
Process requirements 284
Process tables 249–53, 250t
Process variability 564, 564f
Product certifications...282, 283

Product development 592–93
Product failure, cost of..585
Productivity
   of furnaces 579–87
   vs. quality... 230f
   uptime 497
Product life cycle 592f
Product requirements 283
Programmable logic controllers 401
Proof load test.. 296t
Proof torque test...325t
Propane 19, 23, 90, 501, 504, 522
Propellers/propeller fans 207, 214, 214t, 410
Proportional limit...329
Protection tubes 432, 433t
Pumps... 206, 214, 401, 464
Purchase, of heat-treatment equipment... 583–86
Purchase orders 280–81, 282
Purging
   air...9, 9f
      exothermic gas in..93
      furnace atmosphere in... 15
      nitrogen.. 546, 604
PVP (polyvinyl pyrrolidone), as quench media..................................... 156, 158, 158f
Pyrometry, standards and requirements for 251, 252–53, 401
Pyrometry-related equipment, calibration of 247

Q

Quality
   cost of 566
   vs. productivity.. 230f
   role of 294
Quality assurance 294
Quality control..294
   reference sources about 694–96
Quenchants. *See* Quench media
Quench chutes.. 146, 215f
   maintenance of 402
Quench flow.. 214–15, 214t
Quenching. *See also specific types of quenching*
   baskets for 628, 628f
   checklist for...387
   direct 130, 206
   of fasteners..650
   fixture..206
   free...206
   fundamental relationships in...130, 131f

gas 127–28, 129f
heat removal mechanisms during 140–43
improper 300f
interrupted 130, 206
Ishikawa diagram of 132f
liquid 127–28, 129f
maintenance guidelines for 402–3, 439–58
principles of 127–33
rapid 127
reference sources about 695
selective 130
slow 127
spray or fog 64, 64f, 128, 130
of steel 55f
time 130
variables in 131, 132f
Quench media. *See also specific quench media*
   CQI-9 requirements for 253
   definition of 127–28
   effect of soot on 500
   for fasteners 636
   heat-transfer mechanism in 129f
   ideal, properties of 143–44
   mixing of 440
   role of 135
   selection of 129–30, 135–36, 137t–138t, 439–40
   types of 127–28, 128f, 135, 439
Quench requirements 385
Quench solenoids 395f
Quench tanks
   agitation in 206–7, 207f, 208f
   in brine/caustic quenching 441–44
   classification of 441
   design of 205–12, 651
   distortion and staining problems in 147
   draft tubes in 207, 208f, 214–15
   fireballs in 524
   maintenance of 440–41
   in oil quenching 150, 447–53, 505
   in polymer quenching 151, 154, 158f, 169, 444–46
   sensors for 150
   water 441–44

R
RA 602 CA 630, 632
Racks, inspection of 388
Radiant-tube leaks 111, 504

tests for	477–79, 482, 504
Radiant tubes	632
alloy	403–4
ceramic	404
effect of carbon on	498
failure of	300f
maintenance of	403–4
wrought	632
Radiographic testing	370t, 373, 373f
Recordkeeping	384
Reduction	17t
definition of	28, 29
Reduction in area	327
Reduction potential	28
Ellingham-Richardson diagram of	21–23, 22f
Reference library, for heat treaters	683–98
Refractive index	169, 444, 445f
Refractometers	251, 444–45, 444f
Refractory, carbon penetration into	498–500, 498f, 501
Refrigerant dryers	91–92
Regulators	464
Residual-stress analysis	338
Resins, for metallurgical sample mounting	705–6
Retaining rings, neutral-hardened	1–2, 2f
Retorts	630, 630f
Retrofitting	527–28
Reverse engineering	595–601
Ring-gears, carburizing of	59, 60f
Rock salt, as brine quenchant	186–87
Rockwell and Rockwell superficial hardness tests	282, 305, 306f, 309–12, 309t, 310f, 311t, 315, 325t
common problems in	317–19
standards for	321
Rod alloy baskets	628
Root-cause and corrective-actions (RCCA)	232, 244, 245
Root cause and effect analysis	581–82
RX® gas. *See* Endothermic gas	

S

SAE/AISI number	279
SAE-ARP-1962	219, 230, 234, 246
SAE International steel grades	
12L14	2–3
1005	490
1008	490
1010	490
1050-1080	177

1075	149
1117	149
4130	186f, 618f
4140	63, 67, 148, 181t
4150	63, 64f, 177, 598
4340	63, 67, 598
4365	177
4620	148
5100 series	177
8620	300f
8620H	60, 61f
52100	1–2, 178
Safety	511–18, 519–26
common hazards	523–25
flame curtains	427
in lean manufacturing	570
proactive program for	520
responsibility for	565
rules and regulations for	519–20
Safety purging	15
Salt disposal	540–41
Salt quenching	175–79, 176f, 177t
high-temperature	454, 455f
interrupted	175
low-temperature	454, 455–56, 456f
maintenance guidelines for	402, 454–57
salt disposal from	540–41
Sample coolers	464
Sample cutoff saw	702f
Sample ports	404–5
Sample preparation, metallurgical	348–51, 350f, 351f, 352f, 699–720
etching	715–20, 716t, 717t
mounting of specimens	704–3, 705f
sampling	699–700
sectioning of specimens	700–703, 701t, 702t
Sample pumps	464
Sand filtration	169
Saws	351f, 702f
SCADA systems	572
Scanning electron microscope (SEM)	344, 362
Scanning transmission electron microscope (STEM)	362–63
Screens, inspection of	388
Seals	396f, 405
Selective catalytic reduction (SCR)	537
Selective noncatalytic reduction (SNCR)	537
Self-audits	231, 233, 245

Sensors

for hydrogen	31
maintenance of	400, 405, 459–74
on oxygen (carbon) probes	110
for quenching	150
system accuracy tests of	258–60
in temperature uniformity surveys	262
on trays	388
Sensor technology	4
Serpentine grids	628
Service factors, in product failure	364, 365t
Shafts	
carburizing of	59f, 60
keyways of	286, 286f, 288, 288f
Shear	358
Shear strength testing	334, 334f
Shear test	296t
Shim-stock test	112, 247, 461, 488, 489–94, 490t, 493t
combustion method	490–91, 491t, 492, 492f, 493
spectrographic analysis method	490
weight-gain method	490, 491–92, 491f, 492t
Shipping tickets	282, 283
Sieve dryers	83
Sight ports	404–5
Sigma-phase embrittlement	614–16, 614f, 620
Silicon bronzes, as fastener material	646
Silicon carbide, eutectic melting of	631
Silicon iron, annealing of	84
Silicon steels, annealing of	93
Single-part processing water quenching method	184
Sintering	68–73
definition of	68–69
in dissociated ammonia atmosphere	86
furnace atmosphere for	68–69, 69t
process tables for	250t
reference books about	696
Sludge	166–67, 541
Smog (ozone)	535
Smoke-bomb test	481–82, 481f, 482t
S-N curve	332, 333f
Sodium chloride, as quench media	186–87
Sodium hydroxide, as quench media	186–87
Sodium polyacrylate (ACR), as quench media	156, 159, 159f
Softening operations/processes, definition of	277
Solution heat treatment, annealing process temperatures for	48t
Soot	18, 19, 498–504
air leak-related	480
in batch furnaces	500–501, 501f

burning temperature of 500
in continuous furnaces 501–4, 503f, 504f
dew point as indicator of 116
as dew-point instrument contaminant 467, 469–70
effect on carburizing rate 19
in exothermic-gas generators 92
interaction with oxygen 20
management of 405–6
in pusher furnaces 503–4, 504f
in quench tanks 500, 501
in rotary-hearth furnaces 504
Special operations/processes 277
    definition of 277
    Nadcap accreditation audits of 243
Specifications
    for commercial heat-treatment shops 279–82
    for in-house heat-treatment shops 277–78
    on purchase orders 280–81
Specific gravity 419
    conversion factors 418f
    of gases 31t
Staining, of parts 165, 166
Stainless steel
    annealing of 85
    liquid-metal embrittlement of 616–17
    as muffle material 631
    nickel, brazing of 23
    reference books about 686
    sigma-phase embrittlement of 614, 620
    sintering atmosphere for 69–70
    sintering of 86
    thread damage to 299f
Stainless steel alloys, metal dusting of 300f
Stainless steel fasteners 635, 638, 644–45
Standards, for heat treatment
    external standards 279–82
    internal standards 277–78
Statistical data/methods
    in in-control process evaluation 292f
    in mechanical testing 338
    in out-of-control process evaluation 293f
Steam atmosphere 8t, 212
Steel
    austenitizing of 55f
    case hardening of 55f
    etchants for 716t, 717t
    quench-and-tempered 279t

quenching of 55f, 175, 206–12, 215, 216f
reference sources about 684–86
tensile strength 326
Steel bars, hardening of..63–64, 64f
Steel silver, exfoliated... 605f
Stop-off paint vapors.. 110
Straightening, combined with martempering 179
Strain..327–28
   definition of..326
   at failure 329
Strength, of materials 327
Stress..327–28
   causes of 288–89
   definition of..326
   triaxial 336
Stress-concentration angles 285, 285f
Stress-corrosion cracking (SCC) 609–16, 644
Stress durability tests..337
Stress relief..48t, 174f, 175f, 192, 424
Stress risers..285, 285f, 286
Stress-rupture tests... 335, 335f
Stress-strain curves..328–31, 328f, 330f
Structural bolts, oil quenching of... 148, 148f
Sulfur, as metal dusting preventive..508
Sulfur dioxide atmosphere 8t
Supervisors, of heat-treatment operations 565, 567
Surface, "bright"/clean 4, 27
Surface-flaw testing.. 374–75
Switches, maintenance of 405
System accuracy tests (SATs) 249, 252, 257–60, 258t

T

Tapered-tube rotameters 416
Temperature-analysis software...261
Temperature-control instruments 251, 252
Temperature profiling 261
Temperature ratings 384
Temperature. *See also under specific heat-treatment processes*
   CQI-9 monitoring guidelines for..248
   effect on embrittlement..609, 616
   effect on flowmeter readings 419–20, 420t, 421t
   effect on furnace-atmosphere reactions 16
   effect on oxidizing/reducing potential................................21–23, 22f, 23f, 29, 29f
Temperature uniformity survey (TUS) 249, 251, 252, 486–88
   of atmosphere furnaces 265–68, 265f, 266f, 268f
   calibration in 263–64
   failure of...261

guidelines for 261–65
with or without load 261
of ovens 269–73, 270f, 272f
test racks for 268, 268f
Tempering 130
baskets for 629
in dissociated ammonia atmosphere 86
immediate 130
temperature for 251
temperature uniformity surveys of 251
of tool steel 86
Tempering furnaces 651
oily parts in 549
TempTab® ceramic shrinkage device 487f
Tensile strength 326, 330, 617, 617f
Tensile tests 296t, 326–31, 327f
Terminology, of heat treatment 653–81
Tertiary diagram, for hydrogen, oxygen, and nitrogen 30, 30f
Testing .... 291-380, 477-498. *See also* Mechanical testing; *names of specific tests*
choice of tests 294
rationale for 291–301
reference sources about 695
standards for 297
Thermal conductivity, of gases 31t
Thermal content, of gases 31t
Thermocouples 429–34, 430f
accreditation audits of 245
in atmosphere furnace surveys 266
calibration of 249, 252
configurations 249
CQI-9 specifications for, ... 249, 258
definition of 430
installation/positioning of 406, 431–32, 432t
for part/load 486, 487f
protection tubes for 432, 433t
replacement of 406
standards for 434
for system accuracy tests 266–67
system accuracy tests of 257–58, 258t
of temperature-profile systems 261
in temperature uniformity surveys 261, 262–64
Thermocouple wire 431, 434
Thermography 370t, 378–79, 379f
Time, as furnace-atmosphere reaction factor 16
Tin-plated electrolytic tough pitch (ETP) copper terminal lugs 604f
Tool steel 86
reference sources about 686–87

Torque/torque-tension testing 296t, 331–32, 332f
Total acid number (TAN) 165
Total cost of ownership (TCO) 579–87
    hidden costs component 583–86
    overall equipment efficiency (OEE) measure 580
Total temperature rise rule 155
Toughness
    definition of 297
    tests of 336
Training, in heat treatment 219–27
    audit preparation courses 244
    course outlines for 222–26
    effect on profitability 584
    in metallographic interpretation 299
    Nadcap accreditation requirements for 246
Transmission electron microscope (TEM) 362
Transverse rupture test 327
Trays (grids) 388, 627–28, 627f
True Equipment Cost (TEC) 584
TS16949 281
Tubes, annealing of 46

U

Ultimate tensile strength 329
Ultrasonic inspection 325t
Ultrasonic leak detection 377, 379
Ultrasonic testing (UT) 370t, 375–76, 375f, 376f
Uptime availability 583
Uptime utilization 585

V

Vacuum atmosphere 8t
Vacuum Heat Treatment (Herring) 179
Vacuum heat treatment, reference sources about 696
Vacuum purging 9
Valves, in gas-sampling systems 464
Valve-seat inserts, sintering of 70, 70f
Vapor-blanket cooling stage 140, 142f, 143f
Vapor-transport cooling stage 141, 142f, 143
Vibration
    excessive 495, 496f
    fasteners' resistance to 643–44, 644f
Vibration meters 411, 447, 495
Vibration tests 337
Vickers indenters 313, 313f
Viscosity, of quench media 209, 444–46
    effect on heat-transfer rates 141

kinematic 169, 444, 445f
of oil quench media 141, 164, 164f, 165
of polymer quench media.. 169
Volumetric-flow devices 414–15

W

Warning signs 520–21
Warranties, for heat-treatment equipment 586
Water
    effect on cooling rate 181, 181t
    in quench oil 164, 168f, 212, 245, 402, 545
    in quench salt... 175, 176f
Water-cooling systems 551–60
Water-gas reactions 17t, 63
Water leaks
    dew point-based indication of 117
    differentiated from air leaks 23–24, 482–86, 484t–486t
Water quality.. 538–40, 539t, 540t, 551
    for cooling towers 556t
Water quenching 128t, 144–45, 212, 403
    intensive 183–84, 184f
    limitations to..182–83
    quench tank maintenance in 441–44
Water vapor... 17t. *See also* Dew Point
    in dissociated-ammonia generators.. 83, 84t
    effect on steel oxidation...29
    as furnace atmosphere component 15
Water-vapor reactions 20–21
Wear.. 358, 358t
Weigh scales 250, 252
Welding
    Nadcap audits of...243
    reference sources about 689–90
Wire patenting... 178
Work instructions 277–78, 279t
Wrought-alloy retorts 630t

Y

Yield strength..326, 329
Young's modulus..326, 329

Z

Zinc
    liquid-metal embrittlement of... 616
    as oxygen (carbon) probe contaminant... 110
    vapor pressure of..52